T0255496

The China Geological Survey Series

This Open Access book series systematically presents the outcomes and achievements of regional geological surveys, mineral geological surveys, hydrogeological and other types of geological surveys conducted in various regions of China. The goal of the series is to provide researchers and professional geologists with a substantial knowledge base before they commence investigations in a particular area of China. Accordingly, it includes a wealth of information on maps and cross-sections, past and current models, geophysical investigations, geochemical datasets, economic geology, geotourism (Geoparks), and geo-environmental/ecological concerns.

Chuanlong Mou • Xiuping Wang •
Qiyu Wang • Xiangying Ge •
Bowen Zan • Kenken Zhou •
Xiaowei Chen • Wei Liang

Lithofacies Paleogeography and Geological Survey of Shale Gas

Chuanlong Mou
Chengdu Center
China Geological Survey
Chengdu, Sichuan, China

Qiyu Wang
Chengdu Center
China Geological Survey
Chengdu, Sichuan, China

Bowen Zan
Chengdu Center
China Geological Survey
Chengdu, Sichuan, China

Xiaowei Chen
Chengdu Center
China Geological Survey
Chengdu, Sichuan, China

Xiuping Wang
Chengdu Center
China Geological Survey
Chengdu, Sichuan, China

Xiangying Ge
Chengdu Center
China Geological Survey
Chengdu, Sichuan, China

Kenken Zhou
Chengdu Center
China Geological Survey
Chengdu, Sichuan, China

Wei Liang
Chengdu Center
China Geological Survey
Chengdu, Sichuan, China

ISSN 2662-4923 ISSN 2662-4931 (electronic)
The China Geological Survey Series
ISBN 978-981-19-8863-9 ISBN 978-981-19-8861-5 (eBook)
https://doi.org/10.1007/978-981-19-8861-5

This Springer imprint is published by the registered company Springer Nature Singapore Pte Ltd.
The registered company address is: 152 Beach Road, #21-01/04 Gateway East, Singapore 189721, Singapore

Preface

According to the concepts of traditional oil and gas geological theory, the exploration of shale has been forbidden. However, with the emergence of micro/nano-pores, development of innovative theories such as those of non-Darcy flows, and industrial application of horizontal wells and multi-stage hydraulic fracturing and other technologies, commercial exploitation of shale gas has been realized. Shale gas represents the "residual gas" in source rock that has not been discharged in time. It exists in the form of adsorbed, free or dissolved gas and can be described as a biogenic gas, thermogenic gas or mixture of both. At present, research, exploration and development of shale gas are being actively conducted worldwide. Moreover, the recognized importance of shale gas in natural gas production has changed the world energy strategies and represents a valuable strategic direction for China's national oil and gas energy development.

In China, after more than a decade of research and development, significant progress has been made in the research, exploration and development of shale gas reservoirs, especially with the discovery of the Fuling shale gas field. Early preliminary investigations and comparative studies with shale gas accumulation conditions in the United States indicated that the Yangtze plate region of South China, central and Eastern China, Northwestern China, Ordos Basin, Tarim Basin and Tibetan regions have valuable prospects for shale gas exploration. In particular, Southern China has enormous shale gas resource potential. At present, with the rapid development of China's economy, the demand for oil and gas resources has been intensified, and a prominent mismatch exists between the oil and gas supply and demand. Development of China's shale gas industry is expected to increase the oil and gas reserves, which can help alleviate the abovementioned mismatch and optimize the energy structure. This aspect appears to be the motivation for the large-scale geological survey of shale gas in China.

The key factors influencing the accumulation of shale gas can be summarized as follows: First, shale as a source rock must be able to produce a large amount of thermogenic or biogenic gas and must have sufficient in situ gas content; second, shale must have enough matrix pores. The basic geological factors affecting shale gas accumulation include the organic carbon content (TOC), organic matter type and maturity (*R*o), gas shale thickness, mineral composition, reservoir characteristics, burial depth and formation pressure. The basic material elements of shale development, such as the organic carbon content, organic matter type and mineral composition, are controlled by the sedimentary environment or sedimentary facies. The sedimentary environment controls the thickness, distribution area, organic matter content and other characteristics of shale and determines the types of sedimentary rocks and mineral composition. Notably, the rock types and mineral composition considerably influence the development of reservoir physical properties, thereby affecting the shale gas accumulation. Therefore, in essence, the sedimentary environment is the fundamental factor that determines the enrichment degree of shale gas.

Considering the rapid development of shale gas research and exploration in China, the following three important tasks must be performed in the geological survey of shale gas: (1) determine the characteristics of source rocks, such as lithologic characteristics, sedimentary environment characteristics, sedimentary microfacies types and characteristics, organic matter content and mineral composition; (2) identify the temporal and spatial distributions of source rocks, including the thickness, burial depth, fine distribution and area; and (3) optimize the prospective and favorable areas for shale gas reservoirs, comprehensively study and delineate the exploration target area, and provide a scientific basis for the final development of shale gas. Practical experience indicates that these three key objectives can be accomplished through lithofacies paleogeographic research and mapping.

In theory, lithofacies paleogeography is an important branch of sedimentary geology, focused on studying the tectonic evolution background, sedimentary basin properties and migration of oil and gas in geological history. In practice, lithofacies paleogeography can guide the exploration and development of oil and gas resources and prediction of prospects associated with sedimentary stratified mineral and water resource exploration. The mapping method, which can preliminarily clarify the internal relationship between the deposition and energy and mineral resource distribution and is an effective method for prospecting, can be extended to shale gas resources. The sedimentary environment determines the characteristics of the basic geological elements of shale gas, such as the content of organic matter, sedimentary thickness and mineral composition. Based on this knowledge, by examining the tectonic setting as the premise and studying the regional fine sedimentary facies, the lithofacies paleogeography can be mapped to define the spatial and temporal distributions of the favorable facies zones of source rocks, thereby providing a foundation and direction for shale gas exploration. In the evaluation of shale gas districts, the fine sedimentary facies, diagenesis and diagenetic evolution must be analyzed to determine the factors influencing the shale gas enrichment. Subsequently, the boundary contour must be mapped, and the appropriate parameters must be selected and superimposed on the lithofacies paleogeographic map. The highly coupled areas are expected to be the key areas for further exploration and development.

In recent years, considerable shale gas research and exploration have been performed in the Sichuan Basin, and shale gas fields such as Fuling, Changning and Weiyuan have been discovered successively. Nevertheless, considerable scope of improvement remains in terms of the shale gas exploration in the Sichuan Basin, as the existing research methods for the fine sedimentary facies of source rocks, lithofacies paleogeographic mapping and shale gas selection evaluation are limited. Moreover, the use of lithofacies paleogeography and mapping technologies as the guiding theory and key techniques for shale gas geological surveys has not been extensively considered yet.

From 2012 to 2014, the authors worked on the projects "Research and selection evaluation of shale gas accumulation conditions of Lower Paleozoic strata in Southern Sichuan Basin", entrusted by the Exploration Southern Branch of Sinopec, and "Basic geological survey of marine shale gas of Lower Paleozoic in Sichuan Basin", entrusted by the Chengdu Geological Survey Center of China Geological Survey. The study area can be considered tectonically equivalent to the backbulge in Inner Yangtze Craton Basin. The research task was to evaluate and select regions with shale gas blocks. By comprehensively analyzing the outcrop, drilling characteristics and seismic profile through the theory and mapping method associated with sedimentology and lithofacies paleogeography and concepts of geochemistry and tectonic geology, a series of geological maps were compiled, including the comprehensive sedimentary histogram of the source rock segment, comprehensive shale gas histogram, fine lithofacies paleogeography of the shale gas development layer group, organic carbon content plane distribution and average brittle mineral content distribution of organic-rich shale, and hydrocarbon generation and exploration potential distributions. Considerable progress was made in shale gas source rock stratigraphic classification and correlation and examination of the fine sedimentary facies, sedimentary environment and lithofacies paleogeography of the hydrocarbon source rocks, hydrocarbon source rock development characteristics and effects of sedimentation on the shale gas sedimentary environment. Moreover, the favorable areas for shale gas were evaluated, and a detailed and systematic method for shale gas basic geological surveys was established. The preliminary study performed considering the black rock series shale gas reservoir in the Ordovician Wufeng and Silurian Longmaxi Formations indicated that the most favorable development area is the deep-water shelf facies area. However, further analyses showed that not all deep-water shelf facies areas are favorable development areas for shale gas reservoirs. When the costs of exploration and development and buried depth are not considered, the most favorable area typically pertains to the carbonaceous (silicon) muddy deep-water shelf facies, and the TOC of black rock series is usually >2.0%. The content of carbonate minerals in the mineral composition is lower than 30%, and the maturity Ro is moderate (2.0–3.0%). In the geological investigation of shale gas in the Ordovician Wufeng and Silurian Longmaxi Formations, lithofacies and paleogeography research and mapping were performed to effectively superimpose the main geological characteristic parameters of shale gas development. Subsequently, the favorable shale gas areas of the Ordovician Wufeng and Silurian Longmaxi Formations in the Southern Sichuan Basin and its periphery were selected. The target areas for commercial development were defined considering the influencing factors, such as the surface conditions and burial depth. According to the exploration and development practice, lithofacies paleogeography research and mapping as the key technical means can facilitate a breakthrough in shale gas geological surveys. In other words, lithofacies paleogeography research can provide guidance for shale gas geological surveys.

Based on the authors' work experience and research results, this book describes the use of the abovementioned method for shale gas geological survey and exploration and development research. To provide scientific, reliable and direct guidance and facilitate the ongoing large-scale shale gas geological survey and exploration and development in China, the Ordovician Wufeng and Silurian Longmaxi Formations in the Southern Sichuan Basin and its periphery are considered examples.

We emphasize the use of lithofacies paleogeography and related mapping methods as the key technology to guide and perform geological surveys of shale gas. The formation and spatial distributions of shale gas are studied at different scales, specifically, macro-, micro- and ultramicroscales. The shale gas enrichment factors are considered to comprehensively examine the enrichment law and selection and evaluation methods.

This book was completed in 2016 and first printed and published in March 2017. The purpose of this book is to illustrate that lithofacies paleogeography can be used as a guide for shale gas geological surveys by elaborating theories and examples to provide theoretical guidance and methodological references for shale gas geological surveys. In 2021, this book will be republished after supplementing and incorporating the domestic research progress in the past five years. In recent years, rapid exploration and development of shale gas have occurred in China, and the breakthrough of shale gas exploration in the Sichuan Basin has demonstrated that lithofacies paleogeography can serve as a guide for shale gas geological investigations.

Chengdu, China

Chuanlong Mou
Xiuping Wang
Qiyu Wang
Xiangying Ge
Bowen Zan
Kenken Zhou
Xiaowei Chen
Wei Liang

Acknowledgments

In the process of project research on which this book is based, we received assistance from Xusheng Guo, Zhihong Wei, Dongfeng Hu, Qingqiu Huang and other experts from the Sinopec Exploration South Branch and strong support from leaders of the Chengdu Geological Survey Center of China Geological Survey. Academician Baojun Liu, Prof. Xiaosong Xu and Prof. Qian Yu from the Chengdu Geological Survey Center of China Geological Survey provided guidance and assistance for the project and the book. We thank Dr. Min Deng, doctoral student Yao Shengyang and master's student Huang Xiaoming for sorting the relevant materials for this book. Dr. Chao Chen, Dr. Qian Hou, Dr. Jianjun Zhang, Dr. Binsong Zheng, Dr. Yuancong Wang, Master Xiaoyong Sun, Master Zhiyuan Tan, Master Penghui Xu, Master Xin Men, Master Lin Tong, and master's students Qingsong Zhang and Muyuan Wang provided valuable help in the project research as well as the drafting of this book. We appreciate the efforts of Director Zhaoguo Wang and Prof. Jianlong Zhang from the Chengdu Geological Survey Center of China Geological Survey for supporting the English translation of this book. Moreover, we thank Dr. Xiaolin Zhou, Dr. Jianglin He, Dr. Jinyuan Huang, Dr. Ankun Zhao and Dr. Songyang Wu for facilitating the English translation of this book to diversify the readership. We express our heartfelt thanks to all the abovementioned individuals.

About This Book

Based on the research and state of exploration of shale gas through geological surveys, this book presents methods to screen and evaluate shale gas based on lithofacies paleogeography and mapping technologies to perform shale gas geological surveys. Based on novel insights into the definition of shale gas, a comprehensive and systematic analysis of the key factors influencing shale gas enrichment indicates that the "sedimentary environment is the fundamental factor determining the degree of shale gas enrichment". By extensively discussing the relationship between lithofacies paleogeography and shale gas enrichment, we demonstrate, for the first time, how lithofacies paleogeography research and mapping can facilitate and guide shale gas geological surveys. We consider the Ordovician Wufeng and Silurian Longmaxi Formations in the Sichuan Basin and its periphery as examples. First, the spatial-temporal distribution of shales rich in organic matter is confirmed from detailed research on lithofacies paleogeography. Subsequently, by effectively superimposing the evaluation parameters of shale gas, the favorable exploration and target areas are predicted. In summary, the research and mapping of lithofacies paleogeography as a basic method and key technology can facilitate the geological surveys of shale gas. In this book, we describe the method and technology, which can guide the realization of large-scale geological surveys and exploration and development of shale gas.

This book has been written based on the authors' scientific research and practical experience, in combination with the research achievements and current state of exploration of shale gas geological surveys at home and abroad. This book can serve as a reference for geologists interested in basic geology, mineral geology, petroleum and natural gas geology and coal field geology and has a high reference value for scientific researchers and teachers in the fields of sedimentary, lithofacies paleogeography, stratigraphy and petroleum geology.

Contents

Introduction

1

Abstract

By summarizing and comparing the basic geological conditions of shale gas development in North America and China, this chapter finds that the development of gas-bearing shales in North America is concentrated, and the resource type is single. In contrast, the development of shale in China involves more layers with more resource types, and basic research has been performed in shales developed in marine–terrestrial transitional facies and terrestrial facies. A comprehensive analysis of the existing research on lithofacies paleogeography and its role in the development of the world's oil and gas industry indicates that lithofacies paleogeography research and mapping methods can preliminarily clarify the connection between the sedimentary environment and distribution of energy and mineral resources.

Keywords

Research status • Shale gas • Lithofacies paleogeography • Comparative study • North America; China

Shale gas is an important unconventional natural gas resource (Ye and Zeng 2008). With the continuously growing social demand for clean energy, increasing natural gas prices, deepening understanding of shale gas reservoir formation conditions, and advancements in drilling technologies, shale gas exploration and development have expanded from North America across the world (Du et al. 2011). In 2018, U.S. shale gas production reached 6669×10^8 m^3, accounting for 63.4% of the total natural gas production, and this high production changed the global natural gas supply pattern (Jiang et al. 2020). Shale gas has emerged as a research hotspot in the exploration and development of unconventional oil and gas resources worldwide, and accelerating the exploration and development of shale gas resources has become a focus of the major shale gas resource countries (Du et al. 2011). With the rising demand for energy, increasing energy pressure and environmental awareness in China, the exploration and development of shale gas resources must be accelerated (Jiang et al. 2020). Consequently, large-scale basic geological investigations of shale gas are being performed in China.

Unlike the USA and Canada, which are characteristic of concentrated shale gas development strata and single type of shale gas resources, China exhibits shale gas development strata that are distributed across geological historical periods, including Cambrian, Ordovician, Silurian, Carboniferous, Permian and Jurassic. Moreover, many types of shale gas resources exist, including marine, marine–terrestrial transitional and terrestrial facies. Compared with North America, Chinese shale gas resources are characterized by large geological ages, large burial depths, high degrees of thermal evolution and complex tectonic and geomorphic conditions (Sun et al. 2020). In the general geological investigations in the 11th and 12th Five-Year Plans, the distribution areas of marine, marine–terrestrial transitional and terrestrial source rocks in China were preliminarily estimated to be as large as 300×10^4 km^2, 200×10^4 km^2 and 280×10^4 km^2, respectively (Zhang et al. 2011). From the perspective of geographical distributions, high-quality marine source rocks are mainly distributed in Southern China, especially the middle-Upper Yangtze region (Liu et al. 2004; Li et al. 2009d), and the lithology mostly includes siliceous shale, black shale, calcareous shale and sandy shale. Most of the high-quality source rocks pertain to siliceous and black shale. This area is a frontier for geological investigation, exploration and development of shale gas in China (Long et al. 2009). The source rocks of the marine–terrestrial transitional facies are mainly distributed in Southern China, Northern China and Junggar in Northwest China. These rocks are scattered but exhibit a high resource potential. The source rocks, which are mostly siliceous shale, coal-bearing black shale, coal-bearing calcareous shale and silty shale, are important types of shale gas resources and gaining

C. Mou et al., *Lithofacies Paleogeography and Geological Survey of Shale Gas*, The China Geological Survey Series, https://doi.org/10.1007/978-981-19-8861-5_1

considerable attention in China. The distribution of terrestrial source rocks is similar to that of marine–terrestrial transitional source rocks, which are distributed in the three major plates in China. Among the terrestrial source rocks, the Jurassic system in Southern China has been the focus of shale gas resource research, followed by the Middle Cenozoic strata in Northern China and a small amount of source rock in the Junggar Basin and the Qinghai-Tibet Plateau. The lithology mainly includes a set of black shales, calcareous shale and silty shales of lacustrine facies.

Since the initial geological survey of shale gas in China and the extensive exploration and development of shale gas in key areas, many geologists have conducted a considerable amount of preliminary research. From 2005 to 2009, the geological conditions of shale gas in Mesozoic hydrocarbon-bearing basins were analyzed with the source rock formations as the research objects, and the geological prospects of shale gas in these Formations were analyzed based on the distribution pattern of Paleozoic source rocks in the basin and outcrop area. Further comparison was performed considering the geological characteristics of the shale gas development Formations in the USA. Based on this research, the shale gas resource prospects in the Upper Yangtze region were analyzed, and the prospective areas were initially selected (Zhang et al. 2003, 2004, 2008a, b; Liu et al. 2004; Zhao et al. 2008; Ye and Zeng 2008; Zhang et al. 2008; Zou et al. 2009; Long et al. 2009; Li et al. 2009d; Cheng et al. 2009; Wang et al. 2009a, b, c). The 12th Five-Year Plan (2010–2015) corresponded to valuable breakthroughs in shale geological investigation and research in China. In the period 2010–2015, considering the basic geological characteristics of shale gas development in China, strategic surveys of shale gas resources were performed nationwide, the national shale gas resource potential was evaluated, and the favorable areas were selected (Zhang et al. 2009, 2010a, 2012b; Zou et al. 2009, 2010a, 2013a; Chen et al. 2009a, 2009b; Pan and Huang 2009; Li et al. 2009d, Wang et al. 2009a, b, c; Huang 2009a; Xu and Bao 2009; Li and Zhao 2009; Nie et al. 2009a, 2011a, 2011b, 2011c; 2012a, 2012b, c; Jiang et al. 2010; Zhang 2011; Duet al. 2011; Zeng et al. 2011; Chen et al. 2011a; Fu et al. 2011; He et al. 2011; Liang et al. 2012; Zhang et al. 2012b, 2013; Wu et al. 2013a; Wang et al. 2013; Zheng et al. 2013a; Li et al. 2014, 2015; Wang 2015; Jiao et al. 2015; Zhang 2015). Moreover, industrial breakthroughs in shale gas exploration and development were achieved through geological investigations and favorable area optimization. With expanding research on shale gas exploration, CNPC, Sinopec, Chongqing, Sichuan and Guizhou have successively established shale gas exploration companies or project departments to conduct basic geological investigations and research on shale gas resources in the Sichuan Basin and adjacent areas. CNPC has selected four favorable blocks: Weiyuan, Changning, Zhaotong and Fushun-Yongchuan (a cooperative block with Shell) in Southern Sichuan and Northern Yunnan-Guizhou and drilled more than 60 shale gas exploration wells, leading to significant exploration breakthroughs. High-yield shale gas flow has been observed in Well Yang101 in Fushun and Well Ning201 in Changning, and Well Wei201 represents the first drilled well for shale gas development. Sinopec has mainly focused on Fuling, Nanchuan and Qijiang in the Southeastern Sichuan Basin and Pengshui outside the basin. The organization has identified more than 50 shale gas exploration wells, achieved exploration breakthroughs in Fuling, Qijiang and Pengshui, established the Fuling National Shale Gas Exploration Demonstration Zone and discovered China's first shale gas field, the Fuling Shale Gas Field, with proven reserves of 106.75×10^9 m^3. In 2015, the Fuling National Shale Gas Demonstration Zone was successfully completed, with a production capacity and output of 5×10^9 m^3 and 31.68×10^8 m^3, respectively, accounting for 71% of China's shale gas production (Cai et al. 2021). In 2016, substantial exploration breakthroughs were made in other prospective areas, and as of 2020, 6 large and medium shale gas fields have been discovered in Fuling, Weirong, Changning, Weiyuan, Zhaotong and Yongchuan in the Upper Ordovician Wufeng Formation and Lower Silurian Longmaxi Formation of the Sichuan Basin (Zou et al. 2017; Cai et al. 2021) (Fig. 1.1), indicating that shale gas exploration in China has entered the industrialized and quasi-industrialized production stages. These achievements demonstrate the prospects for the exploration and development of shale gas in the Sichuan Basin and its peripheral area, which are expected to facilitate the growth of natural gas production in China.

Nevertheless, the shale gas geological investigations are limited, and a guiding ideology and key technical methods are lacking. According to the actual geological investigation results of shale gas in China, the reservoirs are characterized by multiple resource types, widespread distribution and high potential (Zhang 2011). From discovery to final industrial and commercial development, shale gas reservoirs have benefited not only from the development and deepened understanding of oil and gas geological theories, such as the discovery of micro/nanoscale pores and extension of non-Darcy seepage theory but also from the breakthrough and progress of industrial technologies, such as horizontal well development and multistage hydraulic fracturing technologies. In this context, the conduction of basic geological investigations of large-scale shale gas requires targeted knowledge and theoretical guidance of geological disciplines and the application of key technical methods.

Research on the regional geological background and related basic geological characteristics of shale gas development in China, such as the spatial distribution of

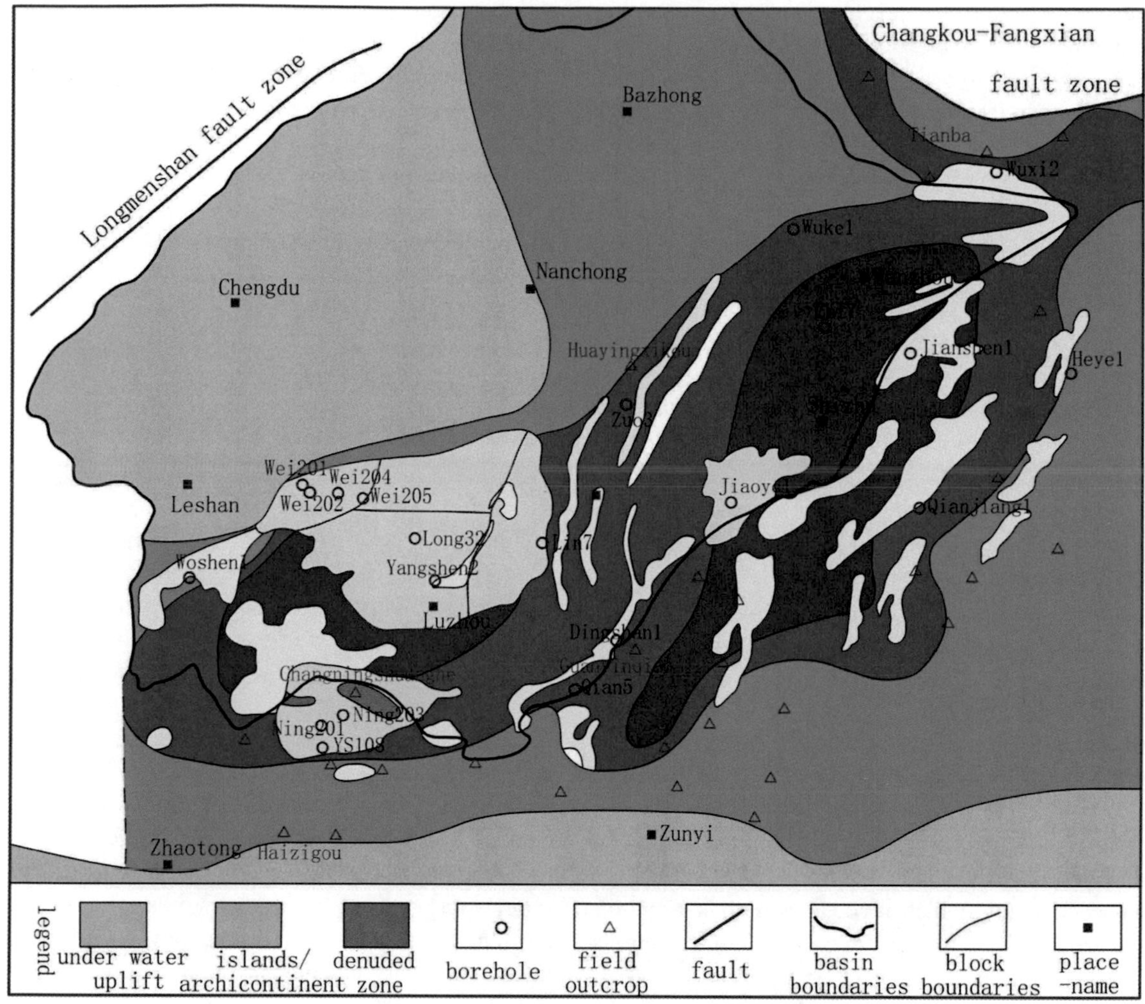

Fig. 1.1 Key drilling spots for shale gas in the Wufeng Formation-Longmaxi Formation in the Sichuan Basin (Zou et al. 2017)

mudstones and shale; sedimentation and diagenesis; and reservoir characteristics, enrichment and accumulation, along with their evaluation and optimization, is still in the lag phase. The level of nationwide geological investigation for shale gas is low. Consequently, except for a few areas in the Sichuan Basin, no major breakthrough has been made in most areas yet. Moreover, except for the Wufeng-Longmaxi Formation, industrial production has been realized, necessitating the expansion of research in other regions. Despite the deepening of shale gas geological investigation and continuous improvement of research and exploration methodologies, the key technical methods have failed to address the requirements for accelerating production development. The main problems can be summarized as follows: Although many researchers have focused on shale gas, most of these studies were performed in certain areas or stratigraphic subdivisions, and only a limited number of studies have been performed on the development characteristics, enrichment and spatial distribution of the complete shale gas reservoir system. The research of sedimentary systems, sequence stratigraphic and petrographic paleogeography with source rocks is not systematic. Moreover, systematic and extensive research on the spatial and temporal distribution characteristics of organic-rich shale as source rocks and reservoirs must be performed using basic and operable key research methods and means. The formation mechanism of shale gas in depositional and diagenetic processes remains to be clarified, as the diagenetic minerals, diagenesis, diagenetic evolution and pore evolution of shale gas reservoirs have not been studied sufficiently. In addition, the formation mechanism of shale gas reservoirs has not been systematically explored.

In conclusion, the extent and research of shale gas geological investigations under the complex geological background of China are limited owing to the lack of integrated, systematic and unified technical approaches and basic geological theories. Through comprehensive comparative research, the authors believe that the key to shale gas geological investigation and development is to understand the basic geological characteristics and spatial distributions of the shale gas reservoir-source rocks, to understand the relevant background and to optimize the prospective, favorable and target areas. In other words, shale gas geological investigations have the following three objectives:

(1) To clarify the characteristics of source rocks, such as lithological characteristics, depositional environment, types and characteristics of sedimentary microfacies (lithofacies), organic matter contents and mineral compositions.
(2) To clarify the spatial and temporal distribution patterns of source rocks, including their thickness, burial depth, fine distribution and area.
(3) To optimize the prospective and favorable areas of shale gas reservoirs to provide a scientific basis for the exploration and development of shale gas.

Theoretically, the key to shale gas geological investigation and research is to focus on basic geological surveys and source rocks, which are the carriers of shale gas development and deposition, especially through lithofacies paleogeography and mapping. Another key aspect is to follow systematic and scientific approaches, use reasonable geological theories and identify critical research methods.

Based on the actual situation, the Southern Sichuan Basin and its adjacent region, which involve the Lower Cambrian Niutitang Formation and Upper Ordovician Wufeng Formation and Lower Silurian Longmaxi Formation, are key areas for shale gas geological surveys in China with promising shale gas development strata and considerable shale gas resource potential. Since 2008, the Ministry of Land and Resources, Chengdu Center of China Geological Survey, CNPC, Sinopec and related research and production units have performed geological investigations and exploration and development work in these regions, with the most significant achievement being the establishment of the first large shale gas field in China. According to the practical explorations, sedimentary (micro) facies, fine petrographic paleogeography and even sequence stratigraphy analyses and geochemistry and petroleum geology concepts can be used to perform the basic geological investigations of shale gas and promote the exploration and development of shale gas in later stages.

Accordingly, based on the authors' work experience and research, this book sublimates the concepts of sedimentology, lithofacies paleogeography, geochemistry and other related theories as guidance to examine shale gas carrier-source rocks. We emphasize that the geological survey of shale gas must be guided by lithofacies paleogeography and its related mapping methods, which can facilitate research on shale gas formation, spatial distribution characteristics, enrichment patterns and optimization of areas at different scales from macroscopic to microscopic to ultramicroscopic. The use of this approach for shale gas geological investigations and exploration and development studies is demonstrated and supported by an example pertaining to the research of gas-bearing formations, namely the Wufeng Formation-Longmaxi Formation in the Southern Sichuan Basin and adjacent areas. This book can provide scientific and reliable guidance for ongoing large-scale shale gas geological investigations and future exploration and development activities in China.

1.1 Current Research Status

According to the definition of petrology, clay rocks refer to sedimentary rocks dominated by clay minerals (content greater than 50%) (Zhao and Zhu 2001; Jiang 2003). Clay rocks that solidify into rock are known as mudstone and shale. Mudstone and shale are rocks with particle sizes of less than 0.0039 mm (i.e., < 4 μm) and composed mainly of clay minerals. In contrast, fine-grained sediments are clay- and silt-grade sediments with particle sizes less than 62 μm and mainly contain clay minerals, silt, carbonate, organic matter, etc. (Schieber and Zimmerle 1998; Aplin and Macquaker 2011). Jiang et al. (2013) referred to sedimentary rocks composed of fine-grained sediments as fine-grained sedimentary rocks or argillaceous rocks. Among these rocks, the rocks with developed and undeveloped lamination are known as shale and mudstone, respectively. The sizes of clay and silt particles are less than 4 μm and between 4 and 62 μm, respective. Since clay and silt particles are difficult to distinguish visually, argillaceous rocks are generally considered to be a mixture of clay and silt. In addition, fine-grained sedimentary rocks are widely distributed, accounting for approximately 2/3 of sedimentary rocks (Macquaker and Adams 2003; Aplin and Macquaker 2011), which is consistent with the proportion of clay rocks in sedimentary rocks reported by Chinese scholars such as Liu (1980) and Jiang (2003). The rocks defined as clay rocks by Chinese scholars may be consistent with the fine-grained sedimentary rocks classified abroad, although they are limited in terms of structural and compositional constraints. Shale gas reservoirs usually include argillaceous rocks, marls, sandstones and carbonate rocks (Zhang et al. 2008b; Bust et al. 2013). Therefore, the term "shale" in shale gas is typically used to represent a geological formation rather than the lithology (Bust et al. 2013). Moreover, shale gas formations actually consist mainly of fine-grained sedimentary rocks rather than muddy shales, but the lithology of shale gas formations is still collectively referred to as "shale" according to present research practice.

The deposition and diagenesis of fine-grained sediments in shale is a weak field of research in sedimentology and even geology due to the small particle size, difficulty of observation and limitations of ultramicroscopic experimental conditions. The study of fine-grained sedimentary rocks in shale gas formations can facilitate the analysis of sedimentary rock genesis and depositional environments and has important petroleum geological significance. In the hydrocarbon system, shale typically acts as a source rock or cap rock, generating hydrocarbons that migrate to reservoirs with superior physical properties or preventing the diffusion of hydrocarbons in the reservoir. The discovery of shale gas indicates that shale has dual characteristics as a source rock and reservoir.

The research and development of shale gas has a history of nearly 200 years. The USA was the first country to conduct shale gas exploration and commercial development. The development of shale gas in the USA can be divided into four major stages: the early exploration and development stage (1821–1975), geological theory and exploration technology research stage (1975–2000), rapid development stage (2000–2006) and rapid increase in production stage (2007 to present) (Wang et al. 2012b). The considerable history of research and exploration experience in the USA, coupled with the gradual enhancement of drilling and extraction technologies in its industrial processes, has transformed shale gas to one of the main energy supplies in the USA at present. According to the U.S. Energy Information Administration (EIA), the shale gas production in the USA in 2020 was 7330×10^8 m^3, accounting for approximately 80% of its total natural gas production. In 2019, the U.S. shale gas production increased by 957×10^8 m^3, accounting for 73% of the global natural gas production growth rate (Zou et al. 2021). The rapid growth of shale gas production in the USA and the satisfactory form of the entire industry are attributable to factors such as the natural gas market prices and long-term support of the state, industry authorities and associations (Wang et al. 2012b). The U.S. Government's advanced leadership pertaining to the "energy independence" strategy is the key to the success of the "shale oil and gas revolution" (Zou et al. 2021). However, from the perspective of shale gas, the innovations and progress in geological theories and methods related to shale gas research have ensured that the USA the only region worldwide to have commercialized the exploration and development.

Owing to the clean energy attributes of shale gas and the bottleneck of conventional oil and gas exploration and development worldwide, the research, exploration and development of shale gas has spread from North America, led by the USA, to the rest of the world, and emerged as a hot spot for energy exploration in the recent decade. The exploitation of shale gas resources is of significance for China, which has scarce oil and gas resources but widely developed organic-rich shales. Although shale gas research in China started relatively late, it has undergone three stages, including fractured reservoir exploration and accidental discovery (before 2005), basic research and technical preparation (2005–2009) and industrial breakthrough (2010) (Zou et al. 2010a; Wang et al. 2012b). At present, considerable shale gas research and exploration is being performed in China, focused on innovations in geological theories, geological investigations, exploration technologies and methods and more refined research at the microscopic scale. Notably, China officially initiated shale gas exploration and development after the USA and Canada. Despite starting at the base level, an annual shale gas production of 100×10^8 m^3 has been achieved in China, followed by a historic growth to 200×10^8 m^3 in two years at depths of 3500 m and a breakthrough discovery at depths of 3500–4000 m, which represent notable accomplishments in the history of China's natural gas development (Zou et al. 2021).

1.1.1 Current Status of Foreign Research

The history of the shale gas industry in the USA can be traced to 1821, when the first natural gas well was drilled by Mitchell Energy in the Devonian Durdirk Shale in Chautauqua County, New York (producing gas from an 8 m thick shale fracture at a well depth of 21 m) (Curtis 2002). Since then, shale gas has been used for home lighting in rural Fredonia. The most developed shale-gas-rich zone in the USA is the Barnett Shale reservoir in the Fort Worth Basin in Texas, the successful development of which has received widespread attention (Li et al. 2011a). Since Mitchell Energy and Development Corporation drilled the first well in the Barnett Shale in 1981, 20 years' worth of effort was expended to extract gas from impermeable shale using hydraulic fracturing techniques. The subsequent rapid decline in gas recovery from vertical wells led to the development of horizontal well technologies (Boyer et al. 2011). In 1992, the first horizontal well was drilled in the Barnett Shale gas reservoir, and the development of Barnett Shale was accelerated by advancements in the hydraulic fracturing techniques (Li et al. 2011a). The application of horizontal wells and hydraulic fracturing technologies has enabled the economic development of unconventional shale gas reservoirs (Rahm 2011). In 2008, Barnett Shale became the largest gas-producing formation in the USA, delivering 7% of the nation's natural gas (Boyer et al. 2011).

Organic-rich shales are widely distributed in 48 states in the USA (all states except Alaska and Hawaii), with gas resources ranging from 1483×10^{12} m^3 to 1859×10^{12} m^3 (Li et al. 2011a). Due to the low clay content and high brittle mineral content of marine shales, which are conducive for hydraulic fracturing, shale gas has been typically produced from Paleozoic silicon-rich marine sedimentary formations (Jenkins and Boyer 2008; Chalmers et al. 2012). Moreover, several researchers have focused on lacustrine organic-rich shale. For example, Zhang et al. (2012c) studied the Wealden Shale of Early Cretaceous lacustrine deposits in NW Germany using organic geochemical methods to reconstruct the depositional environment and performed 3D digital simulations of hydrocarbons to clarify the characteristics of shale gas in different areas.

1. Current status of shale gas exploration and development in North America

According to statistics, in the mid-1970s, the USA entered the stage of large-scale development of shale gas. In 2000, the USA had five shale gas basins: Antrim Shale in the Michigan Basin, Ohio Shale in the Appalachian Basin, New Albany Shale in the Illinois Basin, Barnett Shale in the Fort Worth Basin and Lewis Shale in the San Juan Basin, with geological resources of 12.85 × 1012 –25.14 × 1012 m^3 and proven geological reserves of 6994.3 × 108 m^3 (Curtis 2002). Among these basins, Barnett Shale in the Fort Worth Basin is the largest shale-gas-producing area in the USA. The Newark East shale gas field discovered in the Fort Worth Basin in 1981 has emerged as the second-largest gas field in the USA, with annual shale gas production values of 217 × 108 m^3. By 2007, in addition to the five previously discovered shale-gas-producing basins in the USA, more than 20 basins, such as the Oklahoma Basin (Woodford Shale), Arkoma Basin (Fayetteville Shale) and Williston Basin (Bakken Shale) (Yan et al. 2009; Zhou et al. 2012) has been reported, with the Barnett, Marcellus, Fayetteville, Haynesville, Woodford, Lewis, Antrim and New Albany Shale associated with large-scale production.

With the success of shale gas exploration and development in the USA, regional exploration, investigation and testing of shale gas has also begun in Canada. Early exploration and development focused on the Middle Devonian Horn River Basin and Triassic Montney shale area in Northeastern British Columbia. In recent years, this exploration has gradually expanded to other provinces, such as Ontario and Quebec (Zhou et al. 2012). Preliminary predictions of shale gas resources in the Upper Cretaceous Wilrich Formation and its contemporaries, Jurassic Nordegg/Fernie Formation, Triassic Doig/Doig Phosphate/ Montney Formation, Exshaw/Bakken Formation and Devonian Ireton/Duvernay Formation in the Western Sedimentary Basin (eastern British Columbia and Alberta area), are approximately 2.83×10^8 m^3 (Yan et al. 2009), which demonstrates the potential of shale gas resources. The Canadian Society for Unconventional Gas (CSUG) considers the Colorado Shale, Jurassic and Paleozoic shales in the west (including the Bowser Basin in Northern British Columbia) and Devonian shales in the Southeast to have significant potential for development. In 2007, the shale gas production in Western Canada was approximately 8.5×10^8 m^3, with three horizontal wells showing a high daily production rate (9.4×10^4–14.2×10^4 m^3) (Yan et al. 2009). The Horn River and Montney shale gas play are present two of the largest shale gas reservoirs in Canada (Zhou et al. 2012).

2. Geological background of shale gas in North America

Along with technological innovations in shale gas development, US unconventional oil and gas companies are now focusing on the geological characteristics and regional geological background of shale gas reservoirs to increase the number of high-production wells and achieve a higher economic efficiency (Zhou et al. 2012). Shale gas basins in North America are mainly distributed in the areas in which passive continental margins have evolved into foreland basins and in the Paleozoic Craton terrane areas that are rich in conventional hydrocarbon resources (Montgomery et al. 2005; Pollastro et al. 2007; Li et al. 2009d; Zeng et al. 2011). The gas-bearing shale has various degrees of maturity, gas origins and lithofacies and complex depositional environments. The eastern petroliferous basins, such as the Appalachian Basin, Fort Worth Basin in the Gulf of Mexico and Western Canada Sedimentary Basin, are dominated by black shale, with inferences and interpretations of the depositional environment remaining open to debate (Li et al. 2009d). Loucks and Ruppel (2007) and Algeo and Barry (2008) proposed that the organic-rich black shale of the Devonian–Mississippian Barnett Formation in the Fort Worth Basin and central Appalachian basin corresponds to the restricted deep-water deposition in the foreland basin and was deposited below the storm wave-base and oxygen minimum zone (OMZ) at approximately 120–215 m with anoxic-anaerobic characteristics. These sediments are mainly composed of semipelagic ooze (from neritic shelves) and biological skeletal debris, and the sedimentation was largely accomplished by suspension mechanisms such as turbidity currents, mudflows and density currents, belonging to euxinic slope-basin facies. Biomarker data indicate that the main oil-generating Barnett Shale facies is marine in nature and was deposited under dysoxic, strong upwelling, normal salinity conditions (Hill et al. 2007a). The organic-rich mudstone in unit C of the Lower Jurassic Gordondale Formation in the Western Canada Sedimentary Basin (WCSB) was deposited on gentle slopes at water depths of up to 200 m (Ross and Bustin 2008). Hammes et al. (2011) investigated the geological background, depositional environment, stratigraphic characteristics and shale gas potential of the

Haynesville shale and indicated that the Haynesville shale was deposited in a euxinic and anoxic basin with a restricted environment. Romero and Philp (2012) studied the Woodford Shale in Oklahoma, USA, and concluded that high salinity conditions and water density stratification prevailed during the deposition of this formation.

These examples indicate that most of the black shales in North America were first deposited at high sea levels, and the nutrient-rich upwelling currents carried sufficient nutrients from deep-sea biogenic debris, resulting in high biological productivity and a strong reducing environment (Li et al. 2009d).

The analytical application of sedimentology and sequence stratigraphy is valuable in the regional search for shale gas resources and in predicting the shale gas potential. (Zhou et al. 2012). The sequence stratigraphic structure division of shale is less studied than that of coarse-grained clastic and carbonate formations (Bohacs 1998). Harris (2011) performed the sequence stratigraphic division of the Woodford shale in the Permian Basin based on sedimentological and geochemical characteristics and identified a second-order sea level decline cycle in the lower Woodford Formation, showing a lowstand systems tract with enriched total organic carbon (TOC). In contrast, the transgressive and highstand systems tract of the middle-upper Woodford Formation are not highly enriched in TOC. Hemmesch et al. (2014) studied the sea-level changes and sequence stratigraphic characteristics of the Upper Devonian Woodford organic-rich shale in the Palmyra Basin, West Texas and identified second- and third-order sea-level sequences based on shale layer characteristics. Abouelresh and Slatt (2012) classified the lower Barnett Shale in the east-central Fort Worth area into 1–7 depositional units and the upper Barnett Shale into 8–16 depositional units according to the rising and falling characteristics of sea level. The researchers indicated that the lower Barnett Shale was deposited in a low-energy, deep-water environment, slightly far from a terrigenous source area. In contrast, the upper Barnett Shale was deposited in an oxygenated shallower water environment and may have been influenced by tectonic periodic activities with frequent sea-level fluctuations. These aspects indicate that the division of sequence stratigraphy has a guiding significance for the development of organic-rich shales.

3. Petrological characteristics of gas-bearing shales in North America

(1) Rock types in Barnett Shale

The Mississippi Barnett Formation in the Fort Worth Basin is a classic shale gas system (Loucks and Ruppel 2007) whose geologic characteristics have been extensively studied in the recent decades. The Barnett Shale covers approximately 38 counties in the Fort Worth Basin in Texas, and the main gas-producing areas are located in the Northern and Southern parts of the basin. In the eastern part of the basin, the Barnett Shale unconformably overlies the Ordovician-age Viola Group and is overlain by the Pennsylvanian-age Marble Falls limestone (Jarvie et al. 2007).

Recognition of the different lithofacies is an important step in the evaluation of the gas in place, flow capacity and mechanical properties of Barnett Shale. The lithofacies vary in terms of the petrophysical and mechanical properties and organic content (Hickey and Henk 2007). The Barnett interval comprises a variety of facies but is dominated by fine-grained (clay- to silt-sized) particles. For example, the black shale of the Barnett Formation in the Fort Worth Basin consists mainly of calcareous siliceous mudstone and argillaceous lime mudstone, with intercalated thin beds of skeletal debris. Instead of detrital quartz, clay- to silt-sized microcrystalline quartz is the major component of the siliceous Barnett facies (Loucks and Ruppel 2007). Based on the analyses of porosity and organic geochemistry, a petrographic study of the conventional core Mitchell 2 T.P. Sims from Barnett has led to the identification of the following rock types: organic-rich black shale, fossiliferous shale, dolomite rhomb shale, dolomitic shale, phosphatic shale and concretionary carbonate (Hickey and Henk 2007). Abouelresh and Slatt (2012) analyzed Barnett Shale using cores, thin sections and SEM and reported several microsedimentary features that indicate that these common fine-grained rocks may have been transported and/or reworked by unidirectional currents. Six lithofacies were identified based on vertical facies transitions: (1) massive mudstone, (2) rhythmic mudstone, (3) ripple and low-angle laminated mudstone, (4) graded mudstone, (5) clay-rich mudstone and (6) spicule-rich mudstone. A number of sedimentary structures and textures indicate that a variety of current-related processes were active in sediment transport and deposition, likely including high-density flows, turbidity currents, storm currents and/or contour currents. The current-induced features of these facies likely included millimeter- to centimeter-scale cross- and parallel laminations, scour surfaces, clastic/biogenic particle alignment and normal- and inverse-size grading.

(2) Mineral composition of Barnett Shale

The mineral composition of the gas-bearing black shale is dominated by quartz, followed by clay minerals, carbonate minerals including calcite and dolomite and minor amounts of pyrite, feldspar and rhodochrosite. Moreover, this shale is characterized by a high organic matter content and traces of naturally occurring copper and phosphate minerals (Li et al. 2009d). The brittle mineral content is high, with quartz contents of 20–70% and carbonate mineral contents lower than 20% (Loucks and Ruppel 2007; Jarvie et al. 2007; Ross and Bustin 2008; Milliken et al. 2012). The clay minerals are mainly illite with some monazite (Bowker 2003), and the compositional maturity is relatively high. According to the minerals, textures, organisms and structures, the Barnett Shale can be grouped into three lithofacies: siliceous shale, clay shale and partly muddy carbonate (Loucks and Ruppel 2007; Jarvie et al. 2007; Ross and Bustin 2008).

The mineral components play an indispensable role in shale reservoirs, and the highest Barnett Shale production is associated with zones with 45% quartz and only 27% clay (Bowker 2003). The shale brittleness is correlated with the content of quartz and carbonate (Jarvie et al. 2007). Martineau (2007) suggested that different areas in the Barnett Shale contain different amounts of siliceous, carbonate and clay minerals, resulting in different fracture characteristics. Considering these aspects, Bowker (2007) highlighted that the Barnett Shale is associated with high shale gas yields owing to the high brittleness. The high brittleness allows the material to be effectively fractured hydraulically, and without these mineral fraction characteristics, shale gas extraction from the Barnett Shale would not be successful using the existing extraction techniques.

(3) Petrological characteristics of other shale formations

Other shale gas formations in North America have been extensively investigated in recent years. According to the research on petrological characteristics, the petrology of these formations is similar to those of the Barnett Shale, albeit with certain differences. For example, the Bossier gas-bearing shale reservoir in Western Texas is a mixture of shale, sandstone and siltstone (Jarvie et al. 2007). Hemmesch et al. (2014) identified seven lithofacies for the Woodford organic-rich shale: shale, phosphate nodule-bearing shale, dolomite, chert layer, radiolarian-bearing calcareous laminae, biotite-bearing mudstone and siliceous rock. The bulk mineralogy of Devonian–Mississippian black shales in the Western Canada sedimentary basin is dominated by biogenic quartz, which accounts for 58–93% of the bulk rock. The low siliceous content is attributable to the high content of carbonate minerals, and the clay minerals are mainly illite, with a small amount of unevenly distributed kaolinite (Ross and Bustin 2008). Most researchers have reported consistent results regarding the origin of silica in siliceous shales in North America (Bowker 2003; Loucks and Ruppel 2007; Ross and Bustin 2007). Similar to the Mississippi Barnett Shale in the Fort Worth Basin, other organic-rich shales in North America consist primarily of brittle minerals, with clay mineral contents lower than 50%. The clay mineral content of the well-known Green River Shale in the Uinta Basin is lower than 10%; the Heather Shale in the North Sea contains 53–57% quartz and exhibits a clay mineral content lower than 5% (Hunt 1996). By analyzing the clay minerals, organic matter content, thermal maturity, humidity and pore structure of the Fayetteville Shale, Bai et al. (2013) concluded that the relative proportions of quartz–carbonate–clay minerals influenced the physical properties of the rock, and the rock composition was the most fundamental factor affecting the effectiveness of drilling and hydraulic fracturing.

In general, the gas-bearing shale in North America is mainly composed of organic-rich siliceous shale, with a small amount of calcium, low clay mineral content and a large number of biogenic clasts. Influenced by the biological activity, the source of silica is predominantly biogenic, and it rarely has a terrigenous origin.

4. Organic matter characteristics of North American shales

The gas-bearing shales in North America are important hydrocarbon source rocks. Curtis (2002) compared the geological and geochemical characteristics of shale formations in the five major basins in North America, including the vitrinite reflectance, organic carbon content, favorable shale thickness and adsorbed gas volume. In particular, the organic carbon content of the Barnett Shale ranges from 2.0 to 7.0%, with an average of 4.5%. The organic carbon content of the Antrim Shale and the New Albany Shale is slightly more than 20%, while the Ohio and Lewis Shales exhibit lower organic carbon contents, mostly < 5%. The adsorbed gas content for the Antrim shale and New Albany shale ranges from a 13% to 70%. The genesis of shale formation in the major shale-gas-producing basins in the USA has biogenic, thermogenic and mixed origins, i.e., low thermal maturity shale (e.g., Antrim Shale with R_o = 0.4–0.6%), high thermal maturity shale (e.g., Barnett Shale with Ro = 1.0–2.1%) and mixed high-low thermal maturity shale (e.g., Ohio Shale with Ro = 0.4–1.3% and New Albany Shale with R_o = –1.3%). Moreover, the kerogen type is predominantly type I–II$_1$ (Curtis 2002; Montgomery et al. 2005). The Devonian–Mississippian strata of the Western Canada Basin exhibit a TOC between 0.95.% and 7% (Ross and Bustin 2008). The Woodford Shale in Oklahoma, USA, has a high organic carbon content of 5.01–14.81%, and the organic matter is predominantly type II (Romero and Philp 2012).

Notably, the organic matter content and thermal maturity vary considerably in different shale gas basins. Current shale gas exploration practices in the USA show that the shale maturity is generally greater than 1.3% in shale-gas-producing areas (Martineau 2007; Pollastro et al. 2007), with a maximum maturity of 4.0% in Southern West Virginia in the Appalachian Basin. Moreover, shale gas is produced only in areas with high maturity levels. The maturity of organic matter in shale reservoirs does not considerably influence the shale gas accumulation, although a higher maturity is favorable for shale gas accumulation (Jiang et al. 2010). In addition, clay minerals are hydrophilic, while organic matter is methanophilic (Zhang et al. 2012b). Therefore, the organic matter content considerably influences the shale gas sorption capacity (Chalmers and Bustin 2008a, 2008b).

5. Characteristics of shale gas reservoirs in North America

The pores in shales, which appear to be singular are isolated, are in fact connected by straight and narrow throats, and the pores are characterized by complex internal structures and porous complexes. The porosity of organic-rich shales in the five major shale gas basins in the USA is typically more than 5%, and the New Albany Shale has a porosity higher than 10% and a low permeability (Curtis 2002). The pores in organic shales can be classified as microporous (pore size up to 2 nm), mesoporous (pore size 2–50 nm) and macroporous (pore size greater than 50 nm) (Chalmers and Bustin 2008a, 2008b; Ross and Bustin 2009). The adsorption capacity of shale is also associated with micropores (Ross and Bustin 2009). The micropores in the Barnett Shale include intragranular, intergranular and intergranular micropores of authigenic minerals. The dominant type pertains to intergranular micropores formed by hydrocarbon generation evolution, clay mineral transformation, fossil silicification and framboidal pyrite formation (Loucks et al. 2009; Ross and Bustin 2009).

Research on shale reservoirs is largely dependent on advanced analytical techniques, with scanning electron microscopy, backscattering and two-dimensional (2D) and 3D imaging techniques being widely used. For example, Slatt et al. (2011) used polarized light microscopy and scanning electron microscopy (SEM) to investigate the pore characteristics of favorable formations of the Barnett and Woodford shale gas systems in North America, distinguishing several types of pores: porous floccules, organoporosity, intraparticle pores and microfractures. Curtis et al. (2012) used focused ion beam techniques combined with backscattering or SEM to observe shale gas core samples from different horizons in nine regions of North America and determined the 3D digital features of the corresponding SEM images to visualize the pore size distribution, pore structure characteristics, connectivity and coexisting minerals. The researchers indicated that pores sized approximately 3 to 6 nm have the highest number but contribute less to the total pore volume than micro- to mesopores, indicating that shale gas reservoirs are dominated by micropores and mesopores with a low porosity. Clarkson et al. (2013) investigated the pore structure of North American shale gas reservoirs by applying small-angle and ultrasmall-angle angle neutron scattering techniques combined with low-pressure adsorption and high-pressure mercury intrusion techniques and concluded that the porosity was determined by the pore size.

Various types of pores exist in shale gas reservoirs, and the evolutionary process is extremely complex due to the combined influence of diagenesis and hydrocarbon generation. Chalmers et al. (2012) comprehensively evaluated the development characteristics of pores in shale reservoirs through physical, organic geochemical and compositional analyses of shale gas favorable formation samples from different regions of North America. The researchers highlighted that the porosity was related to the pore size, the micropores contributed the least to the porosity, and the pore evolution was characterized by an increase in the number of micropores and a decrease in the number of mesopores and macropores as the porosity decreased. Moreover, the researchers clarified the relationships among the distribution characteristics of the porosity and pore size versus the content, type and evolution degree of organic matter and mineral compositions. Notably, the mesopores and macropores are mainly intercrystalline pores, intergranular pores or organic matter pores and do not follow the direction of shale laminae. The pore size is inversely proportional to the specific surface area (Beliveau and Honey 1993). Because the micropores occupy a higher specific surface area than mesopores and the smallest specific surface area pertains to macropores, the porosity decreases with the increase in the number of micropores and decrease in the number of mesopores and macropores. Pores develop during the thermal maturation of kerogen and generation of hydrocarbons (Jarvie et al. 2007). As kerogen matures, the porosity of the micropores increases (Chalmers and Bustin 2008a; Ross and Bustin 2009). Because mesopores and micropores are the main constituents of shale pore space, they are of economic importance for shale gas (Keller et al. 2011), which exists in the form of adsorbed gas. The macropores in the Barnett Shale are mainly derived from the thermal degradation of kerogen, in which kerogen undergoes thermal cracking, leading to petroleum generation (Jarvie et al. 2007; Chalmers and Bustin 2007; Loucks et al. 2009; Modiaca and Lapierre 2012; Mastalerz et al. 2013). Mastalerz et al. (2013) studied the evolution of the pores during diagenesis of the New Albany Shale by analyzing the organic matter, mineral composition and physical and gas-bearing properties of the shale. The researchers reported that the pores did not follow a constant trend during the diagenetic process. Nevertheless, with the generation of hydrocarbons, the porosity exhibited several minima and fluctuations. As the maturity increased, the porosity and total pore volume varied with the pore size distributions and pore types. Thus, the variation in the porosity is considerably influenced by the hydrocarbon generation of kerogen and organic matter transformation due to hydrocarbon migration.

Jarvie (1991) noted that the pore-space changes in organic-rich shales occurred owing to the transformation of organic matter during hydrocarbon generation. Peters (1986) suggested that during early diagenesis (*R*o of 0.6%), up to 0.6 wt% hydrocarbons were transformed by the kerogen, while in middle diagenesis, the porosity decreased with the increasing maturity of the organic matter. During the late maturation stage, the number of available open pores decreased, and the fluid flow was restricted as the early pores filled with oil or solid bitumen. The size of the pore throat was noted to be closely related to the rock porosity and permeability (Nelson and Batzle 2006), and Jarvie et al. (2007) suggested that the blocking of the roaring channels by asphaltene residues led to the low permeability. As the thermal evolution progressed, the porosity increased with the conversion of oil and bitumen to dry gas, which created microfractures and allowed the formerly closed pore system to open.

In conclusion, North American shale gas reservoirs contain mostly micropores to mesopores exhibit a high porosity and low permeability, and the macropores are not developed. The porosity is related to the type of pore, i.e., the size of the pores determines the porosity. The generation of and variation in the pores are mainly associated with the diagenesis, hydrocarbon generation and evolution of organic matter.

6. Research on diagenesis

The study of diagenesis is significance for the porosity and permeability analysis of conventional reservoirs, comprehensive evaluation of reservoirs and reservoir and gas production prediction (Yang et al. 2012). For shale gas, diagenesis controls not only the thermal evolution degree of organic matter, but also the mineral composition of shale, especially the composition of clay minerals. Additionally, the intensity of diagenesis considerably influences the development of reservoir space (Liang et al. 2012). Diagenesis influences the mechanical properties of shale, with compaction transforming loose and soft clays to mudstones and shales and cementation of minerals such as carbonates and quartz causing a shift in the mechanical properties from plastic to brittle sedimentary rocks (Bjørlykke and Karre 1997). Research on shale gas diagenesis has gradually attracted attention. Laughrey et al. (2011) comprehensively analyzed the diagenetic history of the Marcellus Formation in the Sullivan area of Pennsylvania and indicated that when the sediments of the Marcellus Formation were buried at a depth of approximately 500 m, early diagenesis was associated with mechanical compaction and mudstone dewatering. As the buried depth increased, the chemical compaction corresponded to quartz cementation and transformation of clay minerals. The organic pores developed significantly during late catagenesis, and this process continued in the metagenesis state at depths greater than 8 km. Milliken et al. (2012) analyzed the porosity, permeability and TOC of Barnett Shale samples with high maturity (*R*o: 1.52–2.15%) in the eastern Fort Worth Basin and noted that the reservoir factors were not correlated with the composition and structural characteristics of the rock due to diagenesis. Compaction and cementation caused the loss of most of the primary intergranular pores. Most of the pores were thus secondary pores filled with asphaltenes and the clastic particles were replaced.

Notably, shale gas reservoirs, as hydrocarbon-producing layers, are subject to both organic and inorganic modifications during burial and diagenesis, and the formation process is complicated. Therefore, the existing studies on shale gas diagenesis are not sufficient, and a comprehensive detailed and systematic study of the diagenesis and diagenetic evolution of shale gas reservoirs and their impact on the reservoir storage space must be conducted. Because shale-gas-bearing shales are both source rocks and reservoirs in hydrocarbon systems, the rock types are mainly muddy shales with high clay mineral contents. The diagenesis of these rocks can be examined using the research methods of hydrocarbon source rocks. For example, in hydrocarbon source rocks, the clay mineral assemblage and diagenetic evolution are clearly influenced by the acidity and alkalinity of the formation fluids and fluid composition (Niu et al. 2000).

7. Characteristics and evaluation of shale gas reservoirs in North America

From a petroleum geological viewpoint, a large amount of natural gas is generated from source rocks through a series of geological conditions and discharged in large quantities under continuous pressure. These gas migrates to permeable strata such as sandstones and carbonate rocks and accumulates in structural or lithologic gas reservoirs. The part remaining in the fine-grained sedimentary rock system forms the shale gas resources (Tian et al. 2005). This model of the generation and storage of shale gas simplifies the reservoir accumulation process and integrates the gas reservoir characteristic analysis and reservoir evaluation. This comprehensive analysis process is different from that of evaluating conventional oil and gas reservoirs.

Shale gas reservoirs in North America are large-scale continuous accumulations and exist in three states: adsorbed gas, dissolved gas and free gas, with most of the gas corresponding to adsorbed and free gas. Shale is derived from biogenic, thermogenic and mixed types of sources, with thermogenic sources being dominant (Du et al. 2011; Xiao et al. 2013). The same set of shale horizons in the same basin, affected by different stages of thermal evolution, exhibit different types of gas reservoirs. For example, the Woodford Shale in the Late Devonian-Early Mississippian of Oklahoma, USA, exhibits different types of biogenic gas and thermogenic gas reservoirs in different stages of thermal evolution of organic matter in different areas (Cardott 2012). Thermogenic shale gas reservoirs are mainly controlled by the thermal maturity of shale, while the main controlling factors for biogenic shale gas reservoirs are the formation water salinity and level of fracturing (Li et al. 2009d).

The analysis and evaluation of shale gas reservoirs, as comprehensive research tools for the exploration and development of shale gas resources, have been performed for each shale gas basin and formation. For example, Ross and Bustin (2007) analyzed the shale gas potential by studying the organic matter content, organic matter maturity and gas-bearing properties of the Early Jurassic Gordondale mudstone in the Peace River region in Northeastern British Columbia, Canada. Bowker (2007) studied the type, thermal evolution and conversion characteristics of organic matter, combined with the adsorbed gas volume and mineral composition of shale and analyzed the shale gas system. Ross and Bustin (2008) comprehensively analyzed the shale gas potential by performing stratigraphic and tectonic studies of the Devonian–Mississippian system in the Western Canada Basin, examining the organic matter and mineral composition and investigating the gas-bearing properties. Chalmers and Bustin (2012b) examined the shale gas potential of the Cretaceous Shaftesbury Formation in Northeastern British Columbia, Canada through organic geochemistry, mineralogy, porosity, and gas content analyses and concluded that the present burial and organic matter maturity of shale formations influences the hydrocarbon generation capacity more notably than the TOC content. The United States Geological Survey (USGS) identified the Lower Cretaceous Pearsall Formation in Southern Texas as a potential shale gas resources through hydrocarbon investigations of the Mesozoic strata in the Northern Gulf Coast. Moreover, Hackley (2012) verified the potential of shale gas reservoirs by analyzing the lithology, stratigraphy and depositional environment of the Pearsall Formation.

The examples of shale gas development in North America demonstrate that shale gas reservoirs are dominated by adsorbed gas, and the shale adsorption capacity determines the amount of adsorbed gas. The adsorption capacity of shale is related to factors such as the mineral composition, organic matter content, kerogen type, formation water content, pore size and structural characteristics and thermal evolution stage of the organic matter. Organic-rich shale with a higher organic matter content, higher thermal evolution level, and lower formation water content corresponds to higher adsorbed gas volumes (Ross and Bustin 2007, 2009; Hao et al. 2013). The effect of the mineral composition on the gas adsorption can be observed by the fact that quartz and carbonate minerals have lower internal surface areas and therefore adsorb a lower amount of gas (Ross and Bustin 2007). Ross and Bustin (2009) reported that in dry conditions, illite and smectite exhibit higher gas adsorption capacities than kaolinite. Schettler and Parmoly (1990) indicated that the main adsorption space in the Devonian shale of the Appalachian Basin is provided by illite, and the contribution of the kerogen to the adsorption space is less significant. Zhang et al. (2012b) noted that in organic-rich shale, the adsorption capacity of the minerals is lower than that of organic matter. Hill et al. (2007b) analyzed the Barnett Shale in the Fort Worth Basin and concluded that the volume of shale gas is related to the organic matter content, thickness and maturity of the shale.

Overall, in shales with a low matrix porosity, the gas-bearing properties and microfracture development characteristics influence the shale gas production capacity (Curtis 2002). The gas-bearing property of shale gas is related to the content and type of organic matter, level of thermal evolution, rock type, mineral composition and physical properties. In fact, the rock mineral composition and organic matter characteristics are the basis of shale gas development. The level of organic matter thermal evolution determines the type of shale gas reservoir and storage space. The mineral composition, organic carbon content and organic matter maturity of shale rocks are the three most important factors for shale reservoir development (Curtis 2002; Jarvie et al. 2005). Therefore, the evaluation of shale gas reservoirs, analysis of shale gas deposits and prediction of potential resources are based on the fundamental understanding and evaluation of shale gas in terms of the petrography, rock composition, organic matter type and maturity and reservoir properties. The diagenesis and original components of shale must be considered when evaluating reservoirs (Ross and Bustin 2007).

While fractures are necessary to ensure high gas production in the Barnett Shale, the macroscopic fractures do not considerably influence the hydraulic fracturing as they are filled with carbonate minerals (Bowker 2003). In the Barnett shale, most of which is a closed system, the organic and inorganic gases produced by hydrocarbon evolution are not immediately released, leading to the generation of high pressures (Jarvie et al. 2007). Gaarenstroom et al. (1993) estimated the oil and gas cracking capacities and suggested that the pressure generated by 1% oil cracking in a closed system could attain the threshold for rock cracking. This finding suggests that the microfractures and migration channels in the Barnett shale are at least partially derived from the generation of early hydrocarbon and nonhydrocarbon gases and from the secondary cracking process of hydrocarbons after oil and gas are generated. Bowker (2007) found that although Newark East has been the largest natural gas field in the Fort Worth Basin since 2001, due to the development characteristics of the facture system, the gas production in different areas of the Fort Worth Basin varies considerably. The Barnett Shale exhibits low gas production near faults and folds, and the structural fractures determine the gas production of the Barnett shale.

Montgomery et al. (2005) believed that the development of shale gas must be based on a comprehensive study of the geological characteristics, geochemical analyses and technological developments. The geological characteristics can be examined to clarify the characteristics of shale reservoirs, and geochemical analyses can clarify the potential productivity of shale and formation pattern of shale gas. After the reservoir and resource potential have been determined, the technological developments determine the productivity of shale. Therefore, the analysis of the geological characteristics and geochemistry of shale gas reservoirs based on the latest hydraulic fracturing and horizontal well technologies is a fundamental and decisive part of shale gas exploration and development, and the study of its sedimentological, petrological and diagenetic effects and organic geochemical aspects eventually determines the effectiveness of shale gas reservoir development.

According to the shale gas production practices of the USA, the favorable reservoir characteristics of thermogenic shale gas can be summarized as follows: TOC $\geq$ 2%, shale thickness $\geq$ 15 m, $1.1\% < Ro < 3\%$ and quartz content $\geq$ 28% (Li et al. 2009d). Shale gas reservoirs with high production and economic benefits correspond to a wide distribution area, moderate burial depth, large thickness (>30 m), organic matter abundance (TOC $\geq$ 2%), kerogen type I or II_1, moderate maturity ($1.1\% < Ro < 2.5\%$), high gas content (3–10 m^3/t), low water production, moderate clay content (<40%), high brittleness (i.e., low Poisson's ratio and high Young's modulus) and surrounding rock, as these aspects facilitate hydraulic fracturing control (both the upper and lower strata are limestones) (Curtis 2002; Montgomery et al. 2005; Pollastro et al. 2007; Li et al. 2009d). Diverse reservoir rock types, dominant free gas, well-developed caprocks and overpressure reservoirs are also characteristics of high-production shale (Xiao et al. 2013). Curtis (2002) compared the geological and geochemical characteristics of shale gas, such as the vitrinite reflectance, organic matter content, favorable shale thickness and adsorbed gas volume, of favorable shale gas strata in five major basins in North America and concluded that the factors affecting shale gas production rates can compensate for one another.

1.1.2 Research History and Status of Shale Gas in China

China is the third country initial to shale gas exploration, which is later than the USA and Canada. The resource of shale gas is rich in China, which is approximately equal to it in the USA (Zhang et al. 2009). It is no doubt that the shale gas will be the new growth point of natural gas in China (Ye and Zeng 2008). The shale gas has been entered into the commercial stage in China since the industrial shale gas flow had been gotten in the well JY-1 in 2012. Although the annual output of shale gas is gradually rapid growth, the shale gas is mainly coming from the middle-shallow burial depth layer (1000–3000 m in depth) in China. There is a big gap between the annual output of shale gas in China and USA (Zhang et al. 2021). In present, the breakthrough of shale gas has been gotten in the marine shale gas in the Southern China; however, there are a few of questions and challenges in the shale gas industry in China now. The enrichment condition of the shale gas has not been well understood (Jiang et al. 2020).

Since the 1960s, the industrial gas flow has been intermittently observed in the fractured shale reservoirs in different basins. However, it has not been paid enough attention. The preliminary research on shale gas resources was started in 2004 by the Strategic Research Center of Oil and Gas Resources, Ministry of Land and Resources, China (SRCOGR, which is presently known as the Ministry of Natural Resources) and China University of Geosciences (Beijing), which was the first time focused on the shale gas in China. Since 2006, the shale gas exploration has been started in China symbolized by the project "The potential evaluation and favorable area prediction for shale gas resource in the key areas of China". This project is belonged to the national major project named "The strategic survey of oil and gas resources and favorable area prediction" which is carried out by the institutes organized by the SRCOGR, such as, China National Petroleum Corporation (CNPC), Sinopec Corporation (SPC), Chengdu Center of Geological Survey, Chongqing Coal Geology Research Institute. The shale gas in China was assessed by dividing the China into five evaluation units, such as the Upper Yangtze and Yunnan-Guizhou-Guangxi unit, Middle-lower Yangtze and South-east unit, the north and Northeast unit, the Northwest unit and the Tibetan unit. Five pilot areas were set up as the Sichuan–Chongqing–Guizhou–Hubei pilot area for shale gas, the Qijiagulong depression in Songliao basin pilot area for shale gas and oil shale, the shale gas-tied gas-coalbed methane pilot area in the Northern Qinshui basin, the lower Yangtze and Anhui-Zhejiang pilot area for shale gas. Several key blocks were selected out too. The companies major in the shale gas exploration were set up by the CNPC, SPC,

China National Offshore Oil Corporation (CNOOC), which symbolizes the beginning of shale gas exploration in China. Meanwhile, the industrial shale gas flow was gotten in some local areas. Furthermore, the shale gas research teams were gradually established in research institutions and private enterprises too. In present, the disciplines boom has been formed in shale gas exploration. The formal shale gas field development was kicked off with the setting up of Fuling shale gas field, in July 2014.

According to the Resource evaluation results of the Ministry of Natural Resources, PRC (formerly Ministry of Land and Resources) in 2015, the shale gas resource is about 121.86×10^{12} m^3 in China. The recoverable resources of shale gas are 21.81×10^{12} m^3. The marine shale gas is 13.00×10^{12} m^3, mainly enrichment in the Sinian, Cambrian, Silurian Formation and so on, in the Upper Yangtze area and Western Tarim basin. The terrestrial shale gas is 3.73×10^{12} m^3 which is enrichment in the Ziliujing Formation in Sichuan Basin, the Yanchang Formation in Ordos basin, the Shahejie Formation in Bohai Bay Basin and the Qingshankou Formation in Songliao basin. 5.08×10^{12} m^3 is enrichment in the marine and continental transitional facies shale, such as the Permian Formation in the middle-Upper Yangtze, the Carboniferous–Permian Formation in the Ordos basin, Junggar Basin, Tarim Basin and so on (Sun et al. 2020). During the 2016–2020, the shale gas has entered in the rapid development stage. A lot of the innovation and breakthrough were gotten in this period. In present, the shale gas has been seeming as an important field for increasing natural gas storage and production in China (Zhao et al. 2019). The shale gas development has been accelerated by CNPC. The production capacity construction has carried out in the shallow shale gas resources (<3500 m in depth) in Changning, Weiyuan and Zhaotong. Up to the end of 2019, the accumulative proved shale gas geological reserves have been $10{,}610 \times 10^8$ m^3. The annual output of CNPC is 80.3×10^8 m^3 and 116.1×10^8 m^3 in 2019 and 2020, respectively. The marine shale gas in Wufeng-Longmaxi Formation has been economically exploited by the Sinopec, in the Fuling area and Weirong area. The accumulative proven shale gas geological reserves amounted to 7255×10^8 m^3 at the end of 2019. The annual outputs of SINOPEC were 73.4×10^8 m^3 and 84.1×10^8 m^3 in 2019 and 2020, respectively (Zou et al. 2021). The Sinopec reported that the accumulative proven shale gas geological reserves were 9408×10^8 m^3, at the end of 2020. The favorable area for shale gas is mainly distributed in Sichuan Basin and its surrounding area (Cai et al. 2021). It is deduced that the annual output of shale gas in China will be 300×10^8 m^3 in 2025, and likely will be 350×10^8 m^3–400×10^8 m^3 in 2030. The shale gas will occupy a major proportion in the increasing of gas in China. The major contributor will be the marine shale gas from the deeper shale strata. The annual output of the marine shale gas likely be 150×10^8 m^3–200×110^8 m^3 in 2030. The shale gas exploitation in Sichuan Basin will accelerate the Sichuan–Chongqing area as the biggest oil and gas production area in China. The Sichuan Basin will be constructed as the "Daqing gas field" (Zou et al. 2021).

The shale gas in China is characterized with different enrichment types, wide distribution and with huge potential. Shale strata were well deposited in the strata of different ages. The shale strata are deposited in various environment, such as the marine, terrestrial and the transitional facies. The organic-rich shale is commonly act as the source rock in the petroliferous basin. The Paleozoic shale in Sichuan Basin is mainly marine sedimentary, with stable regional distribution, large thickness, organic matter enriched and high thermal evolution. A large amount of oil and gas has been observed in Paleozoic shale which is a realizable field for shale gas exploration and development (Zou et al. 2010a). The Paleozoic shale in Southern China has experienced a complex tectonic evolution. It is with the similar geological conditions and tectonic evolution characteristics to the typical shale gas basins in eastern USA (Long et al. 2009). Paleozoic shale gas is an important field to the exploration and production of shale gas in China.

At the beginning of shale gas study, the scholars were focus on the enrichment mechanism (Zhang et al. 2003, 2004), the accumulation condition (Zhang et al. 2008a; Chen et al. 2009a; Wang et al. 2009a, b, c, Nie et al. 2009a), the evaluation of favorable area (Zhang et al. 2008b; Zhao et al. 2008; Cheng and Pi 2009; Pan and Huang 2009; Li et al. 2009d) and so on. The research degree is relatively high in the middle-large petroliferous basin (Zhang et al. 2008; Wang et al. 2009a; Huang, 2009b). The study is the most advanced in the marine shale in south China. Based on the enrichment mechanism study and the comparison of geological conditions of shale gas in China and the USA, Zhang et al. (2009) concluded that the geological conditions are superior in China for shale gas, with the same shale gas potential as the USA. The Shale gas in China is characterized by high abundance of organic matter, high thermal evolution and stronger post-renovation. Meanwhile, the shale gas is characterized by coexistence of marine and continental facies, dominated by the sedimentary zones and complex distribution. For the absent of core samples for shale gas, most of the early studies were based on the conventional oil and gas exploration data, coalbed methane and solid mineral exploration data. The shale samples were coming from the outcrops or the shallow layer. Due to the lack of practices data, many analogies are carried out by referring to foreign materials such as the USA (Xu and Bao 2009, Nie et al. 2009a, Zhang et al. 2004; Zeng et al. 2011; Chen et al. 2011a). Meanwhile, to a certain degree, the progress of the shale gas industry is restricted by the deficiency of the shale gas exploration theories and methods (Li and Zhao 2009). Basing on the geological characteristics and accumulation conditions of Shale gas in China, Li et al. (2012c) concluded that Marine shale gas exploration prospect in China is the best and Sichuan Basin and its surrounding areas are the most realistic. It needs to be verified the potential of shale gas in the marine and continental transitional shale and coal measure shale. Lacustrine shale gas is mainly distributed in the central area of the depression and with a certain exploration potential. In conclusion, in the early stage of shale gas research in China, a side range of interests was focused on the marine shale gas in the middle and Upper Yangtze area. And, the research degree of it is most advanced in China. Since 2009, great progress has been gotten in the shale gas industry in Southern China. The key factors for the trap and enrichment of Chinese-style shale gas are concluded, such as, the shale gas favorable area is dominated by the "sedimentary facies and preservation conditions", the sweet-spot areas are dominated by the "Tectonic types and tectonic processes" (Guo 2016).

The study area of this book is the Southern Sichuan Basin and its surrounding area. The target layer is the Ordovician Wufeng Formation-Silurian Longmaxi Formation. Hence, this work is basing on the modern research of marine shale gas in Southern China, mainly involving Ordovician Wufeng-Silurian Longmaxi Formation and part of Cambrian Niutiantang/Qiongzhusi Formation in Paleozoic.

1. The geological setting of the shale gas in Southern China

There are three regional source rock layers in Paleozoic in south China. The excellent source rock is characterized with the siliceous-shale and the dark shale in the Lower Cambrian Niutiantang (or Qiongzhusi) Formations, the Upper Ordovician Wufeng Formation to Lower Silurian Longmaxi Formation. The high-quality source rocks are mainly argillaceous and siliceous rocks in the Upper Permian Longtan Formation or Wujiaping Formation and Dalong Formation. The organic-poor limestone is non-source rock or poor source rocks (Fu et al. 2011). Similar to shale gas in north America, favorable shale gas formations in China are deposited in the deep-water platform located at the foredeep belt (or depression zone) in the early foreland basin, (Chen et al. 2011a). The Yangtze platform has been in the stage of consecutive thermal deposition since the late Sinian. The structural pattern is characterized with "two basins separated by one platform" in the early Paleozoic. The shale was mainly deposited in the depression and slope of platform margin. The shale mainly deposited in the intraplatform depressions and platform margins, with the restriction of the geographical pattern of passive continental margin during the lower Cambrian Niutiitang Formation depositional period. The shale is characterized the thinner thickness and high organic matter content in the deep-water environment and thicker thickness and low organic matter content in the shallow-water environment. In late Ordovician to early Silurian, the south China plate initial converged with the Yangtze plate. The shale in Wufeng-Longmaxi Formations is deposited in the plate convergence in the early formation of depression. In the depression, the shale is characterized with thicker thickness, high organic matter content. The organic-rich shale is relatively thinner at the outside of the depression (Chen et al. 2011a; Liang et al. 2009).

The evolution of the middle-Upper Yangtze region began with the breakup of Rodinia continent in Nanhua Period. From Sinian to Early Ordovician, the whole middle and Upper Yangtze region was in the background of extension and splitting. A stable middle-Upper Yangtze craton basin was formed within the continental block (Zeng et al. 2011). There is a difference in the spatial and temporal distribution of the Sinian-Silurian cratonic basin in the middle-Upper Yangtze basin. It has undergone two stages from extension to compression, with the evolution process ordinally from rift basin to fissure basin and depression basin. The first stage is from Sinian to early Ordovician. It is characterized by the evolution from early rift to fissure basin in the extensional environment, and the formation is mainly carbonate. The II stage is from late Ordovician to early Silurian. The study area underwent an extrusion stress environment. The basin is a successional compressional depression basin within the craton. The craton margin is generally extruded and uplifted. As a whole, the basin pattern is restricted by uplift segmentation. The sedimentary formations are mainly clastic and mixed type. The lithological association is characterized with the carbonate is gradually drop, the clastic rocks is gradually increasing in the upward (Huang et al. 2011a).

Basing on the outcrops, well test, core samples and the tests of sedimentology, geochemistry, reservoir engineering, Dong et al. (2010) concluded that the distribution of organic-rich black shale in Qiongzhusi and Wufeng-Longmaxi Formations were dominated by the neritic facies to deep shelf sedimentary environment during the Early Paleozoic in the Upper Yangtze Area. The lower Paleozoic formation is the favorable layer for the shale gas. The organic-rich shale in Longmaxi Formation was deposited during the transgressive systems in Early Silurian (Chen et al. 2009a, Huang et al. 2011a, Li et al 2012c). Meanwhile, the organic-rich shale is deposited in the area with slow regressive system, where is favorable to the shale gas formation. Then the sea level began to decline slowly and entered the high stand systems tract, which was mainly semi-deep water shelf deposition. The sandy content is gradually increased with the increasing of silty shale, argillaceous siltstone and siltstone in upward, which composed the regressive sedimentary sequence.

2. Petrological characteristics

(1) Lithofacies and mineral composition

There is a huge difference between different scholars on the lithofacies division of the Longmaxi Formation in Sichuan Basin and its surrounding area. The petrology has rapidly developed in the last ten years, with a lot of study on the lithofacies (sedimentary microfacies) had been carried out. There are many lithofacies division schemes. It can be roughly divided into three types as following,

(1) The lithofacies are classified mainly according to the mineral composition, assisted with the sedimentary structure. However, no unified standard has been established. For example, one division scheme is proposed by Zeng et al. (2011), which divided the Longmaxi Formation into three lithofacies, such as the carbonaceous shale facies, silty shale facies, marl facies. Liang et al. 2012 proposed that there are five lithofacies in Longmaxi Formation, which is carbonaceous shale, siliceous shale, silty shale, calcareous shale, ordinary shale. Zhang et al. (2012b) concluded that the Longmaxi shale is composed by black, gray-black and dark gray calcareous and siliceous shales, sandy shales or thin siltstones. The laminated structures are commonly observed in the Longmaxi shale. Wang et al. (2014b) proposed that there are eight lithofacies in Longmaxi Formation, which is siliceous shale, calcium siliceous shale, micritic calcareous shale, calcareous laminate shale, shell marl, wavy bedding shale, dolomitic shale, phosphorus shale. Zhao et al. (2016) proposed that Longmaxi shale is composed by the siliceous shales, clayey shales, silty shales, calcareous shales, core-bearing argillaceous limestone/calcareous mudstone, siltstone-fine sandstone and bentonite. Another classification scheme is proposed by Wu et al. (2016). Basing on siliceous mineral (quartz and feldspar), carbonate minerals (calcite and dolomite) three mineral and clay mineral, it can divide the shale into siliceous shale, calcareous shale, clay shale and mixed type of shale. The shale can be classified into more than 30 types according to the contents of the three components.
(2) The shale can be classified in different types, according to TOC content and mineral composition. Jiang et al. (2016) classified the Longmaxi shale in Weiyuan area into 11 types, for example the organic-rich siliceous shale facies and organic-rich carbonate-siliceous shale and so on.
(3) The shale was divided into different types, according to the laminar and mineral composition characteristics too. Firstly, Liu et al. (2011) divided the Wufeng-Longmaxi shale in east Sichuan Basin into eight lithofacies, including the stratified-non-stratified mud/shale dolomitic siltstone, stratified calcareous mud/shale argillaceous siltstone, stratified-non-stratified silty mud/shale, silty-fine-grained sandstone, calcareous nodules, organic-rich shale without lamellation. The mineral mainly composed by quartz or carbonate. This method was optimized by Ran et al. (2016) by taking the silica content into account. The Longmaxi shale can be divided into 9 types, such as the silicon-poor non-parallel laminated shale, silicon-rich parallel laminated shale. Basing on the core samples observation and thin sections identification of Well CX-1, Chen et al. (2013a, b) divide the Longmaxi shale into lamellar shale, lamellar calcareous shale, lamellar dolomitic shale, lamellar carbonate shale and lamellar silty shale. The lamellar is composed by the mud grade of quartz, feldspar, clay, organic matter, silt grade of quartz, feldspar, metasomatic origin of calcite, dolomite and pyrite (local, small) content varies.
(4) Recently, Shen et al. (2021) classified the Wufeng-Longmaxi Formation into four types of shale facies combinations, based on the comprehensive consideration of mineral composition, reservoir physical properties, bedding fractures and the influence of bedding on fracture network formation. This scheme is focused on the fracturing ability of shale gas reservoir to form fracture network. According to hydrodynamic genesis of shale deposition, eight lithofacies shale has been classified into strong and weak hydrodynamic zones by Wang et al. (2014b). And, he pointed out that siliceous shale, calcium-bearing siliceous shale and micritic calc shale under the weak hydrodynamic zone have higher organic carbon content. On the other hand, scholars have also done some research on the lithofacies of Guanyinqiao Formation. Liang et al. (2022) divided the Guanyinqiao lithofacies according to the biological characteristics of the Guanyinqiao Formation. Wang et al. (2016) had discussed the significance of mineral composition and distribution characteristics for shale gas exploration. By referring to the lithofacies division scheme of Wufeng-Longmaxi Formation by Wu et al. (2016), Wei (2020) classified the lower Cambrian shale in Western Hubei into four main lithofacies combinations according to the three-end element of siliceous minerals (quartz + feldspar), carbonate minerals and clay minerals. The shale facies are mainly silica-clay-bearing shale facies, ash/silica mixed shale facies, clay/silica mixed shale facies and

argy-rich/ash-mixed shale facies. The organic matter is enrichment in the siliceous shale, and the TOC in the clayey shale is generally smaller than 2.0%.

The mineral composition is obviously affected by the sedimentary environment. For example, the organic-rich shale deposited in the depression environment of Marine platform is generally characterized with enrich siliceous and calcareous organisms, with little or no clay minerals. Siliceous and calcareous minerals are mainly the remains or detritus of various hydrocarbon-forming organisms which have been buried and evolved through various diagenesis (Qin et al. 2010a). However, the mud shales formed in the sea–land interaction or continental lacustrine basin are usually enrich the clay minerals. The organic matter is saved in the form of organic clay mineral aggregate in shale by the adsorbed of the clay minerals (Lu et al. 1999; Cai et al. 2006, 2007; Li et al. 2006; Yu 2006). A large number of studies reported that the mineral composition of Wufeng-Longmaxi Formation is mainly quartz, followed by clay minerals, and other components also include feldspar and a small amount of carbonate minerals. Liu et al. (2015) found the mineral composition of Longmaxi Formation shale in Southeast Sichuan is quartz (avg. 36.07%), clay minerals (avg. 41.55%) and little carbonate mineral. Quartz content is the highest at the bottom and decreases upward, while clay content increases. This conclusion has been observed in the other shale gas well too, such as the well XY-1, YY-1 and the well in the Jiaoshiba area, where the Wufeng-Longmaxi shale is mainly composed by the quartz and clay minerals, with a small amount of plagioclase, potash feldspar, calcite, dolomite and pyrite (Wu et al. 2015).

The lithology of Longmaxi Formation in Sichuan Basin and its periphery is similar to that of North American shale. Both of them are mainly consisted with carbonaceous shale, siliceous shale, silty shale, and a certain amount of silty fine grained clastic rocks such as fine siltstone, argillaceous siltstone. The mineral composition is mainly quartz, with big content of organic matter, maldistribution of calcium and extensive distribution of pyrite. However, it is different from the bio-siliceous shale in North America that the quartz in the Longmaxi Formation is mainly terrigenous and relatively little diagenetic metasomatism in the study area (Zeng et al. 2011; Chen et al. 2011a ; Liu et al. 2011; Zhang et al. 2012b; Liang et al. 2012). Basing on the petrology, mineralogy and biological characteristics analysis on the marine source rock in the middle-Upper Yangtze, Qin et al. (2010a, b) and Fu et al. (2011) pointed out that the siliceous rocks were mainly biogenic. The mainly minerals of the key source rocks in south China were mainly from the benthic siliceous or calcareous frameworks after burial evolution. The source rocks are mainly biogenic in south China. The organic matter is partly saved in the siliceous and calcium minerals in the marine shale. The shape of these minerals is irregular, and with the characteristics of raw chips, multitudinous elements (especially with some little or other trace elements). There is an obviously differences in smoothness and color of the minerals surface. It is indicating that the siliceous and carbonate minerals in the organic-rich shale is not from terrigenous input (Qin et al. 2010a). Zhang et al. (2012b) found the radiolarians and siliceous sponge spicules in well YS-1, the Sanhui outcrops in south Sichuan province, the Hegou outcrops in the Shizhu county and other areas. Meanwhile, the content of animal organic debris was particularly prominent in the muddy deep water shelf environment, with the highest relative 37% and an average 14%. It likely as an important component for hydrocarbon generation. The organic-rich siliceous rocks are commonly resulted by the combined action of hydrothermal activity and biological deposition. Quartz is mainly from biological deposition and SiO_2 chemical precipitation caused by hot water. This may help us understand the reason that dark bedded siliceous rocks are commonly organic-rich rock, while the organic matter content is commonly very low in the flint which is merely formed by the chemical precipitation (Fu et al. 2011).

Lithology and rock mineral composition are double key internal factors for the mechanical properties of rock and even fracture feasibility (Sui et al. 2007), which are the fundamental factors affecting the pore structure of shale gas reservoir too (Chen et al. 2013). Hence, the mineral composition and brittleness index are two important indicators for the description and evaluation of organic-rich shale reservoirs in Longmaxi Formation now (Liu et al. 2012). Most of the Longmaxi Formation profiles in Sichuan Basin is characterized that the clay and carbonate minerals are increasing in the upward, while the quartz decreasing (Liang et al. 2009; Chen et al. 2011a; Liu et al. 2012). The brittleness index of the organic-rich shale is about 50–75% in the depositional center. As far from the land, the content of carbonate minerals is gradually increasing, while the terrigenous clastic minerals are gradually decreasing, in the limited shallow sea facies (Liu et al. 2012).

Furthermore, the deposition of pyrite is related to biological processes too. A lot of H_2S is formed by the sulfate reacts with organic matter with the action of bacteria in the anoxia water. In the early diagenetic stage, H_2S combines with iron ions to form sulfide. Then, the sulfide reacts with active functional groups in organic matter to form organic sulfur. However, metal ions are more easily combined with H_2S. Once the iron ions present in the environment, the formation of pyrite is prior to that of organic sulfur (Zhang et al. 2013a).

Clay mineral characteristics

Clay minerals as a general term for finely dispersed water-bearing layered silicate and water-bearing amorphous silicate minerals, which are the most abundant minerals in strata (Li et al. 2012c). The formation, transformation and disappearance of clay minerals in mudstone are influenced by many factors such as the depositional environment, diagenesis and source rock (Zhao and Chen 1988). The key factors dominated the formation of clay mineral are commonly different in different regions and layers, which resulted the difference in the distribution and types of clay mineral. It is helpful to analyze the paleoenvironment and diagenesis of clay minerals by analyzing their types, occurrence, content and variation characteristics. For a long time, clay minerals have been seemed as a favorable tool to oil and gas exploration. Sedimentary rocks are rich in montmorillonite, which is conducive to the generation and migration of oil and gas through diagenesis Daoudi et al. (2010). There is a law of co-evolution between the ratio of illite/montmix and organic matter. A lot of organic matter is adsorbed in the clay minerals, which have a strong catalytic effect on the process of hydrocarbon generation (Zhang et al. 2013a). The type and content of illite in shale are important factors and indicators for the content of hydrocarbons.

Little previous work was focused on the clay minerals about the shale gas. Basing on the analysis of clay mineral in the dark shale, Li et al. (2012c) discussed the influence of clay minerals on the reservoir property of shale. It is concluded that the composition, distribution and the formation mechanism of the clay mineral is not only showing the characters of the depositional and diagenetic environment, but also with a certain influence on the porosity and permeability. Therefore, the study on clay minerals should be emphasized in shale gas reservoir.

3. Characteristics of organic matter and its relation to mineral composition

(1) Characteristics of organic matter

There were two hydrocarbon generation centers for Longmaxi Formation in Sichuan Basin and its surrounding area, which are the Wanxian-Shizhu of hydrocarbon generation center in the eastern Sichuan Basin and the Zigong-Luzhou-Yibin hydrocarbon generation center in the Sichuan Basin (Wang et al. 2009a). The burial depth of Silurian source rocks is deeper in the eastern and Northern of Sichuan Basin than the south Sichuan Basin, while the thermal evolution of the shale is similar to each other of the two areas (Zeng et al. 2011).

The organic matter is enriched in the dark Longmaxi shale. Under a microscope, the shale is opaque black and gray black. Several different occurrence forms of organic matter are observed in the microscopic identification and electron probe analysis (Zhang et al. 2013a). The organic matter abundance is an important index for source rocks evaluation. Three conventional indexes are applied to organic matter abundance of source rocks, such as the total organic carbon (TOC), hydrocarbon generation potential ($S_1 + S_2$) and chloroform bitumen “A”. In consider of the high to over thermal evaluation of Paleozoic source rocks in Sichuan Basin, chloroform bitumen “A” and hydrocarbon generation potential can no longer be availably showing the hydrocarbon generation capacity of high-overmature source rocks. Hence, TOC has been regarded as the key index to the evaluation of hydrocarbon generation intensity in Paleozoic mud/shale in Sichuan Basin (Wu et al. 2013a). Therefore, total organic carbon content is an important index for evaluating the abundance of source rocks, as well as an important parameter to show the intensity and amount of hydrocarbon generation. According to the definition of shale gas, organic matter in shale is the parent material of hydrocarbon, as well as an important adsorption medium and carrier of shale gas (Li et al. 2007; Zhang et al. 2013a).

The Longmaxi Formation is characterized with high organic matter content and stronger hydrocarbon generation potential, which is similar to shale gas in North America. According to the TOC tests from the outcrops and core samples, the TOC is in the range of 0.35–18.40%, with an average of 2.52% in the south Sichuan Basin. In the 261 samples, the TOC is ranging from 0.50 to 8.75%, with an average of 2.53%. About 45% of them are with TOC bigger than 2%, which are mainly located at the lower Longmaxi Formation (Huang et al. 2012). Basing on the organic geochemistry analysis, Zhu et al. (2010) pointed out that the TOC is 1.2–5.6% (avg. 3.1%) in Longmaxi Formation in Sichuan Basin. In Sichuan Basin, the TOC is 5.1–6.8% (avg. 3.6%) in Wufeng shale. In the periphery of Sichuan Basin, the value of TOC in Wufeng shale is smaller than it in the Sichuan Basin, such as the TOC is 2.6–4.2% (avg. 3.2%) in well PY-1 (Zhao et al. 2016).

The kerogen in the Longmaxi shale is mainly in the type of Sapropel in Sichuan Basin. Kerogen is amorphous and derived from lower aquatic organisms (Huang et al. 2012). Kerogen was flocculent under scanning electron microscope. In organic matter, the content of algae-amorphous group is 58.9–78.3% (avg. 71.2%). The content of animal organic debris group is 7.5–26.4% (avg. 15.9%). The content of secondary group is 11.2–14.8% (avg. 12.7%) (Zhu et al. 2010). In consider the H/C and O/C ratio of rock pyrolysis analysis cannot be used to determine the kerogen type for the high thermal evolution sample, $\delta^{13}C$ values are commonly used to determine the types of organic matter in Sichuan Basin, for the carbon isotopic composition of kerogen is less affected by thermal evolution (Hao and Chen 1992; Huang et al. 1997). The $\delta^{13}C$ values of the I-II_1 type kerogen are ranging from −32.04 to −28.78% (avg. −30.23%) (Wang et al. 2000). According to the maceral of kerogen, Chen Wenling et al. (2013) concluded that the kerogen of the shale at the bottom of Longmaxi Formation in Well CX-1 is Type I, with the algal particles 7.7–11.2% (avg. 9.53%), carbon bitumen 2.4–6.8% (avg. 4.0%), microsomal content 4.6–9.4% (avg. 6.65%), animal body contents 2.1–11.6% (avg. 7.76%). According to the kerogen microscopic analysis of core samples in well JY-1, Guo and Liu (2013) pointed out the Longmaxi shale is the I type kerogen, with $\delta^{13}C$ (PDB) −29.2 to −29.3%. The carbon isotope of natural gas is obviously inversed, as $\delta^{13}C_1$ (PDB) is −29.2%, while $\delta^{13}C_2$ (PDB) is −34.05%. It can be concluded that kerogen is mainly the type I kerogen and partly with the II_1 kerogen in the Longmaxi shale in Sichuan Basin (Wang et al. 2014b; Guo et al. 2014). The gas generation potential of Longmaxi shale is better than North American shale in which the kerogen is mainly composed by the II kerogen.

The thermal evolution of the Longmaxi shale is higher than north America. The shale gas discovered in China is thermal genesis and with the similar enrichment conditions to USA. However, there are some obviously differences between China yet (Li et al. 2012c). The thermal evolution history is showing that the Longmaxi Formation in Sichuan Basin was reach the low maturity stage at the end of Early Permian (*Ro* is 0.5–0.7%), reached the peak of hydrocarbon generation at the end of Triassic (*Ro* is 0.9–1.1%) and entered wet gas–gas condensate stage at the end of Early Jurassic (*Ro* is about 1.3%). In present, it is in the late stage of hypermaturity, and all liquid hydrocarbons are cracked into dry gas (Huang et al. 2012). It is reported that the burial depth of Longmaxi shale is distributed in a wide range, especially it is deeper than 5000 m in local area, while the R_o is merely in the range of 2.2 to 4.0% (Wang et al. 2009a; Liu et al. 2009, 2016; Zhu et al. 2010; Chen et al. 2013a, b; Wang et al. 2016; Guo et al. 2014).

(2) The relationship between organic matter and mineral composition

The laboratory studies showing that there is a positive correlation between the TOC and the gas generation rate and adsorbed gas volume of shale gas (Wang et al. 2009a, 2012b). The mineral composition is one of the key factors for the total gas content of shale (Ross and Bustin 2008). Therefore, it is likely with a certain correlation between the mineral composition and organic carbon content in shale. For example, Zhang et al. (2013a) concluded that the silicification of rocks may relate to the organic matter co-existed with cryptocrystalline quartz and illite, in the late filling veinlet. A lot of organic matter is adsorbed on the clay mineral and cryptocrystalline ultrafine quartz, which catalyzes the hydrocarbon generation. Furthermore, a lot of organic matter is observed in cleavage cracks or fissures of primary minerals and secondary enlarged edges, which is indicating that multiple hydrocarbon migration may have occurred during diagenesis in this area.

The Paleozoic shale in middle-Upper Yangtze is mainly composed by the clay minerals content, brittle minerals (quartz and feldspar), carbonate minerals (calcite and dolomite) and a small amount of sulfate minerals and sulfide (Fu et al. 2011; Qin et al. 2010a, b). Basing on the clay minerals and sulfides analysis, Fu et al. (2011) point out that the source rocks deposited in different environment can be with similar abundance of organic matter but with different mineral compositions. However, basing on the systematic X-ray diffraction analysis of organic-rich shale (TOC > 1.5%) sampled from diachronous marine strata in south China, Qin et al. (2010b) pointed out that there is a certain correlation between the total organic carbon content and mineral content of marine shale in South China. There is a negative correlation between the TOC and the clay mineral content, while a positive correlation between the TOC and the content of quartz. This relationship may be caused by the stable environment of platform basin, platform depression and lagoon, which is favorable for the formation of marine organic-rich shale, and is not conducive to the transport and deposition of terrigenous clastic clay with water. In addition, there is a certain correlation between mineral composition and kerogen type, which may be effect by the sedimentary environment, provenance and other factors (Fu et al. 2011).

However, in middle-Upper Yangtze, there is no obvious correlation between the thermal maturity and the content of clay mineral and quartz in the Paleozoic marine shale. There is an obviously linear relationship between the burial depth and the content of illite and illite/smectite formation in the clay mineral. The content of illite is obviously increased with the increase of the depth, while the illite/smectite formation is decreasing with the increasing of the depth (Fu et al. 2011). Hence, it can be concluded that the composition of the mineral is obviously correlate to the TOC. The thermal evolution of the organic matter is complemented with the transformation of clay minerals. I can be deduced that the sedimentation had influenced the enrichment of organic matter and the mineral components of shale (Li et al. 2009d). The diagenesis is mainly present by the transformation of clay minerals (Lei et al. 1995).

The organic-rich layer located at the bottom of Ordovician Wufeng Formation and Silurian Longmaxi Formation is the most favorable formation for the enrichment of shale gas

(Zhang et al. 2013a). The "deep water shelf-benthic algal mat model" is the typical model of the marine source rocks formation in Southern China (Liang et al. 2009). In marine environment, the source rock is favorable to deposited in the undercompensated environments, such as shallow-deep water basin, deep water shelf basin, deep water shelf facies (Chen et al. 2006a; Qin et al. 2009).

4. Reservoir characteristics

cMicropores and nano-pores are commonly observed in the marine organic-rich shale in China, in the forms of intergranular pores, intragranular pores and organic pores. Especially, the nano-pore throats formed after the hydrocarbon generation of organic matter, which are the main space for shale gas enrichment (Nie et al. 2009a). The nano-pore was firstly observed by Zou et al. (2011c), and was divided into organic nano-pores, nano-pores within particles and micro-fractures and so on. He has opened the prelude to the visual study of shale gas reservoir space in China.

In the study area, the porosity, permeability and pore types of the Longmaxi shale are similar to those in north America. However, due to the differences in sample collection and analysis methods, the reservoir physical tested results of the Longmaxi shale are significantly different among different scholars, although the pore types are similar. Zeng et al. (2011) pointed out that Longmaxi shale in Sichuan Basin is a compact lithology in which the porosity is often less than 2%. The micropores of shale reservoir are mainly illite flake micropores, microfractures, matrix micropores, intragranular dissolution micropores, berrylike pyrite intergranular micropores and so on. According to the helium pycnometry and pressure pulse decay experiments, the average porosity and permeability of Wufeng-Longmaxi shale are 2.11–12.46% and 0.0063–104.41 × 10^{-3} μm^2 in Sichuan Basin and its surrounding areas, respectively. The reservoir physics of the Wufeng-Longmaxi shales is good, except the south Sichuan Basin. The scanning electron microscopy (SEM) analysis showed that the interlayer microcracks and secondary micropores are in the organic-rich layer too. The width of the fractures is in the range of 0.5–60 μm. The diameter of secondary pore is ranging 1 μm to 40 μm. The average face rate is 8.77% (Wang et al. 2012b). According to the test results of 33 core samples from Longmaxi Formation conducted by Weatherford, the porosity of Longmaxi shale is 1.15% to 5.80% (avg. 3.00%), and about 80% samples are bigger than 2.00%. Especially, basing on the 8 core samples of Longmaxi shale from the shale gas well with shale gas breakthrough in Weiyuan area, the porosity is 1.7–5.8% (avg. 4.2%), permeability is 0.00025–1.73700 × 10^{-15} m^2 (avg. 0.421 × 10^{-15} m^2) (Huang et al. 2012). Basing on the high-power SEM analysis of 69 samples from south Sichuan, Wang et al. (2012b) pointed out that the pore types in shale were diverse. A large number of pores and natural fractures are formed in the organic-rich shale. The pores can be divided into four types of matrix pores, such as residual primary pores, organic pores, interlamellar micropores of clay minerals, unstable mineral dissolution pores. The organic pores and the interlayer micropores between the clay minerals are the main contributors to shale reservoir space. The facture densest section is located at the lower part of Longmaxi Formation. Organic pores and interlayer micropores of clay minerals are the major component of matrix pores in shale, which is the significant difference between shale and sandstone reservoirs, and both of them are formed by the diagenesis. Basing on the reservoir physical tests of 159 core samples from the organic-rich shale section with 38 m thickness in well JY1, Guo et al. (2013) concluded that the porosity is 2.78–7.08% (avg. 4.80%), and the permeability is 0.0016–216.601 × 10^{-3} μm^2(avg. 0.16 × 10^{-3} μm^2). The reservoir space types are diverse too, including organic pores, intergranular pores, intergranular pores, dissolution pores, organic shrinkage pores, structural fractures and cleavage fractures. The permeability of the shale without fractures is commonly less than 0.01 × 10^{-3} μm^2. Based on the alcohol method and gas method for porosity and permeability respectively in the core sample of Longmaxi Formation from well CX-1, Chen et al. (2013a, b) pointed out the porosity is 1.92–10.64% (avg. 5.68%), and the permeability (including fracture permeability) is 2.36–32.37 × 10^{-3} μm^2. Basing on the characteristics of the SEM analysis of the sample from Qilong outcrops in Xishui County, Ran et al. (2014) divided the pores in shale into three types, such as micropores ($\leq$10 nm), mesoporous (10–1000 nm) and macropores ($\geq$1000 nm). Up to 10 parameters was concluded by Tu et al. (2014) to the reservoir evaluation of shale gas, including organic carbon content, organic matter maturity, effective thickness, reserve abundance, porosity, gas content, adsorbed gas content, reservoir pressure, burial depth, clay mineral content and so on. Liu et al. (2021) pointed out that the morphology of nano-pores in Wufeng-Longmaxi shale is controlled by pore location (pore type), kerogen type, burial depth, thermal maturity and pore size.

As mentioned above, the reservoir space of Wufeng-Longmaxi shale in Sichuan Basin is controlled by the mineral composition, lithofacies type, organic carbon content, maturity of organic matter and diagenesis. There are diverse types of reservoir space in Longmaxi shale, which is mainly component by organic pores, interlayer micropores of clay minerals and intergranular pores, secondary the dissolution pores and intergranular pores. Microfractures are well formed in shale, which is conducive to the reservoir physical properties and hydraulic fracturing treatments. It is reported that the organic matter in Wufeng-Longmaxi shale is composed by bitumen and kerogen. Xie et al. (2021) pointed out that there are a lot of pores in the pyrobitumen, mainly in the form of bubbles and spongy pores. However, little or no pores are observed in algal fragments, and no irregular pores are observed in bacterial aggregates. Zhu et al. (2016) pointed out that the paleontological fossils can form organic pore network system which is contribute to the pore space in shale. In Longmaxi Formation, graptolites are preserved in multi-layer carbonized films, with a few of interlayer pores and pore space in graptolites. The porosity of the graptolites shale is bigger than 2%, while the permeability is commonly smaller than 0.5×10^{-3} μm^2. The microfractures in shale are conducive to the increasing of permeability which plays a significant role in hydrocarbon migration and shale gas exploitation. There is an obvious correlation between the reservoir characteristics and mineral composition in organic-rich shale. For example, Chen et al. (2011a, b, c, d) found that the porosity of Longmaxi Formation in Southern Sichuan Basin increases with burial depth. There is a positive correlation between porosity and the content of brittle minerals such as quartz, while a negative correlation between porosity and the content of clay. Quartz is characterized with stability mechanical, and it can act as a good supporting role in the pores. The correlation coefficient is big between the porosity and the content of quartz (Hu et al. 2021).

5. Study on diagenesis

Although the shale gas exploration in China is started much later than USA, it is based on the successful practices experience of shale gas in North America. A rapid progress has gotten in shale gas production in China. However, as the Shale gas exploration in North America, little work has carried out on the reservoir diagenesis. Previous studies on the diagenesis of shale gas reservoirs are focus on the organic geochemistry such as the TOC, thermal maturity of the source rock (Liu et al. 2011; Wang et al. 2012b; Huang et al. 2012; Zuo et al. 2012; Nie et al. 2012a, b; Li et al. 2012c; Wang et al. 2012b). The diagenesis has been concerned too (Chen et al. 2011a, b, c, d; Liang et al. 2012; Wang et al. 2012b; Li et al. 2012c). Basing on the characteristics and transformation of clay minerals in organic-rich shale in Southeastern Chongqing and their influence on reservoir physical properties analysis, Li et al. (2012c) deduced that a lot of micro-fractures were formed in the illitization process of smectite produced micro-fractures, which is a part of the storage and permeability space in shale reservoirs. Wang et al. (2012b) pointed out that diagenetic stage is closely related to the brittleness of shale reservoir. During the late maturity or metamorphosis stage, the minerals in shale are gradually transformed into brittle and stable minerals. A large-scale reservoir conditions is an indispensable condition to the enrichment and high yield of shale gas. With the effect of compression, little primary pore was resided in the shale. The reservoir space in shale is mainly composed by the secondary pores and fractures which are formed by the hydrocarbon generation, transformation of clay minerals and the dissolution of unstable minerals (Liu et al. 2011; Wang et al. 2012b). In consider of the influence of mineral formation (e.g., quartz, dolomite and pyrite) on the shale gas reservoir condition during the diagenetic stage, Wang et al. (2021) pointed out that hyomorphic pyrite, bioquartz and microbial dolomite were mainly formed in the early stage of syngenic-early diagenetic stage, which is with the destructive and constructive dual effects on the preservation of original pores in shale. It is supported that the framework dominated the formation of high-quality shale reservoir. The rigid support framework composed of such minerals and terrigenous debris is beneficial to the preservation of original pores and later fracturing. Zhou et al. (2021) concluded that the carbonate minerals Wufeng -Longmaxi shale are derivatives of methanogen metabolism at the early burial stage of sediments. The carbonate mineral is benefit to the brittleness of shale which improves drillability and fracturing effect of shale reservoir. As for the division of diagenetic stages, Chen et al. (2011a) pointed out that the Longmaxi shale is characterized with high illite content stable over the Sichuan Basin, which indicated that the Longmaxi shale has enter into the late diagenetic stage. The diagenetic degree of Longmaxi shale in Southern Sichuan Basin is appropriate to the enrichment of shale in consider of the thermal evolution conditions. As the source rock in Sichuan Basin, the shale gas reservoir was experienced by organic and inorganic processes diagenesis during burial. The formation process of shale gas reservoir is relatively complicated. Basing on the analysis of pores influenced by the diagenesis of organic-rich shale in lower Qiongzhusi Formation in well NJ-1, Fu et al. (2015) pointed out that some pores were cemented and filled by the dissolved silicon recrystallization during diagenesis. The pores are gradually becoming closed or semi-closed pores with the increase of quartz content, and the porosity is decreased. Hence, the diagenesis study in shale gas discipline (especially shale gas in China) has still in the primary stage now. Up to now, few comprehensive and systematic study has been carried out on the diagenesis and diagenetic evolution of shale gas reservoirs.

6. The characteristics and evaluation of gas reservoirs

According to the tested data from the core samples from well CX-1 in south Sichuan Basin, the commercial value of shale gas is dominated by the content of free gas and adsorbed gas. The content of free gas is closely related to its structure. However, the adsorption gas is affected by environmental factors such as temperature, pressure and so on. In the same condition, the adsorption volume is positively correlated with the organic matter content (Wang et al. 2009a, b, c). The organic carbon content is positively correlated with gas content in well CX-1. It can be concluded that the richer of organic matter the bigger total gas content in the shale. Longmaxi Formation is favorable for the enrichment of shale gas for the characteristics of it such as, big TOC, high thermal maturity, with micro-fractures, appropriate burial depth. Therefore, the shale gas reservoir was mainly evaluated according to the TOC at the initial stage of shale gas in China.

Kerogen and clay minerals in shale is an important carrier for the adsorption hydrocarbon. The characteristics of pores in shale and its gas adsorption capacity is not only controlled by the type of clay minerals, but also the thermal maturity and the petrogenetic of shale. The specific area of shale is influenced by the content of micro pores in shale to, which is dominated the adsorb ability of the shale (Ji et al. 2012). Hence, the shale gas reservoir evaluation should be based on petrological study. In addition to petrology and physical properties studies as the evaluation of conventional reservoirs, the evaluation of shale reservoirs should take the total gas content and the technical feasibility into consider (Zhu et al. 2009). Longmaxi shale has been in the hypermaturity stage (Zou et al. 2010a). The gas in the shale is mainly in the form of free and adsorbed state, little is in the dissolved state (Wang et al. 2011a). Basing on the simulation study of lower Cambrian shale in Western Hubei, Fang et al. (2021) pointed out that the shale gas is mainly component by the free gas, secondary adsorb gas. Hence, the study of pores and the composition of free and adsorbed gas are with great significance to revealing the gas-bearing characteristics of shale. Under a certain temperature–pressure condition, the adsorption gas content is positively correlated to the organic carbon content. Meanwhile, the content of adsorb gas is also significantly correlated with the type of organic matter, thermal maturity, mineral composition (especially clay minerals), humidity (water content) and pore structure (Hao et al. 2013). Adsorption capacity is one of the important indicators for shale gas reservoir evaluation. The evaluation parameters include lithological association, mineral composition, tectonic structure, TOC, vitrinite reflectance (R_o), kerogen type and diagenesis (Yu et al. 2012). Basing on the analysis of the adsorption characteristics of shale and the difficulties would be faced in shale gas exploration in China, Hao et al. (2013) pointed out that the adsorption capacity of the shale is influenced by the characteristics of organic matter (enrichment degree, type and maturity), mineral composition, pore size and structure, water content and regional temperature–pressure condition. Zhang et al. (2012b) found that the higher thermal maturity the more adsorbed gas in shale. In organic-rich shale, the adsorption capacity of minerals to gas is weaker than that of organic matter. In the laboratory, the gas adsorption capacity is directly proportional to pressure and inversely proportional to temperature (Zhang et al. 2012b). In practices, the adsorption capacity of shale is affected by pressure and increases with the increase of burial depth but decreases with the increasing of temperature when it reaches a certain extent (Hao et al. 2013). Basing on the analysis of state and content of gas in shale influenced by the composition of shale, tectonic and geological conditions, Jiang et al. (2016a, b) and Zou et al. (2015, 2016) pointed out that the gas in shale is mainly in the state of adsorption and free state, which is closely related to the temperature–pressure (depth) condition. The composition ratio of adsorbed and free gas is a direct impact on the occurrence characteristics of shale gas, which should be paid attention.

The experimental and analogical studies on the enrichment of shale gas have been carried out by the Chinese scholars, basing on the experience of shale gas production of USA and the lithologic characteristics, hydrocarbon generation capacity, hydrocarbon accumulation capacity of shale in China (Li et al. 2009d; Jiang et al. 2010; Liang et al. 2012). Zhong et al. (2019) pointed out that the shale gas reservoirs in the Wufeng-Longmaxi Formation are with six features, such as organic geochemical characteristics, gas content, lithology and mineral composition, brittleness (fractured ability), physical properties and heterogeneity of the gas reservoirs. Liang et al. (2016) concluded the key factors influenced the enrichment of shale gas in Sichuan Basin are included: (1) a huge thick organic-rich shale which provides a hydrocarbon basis for shale gas; (2) an enclosed environment which is conducive to the preservation of shale gas; (3) a huge number of organic pores which provides space for the reservoir of shale gas; (4) the intensive fractures in brittle sections which is contribute to the reservoir space of shale; (5) the dense high-angle fractures are beneficial to the formation of complex fracture networks in later fracturing treatment. (6) compressional faults are favorable for shale gas preservation for its good plugging ability. One typical feature has been observed in the Wufeng-Longmaxi Formation, which is the total gas content are commonly different in different areas (Fig. 1.2). Basing on the statistical analysis more than 600 shale samples from Wufeng-Longmaxi Formations, Qiu et al. (2019) pointed out that the gas content is obviously varied in vertical. The smaller one is almost on gas content, while the biggest one is $9.0m^3/t$. There are great differences in gas content in favorable (enrichment) and sweet spots, according to the comparative analysis of different regions such as Weiyuan, Changning, Fuling and Wuxi are the main areas. An important feature of Wufeng-Longmaxi shale gas is the pressure coefficient of it (Fig. 1.2). According to the classification criteria of natural gas reservoirs, shale gas reservoirs can be divided into low-pressure gas reservoirs (pressure coefficient < 0.9), atmospheric pressure gas reservoirs (0.9 $\leq$ pressure coefficient < 1.3), high-pressure gas reservoirs (1.3 $\leq$ pressure coefficient < 1.8) and ultra-high-pressure gas reservoirs (pressure coefficient $\geq$ 1.8). The overpressure is commonly observed in Sichuan Basin, such as the Jiaoshiba and Changning-Weiyuan shale gas field where the shale gas has been commercially exploited (Fig. 1.2). Guo et al. (2014) pointed out that the shale gas enrichment and the total resources of shale gas are dominated by the porosity and formation pressure. Hu et al. (2014) concluded that the formation pressure coefficient is a good indicator for preservation conditions the of shale gas. The over pressure is commonly indicted a good preservation condition of shale gas. The formation mechanism of overpressure has been studied Zhang et al. (2016). It is concluded that the bigger TOC is the basic condition for overpressure in shale strata. The higher thermal evolution the more gas is cracked, which is more benefit to formation the overpressure. The preservation of overpressure is closely correlated to the uplifting process too. In conclusion, the shale gas in Wufeng-Longmaxi Formations is characterized with high thermal evolution and ultra-overpressure. The gas is mainly in the free gas state in the overpressure area. The enrichment of shale gas is influenced by the adsorption of shale, the trap of capillary pressure and slowly diffusion, which is named "micro residual enrichment model" by (Tenger et al. 2017).

The atmospheric pressure shale gas reservoirs are commonly observed in Sichuan Basin (Fig. 1.2). According to the structural conditions, shale quality and production characteristics in Sichuan Basin and its surrounding area, Nie et al. (2019) divided the pressure shale gas reservoirs in Wufeng-Longmaxi Formations into four types, such as organic-rich eroded away or denudation type, early diffusion type, fracture or fault damaged type and residual syncline type. Meanwhile, Nie et al. (2019) deduced that although atmospheric pressure shale gas reservoirs are characterized with poor-middle enriched and medium–low quality shale, large total resources and huge reserves may be preserved in some area. Guo et al. (2020) proposed that favorable atmospheric shale gas reservoirs can be predicted with the following parameters, such as TOC > 3%, porosity > 3% and total gas content > 3m^3/t in the complex structural areas outside the basin.

In the primary stage, the favorable area for shale gas had been predicted by the analogy analysis between China and the five shale gas production areas in USA. Most of the scholars predicted the favorable area for shale gas by taking into account of the shale thickness, buried depth, TOC and R_O. Partly scholars took the isothermal adsorption capacity in to account too, which optimized the parameters for predicting the favorable target area for shale gas. However, little total gas content data can be obtained in the pre-drilling stage. Hence, in the primary stage, the prediction of favorable areas for shale gas in China actually was a prediction on the organic-rich shale, which has not entered into the optimization stage of shale gas. According to the primary stage of shale gas exploration in China, the shale gas resources are widely distributed in China. Marine strata in south China are with superior geological conditions for shale gas accumulation and abundant shale gas resources, which is expected to become an important strategic replacement area for oil and gas resources in China. Especially, the Sichuan Basin deserves the first attention (Zhu et al. 2010).

There are many factors affecting the scale exploitation of shale gas. The key factor is the reliability of geological evaluation criteria for core area. The determination of the core area is related to whether the target of maximum enrichment of shale gas can be identified in the early stage of shale gas exploration. If the core area is predicted reliability, the shale gas flow will be obtained which is the basement to achieve large-scale economic exploitation in later period (Wang et al. 2012b). The favorable areas and its geological characteristics are commonly different among different studies, for the geological evaluation criteria for shale gas is inconformity among different scholars. For example, Wang et al. (2012c) proposed that the core area for shale gas in Wufeng-Longmaxi Formations in Sichuan Basin should be with big thickness (mostly 40–100 m black shale thickness), big TOC (> 1.5% in most areas), high thermal maturity (*R*o. avg. 1.49–3.135) and big total gas content (avg. > 2 m^3/t). Zuo et al. (2012) proposed that there are seven key geological factors for predicting the favorable area for shale gas in Southeast Sichuan Basin, such as abundance of organic matter, thermal maturity, organic matter type, brittle mineral content, shale thickness, burial depth and structural configuration. The evaluation criteria for favorable exploration areas are as follow, (1) thickness of organic-rich shale is > 100 m; (2) the top burial depth of Longmaxi shale is shallower than 2400 m; (3) TOC > 2%; (4) *R*o > 1.3%; (5) regional tectonic strength is weak. Huang et al. (2012) pointed out that the favorable areas for shale gas in Longmaxi Formation in Southern Sichuan Basin is characterized with TOC (0.50–8.32%, avg. 2.53%), effective thickness (20–260 m) at the lower section of Longmaxi Formation, sapropelic-type kerogen, Ro (>1.8%), in high to hypermaturity stage, with nanoscale pores, porosity (1.15–5.80%, avg. 3.00%), enrich brittle minerals, big total gas content (0.3–5.1 m^3/t, avg. 1.9 m^3/t), total hydrocarbon is abundant. The areas may be great shale gas resource potential with buried depth (<3600 m).

The preliminary geological evaluation standard for China's shale gas core area is established by Wang et al. (2012b),

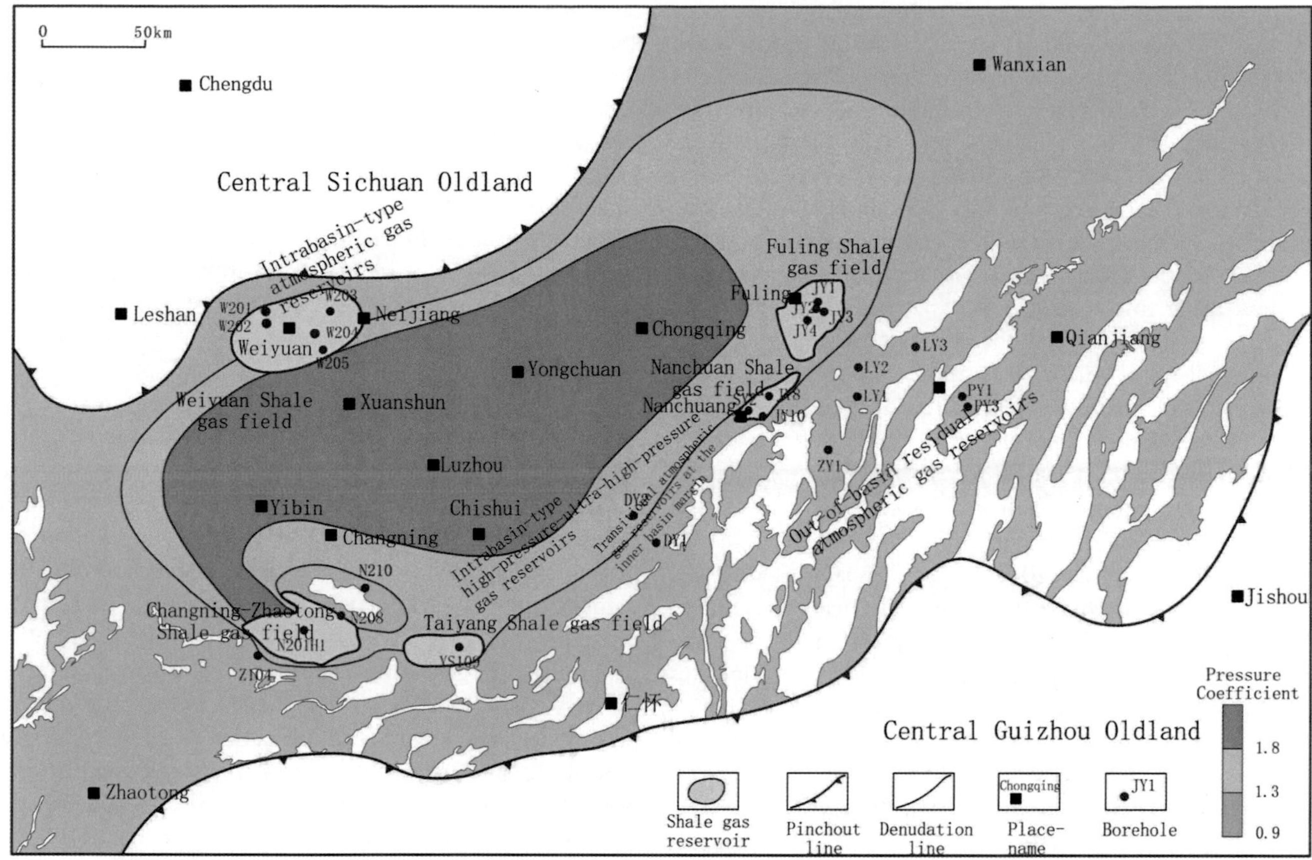

Fig. 1.2 Distribution characteristics of pressure coefficient in Wufeng-Longmaxi Shale gas reservoirs in Sichuan Basin and its surrounding area

basing on the experience of shale gas production of North American and the research progress of China shale gas. The core areas for shale gas have been selected out in Southern Sichuan Basin. Meanwhile it concluded that the Southern Sichuan Basin is the most realistic shale gas development zone in China. Li et al. (2013b) established a set of evaluation index system suitable for high-evolution marine shale gas in Southern Sichuan Basin based on the research results of various indicators. Good results have been gotten by applied this system to the core area selection and potential evaluation of Longmaxi shale gas in Southern Sichuan Basin.

Furthermore, the terrestrial shale gas study has been started early. For example, the enrichment condition and the exploration potential has been studied by Ye and Zeng (2008). They found that the favorable shale in Xujiahe Formation is the dark shale mainly deposited in littoral lake swamp and inland lake swamp facies. Basing on the analysis of shale gas geological conditions, Zhang et al. (2008) pointed out there might be a huge amount of shale gas resources in paleogene Shahejie Formation in Jiyang Depression. Zhou et al. (2011) has discussed the reservoir-forming conditions and exploration direction of Mesozoic lacustrine sedimentary shale gas in Fuxian block, Ordos Basin. Wang et al. (2011a, b) has evaluated the shale gas condition in the Ordovician Marine shale, carboniferous-Permian Marine and continental transitional shale and upper Triassic lacustrine shale in Ordos Basin and pointed out that the latter two layers are with the potential to form shale gas reservoirs. Yang et al. (2013) has studied the characteristics of micro pores in shale and its geological significance to the enrichment of shale gas in continental shale in Ordos Basin. He pointed out the enrichment and migration of shale gas is mainly contributed by the intergranular pore in clay minerals content aggregation and intragranular pore between the layers. The two types of intergranular pores are main control factors of reservoir permeability anisotropy, secondary the intergranular pore and dissolution pore. Organic pores may contribute less due to it is very rare. At the same time, the influence of microcracks cannot be ignored, which is the main microchannel connecting the macropores and mesopores. Lin et al. (2013a, b) analyzed the geological conditions and organic geochemical parameters of terrestrial organic-rich shale, by concluding the distribution of terrestrial organic-rich shale in China. It is concluded that the continental organic-rich shale in China is mainly distributed in the Middle Cenozoic in north and Northeast China, Northwest China and some parts of south China. On the whole, the terrestrial shale gas enrichment in China is characterized with large cumulative thickness, diverse types of organic matter, with a lot of reservoir space, good preservation conditions, high abundance of resources and convenient surface conditions. Moreover, the terrestrial organic-rich shale in China mostly overlaps on the conventional oil and gas fields which with the data and equipment advantages. Therefore, it is likely with better economic recovery. At the same time, the research on shale gas in China is mainly focused on basic theory. The geophysical method using to the shale gas exploration has still in the primary stage. Basing on the analysis of geological and geophysical response characteristics of organic-rich layer in Southern Sichuan Basin, Li et al. (2011a) has established a set of geophysical technical process for shale gas exploration, through the research for seismic data acquisition, processing and interpretation. And, some new progress has been obtained in the seismic exploration of shale gas. In addition, Chen et al. (2011a) systematically studied the formation conditions of shale gas in Hetaoyuan Formation of Cenozoic lacustrine basin-Biyang Depression of Nanxiang basin for the first time, by referring to the evaluation indexes of marine shale gas in the USA.

1.2 Current Status of Lithofacies Paleogeography

Lithofacies paleogeography is an important branch of modern sedimentary geology, which originated from sedimentology. Its development has a close relationship with the development of sedimentology. Lithofacies paleogeography can become a branch of discipline, which is related to the development of sedimentary petrology into sedimentology, and this learning process is inseparable. In a sense, sedimentology is the main basis for the study of lithofacies paleogeography. Now, however, its research scope has already gone beyond sedimentary petrology. Based on a high integration and intersection of various geological disciplines, and supported by the theory of tectonic activity and dynamic transformation, it focuses on investigating and studying the reorganization of oceans and land, distribution of land and sea, nature of basins, ecological environment and allocation of mineral resources from a global perspective (Mou et al. 2016a).

1.2.1 Development History of Foreign Lithofacies Paleogeography

Due to lack of theoretical guidance, the original sedimentology basically describing the appearance of sedimentary rocks, focusing on studying and describing their external features and trying to explain their causes. Until 1830, Lyell proposed the idea of "Uniformity Theory", including the "Walther law of facies" proposed by Walter at the same time. For a long time in the future, research in the geological community will be based on these guiding ideologies (Hua and Zhang 2009).

Until 1939, Twenhofel published the book of the Principles of Sedimentation and Trask published the book of the Modern Marine Sedimentation. They explored the characteristics of modern sedimentary environment in methodology and provided necessary tools for explaining ancient geological history (Hua and Zhang 2009). This is also one of the most important and commonly used principles and methods in the development of sedimentology, that is, to speculate the formation environment of ancient sedimentary rocks according to the modern sediment forming environment, conditions and climates of modern sediments can be summarized as "present being a key to past". In addition, the principle and method of particle size classification of debris particles proposed by Udden, Wentworth, etc., conforming to the laws of Fluid Mechanics and the Normal Distribution of particles; Boggs researched on micro-petrology of sedimentary rock and Milner applied mineral research on provenance and stratigraphic correlation division, etc. These typical examples represent the origin stage of sedimentary-based lithofacies paleogeography. It could be said that the lithofacies paleogeography and related mapping works (methods) at the origin stage have little relation with mineral resources and energy resources.

The World War II and the reconstruction for various countries after the war intensified demand for energy and minerals, which indirectly contributed to the rapid development of sedimentology and lithofacies paleogeography. Among them, the most important is a large number of new technologies and methods to continuously used in sedimentology, which makes the comprehensive development of various testing technologies in this period become mainstream, such as the application of X-ray diffraction and mathematical statistics in particle size analysis (Hua and Zhang 2009). Typical examples are as follows: In 1935, based on the study of sedimentary mineral components, Peitizhuang compiled the contour map of sedimentary mineral components for the first time according to the content and characteristics of different mineral components. In 1949, Sedimentary Rocks was published, which studied the classification of sedimentary rocks and discussed the strata and tectonic environments. It was the first research work on systematic classification in the history of sedimentology. The main task of sedimentology is to infer the paleoenvironmental characteristics of rock strata. In 1945, Krubin not only obtained the paleosedimentary environment of the research object through a great deal of research, but also made a quantitative study on the material expression (rocks, etc.) of the sedimentary environment and concluded that "boundary conditions, particles and hydrodynamic conditions (energy) are the three main factors in the sedimentary system", which enriched the relevant knowledge of sedimentology. In the research history of sedimentary basin analysis, Pettijohn and Crews (Cloos) also systematically used paleocurrent direction and rock correlation to conduct a preliminary analysis, which is the first time in the field of sedimentology. The innovation and application of these theories and methods not only directly promoted the development of sedimentology itself, but also accelerated the development of lithofacies paleogeography and gradually deepened the relationship between this discipline and mineral and energy resources. This is the stage of development of sedimentology and lithofacies paleogeography.

Sedimentology and lithofacies paleogeography entered the modern research stage after 1950, which was marked by the most famous, representative and innovative theory, "Turbulence Theory", which was a milestone in the development history of sedimentology and lithofacies paleogeography. Subsequently, under the guidance of Kuenen, Bouma further proposed the "Bauma sequence" model. The next 20 years or so also witnessed the rapid development of sedimentology and lithofacies paleogeography, and a large number of high-level summative and innovative monographs were published: Sedimentary Petrology and Sedimentology by Douglas, Analysis of Paleocurrent and Basin by Pettijohn and Porter, etc. Among them, there are several works and theories of great significance: the first is the classification of limestone published by Fok (1959, 1962), which is a very important breakthrough in the study of sedimentary facies of carbonate rocks; Secondly, mechanical concepts such as fluid flow pattern, Reynolds and Froude numbers were introduced into sedimentology, which greatly promoted the hydraulic interpretation of sedimentary structures and the study of formation mechanism. The flume experiment of Simons and Richardson (1961), in which the concept of water flow dynamics was formulated to explain the sequence of sedimentary structures, has since established the basis for sedimentology and lithofacies paleogeography. Then, there is the theory of plate tectonics developed from Wegener's continental drift theory and the seafloor spreading theory of American seismic geologists Dietz and Holden (1970) and Heirtzler (1969). Mckenzie and Parker, Princeton University's Morgan and Lamont Observatory (France) Pichon and others jointly put forward, it is a specific extension of the theory of seafloor spreading and finally published by Morgan and Series. In the same year, French geologist Pichon divides the earth's rock strata into six plates, namely the Pacific plate, the Eurasian plate, the American plate, the Indian Ocean plate, the African plate and the Antarctic plate. This theory provides a basis (theoretical basis) for understanding the large-scale distribution of sedimentary facies and biota, as well as the migration of the crust, thus prompting sedimentologists and paleogeographers to consider the influence of structure and plate movement on sedimentation. Therefore, a new discipline, basin analysis, has also been developed.

In short, from 1950s to mid-1970s, because of the background of global geological research, many geologists found that although sedimentary rocks contained a large variety of extremely important mineral resources, some of them only occurs in sedimentary rocks. In order to study the occurrence and distribution of these minerals and petroleum and further expand their exploitation scope, many researchers began to pay attention to the origin and nature of sedimentology and lithofacies paleogeography, as well as their deep relationships with minerals and petroleum and relationships including causal relationships, spatial relationships and so on. The results prove that the paleoenvironment or paleogeography not only controls the formation of sedimentary layered deposits, oil, gas, etc., but also controls the formation and enrichment of many-layered deposits. Therefore, it is very important to study the natural geographical environment when sedimentary rocks were formed. Essentially, the continuous innovation of theories and methods related to sedimentology and lithofacies paleogeography has also fundamentally guides the deepening of research and exploration in the field of mineral and the petroleum industry and made great achievements. This knowledge should be very important in the development of sedimentology and lithofacies paleogeography in this period.

From the late 1970s to the 1980s, the development of sedimentology and lithofacies paleogeography should have matured. Many sedimentologists not only supplemented and improved the various theories and methods of their predecessors, but also put forward many more constructive theories and methods. The development of stratigraphic stratigraphy based on seismic stratigraphy not only enriched the theory of sedimentology, but more importantly, provided the basis and methods for the preparation of more accurate paleogeographic maps, with representative works such as seismic stratigraphy in oil and gas exploration (Vail 1977) and principles of stratigraphy (Vail and Posamentier 1988). The global ocean anoxic event discovered (proposed) by Schlanger and Jenkyns (1976) according to the plate tectonic theory provides an excellent foundation and methodology for global research. It is worth noting that the theoretical basis and methods are put forward for mineral resources and oil and gas resources. The former being based on the oil and gas industry foundation, then researched and put forward and finally served for the research, exploration and development of oil and gas; the latter having mineral and hydrocarbon resources due to the anoxic seabed, where the sediments are black and rich in organic carbon, generally not disturbed by benthos and often form marine striped sediment containing pyrite and heavy metals. Generally speaking, the development of sedimentology and lithofacies paleogeography in this period is inextricably linked with mineral resources and energy resources (such as hydrocarbons) and guides the research, exploration and development of mineral resources and the hydrocarbon industry.

Since 1990s, sedimentology and lithofacies paleogeography have entered a period of modern comprehensive development. Information from the International Association of Sedimentologists (IAS) (the 12th to 19th, 1986–2014) from 1986 to 2014 shows that contemporary sedimentologists have not only placed greater emphasis on basic sedimentological and lithofacies paleogeographic research (summarised as the process of refining the theoretical system of sedimentology), but also strengthened new sedimentological research methods with a view to increasing the understanding of the sedimentary environment of "sedimentary rocks" from a more detailed macro- and microscopic and integrated perspective, thus better serving the recovery of the paleoenvironment and paleogeography, and thus enabling accurate predictions of the genesis and spatial distribution of mineral and energy resources. On this basis, we will seize contemporary hot issues and conduct targeted research. The specific mainstream development directions are: high-precision sequence stratigraphy research (covering the compilation method of high-precision lithofacies paleogeographic map, instantaneous lithofacies paleogeographic map, global sea level change (curve) research, etc.) sedimentary basin analysis (especially sedimentary basin analysis in complex orogenic belts), dolomite genesis research, deep-water sedimentation and sedimentation research, sedimentary structure evolution and sedimentary response research, climate, environmental change and resource sedimentology research, etc. Another particularly important research status and the trend is the comprehensive research of interdisciplinary and cross-penetration. Nowadays, the concept of time coordinates has not only been introduced into the understanding of sedimentological laws, but also closely integrated with tectonic theory, earthquake-sequence stratigraphy, geophysics, geochemistry and computers, etc., using "new methods and new technologies". Begin to explore the regularity of sediment movement in four-dimensional space (Jiang et al. 2014), to better serve basic geological research, mineral and energy industry development, etc. This will be the development trend and direction of the contemporary and future sedimentology research model with "sedimentology, supported by multidisciplinary integration and cutting-edge technology" (Liu et al. 2006; Zhang and Xin 2006; Jiang et al. 2007; Hua and Zhang 2009; Zheng et al. 2013a, b; Jiang et al. 2014).

It is not difficult to see that the development history of sedimentology and lithofacies paleogeography abroad (Table 1.1) is also the development history of mineral and energy industries to some extent. Among them, oil–gas sedimentology, sequence stratigraphy derived from sedimentology, etc., combined with geophysics, geochemistry and other comprehensive disciplines, have achieved good results in the research of mineral resources and energy resources. They not only fully display sedimentology and lithofacies paleogeography at macro- and micro-levels, but also make the research more comprehensive and accurate and has been applied to the exploration and development of mineral resources and energy resources. In a word, the theories and technologies of sedimentology and lithofacies paleogeography accompany and guided the rapid development of the mineral and energy industries.

1.2.2 Research Status of Lithofacies Paleogeography in China

The study of lithofacies paleogeography in China started early. Although the related theories and methods are relatively lacking in innovation, most of them are the basis for the theories and methods of sedimentology and lithofacies paleogeography abroad. In the past, theoretical supplements or fruitful innovations were made according to Chinese geological conditions, but considerable progress and remarkable achievements were made.

In the aspect of restoring paleoenvironment or paleogeography, the lithofacies paleogeography research in different periods in China have produced different mapping guiding ideologies, principles and methods and formed corresponding representative monographs or achievements (Mou et al. 2016b). In chronological order, in the 1950s, Mr. Liu (1955) compiled the earliest large-scale paleogeographic atlas of lithofacies in China, "Paleogeographic Map of China" based on stratigraphy. In 1965, Lu Yanhao, a paleontologist, compiled Cambrian Paleogeography from the perspective of paleobiogeography (Lu et al. 1965). In 1984, Guan Shicong compiled "Sedimentary Facies and Oil and Gas in Chinese Sea-Continent Change Sea Areas", taking sedimentology as the theoretical basis and combining the knowledge of geotectonics, stratigraphy, petroleum and natural gas geology and other disciplines), which is the first attempt in China to apply sedimentology and lithofacies paleogeography to oil and gas energy resources. In 1985, Professor Wang Hongzhen and his team kept pace with the times, adopted the idea of "tectonic activity theory" and "geological historical evolution stage theory" and compiled the "Chinese Paleogeographic Atlas" (Wang et al. 1985). It not only has some innovations and breakthrough in mapping idea, but also is ingenious in map expression: although the overall shape of the map is still plotted with the present latitude and longitude, the main continental blocks in China and the tectonic boundary with the orogenic belt are used in the crust. Docking subduction zones and crustal

Table 1.1 Time and significance of important theories (events) in the development of sedimentology (after Jiang et al. 2014)

Stage	Time	Theory (event)	Representative	Significance
Budding in its initial formative stages	1830–1839	Geological theory, the past and the present	Lyell CC, Trask	The realist principle of discussing the present and the past
	1884	Deep-sea sediments	Murray J	Classification and description of deep-sea sediments
	1894	Introduction to geology for historical sciences	Walther J	Propose the "phase sequence" law
	1913	Sedimentary petrology	Hatch FH	Sedimentary petrology was separated from stratigraphy and gradually became a unique part of the earth sciences The sub-discipline was established; and the study of sedimentation began to be emphasized
	1922	Introduction to sedimentary petrology	Milner HB	
	1926	Sedimentation tutorial	Twenhofel WH	
	1931	Journal of Sedimentary Petrology	Founding of the American SEPM Society	Marks sedimentology as a separate discipline
	1932	Phase theory	Налцвкцй Д. В	
	1932	Sedimentary petrology	Швецов	From qualitative research to semi-quantitative research, more emphasis is placed on sedimentation and deposition Study on the formation mechanism of sedimentary petrology, an important sign of the maturity of sedimentary petrology
	1939	Principle of sedimentation	Twenhofel WH	
	1935, 1949	Sedimentary rocks and sedimentary minerals research	Pettijohn FJ	The first contour map of sedimentary mineral composition, an important sign of sedimentary petrology maturity
	1950	Turbidity current is the reason for the formation of graded bedding, turbidity current theory	Kuenen PH	Opened a new chapter in the study of turbidity current, a milestone in the history of sedimentology and lithofacies paleogeography research and development
Sedimentary petrology	1951	"Sedimentary Petrology", "Theoretical Basis of Sedimentary Genesis", "Analysis of Paleoflow Water and Basin", "Genesis of Sedimentary Rocks" by H.Blattetal, "Sedimentary Facies and Sedimentary Environment"	Deoglas DJ, Trahov, Petty Zhuang and Porter, Blattetal H, Reading	Sedimentology is a new stage in the development of sedimentary petrology
	1952	International sedimentological society founded (IAS)		Publish the latest research results in the field of sedimentology, promote international cooperation and exchanges, advocate Guide multi-disciplinary cross-integration and promote global sedimentology research and exchange services
	1959, 1962	Practical petrological classification of limestone, mechanical concepts such as fluid state concept, Reynolds and Froude numbers are introduced into sedimentology	Simons and Richardson	Carbonatite research has entered a new stage, and the study of sedimentology and lithofacies paleogeography has a basis for sedimentary dynamics
	1961, 1968	Development of the theory of plate tectonics from the theory of continental drift and seafloor spreading	Alfred Lothar Wegener, Dietz R, Hess H, Mckenzin DP, Parker RL, Morgan WJ, Lepichon X	This has provided the basis (theoretical basis) for understanding the widespread spread of sedimentary phases and biogenic assemblages, as well as the transport of crustal material, which has led sedimentologists and paleogeographies to consider the effects of tectonism and plate movement on sedimentation and paleogeographic change, and thus to the development of a new discipline, basin analysis
	1961	Turbidite	Bouma AH	Propose the famous "Boma sequence'
	1962	Sedimentary rock research methods	Liu Baojun	The beginning of the development of sedimentary petrology in China and the establishment of a theoretical system of sedimentology
	1962	Sedimentary petrology	Wu Chongjun	
	1964	Structural genetic classification of carbonatite	Ye Zhizheng, He Qixiang, etc	The beginning of the modern concept of carbonatite research in China
	1970	The beginning of the modern concept of carbonatite research in China	Pieson SJ	First to use logging for oilfield sedimentology studies

(continued)

Table 1.1 (continued)

Stage	Time	Theory (event)	Representative	Significance
All-round development of sedimentology	1977, 1978	Seismic stratigraphy, comprehensive analysis of sea level change, "Application of seismic stratigraphy in oil and gas exploration" Global oceanic hypoxia event	Vail PR, Schlanger S	Combine sedimentary facies analysis with seismic data, introduce the concept of sequence and propose the concept of isochronous stratigraphic framework The global oceanic anoxic event proposed by Schlanger, based on the theory of plate tectonics, provides a good foundation and method for global research
	1984	Principles of sedimentary basin analysis	Miall AD	Synthesis of structural geology and sedimentology
	1988	"Principles of Sequence Stratigraphy"	Vail PR	The development of stratigraphic stratigraphy based on seismic stratigraphy has not only enriched the theory of sedimentology, but more importantly has provided the basis and methodology for the preparation of more accurate paleogeographic maps
	1986–2010	12th International Congress of Sedimentology	Hosted by the International Sedimentary Society (IAS)	Emphasis on the basic research of sedimentology, while strengthening new research methods; sequence stratigraphy and sedimentary basin analysis, global sea level change, carbonate rock and dolomitization research become mainstream; high-precision sequence stratigraphy, deep water Sedimentation and sedimentation, diagenesis, sedimentary tectonic evolution, climate, environmental changes, and resource sedimentology continue to be the focus of research, representing the development direction of sedimentology
		Thirteenth International Congress of Sedimentology		
		14th International Congress of Sedimentology		
		15th International Congress of Sedimentology		
		16th International Congress of Sedimentology		
		17th International Congress of Sedimentology		
		18th International Congress of Sedimentology		
	2010–present	The main trend in the development of sedimentology is the cross-pollination and integration of various disciplines. This is reflected in the introduction of the concept of time coordinates in the understanding of sedimentological laws, but also in the close integration with geotectonic theory, seismic-stratigraphic stratigraphy, geophysics, geochemistry and computers, etc., and the use of "new methods and techniques" to begin to explore the regularity of sediment movement in four dimensions. The aim is to better serve basic geological research, mineral, and energy industry development, etc. This will be the development trend and direction of the contemporary and future sedimentology research model based on "sedimentology, supported by multidisciplinary integration and frontier technology"		

superimposed subduction zones represent their tectonic divisions and related properties, which are consistent with the idea of "tectonic activity theory" in the idea of compiling maps; the second is the evolutionary nature of the ocean-continent transition process between continental blocks in China. The final description of the orogenic form between them is consistent with the "geological historical evolutionary stage theory" in the idea of compiling maps. This is the first generation of lithofacies paleogeography atlas in China. It is not only the Chinese sedimentology, but also the breakthrough and innovation of the whole geoscience research, which has important guiding and enlightening significance.

After the 1990s, sedimentology and lithofacies paleogeography in China entered a period of development in which a hundred schools of thought competed, drawing on and absorbing the guiding ideas of the "activity theory" from abroad, and the "geological activity theory" in China, which was used to compile "fixed theory" lithofacies paleogeographic maps within the present-day coordinates of latitude and longitude in China, with emphasis on "petrographic or sedimentary" elements (Mou et al. 2016b). Representative figures and works include Liu Baojun and Xu Xiusong, who proposed the idea of "tectonic control of basins and basin control of phases" in the compilation of maps and compiled the Atlas of South China Earthquake - Triassic Lithofacies Paleogeography (Liu et al. 1994). In summary, this paleogeographic atlas has the following two features: (1) the idea of "structure controlling basin and basin controlling facies" was adopted for the first time in China to compile lithofacies paleogeographic map set, and the compiled paleogeographic map fully considered the crustal evolution background of each fault stratum in each geological history period and reflected its structure and basin nature in the form of corner maps. The paleogeographic map is a preliminary attempt at dynamic paleogeographic recovery (Mou et al. 2016b); (2) it focuses on dissecting the tectonic and basin-embedded properties of the sediments, emphasizing them as a comprehensive causal body of tectonic subsidence, sea-level rise and fall and material supply rates and also focuses on the effects of event deposition and continental margin evolution, etc. Another example is the quantitative lithofacies paleogeographic mapping method pioneered by Feng Zengzhao, which adopts a "single-factor analysis and multi-factor synthesis" approach to compile a series of quantitative lithofacies paleogeographic maps of various regions and selected geological periods in China (Feng and Wu 1988; Feng 1991; Feng et al. 1994, 1997a, b, 1999), such mapping methods have been widely applied in the field of oil and gas exploration and development in China (Mou et al. 2016b), a typical example of the application of Chinese lithofacies paleogeography to the oil and gas industry. Of course, in the early research stage of the oil and gas industry, there are many typical examples of lithofacies paleogeography and its mapping methods as the main guiding ideology, for example, the application of lithofacies paleogeography and its mapping methods played a key role in the discovery of the Puguang gas field in China. The collaborative team of Ma et al. (2002, 2005; Ma and Cai 2006; Ma et al. 2009), Mou et al. (2003) and others compiled a more detailed lithofacies paleogeography map of the Changxing Formation in the Northeast Sichuan area after detailed studies of the sedimentary phases and sedimentary micro levels, which was a key guide for oil and gas field companies in good deployment, etc. As a result of the sedimentary paleoenvironment (paleogeography). The different views on sedimentary paleoenvironment (paleogeography) have led other oil and gas companies to give up other promising fields. This simple example not only proves that lithofacies paleogeography and its mapping methods have a fundamental and critical guiding role in the pre-hydrocarbon industry studies and even further exploration and development processes, but also shows that Chinese sedimentologists have also been actively exploring ways to enhance the utility of lithofacies paleogeography (Mou et al. 2016b), i.e., to integrate sedimentary lithofacies paleogeography studies with mineral and energy resource prediction. This is also an important aspect of the study of lithofacies paleogeography in China. This is also a major feature and development trend of lithofacies paleogeography in China since the end of 1990s. Another main feature and development trend is to explore how to improve the isochronism, instantaneity and objectivity of lithofacies paleogeography research (Mou et al. 2016b). It should be said that since Vail (1977) combined sedimentary facies analysis with seismic data, introduced the concept of sequence and proposed the concept of isochronous stratigraphic framework, how to draw up the lithofacies paleogeographic maps of system tracts or related interfaces in different sequences, so as to more truly reflect the sedimentary evolution of sedimentary basins, and more objectively and dynamically reflect the filling sequence and historical changes of sedimentary basins (Mou et al. 1992; Mou 1993, Mou et al. 1999, 2016b; Ma et al. 2002, 2005; Ma and Cai 2006; Ma et al. 2009; Jiang et al. 2007; Hua and Zhang 2009), and finally, the relationship between them and mineral and energy resources (time and space, genetic mechanism) will be clarified more precisely, so as to better serve the mineral and energy industries, this has always been the goal of sedimentologists. In China, the instantaneous mapping method of stratigraphic sequences developed by Mou et al. (1992) was the pioneer and representative of this method. In these two works, not only the theory and methodology of stratigraphic paleogeography are discussed in-depth, but also the practice and application of this methodology are discussed. These works are closely related to oil and gas exploration, development and applications. The theory provides a fundamental basis for the sedimentary environment and spatial and temporal distribution of sediments, minerals and energy resources, while the preparation of medium- and large-scale stratigraphic paleogeographic maps at the basin scale guides the exploration and development of oil and gas and achieves practical results, reflecting its application value. Up to now, many sedimentologists have explored and expanded the study of stratigraphic paleogeography and more elaborate mapping methods and achieved a great deal of results, such as Deng et al. (1997) and Zheng et al. (2000, 2001, 2002, 2008, 2010b) have been trying to apply the relevant principles and methods to the study and prediction of terrestrial sedimentary environments. Ma et al. (2009) compiled and published a tectonically stratified lithofacies paleogeographic map of the South China.

Just like the research and development history of sedimentology and lithofacies paleogeography in foreign countries, the development history of China is also accompanied by the development of the mineral and energy industries. Every innovation in theory and related technical methods is closely related to the demand of mineral and energy industries. The development of the latter cannot be separated from the key guidance of the former.

In China, oil and gas sedimentology developed from lithofacies paleogeography has experienced the stage of learning from abroad to practice (1949–1970), the stage of enriching and perfecting oil and gas sedimentology theory (1970–1990), and the stage of production and practice of combining oil and gas sedimentology with stratigraphic stratigraphy, seismology, logging, experiment and new computer technology (after 1990) (Gu and Zhang 2003). Nowadays, from the perspective of basic research and theoretical innovation, although the paleogeographic research of the activity theory and the research on the distribution of oceans, seas and continents on the paleolatitude coordinates are still the goals that sedimentary geologists strive and pursue, they are also the goals of sedimentary paleogeography today. Frontier areas of research. However, in terms of its practical applicability, just as the development of sedimentology and lithofacies paleogeography has become more integrated, the development of paleogeography cannot ignore the needs of the people it serves and still needs to persist in exploring appropriate guiding ideas and mapping methods in order to better serve the mineral and energy resources industries that are closely related to our lives. In recent years, Mou and Xu (2010) put forward the idea of "structural control of basin, basin control of facies, facies control of oil and gas basic geological conditions", and made full reference to mineral and energy resources, comprehensive application of oil and gas geology and other disciplines, the compilation of the lithofacies and paleogeography Atlas of China (Ediacaran-silurian) (Mou et al. 2016b) is an attempt to integrate basic research, theoretical innovation and practical application and finally achieve better results. This is of great significance not only for the study of sedimentology and lithofacies paleogeography, but also for the research and exploration of energy and mineral resources.

In recent years, as shale gas has become a research hotspot in the field of energy, the concept of shale gas sedimentology has emerged. Organic-rich shale, as the host of shale gas reservoir development, has its unique petrological characteristics. Identifying different petrological features is the key to evaluating the formation conditions, in-situ gas content and resources of shale gas reservoirs (Yang et al. 2010). However, at present, most scholars at home and abroad have studied the content and type of organic carbon, maturity, mineral composition, fracture system, temperature and pressure, structural preservation conditions and other aspects and deeply studied the adsorption mechanism (Zhang et al. 2003; Jarvie et al. 2007; Travis et al. 2008; Nie et al. 2009a; Zou et al. 2010c, 2011a). However, the petrological characteristics, depositional environment and depositional model of the organic-rich mud shale developed in shale gas reservoirs are less involved, which in turn determine the material basis for the development of shale gas, and the basic geological characteristics of shale gas investigation are more important.

Work in fine-grained sedimentology and paleogeography, including mudstone and shale, dates back to the 1940s (Dapples and Rominger 1945), with an initial focus on macroenvironmental and lithofacies studies, such as submarine channels and lithography. Submarine fans (Dill et al. 1954; Sullwold and Harlod 1960), turbidity currents and turbidites (Kuenen 1950), tidal currents and tidal rocks (Straaten and Kuenen 1958) and mud in Salt Lake (Dellwig 1955) environments and wind dust deposition in marine environment (Radczewski 1939). In the 1970s and 1980s, research expanded into sedimentary petrology (Picard, 1971; Spears et al. 1980) and sedimentation modelling, creating models for deep-sea suspended sedimentation (Schubel and Carter 1984), turbidity and isobath sedimentation (Stow and Piper 1984), tidal current sedimentation (McCave 1971), explored the genesis of fine-grained sedimentation in terrestrial lake environments (McCave 1971), conducted experimental (Kranck 1980) and simulation studies (McCave and Swift 1976; Stow and Bowen 1978), analysed the mechanisms of fine-grained sedimentation (McCave 1984), noting modeling studies of organic-rich fine-grained sediments (Sageman et al. 1991). In the 1990s, people gradually paid attention to the study of fine-grained sedimentary microstructure (Hulbert 1991) and its controlling factors. With the rapid development of the unconventional oil and gas industry, especially since the beginning of the 21st Century, with the development of the shale gas industry dominated by the USA, the study of fine-grained sediments and tight facies has received renewed attention (Jia et al.

2014). At the same time, for shale gas, organic-rich shale deposition modeling and mechanism research has attracted much attention (Loucks and Ruppel 2007). Various genetic models have been established for source rocks, mainly including upwelling models (Sageman et al. 1991; Su et al. 2007; Li et al. 2008; Zhang 2015), oceanic hypoxia event model (Sageman et al. 1991; Cappellen and Ingall 1994; Erbacher et al. 2001; Su et al. 2007; Li et al. 2008; Zhang 2015), pelagic suspended sedimentary model (Stow and Tabrez 1998; Stow et al. 2000), Black Sea model (Wijsman et al. 2002), continental margin slope-basin model (Loucks and Ruppel 2007), shallow-water shelf models (Leckie et al. 1990; Egenhoff and Fishman 2013), peripheral foreland basin depositional models (Lehmann et al. 1995) and inland lake basin depositional models (Piasecki et al. 1990; Gilbert et al. 2012; Xu et al. 2015a, b). At present, most of the evaluation methods for shale gas reservoir source rocks adopt the inversion method, that is, using residual organic matter in strata to evaluate source rock. However, some scholars use the emerging geobiological theory to evaluate the source rocks by forward modeling (Xie et al. 2006; Yin et al. 2008; Wu 2012).

1.2.3 Application Status of Lithofacies Paleogeography in Shale Gas Industry

The development history and research status of global sedimentology and lithofacies paleogeography show that, on the one hand, the development of sedimentology and lithofacies paleogeography in the world is, to a certain extent, driven by the development and demand of the mineral and oil and gas industries. On the other hand, lithofacies paleogeography not only has important basic theoretical significance for humans to study the tectonic evolution history of the earth, the control of tectonic activity on the attributes of sedimentary basins, the restoration of ocean-continent patterns and the transformation of geological historical periods, etc. In the research of oil and gas resources, exploration and development, prospect prediction of sedimentary layer-controlled minerals, and water resources exploration also play a very critical role (Mou et al. 2016b). In layman's terms, different sedimentary systems and sedimentary facies control the most primitive exploitation conditions of mineral resources and oil and gas resources in the geosphere; that is, different paleoenvironment or paleogeography will form different mineral resources and oil and gas resources. The rule is summarized as "facies control oil and gas, basic geological conditions" (Mu and Xu 2010). Then, we should be able to preliminarily reveal the internal relationship between sedimentation and the distribution of energy and mineral resources through lithofacies paleogeography and mapping methods. If the knowledge from other disciplines is integrated on this basis, we will be able to reveal the causes of energy and mineral resources more deeply.

Therefore, it is concluded that lithofacies paleogeography is not only a theoretical guide to the study of mineral and energy resources, revealing their intrinsic connections, which has the property of 'guidance', but also its mapping method is itself a method for finding minerals and oil and gas, which can be directly applied to the exploration and development of mineral and energy resources (Mou et al. 2011, 2016b), which has the property of 'key technology'.

As early as 70 years ago, Mr. Xie Jiarong put forward the forward-looking view of paleogeography as a guide for prospecting work (Xie 2001). He was the first geologist to openly put forward paleogeography as a guide to prospecting. The first application of facies paleogeography and its mapping method in oil, gas and mineral exploration and development. Xie Jiarong (2001) proved the key role of paleogeographic interpretation by the discovery of new coalfields and coalfields in Huainan and considered that paleogeographic conditions controlled the physical and chemical properties of coal seams and the economic value of exploration and development. The distribution of bauxite and phosphate deposits is also controlled by paleogeography. He further pointed out that the exploration of copper and iron ore still needs a correct understanding of paleogeographic characteristics. Under the guidance of this theory, lithofacies paleogeography of different scales is a method to search for mineral resources such as coal seams, bauxite, phosphate rock, copper and iron ore. In addition, sedimentology, lithofacies paleogeography and its mapping methods have been playing a very important role in conventional oil and gas exploration (Feng and Wu 1988, Feng et al. 1991, 1994, 1997b, 1999, 2000; Mou et al. 1992, 1997b, 2000, 2010, 2011, 2014, 2016b; Tian and Wan 1993, 1994; Chen and Wang (1999), 2009a; Ma et al, 2002, 2005; Ma and Cai 2006; Ma et al. 2009). Tian and Wan (1993) guided by lithofacies paleogeography, studied the distribution, sequence and lithofacies of the Jurassic in China according to the Chinese tectonic environment, paleogeographic outline, sedimentary lithofacies and stratigraphic sequence characteristics and analyzed the hydrocarbon-bearing geological conditions of the Jurassic in China. Feng and Wu (1988) studied the petrology and lithofacies paleogeography of the Qinglong Group in the lower Yangtze region, and from the isopach map of dark rock group of the Qinglong Group, the distribution law of oil layers can be seen historically and comprehensively; Granular limestone formed in shoal environment, quasi-contemporaneous dolomite and limestone with structural cracks are all favorable oil and gas reservoirs; Marl can also be used as oil and gas caprock. He studied the hydrocarbon generation, storage and caprock conditions of the Qinglong Group from the perspective of lithofacies paleogeography. In his monograph "China Sedimentology", "Lithofacies Paleogeography and Jingbian Gas

Field", "Lithofacies Paleogeography and Hudson Oilfield in the Tarim Basin", "Lithofacies Paleogeography and the Hudson Oilfield in the Tarim Basin", "Reef Gas in the Upper Permian Changxing Formation in the Sichuan Basin" Reservoir exploration", "Lithofacies paleogeography research has led to a major breakthrough in the exploration of oolitic shoal gas reservoirs in the Feixianguan Formation in the Sichuan Basin", and "the important role of sedimentary facies in the breakthrough of oil and gas exploration in the Yanchang Formation", etc. The guiding role of geography or sedimentary facies in oil and gas exploration (Feng et al. 1991, 1994, 1997b, 1999, 2013), these are typical applications of quantitative lithofacies paleogeography and mapping (methods) in oil and gas exploration and development. Liu and Xu (1994) discussed the prospect of oil and gas by studying lithofacies paleogeography in Southern China, laying a foundation for the application of lithofacies paleogeography research and mapping (methods) in oil and gas exploration in Southern China. During the "Seventh Five-Year Plan" period, in the project of "Prospecting Foreground Prediction of Lithofacies Paleogeography, Sedimentation and Stratum Control in South China" presided by Academician Liu Baojun, Mou et al. (1992) used the theory and method of sequence stratigraphy to combine the Combined with this, the method of sequence lithofacies paleogeography was first proposed in China, and the ore-controlling mechanism of the sedimentary system tract was discussed by taking a Devonian sedimentary layer-controlled ore deposit in South China as an example. And, it is the first application of sequence stratigraphy in oil and gas minerals. In future research, it is further considered that "the division of sedimentary sequence sedimentary system tract and the compilation of the sequence lithofacies paleogeographic map can be used for the research and analysis of the spatial configuration of the oil and gas source-reservoir-cap rock", and it is pointed out that this method provides research ideas and working modes for "selection and evaluation from sequence stratigraphy-sequence lithofacies paleogeography-oil and gas source-reservoir-cap rock", and the Permian in Hunan, Hubei and Jiangxi are taken as the research objects to compile corresponding sequence sedimentary system. Geographical map, combined with the sedimentary characteristics of each system tract, analyzed horizontally and vertically and discussed the spatial configuration of oil and gas source-reservoir cap rocks in the Hunan-Hubei-Jiangxi region (Mou et al. 1992; Mou 1994; Mou et al. 1997b, 2000). Guidance and application of facies paleogeography [especially sequence lithofacies paleogeography and its mapping (methods)] in the oil and gas geological survey in the Hunan, Hubei and Jiangxi regions. The idea and method of applying lithofacies paleogeography theory, lithofacies paleogeography or sequence lithofacies paleogeography has been widely studied and applied in national oil and gas geological surveys and even used to guide oil and gas (Mou et al. 1992; Mou 1994; Mou et al. 1997b, 1999, 2007, 2010, 2011, 2014; Xu et al. 1994; Mei et al. 1996, 2005, 2006, 2007; Li et al. 1998; Chen and Wang 1999, 2000, Chen and Ni (2005), 2009; Tian et al. 1999, 2000, 2008, 2010; Zheng et al. 2000, 2001, 2002, 2009; Ma and Cai 2006; Ma and Chu 2008; Zhou et al. 2006; Wang et al. 2007, Zhu et al. 2008; Lu and Ji 2013). The most typical application is the discovery of the Puguang gas field in the Sichuan Basin. Ma et al. (2010) believed that the innovation of geological knowledge, explore ideas and exploration technology had brought about the discovery of the Puguang gas field. The main contribution of the distribution pattern of salt rock reservoirs is the study of lithofacies and paleogeography and the compilation of related maps. On the basis of detailed analysis of the sedimentary characteristics of the depositional system of the Permian Changxing Formation, the time and the spatial distribution of each sedimentary facies belt were defined, and the corresponding lithofacies paleogeographic maps were compiled. It is believed that the reefs of the Changxing Formation are reefs on the gentle slopes of the carbonate platform margins, which are intermittently distributed along the platform margins. It is pointed out that the shoals and reefs on the platform margin. It is the most favorable facies belt for oil and gas reservoirs, and reef dolomite and granular dolomite are the most favorable microfacies for reservoirs (Mou et al. 2016b). In fact, the reef gas reservoir of Changxing Formation in the upper Permian in the Sichuan Basin has experienced a difficult geological survey and exploration process. During this process, under the misleading of two wrong geological understandings, the exploration of the reef gas reservoirs at that time was stagnant (Feng et al. 2008). To make a breakthrough, it is necessary to re-understand the genesis of reef gas reservoirs (shoals and reefs at the edge of the platform) and geological distribution patterns (lithofacies and paleogeographic features). By studying the characteristics of reefs and reef gas reservoirs, it is considered that reef gas reservoirs are obviously controlled by lithofacies paleogeography and sedimentary facies belt, and lithofacies paleogeography controls the distribution of reefs and reef gas reservoirs. The exploration success rate of reef gas reservoirs in the Changxing Formation of Upper Permian in the Sichuan Basin has been improved, and a major breakthrough has been made in reef gas reservoir exploration (Feng et al. 1997a, b, 1999, 2013; Mou et al. 2003, 2005; Ma et al. 2002, 2006, 2007, 2010, 2014).

In a word, the theoretical development of lithofacies paleogeography and the progress of related technologies, such as the innovation and development of mapping technology of sequence lithofacies paleogeography, have played an important role in the geological investigation and exploration and development of sedimentary minerals and conventional oil and gas resources in the past decade. It plays a very important role in developing new oil exploration fields such as rivers, deltas and slump turbidite fans, deep-water gravity flow deposits, beach bars, reefs and carbonate uplifts and promoting the exploration and development of oil and gas resources. It has played an irreplaceable role in theoretical guidance (Zhu 2008) and is a key technical method.

Shale gas reservoirs are a kind of oil and gas resource. However, the guiding role of lithofacies paleogeography theory and mapping methods in shale gas geological survey and exploration and development, as well as the status of key technologies and methods, have not been fully understood. Geological survey, exploration and development of shale gas in China have made remarkable achievements in a short time, but it is still in the initial stage. It is moving from theoretical research to practice, from investigation in key fields to multi-field and multi-type evaluation and from local experiments to scale investment. Originally, the research work on shale gas geology in China has drawn on the experience of the USA from the very beginning, while the geological conditions of shale gas in the USA are relatively simple and its basic research work is relatively maturity, thus the basic geological background such as lithofacies paleogeography distribution and mapping methods in shale gas geological survey and exploration and development of the application of less attention and research.

It is considered that through basic geological research (such as sedimentary facies, lithofacies and paleogeography) and the determination of source rocks, especially the determination of high-quality source rocks, the potential of oil and gas resources in petroliferous basins can be scientifically evaluated, and the generation, migration and migration of oil and gas can be deeply recognized. The laws of migration, aggregation and enrichment are of great significance (Lu et al. 2006; Hou et al. 2008; Zhang et al. 2010a, b). Among them, the shale gas reservoir with the integration of source and reservoir is especially true, and it is particularly important to study its material basis and controlling factors in depth. As mentioned earlier, predecessors have done some research on the sedimentary model and enrichment mechanism of shale gas reservoir source rocks and achieved some results. However, the research on the relationship between source-reservoir allocation, favorable sedimentary environment and main controlling factors of source rocks is still insufficient, especially the application of sedimentology and lithofacies paleogeography in shale gas has not been studied. Fortunately, they have already participated in it: Liang et al. (2011) studied the influence of the Niutitang Formation depositional environment on the enrichment of organic matter in Northern Guizhou and analyzed the Wufeng and Longmaxi Formations through electron microscope observation and mineral content analysis. The mineralogy, lithofacies and storage space types of the shale in this group were discussed, and the controlling factors for the performance of related shale reservoirs were discussed. Many factors control the distribution of the hydrocarbon source rock (Huang 2011), such as the paleoclimate, paleostructure, paleoenvironments, it is impartial to put forward that "anoxic conditions" alone is the main control factor for effective source rocks, and "paleoproductivity" has a good correlation with organic-rich marine source rocks. As for the formation environment of source rocks, the high organic matter productivity of surface water is more important than the anoxic environment of bottom water. As long as the organic matter productivity is high enough, source rocks can also be formed in an oxygen-containing non-reducing environment. Gao et al. (2012) studied the shale of the Sargan Formation and the shale of the Yingan Formation in the Tarim Basin and concluded that the organic matter content and density of the source rocks of the two groups are consistent with the rise and fall of sea level. Corresponding relationship, all have the characteristics of sea-level rise, organic matter abundance increase, mud shale density decrease, and sea-level decrease, organic matter abundance decrease and mudstone density increase. And, it is pointed out that the sea level change is influenced by the comprehensive control of tectonic activities and changes in the sedimentary environment. Xie et al. (2015) further analyzed the biological composition and organic carbon content according to the idea of depositional environment-biological composition-source rock formation, and then discussed the relationship between depositional environment and source rock development and concluded that the former absolutely controls the latter. Lin et al. (2015) also believed that the structure and paleoclimate can control the formation and disappearance of the accommodation space, and the supply of water and sediments, the products of the sedimentary environment, affects the development and evolution of the basin and also affects the physical and chemical properties of sedimentary water bodies. These properties and biological development ultimately control the development characteristics of source rocks by influencing preservation conditions, paleoproductivity and sedimentation rate. In a word, the source rock is the material base of shale gas exploration, and its forming environment and its control on the sedimentological characteristics and processes of the formation and preservation of the source rock are studied. It is of great significance to the shale gas geological survey and further exploration and development in the future and deserves the attention of many shale gas researchers.

In the research on the characteristics of shale gas reservoirs, most of the predecessors studied shale as a source rock, but relatively few studied it as a reservoir rock. In North America, significant progress has been made in the study of shale gas pore characteristics by studying the microscopic characteristics of shale reservoirs represented by micron-scale and nano-scale microporosity, which plays an important role in guiding the exploration and development of shale gas. In China, Zou et al. (2010b) discovered nano-scale pores in mud shale for the first time through nano-CT technology, which opened the prelude to the research on nano-scale pores in oil and gas reservoirs. The same is true for shale gas reservoirs. Many scholars have studied the characteristics, types and forming conditions of shale reservoirs and put forward the evaluation parameters for shale reservoirs (Curtis 2002; Jiang et al. 2010; Zou et al. 2010b; Guo et al. 2011). Because of the importance of shale gas reservoir characteristics, the study of its controlling factors has been paid more and more attention. The study found that there is a close relationship between sedimentary (micro)facies and shale reservoir lithofacies, characteristics, reservoir properties and distribution, etc. However, is this relationship a positive correlation under any conditions? It is still a positive correlation under certain conditions, or even a negative correlation, and there is no clear conclusion yet. Relevant research is being carried out gradually, and many researchers have begun to work in this area in the past two years. For example, Guo et al. (2015) studied the shale reservoir rocks of the Shanxi Formation through parameters such as logging data, logging data, core observation description, thin section analysis, particle size analysis, X-ray diffraction whole-rock analysis and total organic carbon content. And, the sedimentary microfacies controlling the distribution of shale reservoirs in the Shanxi Formation are described in detail by facies and sedimentary microfacies types. Through the study of lithofacies paleogeography and mapping technology, the spatial distribution law of favorable sedimentary microfacies is recognized. Application of lithofacies and paleogeography research and mapping in shale gas geological survey and in-depth study of reservoirs. Xu et al. (2015a, b) also conducted a comprehensive analysis and research on the reservoir characteristics of shale formations in the central and Southeastern Hunan depression by using X-ray diffraction, scanning electron microscopy and isothermal adsorption technology under the background of regional deposition. It is another practice of lithofacies paleogeography in shale gas geological investigation and in-depth study to study the distribution of favorable reservoirs by geographic mapping.

To sum up, the conventional oil and gas geological survey and exploration and development practice in China for decades, as well as the shale gas geological survey and exploration and development practice in recent decades, show that the distribution of source rocks in China, that is, the sedimentary environment of the potential shale gas distribution area and the later multi-stage structural transformation, has not been systematically studied up to now. The relationship between organic geochemical parameters (organic carbon content, type of organic matter, thermal evolution degree of organic matter, etc.) and mineral composition of source rocks and tectonic background, sedimentary environment and paleogeographic distribution has not been systematically or fully studied in China, and only a few reports are available (Li et al. 2013a, b; Wang et al. 2013; Wang (2015)). Sedimentology and lithofacies paleogeography are the theoretical guidance for shale gas geological survey, and lithofacies paleogeography mapping technology is the key technology to optimize shale gas prospect areas, favorable areas and target areas. The related theories, methods and technologies have not been systematically developed. Mou et al. (2016a) discussed the role of lithofacies paleogeography and its mapping method in shale gas survey in relation to the study of shale gas in the Longmaxi Formation in Southern Sichuan Basin and adjacent areas, it is the first time that the study of lithofacies paleogeography can be used as a guide and a key technical method for shale gas geological survey. In fact, in the face of extremely complex geological conditions of shale gas in China, the extensive and relatively lagging theoretical research on shale gas geology and the weak research foundation, in the geological survey of shale gas in China, attention should be paid to the research of key technologies and methods of basic geology and geological survey, and the main research directions should be the research of source rock lithofacies paleogeography, mapping method and key technologies and methods of geological survey. The practice of shale gas exploration and development in China also proves that lithofacies paleogeography controls the combination of basic elements of shale gas accumulation and the distribution of shale gas resources. It is necessary to strengthen the study on the controlling effect of sedimentation on the distribution law of shale section, that is, source rocks. It is necessary to strengthen the superposition of multi-information, multi-scale, diversified and digitized paleogeography and shale gas geological conditions parameters based on mapping and explore the methods of optimizing the prospect, favorable area and target areas.

References

Abouelresh MO, Slatt RM (2012) Lithofacies and sequence stratigraphy of the Barnett Shale in east-central Fort Worth Basin, Texas. AAPG Bull 96(1):12–22

Algeo TJ, Barry MJ (2008) Tracemetal covariation as a guide to water-mass conditions in ancient anoxic marine environments. Geosphere 5(4):872–887

An XX, Huang WH, Liu SY et al (2010) The distribution development and expectation of shale gas resource. Res Indus 12(2):103–109

Aplin AC, Macquaker JSH (2011) Mudstone diversity: origin and implications for source, seal, and reservoir properties in petroleum systems. AAPG Bull 95(12):2031–2059

Bai BJ, Elgmati M, Zhang H et al (2013) Rock characterization of Fayetteville shale gas plays. Fuel 105:645–652

Beliveau D (1993) Honey, I shrunk the pores!: J Can Petrol Technol 32 (8):15–17

Bjørlykke K, Kaare H (1997) Effects of burial diagenesis on stresses, compaction and fluid flow in sedimentary basins. Mar Pet Geol 14 (3):267–276

Boggs S (1992) Petrology of sedimentary rocks. Springer, Netherlands

Bohacs KM (1998) Introduction: mud rock sedimentology and stratigraphy—challenges at the basin to local scales. In: Schieber J, Zimmerle W, Sethi P (eds) Shales and mudstones: basin studies, sedimentology and paleontology: Stuttgart, Schweizerbart'sche Verlagsbuchhandlung, vol 1, pp 13–20

Bouma AH (1962) Sedimentology of some flysh deposits: a graphic approach to facies interpretation. Elsevier, pp 166–169

Bowker KA (2003) Recent development of the Barnett Shale play, Fort Worth Basin: West Texas. Geol Soc Bulletin 42(6):1–11

Bowker KA (2007) Barnett Shale gas production, Fort Worth Basin: issues and discussion. AAPG Bull 91(4):523–533

Boyer C, Clark B, Jochen V et al (2011) Shale gas: a global resource. Oilfield Rev 23(3):28–37

Bureau of Geology and Mineral Resources of Sichuan Province (1991) Regional geology of Sichuan Province. Geological Publishing House, Beijing, pp 1–732

Bust VK, Majid AA, Oletu JU et al (2013) The petrophysics of shale gas reservoirs: technical challenges and pragmatic solutions. Pet Geosci 19:91–103

Cai GJ, Xu JL, Yang SY (2006) The fractionation of an argillaceous sediment and difference in organic matter enrichment in different fractions. Geol J China Universities 12(2):234–241

Cai JG, Bao YJ, Yang SY, et al (2007) Occurrence and enrichment mechanism of organic matter in argillaceous sediments and mudstones. Sci China (Seri. D): Earth Sci 37(2):24−253

Cai XY, Zhao Pei R, Gao B et al (2021) Sinopec's shale gas development achievements during the "Thirteenth Five-Year Plan" period and outlook for the future. Oil Gas Geol 42(1):16–27

Cappellen PV, Ingall ED (1994) Benthic phosphorus regeneration, net primary production, and ocean anoxia: a model of the coupled marine biogeochemical cycles of carbon and phosphorus. Paleoceanography 9(5):677−692.

Cardott BJ (2012) Thermal maturity of woodford shale gas and oil plays, Oklahoma, USA. Int J Coal Geol 103:109–119

Chalmers RL, Bustin RM (2007) The organic matter distribution and methane capacity of the Lower Cretaceous strata of northeastern British Columbia, Canad. Int J Coal Geol [J] 70:223–239

Chalmers GR, Bustin RM (2012) Light volatile liquid and gas shale reservoir potential of the Cretaceous Shaftesbury Formation in Northeastern British Columbia, Canada. AAPG Bull 96(7):1333–1367

Chalmers RL, Bustin RM (2008a) Lower Cretaceous gas shale in northern British Columbia: Part I. Geological controlon methane sorption capacity: Bull Can Petrol Geol 56:1–21

Chalmers RL, Bustin RM (2008b) Lower Cretaceous gas shale in northern British Columbia: part II. Evaluation of regional potential gas resources. Bull Can Petrol Geol 56:22–61

Chalmers RL, Bustin RM, Bustin AM (2011) Geological controls on matrix permeability of the Doig–Montney hybrid shale-gas-tight-gas reservoir. Northeastern British Columbia (NTS 093P). In: Geoscience British Columbia Summary of Activities. Geoscience British Columbia, Report 2012 (1), pp 87–96

Chalmers GR, Bustin RM, Power IM (2012a) Characterization of gas shale pore systems by porosimetry, pycnometry, surface area, and field emission scanning electron microscopy/transmission electron microscopy image analyses: Examples from the Barnett, Woodford, Haynesville, Marcellus, and Doig units[J]. AAPG Bull 96(6):1099–1119

Chen HD, Wang CS (1999) Permian sequence stratigraphy and basin evolution in South of China. Acta Sedimentol Sin 17(4):529–525

Chen HD, Ni XF (2005) Sediment accumulation on the steep slope zone of the continental down-warped lacustrine basin and dynamics of sequence filling: an example from the Triassic Yanchang Formation in the Ordos Basin, Eastern Gansu. Abstract of the symposium on Meso-Cenozoic evolutionary dynamics and its effect on resources and environment in Ordos Basin and its periphery

Chen B, Lan ZK (2009) Lower Cambrian shale gas resource potential in Upper Yangtze region. China Petrol Explo 3:15–20

Chen B, Pi DC (2009) Silurian Longmaxi Shale Gas potential analysis in middle & upper Yangtze region. China Petrol Expl 3:15–19

Chen HD, Qin JX, Tian JC et al (2000) Source-reservoir-cap rock association within sequence framework and its exploration significance. Acta Sedimentol Sin 18(2):215–220

Chen JF, Zhang SC, Bao ZD et al (2006) Main sedimentary environments and influencing factors for development of marine organic-rich source rocks. Marine Origin Petrol Geol 11(3):49–54

Chen HD, Huang FX, Xu SL et al (2009a) Distribution rule and main controlling factors of the marine facies hydrocarbon substances in the middle and upper parts of Yangtze region, China. J Chengdu Univ Tech (Sci Tech Edition) 36(6):569–577

Chen HD, Zhong YJ, Hou MC et al (2009b) Sequence styles and hydrocarbon accumulation effects of carbonate rock platform in the Changxing−Feixianguan formations in the Northeastern Sichuan Basin. Oil Gas Geol 30(5):539–547

Cheng KM, Wang SQ, Dong DZ et al (2009c) Accumulation conditions of shale gas reservoirs in the Lower Cambrian Qiongzhusi formation, the Upper Yangtze region. Nat Gas Ind 29 (5):40–44

Chen SB, Zhu YM, Wang HY et al (2010) Research status and trends of shale gas in China. Acta Petrolei Sinica 31(4):689–694

Chen B, Guan XQ, Ma J (2011a) Early Paleozoic shale gas in upper Yangtze region and North American shale gas potential Barnett contrast[J]. J Oil Gas Technol (J Jianghan Petrolum Inst) 33(12):23–27

Chen SB, Zhu YM, Wang HY et al (2011b) Shale gas reservoir characterisation: a typical case in the Southern Sichuan Basin of China. Energy 36:6609–6616

Chen SB, Zhu YM, Wang HY et al (2011c) Characteristics and significance of mineral compositions of Lower Silurian Longmaxi Formation shale gas reservoir in the Southern margin of Sichuan Basin. Acta Petrolei Sinica 32(5):775–782

Chen X, Yan YX, Zhang XW et al (2011d) Generation conditions of continental shale gas in Biyang Sag, Nanxiang Basin. Petrol Geol Exp 33(2):137–141

Chen HD, Ni XF, Tian JC et al (2006) Sequence stratigraphic framework of marine lower assemblage in South China and petroleum exploration. Oil & Gas Geol 27(3):370−377

Chen SB, Xia XH, Qin Y et al (2013a) Classification of pore structures in shale gas reservoir at the Longmaxi Formation in the south of Sichuan Basin. J China Coal Soc 38(5):760–765

Chen WL, Zhou W, Luo P et al (2013b) Analysis of the shale gas reservoir in the Lower Silurian Longmaxi Formatio, Changxin1well, Southeast Sichuan Basin, China. Acta Petrologica Sinica 29(3):1073–1086

Cheng LX, Wang YJ, Chen HD et al (2013) Sedimentary and burial environment of black shales of Sinian to Early Palaeozoic in Upper Yangtze region. Acta Petrologica Sinica 29(8):2906–2912

Christiansen FG, Olsen H, Piasecki S et al (1990) Organic geochemistry of upper palaeozoic lacustrine shales in the east greenland basin. Org Geochem 16(1–3):287–294

Clarkson CR, Solano N, Bustin RM et al (2013) Pore structure characterization of North American shale gas reservoirs using USANS/SANS, gas adsorption, and mercury intrusion. Fuel 103:606–616

Cloos E, Pettijohn FJ (1973) Southern Border of the Triassic Basin, West of York, Pennsylvania: fault or overlap? Geol Soc Am Bull 84 (2):523–535

Curtis JB (2002) Fractured shale-gas systems. AAPG Bull 86 (11):1921–1938

Curtis ME, Sondergeld CH, Ambrose RJ et al (2012) Microstructural investigation of gas shales in two- and three-dimensions using nanometer-scale resolution imaging. AAPG Bull 96(4):665–677

Daoudi L, Ouajhin B, Rocha F et al (2010) Comparative influence of burial depth on the clay mineral assemblage of the Agadir-Essaouira basin (Western High Atlas, Morocco). Clay Miner 45:453–467

Dapples EC, Rominger JF (1945) Orientation analysis of fine-grained clastic sediments: a report of progress. J Geol 53(4):246–261

Dellwig IF (1955) Origin of the salina salt of michigan. J Sediment Res 25:83–110

Deng HW, Wang HL, Li XM (1997) Application of base level principle in prediction of Lacustrine reservoirs. Oil Gas Geol 18(2):90–95

Dietz RS, Holden JC (1970) The breakup of Pangaea. Sci Am 223 (4):30–41

Dill RF, Dietz RS, Stewart H (1954) Deep-sea channels and delta of the monterey submarine canyon. Geol Soc Am Bull 65(2):191–194

Dong DZ, Cheng KM, Wang YM et al (2010) Forming conditions and characteristics of shale gas in the Lower Paleozoic of the Upper Yangtze region, China. Oil Gas Geol 31(3):288–299

Dorrik VS, Anthony JB (1980) A physical model for the transport and sorting of fine-grained sediment by turbidity current. Sedimentology 27:31–46

Du JH, Yang H, Xu CC et al (2011) A discussion on shale gas exploration and development in China. Nat Gas Ind 31(5):6–8

Egenhoff SO, Fishman NS (2013) Traces in the dark-sedimentary processes and facies gradients in the upper shale member of the upper devonian-lower mississippian bakken formation, williston basin, north dakota, U. S. A. J Sediment Res 83(9):803−824

Erbacher J, Huber BT, Norris RD et al (2001) Increased thermohaline stratification as a possible cause for an ocean anoxic event in the cretaceous period. Nature 409(6818):325–327

Fang RH, Liu XQ, Zhang C et al (2021) Molecular simulation of shale gas adsorption under temperature and pressure coupling: case study of the Lower Cambrian in Western Hubei Province. Nat Gas Geosci 32(11):1–15

Feng ZZ (1991) Listhofacies paleogeography of Permian Middle and Lower Yangtze Region. Geological Publishing House, Beijing

Feng ZZ (2004) Cambrian and Ordovician lithofacies paleogeography of China. Petroleum Industry Press, Beijing

Feng ZZ (2005) Discussion on petroleum exploration of marine strata in South China from quantitative lithofacies palaeogeography. J of Palaeogeography (Chinese Edition) 79(1):1–11

Feng ZZ (2008) Formation, development, characteristics and prospects of paleogeography in China. In: National Conference on Palaeogeography and Sedimentology

Feng ZZ, Wu SH (1987) Studing and mapping lithofacrs paleogeography of Qinglong Group of Lower-Middle Triassic in The Lower Yangtze Valley. Acta Sedimentol Sin 5(3):40–58

Feng ZZ, Wu SH (1988) Potential of Oil and Gas of Qinglong Group of Lower-middle Triassic in Lower Yangtze River Region from the Viewpoint of Lithofacies Paleogeography [J]. Acta Petrol Sinica 9 (2):1–11

Feng ZZ, Wu SH, He YB (1993) Listhofacies paleogeography of Permian Middle and Lower Yangtze Region. Acta Sedimentol Sin 11(3):13–24

Feng ZZ, Wang YH, Liu HJ et al (1994) Chinese sedimentology. Petroleum Industry Press, Beijing

Feng ZZ, Yang YQ, Jin ZK et al (1999) Potential of oil and gas of the carboniferous in South China from the viewpoint of lithofacies palaeogeography. J Palaeogeography (Chinese Edition) 1(4):86–92

Feng ZZ, Bao ZD, Yang YY et al (1997a) Why an important breakthrough in marine oil-gas exploration in South China so far having not been made. Marine Origin Petrol Geol 2(4):4–7

Feng ZZ, Li SW, Yang YQ et al (1997b) Potential of oil and gas of the Permian of South China from the viewpoint of lithofacies paleogeography. Acta Petrolei Sinica 18(1):10–17

Feng ZZ, Jin ZK, Bao ZD et al (2000) Theory of Marine carbonate oil and gas in Southern China from quantitative lithofacies paleogeography. Marine Origin Petrol Geol Z1:123–123

Feng ZZ, Bao ZD, Shao YL et al (2013) Sedimentology in China (2nd Edition). Petroleum Industry Press, Beijing

Folk RL (1959a) Spectral subdivision of limestone types. AAPG Special Volumes 38(1):62–84

Folk RL (1959b) Practical petrographic classification of limestones. AAPG Bull 43(1):1–38

Fu XD, Qin JZ, Tenger, et al (2011) Mineral components of source rocks and their petroleum significance: A case from Paleozoic marine source rocks in the Middle-Upper Yangtze region. Petrol Explo Develop 38(6):671−684

Fu CQ, Zhu YM, Chen SB (2015) Diagenesis controlling mechanism of pore characteristics in the Qiongzhusi Formation Shale, East Yunnan. J China Coal Soc 40(S2):439–448

Gaarenstroom L, Tromp AJ, Jong MC, et al (1993) Overpressures in the central North Sea: implications for trap integrity and drilling safety. In: Parker JR (ed) Petroleum geology of Northwest Europe: Proceedings from the 4th Conference, The Geological Society. London, pp 1305–1313

Gao ZY, Zhang SC, Ye L et al (2012) Relationship between high-frequency sea-level changes and organic matter of Middle-Upper Ordovician marine source rocks from the Dawangou section in the Keping area, Xinjiang. Acta Petrolei Sin 33(2):232–240

Gilbert R, Desloges JR (2012) Late glacial and Holocene sedimentary environments of Quesnel lake, British Columbia. Geomorphology 179:186–196

Gilbert JA, Steele JA, Caporaso JG, et al (2012) Defining seasonal marine microbial community dynamics. ISME J 6(2):298–308

Gu JY, Zhang XY (2003) Development review and current application of petroleum sedimentology. Acta Sedimentol Sin 21(1):137–141

Guan SC, Yan HY, Qiu DZ et al (1980) Investigations on the marine sedimentary environmental model of China in Late Proterozoic to Triassic periods. Oil Gas Geol 1(1):2–17

Guo TL (2013) Evaluation of highly thermally mature shale-gas reservoirs in complex structural parts of the Sichuan Basin. J Earth Sci 1(24):863–873

Guo TL (2016) Key geological issues and main controls on accumulation and enrichment of Chinese shale gas. Pet Explor Dev 43 (3):317–326

Guo TL, Liu RB (2013) Implications from marine shale gas exploration breakthrough in complicated structural area at high thermal stage: Taking Longmaxi Formation in Well JY1 as an example. Nat Gas Geosci 24(4):643–651

Guo TL, Zhang HR (2014) Formation and enrichment mode of Jiaoshiba shale gas field, Sichuan Basin. Pet Explor Dev 41(1):28–36

Guo YH, Li ZF, Li DH et al (2004) Lithofacies palaeogeography of the early Silurian in Sichuan area. J Palaeogeography (Chinese Edition) 6(1):20–29

Guo L, Jiang ZX, Jiang WL (2011) Formation condition of gas-bearing shale reservoir and its geological research target. Geol Bull China 31(2):385–392

Guo XS, Hu DF, Wen ZD et al (2014) Major factors controlling the accumulation and high productivity in marine shale gas in the Lower Paleozoic of Sichuan Basin and its periphery: a case study of the Wufeng-Longmaxi Formation of Jiaoshiba area. Geology in China 41(3):893–901

Guo W, Liu HL, Xue HQ et al (2015) Depositional facies of Permian Shanxi Formation gas shale in the Northern Ordos Basin and its impact on shale reservoir. Acta Geol Sin 89(5):931–941

Guo TL, Jiang S, Zhang PX et al (2020) Progress and direction of exploration and development of normally-pressured shale gas from the periphery of Sichuan Basin. Pet Geol Exp 42(05):837–845

Hackley PC (2012) Geological and geochemical characterization of the Lower Cretaceous Pearsall Formation, Maverick Basin, South Texas: a future shale gas resource? AAPG Bull 96(8):1449–1482

Hammes U, Hamlin HS, Ewing TE (2011) Geologic analysis of the Upper Jurassic Haynesville shale in East Texas and West Louisiana. AAPG Bull 95(10):1643–1666

Hao F, Chen JY (1992) The cause and mechanism of vitrinite reflection anomalies. J Pet Geol 15:419–434

Hao F, Zou HY, Lu YC (2013) Mechanisms of shale gasstorage: implications for shalegas exploration in China. AAPG Bull 97 (8):1325–1346

Harris NB (2011) Expression of sea level cycles in a black shale: woodford shale, Permian basin. In: Abstracts volume of AAPG Annual Convention Exhibition, April 10-13, Houston, Texas, USA.

He JX, Duan Y, Zhang XL et al (2011) Geologic characteristics and hydrocarbon resource implication of the black shale in Niutitang Formation of the Lower Cambrian, Guizhou Province. J xi'an Shiyou University (Natural Science Edition) 26(3):37–42

Heirtzler JR (1969) The theory of sea floor spreading. Sci Nat 56 (7):341–347

Hemmesch NT, Harris NB, Mnich CA et al (2014) A sequence-stratigraphic framework for the Upper Devonian Woodford Shale, Permian Basin, West Texas. AAPG Bull 98(1):23–47

Hickey JJ, Henk B (2007) Lithofacies summary of the Mississippian Barnett Shale, Mitchell 2 T.P. Sims well, Wise County, Texas[J]. AAPG Bull 91(4):437–443

Hill RJ, Jarvie DM, Zumberge J et al (2007a) Oil and gas geochemistry and petroleum systems of the Fort Worth Basin. AAPG Bull 91 (4):445–473

Hill RJ, Zhang E, Katz BJ et al (2007b) Modeling of gas generation from the Barnett Shale, Fort Worth Basin, Texas. AAPG Bull 91 (4):501–521

Hou YJ, Zhang SW, Xiao JX et al (2008) The excellent source rocks and accumulation of stratigraphic and lithologic traps in the Jiyang depression, Bohai Bay Basin, China. Earth Sci Front 15(2):137–146

Hu DF, Zhang HR, Ni K et al (2014) Main controlling factors for gas preservation conditions of marine shales in Southeastern margins of the Sichuan Basin. Nat Gas Ind 34(6):17–23

Hu DG, Wan YQ, Fang DL et al (2021) Pore characteristics and evolution model of shale in Wufeng-Longmaxi Formation in Sichuan Basin. Periodical of Ocean University of China 51 (10):80–88

Hua X, Zhang QQ (2009) Present situation and prospect of lithofacies paleogeography. Sci Tech Info 33:436

Huang JZ (2009a) Exploration prospect of shale gas and coal-bed methane in Sichuan Basin. Lithologic Res 21(2):116–120

Huang JZ (2009b) The pros and cons of paleohighs for hydrocarbon reservoiring: A case study of the Sichuan basin. Nat Gas Ind 29 (2):12–17

Huang JZ, Chen JJ, Song JR, et al (1997) Hydrocarbon source systems and formation of gas fields in Sichuan Basin: Science in China series D-Earth Sci 40:32−42

Huang FX, Chen HD, Hou MC et al (2011a) Filling process and evolutionary model of sedimentary sequence of Middle-Upper Yangtze craton in Caledonian (Cambrian-Silurian). Acta Petrologica Sin 27(8):2299–2317

Huang YR, Zhang ZH, Li YC et al (2011b) Development regularity and dominant factors of hydrocarbon source rocks in worldwide deepwater petroliferous basins. Marine Origin Petrol Geol 16 (3):15–21

Huang JL, Zou CN, Li JZ, et al (2012) Shale gas accumulation conditions and favourable zones of Silurian Longmaxi Formation in south Sichuan Basin, China[J]. J China Coal Soc 37(5):782–787

Hulbert MH (1991) Microstructure of fine-grained sediments. Springer

Hunt JM (1996) Petroleum geolgoy and geochemistry. Freeman Company, San Fransisco

Jarvie DM (1991) Total organic carbon (TOC) analysis. In: Merrill RK (ed) Source and migration processes and evaluation techniques: AAPG Treatise of Petroleum Geology. Handbook Petrol Geol, 113–118

Jarvie MD, Hill JR, Pollastro MR (2005) Assessment of the gas potential and yields from shales: the Barnett Shale model. In: Cardott BJ (ed) Unconventional energy resources in the southern mid-continent, conference. Oklahoma Geol Surv Circular 110:34

Jarvie DM, Hill RJ, Ruble TE et al (2007) Unconventional shale−gas systems: the Mississippian Barnett Shale of North−central Texas as one model for thermogenic shale−gas assessment. AAPG Bull 91 (4):475–499

Jenkins CD, Boyer CM (2008) Coalbed- and shale-gas reservoirs. Distinguished author series. J Petrol Techn, February Issue, 92−99, SPE103514

Ji LM, Qiu JL, Zhang TW (2012) Experiments on methane adsorption of common clay minerals in shale. Earth Sci (J China University of Geosciences) 37(5):1043–4050

Jia CZ, Zheng M, Zhang YF (2014) Four important theoretical issues of unconventional petroleum geology. Acta Petrolei Sinica 35 (1):10

Jiang ZX (2003) Sedimentology. Petroleum Industry Press, Beijing

Jiang WH, Dong CM, Yan JN (2007) Study on present situation and trend of lithofacies paleogeography. Fault−Block Oil & Gas Field 14(3):1−3

Jiang YQ, Dong DZ, Qi L et al (2010) Basic features and evaluation of shale gas reservoirs. Nat Gas Ind 30(10):7–12

Jiang ZX, Liang C, Wu J et al (2013) Several issues in sedimentological studies on hydrocarbon−bearing fine−grained sedimentary rocks. Acta Petrolei Sinica 34(6):1031–1039

Jiang WC, Zhang XR, Fang S et al (2014) Overview of development of sedimentology and hot issues in oil and gas exploration. Geol Res 23(4):398–407

Jiang YQ, Song YT, Qi L et al (2016a) Fine lithofacies of China's marine shale and its logging prediction: a case study of the Lower Silurian Longmaxi marine shale in Weiyuan area, Southern Sichuan Basin, China. Earth Sci Front 23(01):107–118

Jiang ZX, Tang XL, Li Z et al (2016b) The whole−aperture pore structure characteristics and its effect on gas content of the Longmaxi Formation shale in the Southeastern Sichuan Basin. Earth Sci Front 23(02):126–134

Jiang ZX, Song Y, Tang XL et al (2020) Controlling factors of marine shale gas differential enrichment in Southern China. Pet Explor Dev 47(3):617–628

Jiao FZ, Feng JH, Yi JZ et al (2015) Direction, key factors and solution of marine natural gas exploration in Yangtze Area. China Petrol Explor 20(2):1–8

Jin JL, Zou CN, Li JZ et al (2012) Shale gas accumulation conditions and favorable zones of Silurian Longmaxi Formation in south Sichuan Basin, China. J China Coal Soc 37(5):782–787

Keller L M, Holzer L, Wepf R, et al (2011) 3−Dgeometry and topology of pore pathways in Opalinus clay: implications for mass transport. Appl Clay Sci 52:85–95

Kinley TJ, Cook LW, Breyer JA et al (2008) Hydrocarbon potential of the barnett shale (mississippian), Delaware Basin, West Texas and Southeastern New Mexico. AAPG Bull 92(8):967–991

Kranck K (1980) Experiments on the significance of flocculation in the settling of fine−grained sediment in still water. Can J Earth Sci 17 (11):1517–1526

Krumbein WC (1945) Modern sedimentation and the search for petroleum. Tulsa Geological Society.

Kuenen PH (1950) Turbidity currents as causes of graded bedding. J Geol, 58

Laughrey CD, Lemmens H, Ruble TE, et al (2011) Black shale diagenesis: insights from integrated high-definition analyses of post-mature Marcellus Formation rocks, Northeastern Pennsylvania. Abstracts Volume of AAPG Annual Convention and Exhibition, April 10−13, Houston, TX, USA

Leckie D, Singh C, Goodarzi F et al (1990) Organic−rich, radioactive marine shale; a case study of a shallow−water condensed section, cretaceous shaftesbury formation, Alberta, Canada. J Sediment Res 60(1):101–117

Lehmann D, Brett CE, Cole R et al (1995) Distal sedimentation in a peripheral foreland basin: Ordovician black shales and associated flysch of the Western Taconic Foreland, New York state and Ontario. Geol Soc Am Bull 107(6):708–724

Lei HY, Shi YX, Fang X et al (1995) Influences of the diagenetic evolution of aluminosilicate minerals on the formation of the transitional zone gas. Acta Sedimentol Sin 13(2):22–33

Li R (1998) The distribution of Cand Oisotopes in sequence stratigraphic framework of Meso-and Neo-proterozoic in the Ming Dynasty Tombs of Beijing. Acta Geol Sin 72(2):190–191

Li GF, Zhao PD (2009) Application of geo-anomaly pre-prospecting theory to shale gas exploration. Nat Gas Ind 29(12):119–124

Li SY, Yu BS, Hai LD, et al (2006) The mineralogical phase and preservation of organic matter in sediments of the Qinghai Lake. Acta Petrologica Et Mineralogica 25(6):4, 93–498

Li XJ, Hu SY, Cheng KM (2007) Suggestions from the development of fractured shale gas in North America. Pet Explor Dev 37(4):392–400

Li SJ, Xiao KH, Wo YJ et al (2008) REE geochemical characteristics and their geological signification in Silurian, West of Hunan Province and North of Guizhou Province. Strategic Study of CAE 22(2):273–282

Li DH, Li JZ, Wang SJ et al (2009a) Analysis of controls on gas shale reservoirs. Nat Gas Ind 29(5):22–26

Li JZ, Dong DZ, Chen GS et al (2009b) Prospects and strategic position of shale gas resources in China. Nat Gas Ind 29(5):11–16

Li XJ, Lv ZG, Dong DZ et al (2009c) Geologic controls on accumulation of shale gas in North America. Nat Gas Ind 29(5):27–32

Li YX, Nie HK, Long PY, et al (2009d) Development characteristics of organic-rich shale and strategic selection of shale gas exploration area in China[J]. Nat Gas Ind 29(12):115–118

Li X, Duan SK, Sun Y, et al (2011a) Advances in the exploration and development of U.S. shale gas. Natural Gas Indus 31(8):124−126

Li YX, Qiao DW, Jiang WL et al (2011b) Gas content of gas−bearing shale and its geological evaluation summary. Geol Bull China 30(2–3):308–317

Li ZR, Deng XJ, Yang X et al (2011c) New progress in seismic exploration of shale gas reservoirs in the Southern Sichuan Basin. Nat Gas Ind 31(4):40–43

Li JZ, Li DH, Dong DZ, et al (2012a) Comparison and enlightenment on formation condition and distribution characteristics of shale gas between China and U. S. Strategic Study of CAE 14(6):56–63

Li J, Yu BS, Liu CE et al (2012b) Clay minerals of black shale and their effects on physical properties of shale gas reservoirs in the Southeast of Chongqing: a case study from Lujiao Outcrop section in Pengshui, Chongqing. Strategic Study of CAE 26(4):732–740

Li YF, Fan TL, Gao ZQ et al (2012c) Sequence stratigraphy of Silurian Black shale and its distribution in the Southeast Area of Chongqing. Nat Gas Geosci 23(2):299–306

Li YX, Zhang JC, Jiang SL et al (2012d) Geologic evaluation and targets optimization of shale gas. Earth Sci Front 19(5):332–338

Li SZ, Jiang WL, Wang Q et al (2013a) Research status and currently existent problems of shale gas geological survey and evaluation in China. Geol Bull China 32(9):1440–1446

Li YJ, Liu H, Zhang LH et al (2013b) Lower limits of evaluation parameters for the lower Paleozoic Longmaxi shale gas in Southern Sichuan Province. Scientia Sin (Terrae) 43(7):1088–1095

Li H, Bai YS, Wang BZ et al (2014) Preservation conditions research on shale gas in the lower Paleozoic of Western Hunan and Hubei area. Petrol Geol Recov Eff 21(6):22–25

Li WY, Sang SX, Wang R (2015) The analysis of shale gas preservation conditions of the Lower Cambrian Hetang Formation in the Northeastern of Jiangxi. Science Tech Eng 15(11):25–30

Liang DG, Guo TL, Bian LZ et al (2009) Some progresses on studies of hydrocarbon generation and accumulation in marine sedimentary regions, Southern China (Part 3): controlling factors on the sedimentary facies and development of palaeozoic marine source rocks. Marine Origin Petrol Geol 14(2):1–19

Liang C, Jiang ZX, Guo L et al (2011) Characteristics of black shale, sedimentary evolution and shale gas exploration prospect of shelf face taking Weng'an Yonghe profile Niutitang group as an example. J Northeast Petrol Univ 35(6):11–22

Liang T, Yang C et al (2012) Characteristics of shale lithofacies and reservoir space of the Wufeng−Longmaxi Formation, Sichuan Basin. Pet Explor Dev 39(6):691–698

Liang F, Bai WH, Zou CN et al (2016) Shale gas enrichment pattern and exploration significance of Well Wuxi−2 in Northeast Chongqing, NE Sichuan Basin. Pet Explor Dev 43(03):350–358

Liang PP, Guo W, Wang N et al (2022) Lithofacies and sedimentary environments of the uppermost Ordovician Kuanyinchiao Formation from the wells in Weiyuan-Luzhou, Southern Sichuan. Chinese J Geol (Scient Geol Sin) 57(01):115–126

Lin JF, Hao F, Hu HY et al (2013a) Depositional environment and controlling factors of source rock in the Shahejie Formation of Langgu sag. Acta Petrolei Sinica 36(2):163–173

Lin LM, Zhang JC, Tang X et al (2013b) Conditions of continental shale gas accumulation in China. Nat Gas Ind 33(1):35–40

Liu HY (1955) Map of ancient geography of china. Science Press, Beijing

Liu BJ (1980) Sedimentary petrology. Geological Publishing House, Beijing

Liu BJ, Zeng YF (1985) Basic and working methods of lithofacies paleogeography. Geological Publishing House, Beijing

Liu BJ, Xu XS (1994) Lithofacies palaeogeography atlas of South China (Sinian−Triassic). Geological Publishing House, Beijing

Liu BJ, Xu XS, Xia WJ et al (1994) Lithofacies palaeogeography atlas of South China (Sinian-Triassic) [M]. Beijing, Geological Publishing House

Liu BJ, Zhou MK, Wang RZ (1990) Paleogeographic outline and tectonic evolution of early Paleozoic in Southern China. Acta Geoscientica Sinica 1:97–98

Liu BJ, Xu XS, Pan XN et al (1993) Crustal evolution and mineralization of the ancient continental deposits in South China. Science Press, Beijing

Liu CL, Li JM, Li J et al (2004) The resource of natural gas in China. J Southwest Petrol Univ (Sci & Technology Edition) 26(1):9–12

Liu BJ, Han ZZ, Yang RC (2006) Progress, prediction and consideration of the research of modern sedimentology. Special Oil Gas Reserv 15(5):1–9

Liu SG, Zeng XL, Huang WM et al (2009) Basic characteristics of shale and continuous−discontinuous transition gas reservoirs in Sichuan Basin, China. J Chengdu Univ Techn (Science & Technology Edition) 36(6):578–592

Liu SG, Ma WX, Luba J et al (2011) Characteristics of the shale gas reservoir rocks in the Lower Silurian Longmaxi Formation, East Sichuan basin, China. Acta Petrologica Sinica 27(8):2239–2252

Liu W, Yu Q, Yan JF et al (2012) Characteristics of organic−rich mudstone reservoirs in the Silurian Longmaxi Formation in Upper Yangtze region. Oil Gas Geol 33(3):346–352

Liu YX, Yu LJ et al (2015) Mineral composition andmicroscopic reservoir features of Longmaxi shales in Southeastern Sichuan Basin. Pet Geol Exp 37(03):328–333

Liu SG, Deng B, Zhong Y et al (2016) Unique geological features of burial and superimposition of the Lower Paleozoic shale gas across the Sichuan Basin and its periphery. Earth Sci Front 01:11–28

Liu SG, Jiao K, Zhang JC et al (2021) Research progress on the pore characteristics of deep shale gas reservoirs: An example from the Lower Paleozoic marine shale in the Sichuan Basin. Nat Gas Ind 41 (01):29–41

Long PY, Zhang JC, Li YX et al (2009) Potentials of the Lower Palaeozoic shale gas resources in Chongqing and its periphery. Nat Gas Ind 29(12):125–129

Loucks RG, Ruppel SC (2007) Mississippian Barnett Shale: Lithofacies and depositional setting of a deep−water shale−gas succession in the Fort Worth Basin, Texas. AAPG Bull 91(4):523–533

Loucks RG, Reed RM, Ruppel SC et al (2009) Morphology, genesis, and distribution of nanometer−scale pores in siliceous mudstones of the Mississippian Barnett shale. J Sediment Res 79:848–861

Lu LL, Ji YL (2013) Sequence stratigraphic framework and palaeogeography evolution of the Cambrian in Lower Yangtze area. J Palaeogeography (Chinese Edition) 15(6):765–776

Lu YH, Zhu ZL, Qian YY (1965) Cambrian palaeogeography and litho-facies of China. Acta Geol Sin 04:3–11

Lu YH, Zhang WT, Zhu ZL et al (1994) Suggestions for the establishment of the Cambrian Stages in China. J Stratigr 18(4):1

Lu XC, Hu WX, Fu Q et al (1999) Study of combination pattern of soluble organic matters and clay minerals in the immature source rocks in Dongying Depression, China. Chinese J Geol (Scientia Geol Sinica) 34(1):69–77

Lu JC, Li YH, Wei XX et al (2006) Research on the depositional environment and resources potential of the oil shale in the Chang 7 Member, Triassic Yanchang Formation in the Ordos Basin. J Jilin Univ (Earth Science Edition) 36(6):928–932

Ma YS, Cai XY (2006) Exploration achievements and prospects of the Permian−Triassic natural gas in Northeastern Sichuan basin. Oil Gas Geol 27(6):741–750

Ma YS, Chu ZH (2008) Building-up process of carbonate platform and high-resolution sequence stratigraphy of reservoirs of reef and oolitic shoal facies in Puguang gas field[J]. Oil Gas Geol 29(5):548–556

Ma YS, Guo TL, Fu XY et al (2002) Petroleum geology of marine sequences and exploration potential in Southern China. Marine Origin Petrol Geol 7(3):19–27

Ma YS, Fu Q, Guo TL et al (2005) Pool forming pattern and process of the Upper Permian−lower Triassic, the Puguang Gas Field, Northeast Sichuan Basin, China. Pet Geol Exp 27(5):455–461

Ma YS, Chen HD, Wang GL et al (2009) Sequence stratigraphy and paleogeography of Southern China. Science Press, Beijing

Ma YS, Cai XY, Zhao PR et al (2010) Distribution and further exploration of the large−medium sized gas fields in Sichuan Basin. Acta Petrolei Sinica 31(3):347–354

Ma YS, Feng JH, Mou ZH et al (2012) The potential and exploring progress of unconventional hydrocarbon resources in SINOPEC. Strategic Study of CAE 14(6):101–105

Ma YS, Cai XY, Zhao PR (2014) Characteristics and formation mechanisms of reef−shoal carbonate reservoirs of Changxing −Feixianguan formations, Yuanba gas field. Acta Petrolei Sinica 35(6):1001−1011

Macquaker JHS, Adams AE (2003) Maximizing information from fine grained sedimentary rocks: an inclusive nomenclature for mudstones. J Sediment Res 73(5):735–744

Martineau DF (2001) Newark East, Barnett Shale field, Wise and Denton counties, Texas; Barnett Shale frac-gradient variances (abs.): AAPG Southwest Section Meeting, March 1–4, 2003, Fort Worth, Texas, http://www.searchanddiscovery.com/documents/abstracts/southwest/index.htm

Martineau DF (2003) Newark East, Barnett Shale Field, Wise and Denton Counties, Texas; Barnett Shale Frac Gradient Variances, 1−4

Martineau DF (2007) History of the Newark East field and the Barnett Shale as a gas reservoir[J]. AAPG Bull 91(4):399–403

Mastalerz M, Schimmelmann A, Drobniak A et al (2013) Porosity of Devonian and Mississippian New Albany Shale across a maturation gradient: insights from organic petrology, gas adsorption, and mercury intrusion. AAPG Bull 97(10):1621–1643

Mccave IN (1971) Mud layers and deposition from tidal currents; discussion of a paper by g. de v. klein, "tidal origin of a precambrian quartzite; the lower fine−grained quartzite (middle dalradian) of islay, scotland". J Sedimentary Res 41(4):1147−1148

Mccave IN (1984) Mechanics of deposition of fine−grained sediments from nepheloid layers. GeoMarine Lett 4(3):243–245

Mccave IN, Swift SA (1976) A physical model for the rate of deposition of fine−grained sediments in the deep sea. Bull Geol Soc Amer 87(4):541–546

Mei MX, Xu DB (1996) Cognition of several problems on cyclic records of Sedim Entary Strata—discussion on working method Ofoutcrop sequence stratigraphy. Strategic Study of CAE 10(1):85–92

Mei MX, Ma YS, Deng J et al (2005) Sequence−stratigraphic framework for the early palaeozoic of the Upper−Yangtze region. Geoscience 19(4):551–562

Mei MX, Zhang C, Zhang H et al (2006) Sequence−stratigraphic frameworks and their forming backgrounds of paleogeography for the lower Cambrian of the Upper−Yangtze region. Geoscience 20 (2):195–208

Mei MX, Ma YS, Zhang H et al (2007) Sequence−stratigraphic frameworks for the Cambrian of the Upper−yangtze Region: ponder on the sequence stratigraphic background of the Cambrian biological diversity events. J Stratigr 31(1):68–77

Milliken KL, Esch WL, Reed RM et al (2012) Grain assemblages and strong diagenetic overprinting in siliceous mudrocks, Barnett Shale (Mississippian), Fort Worth Basin, Texas. AAPG Bull 96(8):1553–1578

Milner HB, Part GM (1916) Methods in practical petrology: hints on the preparation and examination of rock slices. Nature 97(2435):361

Modica CJ, Lapierre SG (2012) Estimation of kerogen porosity in source rocks as a function of thermal transformation: example from the Mowry Shale in the Powder River Basin of Wyoming. AAPG Bull 96:87–108

Montgomery SL, Jarvie DM, Bowker KA et al (2005) Mississippian Barnett Shale, Fort Worth Basin, northcentral Texas: gas shale play with multitrillion cubic foot potential. AAPG Bull 89(2):155–175

Morgan WJ, Series CR (1968) Rises, trenches, great faults, and crustal blocks. American Geophysical Union

Mou CL (1990) Evolution model from passive margin to foreland basin −Also on the ore−controlling role of foreland basin. Sedimentary Facies and Palaeogeography 2:56–62

Mou CL (1993) Sequence Stratigraphy and ore controlling research of the Devonian Strata in Hunan [D]. Chengdu Univ Technol

Mou CL (1994) Exposure Sequence Stratigraphy of the Devonian Strata in Hunan [J]. lithofacies palaeogeography 14(2):1–9

Mou CL, Xu XS (2010) Sedimentary evolution and petroleum geology in South China during the Early Palaeozoic. Sedimen Geol Tethyan Geol 30(3):24–29

Mou CL, Xu XS, Lin M (1992) Sequence stratigraphy and compilation of lithofacies and palaeogeographic maps−an exampie from the Devonian Strata in Southern China. Sedimen Facies Palaeogeography 4(1):1–9

Mou CL, Zhu XZ, Xing XF (1997a) Sequence stratigraphy of marine Volcano−sedimentary Basins: an example from the Devonian Strata in the Ashele−chonghur Region, Xinjiang. Sedimen Facies Palaeogeography 17(3):11–21

Mou CL, Qiu DZ, Wang LQ et al (1997b) Permian sedimentary basins and sequence stratigraphy in the Hunan−Hubei−Jiangxi region. Lithofacies Palaeogeography 17(5):1–22

Mou CL, Wang J, Yu Q et al (1999) The Evolution of the sedimentary basin in Lanping area during Mesozoic−cenozoic. Mineral Petrol 19 (3):30–36

Mou CL, Qiu DZ, Wang LQ et al (2000) Sequence lithofacies paleogeography and oil and gas of Permian in Hunan, Hubei and Jiangxi. Geological Publishing House, Beijing

Mou CL, Tan QY, Wang LQ et al (2003a) Late Permian bioherm oil paleo−pool discovered in Panlongdong, Xuanhan County, Sichuan. Geological Review 49(3):60–64

Mou CL, Tan QY, Yu Q et al (2003b) The late Permian organic reefal oil pool section in Panlongdong, Xuanhan, Sichuan. Sedim Geol Tethyan Geol 23(3):60–64

Mou CL, Tan QY, Yu Q et al (2004) The organic reefs and their reef −forming model for the Upper Permian Changxing Formation in Northeastern Sichuan. Sedim Geol Tethyan Geol 24(3):65–76

Mou CL, Ma YS, Yu Q et al (2005) The oil−gas sources of the Late Permian Organic Reefal Oil−gas pools in the Panlongdong, Xuanhan, Sichuan. Pet Geol Exp 27(6):570–574

Mou CL, Ma YS, Tan QY et al (2007) Sedimentary model of the Changxing—Feixianguan Formations in the Tongjiang—Nanjiang —Bazhong area, Sichuan. Acta Geol Sin 81(6):820–826

Mou CL, Zhou KK, Liang W et al (2011) Early paleozoic sedimentary environment of hydrocarbon source rocks in the middle−upper Yangtze Region and petroleum and gas exploration. Acta Geol Sin 85(4):526–532

Mou CL, Liang W, Zhou KK et al (2012) Sedimentary facies and palaeogeography of the middle−Upper Yangtze area during the Early Cambrian (Terreneuvian−Series 2). Sedimentary Geol Tethyan Geol 32(3):41–53

Mou CL, Ge XY, Xu XS et al (2014) Lithofacies palaeogeography of the Late Ordovician and its petroleum geological significance in Middle−Upper Yangtze Region. J Palaeogeography (Chinese Edition) 16(4):427–440

Mou CL, Wang QY, Wang XP et al (2016a) A study of lithofacies −palaeogeography as a guide to geologi- cal survey of shale gas. Geol Bull China 35(1):10–19

Mou CL, Wang XP, Wang QY et al (2016b) Relationship between sedimentary facies and shale gas geological conditions of the Lower Silurian Longmaxi Formation in Southern Sichuan Basin and its periphery. J Palaeogeo (Chinese Edition) 18(3):457–472

Mou CL, Zhou KK, Chen XW et al (2016c) Lithofacies palaeogeography atlas of China (Ediacaran−Silurian). Geological Publishing House, Beijing

Murray J, Renard A, Murray J (1884) On the nomenclature, origin, and distribution of deep−sea deposits. Proc R Soc Edinb 12:495–529

Nelson PH, Batzle ML (2006) Single−phase permeability, in J. Fanchi, ed., Petroleum engineering handbook: Volume I. General engineering: Richardson, Texas. Society Petrol Eng 1:687–726

Nie HK, Zhang JC (2010) Shale gas reservoir distribution geological law, characteristics and suggestions. J Cent South Univ (Science Technology) 41(2):700–708

Nie HK, Zhang JC (2011c) Types and characteristics of shale gas reservoir: a case study of Lower Paleozoic in and around Sichuan Basin[J]. Petrol Geol Exp 33(3):219–227

Nie HK, Zhang JC (2012c) Shale gas accumulation conditions and gas content calculation: a case study of Sichuan Basin and its periphery in the Lower Paleozoic[J]. Acta Geol Sinica 86 (2):349–361

Nie HK, Tang X, Bian RK (2009a) Controlling factors for shale gas accumulation and prediction of potential development area in shale gas reservoir of South China. Acta Petrolei Sinica 30(4):484–491

Nie HK, Zhang JC, Zhang PX et al (2009b) Shale gas reservoir characteristics of Barnett shale gas reservoir in Fort Worth Basin. Bull Geol Sci Tech 28(2):87–93

Nie HK, Zhang JC, Li YX (2011a) Accumulation conditions of the Lower Cambrian shale gas in the Sichuan Basin and its periphery. Acta Petrolei Sinica 32(6):961–965

Nie HK, He FQ, Bao SJ (2011b) Peculiar geological characteristics of shale gas in China and its exploration countermeasures. Nat Gas Ind 31(11):111–116

Nie HK, Bao SJ, Gao B et al (2012a) A study of shale gas preservation conditions for the Lower Paleozoic in Sichuan Basin and its periphery. Earth Sci Front 19(3):280–294

Nie HK, Zhang JC, Bao SJ et al (2012b) Shale gas accumulation conditions of the Upper Ordovician−Lower Silurian in Sichuan Basin and its periphery. Oil Gas Geol 33(3):335–345

Nie HK, Wang H, He ZL et al (2019) Formation mechanism, distribution and exploration prospect of normal pressure shale gas reservoir: a case study of Wufeng Formation−Longmaxi Formation in Sichuan Basin and its periphery. Acta Petrolei Sinica 40(02):131−143

Niu B, Yoshimura T, Hirai A (2000) Smectite diagenesis in Neogene Marines and stone and mudstone of the Niigata Basin, Japan. Clays Min 48(1):26–42

Pan RF, Huang XS (2009) Shale gas and its exploration prospects in China. China Petrol Explor 3:1–5

Peters KE (1986) Guidelines for evaluating petroleum source rock using programmed pyrolysis. AAPG Bulletin 70:318–329

Pettijohn FJ (1935) Stratigraphy and structure of Vermilion Township, District of Kenora, Ontario. Geol Soc Am Bull 46(12):1891–1908

Pettijohn FJ (1949) Sedimentary rocks. J Geology

Piasecki S, Christiansen FG, Stemmerik L (1990) Depositional history of an Upper Carboniferous organic-rich lacustrine shale from East Greenland. Bull Can Petrol Geol 38(3):273–287

Picard MD (1971) Classification of fine−grained sedimentary rocks. J Sed Res

Pichon XL, Francheteau J, Bonnin J (1973) Plate tectonics. Develop Geotectonics 13(3):349–355

Pollastro RM, Jarvie DM et al (2007) Geologic framework of the Mississippian Barnett shale, Barnett-paleozoic total petroleum system, Bend arch–Fort Worth basin, Texas. AAPG Bull 91 (4):405–436

Posamentier HW, Vail PR (1988) Sequences, systems tracts, and eustatic cycles. American Association of Petroleum Geologists, Tulsa, OK

Potter PE, Pettijohn FJ (1979) Paleocurrents and basin analysis. Earth −Sci Rev 14(4):366–368

Qin JZ, Tenger, Fu XD (2009) Study of forming condition on marine excellent source rocks and its evaluation [J]. Petrol Geol Exp 31 (4):366–372, 378

Qin JZ, Fu XD, Shen BJ et al (2010a) Characteristics of ultramicroscopic organic lithology of excellent marine shale in the Upper Permian Sequence, Sichuan Basin. Pet Geol Exp 32(2):164–171

Qin JZ, Tao GL, Tenger, et al (2010b) Hydrocarbon−forming Organisms in Excellent Marine Source Rocks in South China. Petrol Geol Exp 32(3):262−269

Qiu Z, Zou C, Wang HY et al (2019) Discussion on characteristics and controlling factors of differential enrichment of Wufeng−Longmaxi Formations shale gas in South China. Nat Gas Geosci 31(2):163–175

Radczewski OE (1939) Eolian deposits in marine sediments, Part 6: special fratures of sediments[J]. AAPG Special Vol (142):496–502

Rahm D (2011) Regulating hydraulic fracturing in shale gas plays: the case of Texas. Enegy Policy 39:2974–2981

Ran B, Liu SG, Sun W et al (2016) Lithofacies classification of shales of the Lower Paleozoic Wufeng−Longmaxi Formations in the Sichuan Basin and its surrounding areas, China. Earth Sci Front 23(02):96–107

Romero AM, Philp RP (2012) Organic geochemistry of the Woodford Shale, Southeastern Oklahoma: how variable can shales be? AAPG Bull 96(3):493–517

Rong JY (1984) The ecological formation evidence and glacier activities for the regreession late Ordovician Epoch in Upper Yangtze region. J Stratigr 8(1):19–29

Ross DJK, Bustin MR (2008) Characterizing the shale gas resource potential of Devonian−Mississippian strata in the Western Canada sedimentary basin: application of an integrated formation evaluation. AAPG 92(1):87–125

Ross DJK, Bustin RM (2007) Shale gas potential of the Lower Jurassic Gordondale Member, Northeastern British Columbia, Canada. Bull Can Pet Geol 55(1):51–75

Ross DJ, Bustin RM (2009) The important of shale gas composition and pore structure upon gas storage potential of shale gas reservoirs. Mar Pet Geol 26:916–927

Sageman BB, Wignall PB, Kauffman EG (1991) Biofacies models for organic−rich facies: Tool for paleoenvironmental analysis, Chapter 5, 542−564

Schettler PD, Parmoly CR (1990) The measurement of gas desorption isotherms for Devonian shale. GRI Devonian Gas Shale Tech Rev 7:4–9

Schieber J, Sedimentary features indicating erosion, condensation, and hiatuses in the Chattanooga shale of central Tennessee: relevance for sedimentary and stratigraphic evolution

Schieber J, Zimmerle W (1998) The history and promise of shale research[C]//Schieber J, Zimmerle W, Sethi P. Shales and Mudstones(V01,1): basin studies, sedimentology and paleontology. Stuttgart: Schweizerbart' sche Verlagsbuchhandlung, 1–10

Schlanger SO, Jenkyns HC (1976) Cretaceous oceanic anoxic events: causes and consequences. Geologie En Mijnbouw 55(3)

Schubel JR, Carter HH (1984) The estuary as a filter for fine−grained suspended sediment. The estuary as a filter, 81−105

Shen C, Ren L, Zhao JZ et al (2021) Division of shale lithofacies associations and their impact on fracture network formation in the Silurian Longmaxi Formation, Sichuan Basin. Oil Gas Geol 42 (01):98–106

Simons DB, Richardson EV (1961) Forms of bed roughness in alluvial channels. Am Soc Civil Eng 87(3):87–105

Slatt RM, O'Brien NR (2011) Pore types in the Barnett and Woodford gas shales: contribution to understanding gas storage and migration pathways in fine−grained rocks. AAPG Bull 95(12):2017–2030

Smith D (1970) Mineralogy and petrology of the diabasic rocks in a differentiated olivine diabase sill complex, Sierra Ancha, Arizona. Contrib Miner Petrol 27(2):95–113

Spears DA, Lundegard PD, Samuels ND (1980) Field classification of fine−grained sedimentary rocks; discussion and reply. J Sediment Res 51(3):1031–1033

Stow DA, Bowen AJ (1978) Origin of lamination in deep sea, finegrained sediments. Nature 274(5669):324–328

Stow D, Piper D (1984) Deep−water fine−grained sediments: facies models. Geol Soc London Special Pub 15(1):611–646

Stow D, Tabrez AR (1998) Hemipelagites: processes, facies and model. Geol Soc London Special Pub 129(1):317–337

Stow D, Mayall M (2000) Deep−water sedimentary systems: new models for the 21st century. Mar Pet Geol 17(2):125–135

Straaten L, Kuenen P (1958) Tidal action as a cause of clay accumulation. J Sediment Petrol 28(4):406–413

Su WB, Li ZM, Ettensohn FR et al (2007) Distribution of black shale in the Wufeng-Longmaxi Formations (Ordovician–Silurian), South China: major controlling factors and implications. Earth Sci: J China Univ Geosciences 32(6):819–827

Sui FG, Liu Q, Zhang LY (2007) Diagenetic evolution of source rocks and its significance to hydrocarbon expulsion in Shahejie Formation of Jiyang Depression. Acta Petrolei Sinica 28(6):12–16

Sullwold, Harold H (1960) Load−cast terminology and origin of convolute bedding: further comments. Geol Soc America Bull 71 (5):635

Sun HQ, Zhou DH, Cai XY et al (2020) Progress and prospect of shale gas development of Sinopec. China Petrol Explor 25(2):14–26

Tenger SBJ, Yu LJ et al (2017) Mechanisms of shale gas generation and accumulation in the Ordovician Wufeng−Longmaxi Formation, Sichuan Basin, SW China. Pet Explor Dev 44(01):69–78

Tian ZY, Wan LK (1993) Lithofacies Palaeogeography and Petroliferous Prospect, jurassic, China. Xinjiang Petrol Geol 14(2):101–115

Tian ZY, Wan LK (1994) Tertiary lithofacies paleoearth and petroleum prospect in China. Petrol Geol Eng 2:1–10

Tian ZY, Zhang QC (1997) Lithofacies paleogeography and oil and gas of petroliferous basins in China. Geological Publishing House, Beijing

Tian JC, Chen HD, Qin JX (2004) Case study of sequence−based lithofacies−paleogeography research and mapping of South China. Journal of Earth Sciences and Environment 26(1):6–12

Tian WG, Jiang ZX, Pang XQ et al (2005) Study of the thermal modeling of magma intrusion and its effects on the thermal evolution patterns of source rocks. J Southwest Petrol Univ (Science & Technology Edition) 27(1):12–16

Tian JC, Li MR, Zhang SP, et al (2010) Diagenetic facies character in sequence stratigraphicframeworks of the steep slope zone in an half fault lake basin-taking Paleogene Shahejie Formation of the north steep slope zone of Dongying depression as example[J]. J Chengdu Univ Technol (Sci Technol Ed) 37(3):225–230

Tian JC, Liu JD, Li Q, et al (1999) High-resolution sequence Stratigraphy of delta-lake sediments– A case study of Sha 2nd Member and Sha 1st Member of Ying61-Ying 87 Block in Dongxin Oilfield [J]. Multipl Oil-Gas Field 3:42–45

Tian JC, Zhang CJ, Zhu YT, et al (2008) Diagenesis and reservoir control of lower Paleozoic in south North China Basin. Abstracts of the 10th National Conference of Palaeogeography and Sedimentology

Tian JC, Zeng YF, Zheng HR, et al (2000) The Research of the Reservoir Characteristics of the Dolostone Intercalated in Mudstone in Terrigenous Oil-bearing Basin—Taking the dolostone on the upper of Sha-III in Dongying sag as an example [J]. J Chengdu Univ Technol (Sci Technol Ed) 27(01):88–92

Trask PD (1939) Organic content of recent marine sediments. In: Trask PD (ed) Recent marine sediments—a symposium. Thomas Murby & Co., London, pp 428–453

Tu Y, Zou HY, Meng HP (2014) Evaluation criteria and classification of shale gas reservoirs. Oil Gas Geol 35(01):153–158

Twenhofel WH (1939) Principles of sedimentation. J Geol 49(8):282–284

Udden JA (1914) Mechanical composition of clastic sediments. Geol Soc Am Bull 25(1):655–744

Vail PR (1977) Seismic stratigraphy and global changes of sea level. Geophys Res Lett 29(22):4–7

Wang HZ (1985) Atlas of ancient geography of China. Geological Publishing House, Beijing

Wang ZG (2015) Preakthrough of fuling shale gas exploration and development and its inspiration. Oil Gas Geol 36(1):1–6

Wang SY, Dai HM, Wang HQ et al (2000) Source rock feature of the South of the Dabashan and Mi−cangshan. Nat Gas Geosci 11(4–5):4–16

Wang DH, Guo F, Ren GX et al (2006) Accumulation and distribution conditions of natural gas from Crom Cretaceous of Southern Songliao Basin. World Geology 25(3):282–290

Wang DH, Meng XH, Luo P, et al (2007) Discussion on the sedimentary environment and origin of the microsparite carbonates of Mesoproterozoic Gaoyuzhuang Formation in Jixian Area Tianjin China. Bull Mineral Petrol Geochem 26(s1):396–403

Wang LS, Zou CY, Zheng P et al (2009a) Geochemical evidence of shale gas existed in the Lower Paleozoic Sichuan basin. Nat Gas Ind 29(5):59–62

Wang SJ, Wang LS, Huang JL et al (2009b) Accumulation conditions of shale gas reservoirs in Silurian of the Upper Yangtze region. Nat Gas Ind 29(5):45–50

Wang SQ, Chen GS, Dong DZ et al (2009c) Accumulation conditions and exploitation prospect of shale gas in the Lower Paleozoic Sichuan basin. Nat Gas Ind 29(5):51–58

Wang FY, He ZY, Meng XH et al (2011a) Occurrence of shale gas and prediction of original gas in−place (OGIP). Nat Gas Geosci 22 (03):501–510

Wang SJ, Li DH, Li JZ et al (2011b) Exploration potential of shale gas in the Ordos Basin. Nat Gas Ind 31(12):40–46

Wang GT, Yang SJ et al (2012a) Shale gas enrichment factors and the selection and evaluation of the core area. Strategic Study of CAE 14 (6):94–100

Wang M, Li JF, Ye JL (2012b) Shale gas knowledge reader. [M]. Science Press, Beijing, 1−69

Wang QB, Liu RB, Li CY et al (2012c) Geologic Condition of the Upper Ordovician−Lower Silurian shale gas in the Sichuan Basin and its periphery. J Chongqing Univ Sci Tech (Natural Sciences Edition) 14(4):17–21

Wang YM, Dong DZ, Li JZ et al (2012d) Reservoir characteristics of shale gas in Longmaxi Formation of the Lower Silurian, Southern Sichuan. Acta Petrolei Sinica 33(4):551–561

Wang Y, Chen J, Hu L et al (2013) Sedimentary environment control on shale gas reservoir: a case study of lower Cambrian Qiongzhusi Formation in the Middle Lower Yangtze area. J China Coal Soc 38 (5):845–850

Wang XP, Mou CL, Ge XY et al (2014a) Study on clay minerals in the Lower Silurian Longmaxi Formation in Southern Sichuan basin and its periphery. Nat Gas Geosci 25(11):1781–1794

Wang YM, Dong DZ, Yang H et al (2014b) Quantitative characterization of reservoir space in the Lower Silurian Longmaxi Shale, Southern Sichuan, China. Scientia Sinica (Terrae) 44(6):1348–1356

Wang XP, Mou CL, Ge XY et al (2015a) Mineral component characteristics and evaluation of black rock series of Longmaxi Formation in Southern Sichuan and its periphery. Acta Petrolei Sinica 36(2):150–162

Wang XP, Mou CL, Ge XY et al (2015b) Diagenesis of black shale in Longmaxi Formation, Southern Sichuan Basin and its periphery. Acta Petrolei Sinica 36(9):1035–1047

Wang YM, Dong DZ, Huang JL et al (2016) Guanyinqiao Member lithofacies of the Upper Ordovician Wufeng Formation around the Sichuan Basin and the significance to shale gas plays, SW China. Pet Explor Dev 43(01):42–50

Wang RY, Hu ZQ, Bao HY et al (2021) Diagenetic evolution of key minerals and its controls on reservoir quality of Upper Ordovician Wufeng−Lower Silurian Longmaxi shale of Sichuan Basin. Pet Geol Exp 43(06):996–1005

Wei SL (2020) Reservoir characteristics and shale gas occurrence mechanism of the lower Cambrian Shuijingtuo shale in the South of Huangling Anticline, Western Hubei. China University of Geosciences

Wentworth CK, Williams H (1932) The classification and terminology of the pyroclastic rocks. Natl Acad Sci Natl Res Council

Wijsman J, Herman P, Middelburg JJ et al (2002) A model for early diagenetic processes in sediments of the continental shelf of the black sea. Estuar Coast Shelf Sci 54(3):403–421

Wu K (2012) Hydrocarbon source rock evaluation methods on shale gas reservoir in high and over−matured areas. J Oil Gas Tech 34 (10):37–39

Wu CJ, Zhang MF, Liu Y et al (2013a) Geochemical characteristics of Paleozoic shale in Sichuan Basin and their gas content features. J China Coal Soc 38(5):794–8799

Wu X, Ren ZY, Wang Y, et al (2013b) Situation of world shale gas exploration and development [J]. Resour Ind 15(5):61–67

Wu Y, Fan TL, Jiang S et al (2015) Mineralogy and brittleness features of the shale in the upper Ordovician Wufeng Formation and the lower Silurian Longmaxi Formation in Southern Sichuan Basin. Petrol Geol Recov Eff 22(04):59–63

Wu LY, Hu DF, Lu YC et al (2016) Advantageous shale lithofacies of Wufeng Formation−Longmaxi Formation in fuling gas field of Sichuan Basin, SW China. Pet Explor Dev 43(02):189–197

Xiao XM, Song ZG, Zhu YM et al (2013) Summary of shale gas research in North American and revelations to shale gas exploration of Lower Paleozoic strata in China South area. J China Coal Soc 38 (5):721–727

Xie JR (1947) The geology and mineral resources of the new coal field and the great Southern Basin. Geol Rev 12(5):317–348

Xie JR, Zhang HY, Shao LY (2001) Palaeogeography as a guide to mineral exploration. J Palaeogeography (Chinese Edition) 04:1–9

Xie SH, Gong YM, Tong JN et al (2006) Stride across from paleobiology to geobiology. Chin Sci Bull 51(19):2327–2336

Xie XM, Tenger QJZ et al (2015) Depositional environment, organisms components and source rock formation of siliceous rocks in the base of the Cambrian Niutitang Formation, Kaili, Guizhou. Acta Geol Sin 89(2):425–439

Xie GL, Liu SG, Jiao K et al (2021) Organic pores in deep shale controlled by macerals: Classification and pore characteristics of organic matter components in Wufeng Formation−Longmaxi Formation of the Sichuan Basin. Nat Gas Ind 41(09):23–34

Xu SL, Bao SJ (2009) Preliminary analysis of shale gas resource potential and favorable areas in Ordos Basin. Nat Gas Geosci 20 (3):460–4565

Xu XS, Mou CL, Lin M (1993) Characteristics and depositional model of the Lower Paleozoic organic rich shale in the Yangtze continental block. Chengdu University of Science and Technology Press, Chengdu

Xu XS, Mu CL, Lin M (1994) Sequence Stratigraphy of Devonian and it controls deposits in intraplate basin, South China. Acta Sedimentologica Sinica 12(1):1–6

Xu FH, Qian J, Yuan HF et al (2015a) Sedimentary mode and reservoir properties of mud shale series of strata in Xiangzhong−Xiangdongnan depression, Hunan, China. J Chengdu Univ Tech (Science Tech Edition) 42(1):80–89

Xu ZY, Jiang S, Xiong SY et al (2015b) Characteristics and depositional model of the lower paleozoic organic rich shale in the Yangtze continental block. Acta Sedimentol Sin 33(1):21–34

Yan CZ, Huang YZ, Ge CM et al (2009) Shale gas: enormous potential of unconventional natural gas resources. Nat Gas Ind 29(5):1–6

Yang ZH, Li ZM, Shen BJ et al (2009) Shale gas accumulation conditions and exploration prospect in Southern Guizhou depression. China Petroleum Exploration 14(3):24–28

Yang ZH, Li ZM, Wang GS, et al (2010) Enlightenment from Petrology Character, Depositional Environment and Depositional Model of Typical Shale Gas Reservoirs in North America [J]. Geol Sci Technol Infor 29(6):59–65

Yang RC, Wang XP, Fan AP et al (2012) Diagenesis of sandstone and genesis of compact reservoirs in the East II Part of Sulige gas field, Ordos Basin. Acta Sedimentol Sin 30(1):111–119

Yang C, Zhang JC, Tang X (2013) Microscopic picpore types and its impact on the storage and permeability of continental shale gas. Ordos Basin 20(4):240–250

Ye J, Zeng HS (2008) Pooling conditions and exploration prospect of shale gas in Xujiahe formation in Western Sichuan depression. Nat Gas Ind 28(12):18–25

Yin HF, Xie SC, Qin JZ, et al, Some exploration of geobiology −biogeology and the biofacies of the earth 2008. discussions on earth biology, biology geology and biologicalphase. Sciencein China (Series D) 38(11):1−8

Yu BS (2006) Particularity of shale gas reservoir and its evaluation. Earth Sci Frontiers 19(3):252−258

Yu Q, Men YP, Zhang HQ, et al (2012) Shale gas resource evaluation and block selection in South Sichuan Basin, internal report.

Yu HZ, Xie JL, Wang XX et al (2006) Organoclay complexes in relation to petroleum generation. Earth Sci Front 13(4):254–280

Zeng XL, Liu SG, Huang WM et al (2011) Comparison of Silurian Longmaxi Formation shale of Sichuan Basin in China and Carboniferous Barnett Formation shale of Fort Worth Basin in United States. Geol Bull China 30(2–3):372–384

Zhang DW (2011) Main solution ways to speed up shale gas exploration and development in China. Nat Gas Ind 31(5):1–5

Zhang WS (2015) The effect of ancient sedimentary environment on source rocks: an example of the Longmaxi Formation in Southeast Sichuan. Inner Mongolia Petrochem Indus 18(1):141–143

Zhang ZH, Xin CB (2006) New progress and development trend of sedimentology. Shanxi Architecture 32(6):91–92

Zhang JC, Xue H, Zhang DM et al (2003) Shale gas and its accumulation mechanism. Geosci- Ence 17(4):466

Zhang JC, Jin ZJ, Yuan MS (2004) Reservoiring mechanism of shale gas and its distribution. Nat Gas Ind 24(7):15–18

Zhang JC, Xu B, Nie HK et al (2007a) Two essential gas accumulations for natural gas exploration in China. Nat Gas Ind 27(11):1–6

Zhang WZ, Yang H, Fu ST, et al (2007b) Studies on the hydrocarbon development mechanism of Chang 91 high−quality lacustrine source rocks of Yanchang Formation, Ordos Basin. Science in China (Series D: Earth Sciences) 37 (Suppl. I):33−38

Zhang LY, Li Z, Zhu RF, et al (2008) Resource Potential of Shale Gas in Paleogene in Jiyang Depression[J]. Nat Gas Ind 28(12):26–29

Zhang JC, Nie HK, Xu B et al (2008a) Geologicalcondition of Shale Gas Accumulation in Sichuan Basin. Nat Gas Ind 28(2):151–156

Zhang JC, Wang ZY, Nie HK et al (2008b) Shale gas and its significance for exploration. Strategic Study CAE 22(4):640–646

Zhang JC, Xu B, Nie HK et al (2008c) Exploration potential of shale gas resources in China. Nat Gas Ind 28(6):136–140

Zhang LY, Li Z, Zhu RF et al (2008d) Resource potential of shale gas in paleogene in Jiyang depression. Nat Gas Ind 28(12):26–29

Zhang JC, Jiang SL, Tang X, et al (2009) Accumulation types and resources characteristics of shale gas in China[J]. Nat Gas Ind 29 (12):109–114

Zhang JC, Li YX, Nie HK, et al (2010a) Geologic setting and drilling effect of the shale cored well Yuye-1, Penshui County of Chongqing [J]. Nat Gas Ind 30(12):114–118

Zhang WZ, Yang H, Xie LQ et al (2010b) Lake−bottom hydrothermal activities and their influences on the high−quality source rock development: A case from Chang 7 source rocks in Ordos Basin. Pet Explor Dev 37(4):424–429

Zhang JC, Bian RK, Tieya J et al (2011) Fundamental significance of gas shale theoretical research. Geol Bull China 30(2–3):318–323

Zhang CM, Zhang WS, Guo YH (2012a) Sedimentary environment and its effect on hydrocarbon source rocks of Longmaxi Formation in southeast Sichuan and northern Guizhou[J]. Earth Sc Front 19(3):136–145

Zhang JC, Lin LM, Li YX, et al (2012b) The method of shale gas assessment: probability volume method[J]. Earth Sci Front 19 (2):184–191

Zhang TW, Ellis GS, Ruppel SC, et al (2012c) Effect of organic matter type and thermal maturity on methane adsorption in shale−gas systems. Organic Geochem 47:120−131

Zhang CM, Jiang ZX, Guo YH et al (2013a) Geochemical characteristics and Paleoenvironment reconstruction of the Longmaxi Formationin Southeast Sichuan and Northern Guizhou. Bull Geol Sci Tech 32(2):124–130

Zhang XL, Li YF, Lv HG et al (2013b) Relationship between organic matter characteristics and depositional environment in the Silurian Longmaxi Formation in Sichuan Basin. J China Coal Soc 38 (5):851–856

Zhang ZS, Hu PQ, Shen J et al (2013c) Mineral compositions and organic matter occurrence modes of Lower Silurian Longmaxi Formation of Sichuan Basin. J China Coal Soc 38(5):766–771

Zhang J, Liu SG, Ran B et al (2016) Abnormal overpressure and shale gas preservation. J Chengdu Univ Tech (Science & Technology Edition) 43(02):177–187

Zhang JC, Tao J, Li Z et al (2021) Prospect of deep shale gas resources in China. Nat Gas Ind 41(1):15–28

Zhao CL, Zhu XM (2001) Sedimentary petrology, 3rd edn. Petroleum Industry Press, Beijing

Zhao XY, Chen HQ (1988) Characteristics of the distribution of clay mine−rals in oil−bearing basins in China and their controlling factors. Acta Petrolei Sinica 9(3):28–37

Zhao Q, Wang HY, Liu RH et al (2008) Lobal development and China's exploration for shale gas. Natural Gas Tech Econ 2(3):11–14

Zhao WZ, Li JZ, Yang T et al (2016) Geological difference and its significance of marine shale gases in South China. Pet Explor Dev 43(04):499–510

Zhao QM, Zhang JC, Liu JG (2019) Status of Chinese shale gas revolution and development proposal. Drilling Engineering 46 (8):1–9

Zhao JH, Jin ZJ, Jin ZK, et al, Lithofacies types and sedimentary environment of shale in Wufeng−Longmaxi Formation, Sichuan Basin. Acta Petrolei Sin 37(5):572−586

Zheng RC, Peng J (2002) Analysis and isochronostratigraphic correlation of high−resolution sequence stratigraphy for Chang−6 Oil reservoir Set in Zhidan Delta, Northern Ordos Basin. Acta Sedimentol Sin 20(1):92–100

Zheng RC, Yin SM, Peng J (2000) Sedimentary dynamic analysis of sequence structure and stacking pattern of base—level cycle. Acta Sedimentol Sin 18(3):369–375

Zheng RC, Peng J, Wu CR (2001) Grade division of base−level cycles of terrigenous basin and its implication. Acta Sedimentol Sin 19 (2):249–255

Zheng RC, Wen HG, Liang XW (2002) Analysis of high−resolution sequence stratigraphy for Upper Paleozoic in Ordos Basin. Mineral Petrol 4(13):66–74

Zheng RC, Wang HH, Han YL et al (2008) Sedimentary facies characteristics and sandbody distribution of Chang 6 member in Jiyuan area of Ordos Basin. Lithologic Reservoirs 03:21–26

Zheng RC, Luo P, Wen QB et al (2009) Characteristics of sequence −based lithofacies and paleogeography, and prediction of Oolitic Shoal of the Feixianguan Formationin the Northeastern Sichuan. Acta Sedimentol Sin 27(1):1–8

Zheng RC, Wen HG, Li FJ (2010a) High resolution sequence stratigraphy. Geological Publishing House, Beijing

Zheng RC, Zhou G, Dong X et al (2010b) The characteristics of hybrid facies and hybrid sequence of Xiejiawan member of Ganxi Formation in the Longmenshan area. Acta Sedimentol Sin 18 (1):33–41

Zheng RC, Li GH, Dai CC et al (2012) Basin−mountain coupling syestem and its sedimentary response in Sichuan analogous Foreland Basin. Acta Geol Sin 86(1):170–180

Zheng HR, Gao B, Peng YM et al (2013a) Sedimentary evolution and shale gas exploration direction of the Lower Silurian in Middle—Upper Yangtze area. J Palaeogeography (Chinese Edition) 15 (5):645–656

Zheng XJ, Bao ZD, Feng ZZ (2013b) Advances of palaeogeography in China in the first 10 years of the 21st century. Bull Min, Petrology Geochem 32(3):301–309

Zheng C, Qin QR, Hu DF et al (2019) Experimental study on "six properties"of shale gas reservoirs in the Wufeng–Longmaxi Formation in Dingshan area, Southeastern Sichuan Basin. Petrol Geol Recov Eff 26(02):14–23

Zhou HR, Mei MX, Luo ZQ et al (2006) Sedimentary sequence and stratigraphic framework of the Neoproterozoic Qingbaikou system in the Yanshan region, North China. Earth Sci Front 13(6):280–289

Zhou W, Su Y, Wang FB et al (2011) Shale gas pooling conditions and exploration targets in the Mesozoic of Fuxian Block, Ordos Basin. Nat Gas Ind 31(2):29–33

Zhou XL, Wang J, Yu Q et al (2012) A review of geological characteristics of shale gas accumulation information from AAPG annual convention. Geol Bull China 31(7):1155–1163

Zhou XF, Li X Z, Guo W, et al (2021) Characteristics, formation mechanism and influence on physical properties of carbonate minerals in shale reservoir of Wufeng–Longmaxi Formations, Sichuan Basin. Natural Gas Geosci 1–14

Zhu XM (2008) Sedimentary petrology. Petroleum Industry Press, Beijing, pp 1–126

Zhu XM, Dong YL, Yang JS et al (2008) Sequence stratigraphic framework and sedimentary system distribution of Paleogene in Liaodong Bay area. Sci China Earth Sci 38(I):1–10

Zhu H, Jiang WL, Bian RK et al (2009) Shale gas assessment methodology and its application: a case study of the Western Sichuan Depression. Nat Gas Ind 29(12):130–134

Zhu YM, Chen SB, Fang JH et al (2010) The geologic background of the Siluric shale–gas reservoiring in Szechwan, China. J China Coal Soc 35(7):1160–1164

Zhu YM, Wang Y, Chen SB et al (2016) Qualitative–quantitative multiscale characterization of pore structures in shale reservoirs: A case study of Longmaxi Formation in the Upper Yangtze area. Earth Sci Front 23(01):154–163

Zobell CE (1939) Occurrence and activity of bacteria in marine sediments: part 6. Special Features of Sediments. AAPG Special Volumes 142:416–427

Zou CN, Shizhen T, Rukai Z et al (2009) Formation and distribution of "continuous" gas reservoirs and their giant gas province: a case from the Upper Triassic Xujiahe Formation giant gas province, Sichuan Basin. Pet Explor Dev 36(3):307–319

Zou CN, Dong DZ, Wang SJ et al (2010a) Geological characteristics, formation mechanism and resource potential of shale gas in China. Pet Explor Dev 37(6):641–653

Zou CN, Li JZ, Dong DZ, et al (2010b) Abundant nano–scale pores have been discovered in shale gas reservoirs in China for the first time. Petrol Expl Develop 37(5)

Zou CN, Guangya Z, Shizhen T et al (2010c) Geological features, major discoveries and unconventional petroleum geology in the global petroleum exploration. Pet Explor Dev 37(2):129–145

Zou CN, Dong DZ, Yang H et al (2011a) Conditions of shale gas accumulation and exploration practices in China. Nat Gas Ind 31 (12):26–39

Zou CN, Tao SZ, Hou LH, et al (2011b) Unconventional petroleum geology, 2nd edn. Geological Publishing House, Beijing, pp 1–162

Zou CN, Zhu RK, Bai B et al (2011c) First discovery of nano–pore throat in oil andgas reservoir in China and its scientific value. Acta Petrologica Sinica 27(6):1857–1864

Zou CN, Yang Z, Tao SZ et al (2012a) Nano–hydrocarbon and the accumulation in coexisting source and reservoir. Pet Explor Dev 39 (1):13–26

Zou CN, Zhu RK, Wu ST et al (2012b) Types, characteristics, genesis and prospects of conventional and unconventional hydrocarbon accumulations: taking tight oil and tight gas in China as an instance. Acta Petrolei Sinica 33(2):173–187

Zou CN, Yang Z, Cui JW et al (2013a) Formation mechanism, geological characteristics and development strategy of nonmarine shale oil in China. Pet Explor Dev 40(1):14–26

Zou CN, Zhang GS, Yang Z et al (2013b) Geological concepts, characteristics, resource potential and key techniques of unconventional hydrocarbon: On unconventional petroleum geology. Pet Explor Dev 40(4):385–399

Zou CN, Dong DZ, Wang YM et al (2015) Shale gas in China: characteristics, challenges and prospects (I). Pet Explor Dev 42 (06):689–701

Zou CN, Dong DZ, Wang YM et al (2016) Shale gas in China: characteristics, challenges and prospects (II). Pet Explor Dev 43 (02):166–178

Zou CN, Zhao Q, Dong DZ et al (2017) Geological characteristics, main challenges and future prospect of shale gas. Nat Gas Geosci 28 (12):1781–1796

Zou C, Zhao Q, Cong lZ, et al (2021) Development progress, potential and prospect of shale gas in China. Nat Gas Indus 41(1):1–14

Zuo ZH, Yang F, Zhang C et al (2012) Evaluation of advantaged area of Longmaxi Formation shale gas of Silurian in Southeast Area of Sichuan. Geol Chem Min 34(3):135–142

2 Analysis of Factors Controlling Shale Gas Enrichment

Abstract

In this chapter, the definition of shale gas is reinterpreted and proposed: Shale gas is the "residual gas" in source rock that has not been discharged in time. Shale gas exists in the form of adsorbed gas, free gas or dissolved gas, which is mainly biogenic gas, thermal gas or a mixture of these two. Lithofacies paleogeography, as a comprehensive reflection of facies and paleosedimentary environment, is the main controlling and basic factors affecting the development of organic-rich shale, and should be the basis of shale gas geological survey, and the related mapping technology should be the key technology in the geological survey process.

Keywords

Shale gas definition • Factors controlling shale gas enrichment • Lithofacies paleogeography and mapping technology • Evaluation of selected block for shale gas; Shale gas geological survey

2.1 What is Shale Gas

The unconventional natural gas usually has been explored by unconventional technologies and is difficult to be explained by the theory of conventional petroleum geology. This type of gas reservoir is generally characterized by low porosity, low permeability and continuous accumulation. Shale gas is one of the most important types of unconventional natural gas and is an important green energy.

The traditional definition of shale gas refers to the natural gas existing in adsorption and free state in low porosity, ultralow permeability, organic-rich dark shale or high-carbon content shale (Zhang et al. 2003, 2004, 2008a, b; Jiang et al. 2010; Xu et al. 2011), it is biogenic, thermogenic or bio-thermal continuous natural gas accumulation, with geological features such as short migration distance, multiple sealing mechanisms, concealed accumulation and high gas content in reservoir formation (Curtis 2002; Zhang et al. 2003; Xu et al. 2011). Generally, shale gas is natural gas extracted from dark shale rich in organic matters (Zou et al. 2010, 2011).

From the aspect of petrological characteristics, Jiang (2003) believed that shale gas reservoirs did not develop in the traditional organic-rich mudstone/shale (clay mineral content over 50%), but developed in the fine-grained sedimentary rock composed of clay and silty fine-grained sediments. With the rapid development of unconventional oil and gas industry, the concept of fine-grained sedimentary rocks and related research start to get the attention from the scholars (Jia et al. 2014). The scholars redefined that the shale gas is natural gas development in fine-grained sedimentary rocks, and believe that the hydrocarbon source rocks generate a large quantity of natural gas through a series of geological conditions, and discharge the gas under the steady pressure. The gas migrated to permeable strata such as sandstone and carbonate rock and accumulated into structural or lithologic gas reservoirs. The natural gas remained in fine-grained sedimentary rock strata formed shale gas resources (Li et al. 2009a, b). In fact, considering the perspective of oil and gas resources, the concept of fine-grained sedimentary rocks is mentioned by many scholars to distinguish the coarse-grained sedimentary rock as an conventional reservoir rock. Currently, fine-grained sedimentary rocks were referred as unconventional reservoir rock with particle size less than 1/16 mm, including siliceous rock, shale, clay and coal rocks. Of course, this concept has greatly enriched the research field of sedimentary petrology, expanded the scope of reservoir research, and developed traditional sedimentology and reservoir geology (Xu et al. 2015). From this perspective, the concept of fine-grained sedimentary rocks should be more specific to unconventional plays as a whole.

Many scholars also believed that the shale gas developed in the intervals which were mainly silty mudstone, argillaceous

C. Mou et al., *Lithofacies Paleogeography and Geological Survey of Shale Gas*, The China Geological Survey Series, https://doi.org/10.1007/978-981-19-8861-5_2

siltstone, siltstone, fine sandstone, coarse sandstone and even thin fine conglomerate interbeds (Zhang et al. 2003, 2004, b). Wu et al. (2013b) also pointed out that shale gas not only existed in the interlayer of mudstone, high-carbon mudstone, shale and silty rock in adsorption or free state, but also existed in fractures, pores and other reservoir spaces as free state, or in kerogen, clay particles and pore surface as adsorbed state. Furthermore, a very small amount is stored in dissolved state in kerogen, asphaltene and petroleum, in interbedded siltstone, silty mudstone, argillaceous siltstone or even sandstone formations. Wang et al. (2012a, b) defined shale gas as natural gas occurring in organic-rich shale and its interlayers. From the perspective of self-generation and self-storage of shale gas reservoirs, the natural gas in non-source rock/lithology should actually belong to other types of conventional or unconventional gas reservoirs in the same accumulation system, such as tight sandstone gas.

Generally, more attention should be paid to the organic-rich shale itself, that is, the source rock itself, in order to accurately understand the definition of shale gas. Based on the geological survey and research of shale gas, the author believes that shale gas is the "residual gas" that has not been discharged in time in the source rock. It exists in the form of adsorbed gas, free gas or dissolved gas, and is mainly biogenic gas, thermogenic gas or a mixture of both. Its meaning emphasizes on two aspects: (1) High-quality hydrocarbon source rocks are shale gas accumulation of the main factor. Shale gas originates in the hydrocarbon source rocks; the carrier rock is the organic matter rich rocks that have certain adsorption storage space or silty shale, rather than organic matter-lack rocks like silt (mass) rocks, carbonate rocks. (2) Shale gas is the "residual gas" which has not been timely discharged in the process of hydrocarbon generation from the relatively high-quality source rock. The shale gas reservoir has the integration features of both source and storage. The hydrocarbon generation residual pore is the main storage space for shale gas and determines the adsorption potential of shale gas and adsorbed gas. Thus, shale gas is a special natural gas resource which cannot be exploited by conventional oil and gas exploitation technology under the current economic and technological conditions. In recent years, as vast progress has been made in horizontal well drilling and fracking, "residual gas" in source rocks, known as shale gas, has begun to be exploited on an industrial basis.

2.2 Geological Characteristics of Shale Gas

Based on the definition of shale gas, shale gas reservoirs should have the three most basic and important geological characteristics as following:

(1) The primary condition and key factor for the shale gas reservoir are that there must be sufficient in situ gas content and sufficient organic matter in mudstone/shale to produce a large amount of biogenic gas and thermogenic gas, which requires that mudstone/shale must be hydrocarbon source rock. Thus, shale gas normally occurs in the source rock.
(2) Integration of source and storage. Shale gas reservoirs have the typical characteristics of in situ accumulation, and the natural gas that cannot be discharged in time remains in the source rock to form gas source accumulation. Moreover, in the long process of hydrocarbon generation, the generation of natural gas is not only to meet the adsorption by particles surface like organic matter and clay minerals in rock, but also need to fully occupy its matrix pore and various types of reservoir space of reserves. When the adsorption gas and dissolved gas reaches saturation, the surplus of natural gas in free state starts to migrate and accumulates the water-soluble gas reservoirs under suitable conditions (Nie et al. 2009a; Wang et al. 2010). Thus, although the shale gas storage carrier rock is a low porosity and ultralow permeability reservoir, it has a wide range of gas saturation.
(3) The development and distribution of shale gas reservoirs are not controlled by structure; there is no obvious or fixed limit trap. It was only controlled by the area of source rock and capping rock. Compared with conventional oil and gas reservoirs, shale gas accumulation belongs to the occurrence and enrichment of natural gas without secondary migration or with very short distance secondary migration and does not depend on conventional traps (Zhang et al. 2011).

Generally, shale gas reservoir is a continuous natural gas reservoir formed by continuous gas supply, continuous accumulation, and it is formed from hydrocarbon source rock as same as gas source rock, reservoir and cap rock (Li et al. 2014). The geological parameters such as lithology, thickness, area, geochemical parameters, physical parameters and mineral composition have direct control of the content of shale gas and determine the accumulation (Wang et al. 2015a; Mou et al. 2016). In fact, the gas content of shale gas indicates the shale gas enrichment, is a direct indicator of the residual gas in organic-rich shale, and is also an important parameter in the calculation of shale gas resources in China (Wang et al. 2013b; Wang et al. 2015a). Thus, in the regional geological survey of shale gas, the analysis of influencing factors of shale gas accumulation is actually the analysis of influencing factors of shale gas content.

2.3 Influencing Factors of Shale Gas Enrichment

The commercial development of shale gas in North America depends on the abundance and proper thermal evolution of organic matter, the content of brittle minerals, preservation conditions, shale thickness, surface topography and hydrology (Martini et al. 1998; Daniel et al. 2007; Gault and Scotts 2007; Martineau 2007; Ross et al. 2008; Li et al. 2014). Key parameters used by foreign oil companies in shale gas favorable area screening (Martini et al. 1998; Daniel et al. 2007; Gault and Scotts 2007; Martineau 2007; Ross et al. 2008) can be divided into two categories, including geological conditions and engineering conditions. The first one controls the generation and enrichment of shale gas, including gas-bearing shale area and thickness, abundance of organic matter, type and maturity, brittle mineral content and hydrocarbon display, etc. The latter one controls the cost of shale gas development, including burial depth, surface topography and road transportation.

In China, scholars (Xu et al. 2011; Chen et al. 2011a, b) believe that there are three key factors controlling the enrichment of shale gas: shale thickness, organic matter content and shale reservoir space (pores and fractures). Wang et al. (2013a, b, c) systematically studied the gas-generating material basis of coal-bearing shale gas in Carboniferous Ceshui Formation in central Hunan from the aspects including black organic-rich shale thickness, organic carbon content, organic matter type and thermal evolution degree. He believes that these three factors are the main key factors affecting shale gas accumulation.

Zhang et al. (2008b) believe that shale gas, as a special type of natural gas accumulation, has lower reservoir-forming threshold (Pang et al. 2004) and consequently leads to a larger distribution area of shale gas. Qiu et al. (2014) believe that shale properties like thickness, burial depth, mass fraction of organic carbon, maturity of organic matter, gas content and preservation conditions are key factors for shale gas accumulation after the in-depth study of Lower Cambrian Marine shale in the Middle Yangtze Region. The previous research (Curtis 2002) shows that shale gas accumulation requires the following geological conditions: The sedimentary strata are mainly mudstone/shale, with the thickness of single layer larger than 10 m, high mud content (the thickness of pure mudstone in mudstone/shale stratum is more than 10%), low organic matter abundance (TOC $\geq$ 0.3%), low maturity threshold (Ro $\geq$ 0.4%) and low porosity (less than 12%). For commercial exploration, the shale gas is required to have a small burial depth (no more than 3 km), development of fractures, high adsorbed gas content (no less than 20%), etc. Mudstone/shale that is still in the stage of gas generation has better favorable shale gas accumulation.

Scholars focus on different shale gas development blocks or different shale formations and have proposed different ideas on key factors affecting shale gas accumulation and enrichment. These factors are mainly thickness of mudstone/shale, organic carbon content, organic matter types, thermal maturity, the brittle mineral content, porosity and permeability of reservoirs, the fracture development, paleo structure, preservation conditions and hydrocarbon content and so on. These are the main factors of gas content and all influencing the shale gas content. Together with environmental factors such as depth, temperature and pressure, they determine whether a block has commercial exploration and development value (Zhang et al. 2011a, b, c; Li et al. 2013a, b). The gas content of shale is the key to evaluate these factors and directly reflects the characteristics of shale gas reservoirs.

Shale gas content refers to the total amount of natural gas existing in rock per ton under the standard temperature and pressure (101.325 kPa, 25 °C), including free gas, adsorbed gas, dissolved gas and so on. At present, the main focus is on adsorbed gas and free gas (Li et al. 2011; Nie et al. 2013); its amount directly determines the economic recoverable value of shale gas reservoirs (Han et al. 2013a, b). Nie et al. (2012) found that shale gas is mainly adsorbed gas, and shale gas content indicates the enrichment degree of shale gas, which is a direct indicator to evaluate the residual gas in shale and an important parameter to calculate the amount of shale gas resources in China (Wang et al. 2013b). Yang et al. (2012) believes that shale gas evaluation is a resource factor, which determines the regional shale gas resource potential and reserves, and is a key indicator for shale gas resource evaluation and favorable block selection. Therefore, the analysis of influencing factors of shale gas enrichment in regional geological survey is actually the analysis of influencing factors of shale gas content.

2.3.1 Organic Carbon Content

Shirley (2001) proposed that the organic carbon content of shale is one of the main factors affecting the adsorbed gas volume of shale, while the abundance, type and evolution degree of organic matter are the main factors affecting the gas generation. Organic matter content determines the hydrocarbon generation capacity, pore space size and adsorption capacity of shale and plays a decisive role in the gas content of organic-rich shale (Li et al. 2011). Generally speaking, the higher the abundance of organic matter and gas generation in source rocks, the higher the enrichment of gas reservoirs (Xu et al. 2011). The successful experience of shale gas exploration and development in North America shows that organic matter content is an important indicator to measure the gas bearing of shale (Curtis 2002; Bowker 2007). Through the study of shale gas in North America, it is found that there is a linear relationship between the content of organic carbon in shale and the gas production rate of shale (Fig. 2.1). The content of organic carbon is an important variable determining the gas production capacity of shale. Due to the adsorption characteristics of organic carbon, its content directly controls the adsorbed gas content of shale. Accordingly, shale adsorbed gas increases (Jarvie et al. 2005). Ross et al. (2007) studied Gordondale formation of Jurassic System in Eastern Canada and found that calcareous or siliceous shale with higher organic carbon content had higher storage capacity for adsorbed shale gas. On the contrary, as organic carbon content decreases, adsorbed gas content also decreases, reflecting the close relationship between shale adsorption capacity and organic carbon content (Bowker 2007).

Domestic scholars also found that organic matter content is the key factor affecting shale gas content. Organic matter content is a key parameter restricting shale gas content and resource potential (Wang et al. 2013b), which determines the amount of hydrocarbon generation and affecting the intensity of hydrocarbon generation (Bai et al. 2011). Organic matter abundance and thermal maturity are the basic conditions for shale gas accumulation (Li et al. 2014). Organic carbon content is also an important indicator to measure the gas bearing of shale (Han et al. 2013b). Studies show that, same as shale in North America, there is a clear linear relationship between organic carbon content and gas production rate of shale in most basins in China (Zhang et al. 2011a, b, c). In the process of actual exploration and research, it is found that the total organic matter content is often positively correlated with the gas content of shale, and the higher the organic matter content in shale, the greater the gas content (Fig. 2.2) (Guo et al. 2013; Wang et al. 2013b; Tu et al. 2014). Organic matter in shale can adsorb shale gas generated by the source rock on its surface, leading to a linear relationship between the shale gas adsorption capacity of organic matter and the total organic carbon content in shale (Bai et al. 2011). Han et al. (2013a) found that the adsorbed gas content of Longmaxi Formation shale in Southeastern Chongqing increased with the increase of organic carbon content and proposed that this was related to the microscopic pore in the organic matter. Xue et al. (2013) measured the organic carbon content and field gas production of Longmaxi Formation shale in Zhaotong block, Sichuan Basin. And the results show that with the increase of organic carbon content, the specific surface area of shale increased, the adsorption capacity enhanced, the saturated adsorption quantity increased, and leading to an increase in gas content. In addition, Nie et al. (2012) analyzed and calculated the gas content of Upper Ordovician-Lower Silurian black shale in the Sichuan Basin and its surrounding areas. The results obtained by tests and calculations are highly consistent. The calculation method is mainly based on the linearity of organic matter content and shale porosity. Shale reservoir porosity is mainly from the organic matter hydrocarbon generation pores, this is under the control of the organic matter content. Thus, it can be further shown that the development of black shale rich in organic matter can be used as the important indexes for evaluation of shale gas. In shale gas exploration, vertical and horizontal permeability increases after hydraulic fracturing, and high organic carbon content also increases in situ permeability. These factors would greatly increase the EUR (Zhang et al. 2011a, b, c).

In general, organic carbon content is the main factor controlling the adsorbed gas content of shale. On one hand, higher organic carbon content leads to higher gas generation potential and higher gas content per unit volume of shale. On the other hand, higher organic carbon content leads to more hydrocarbon generation pores of organic matter, and the more specific surface available for natural gas adsorption,

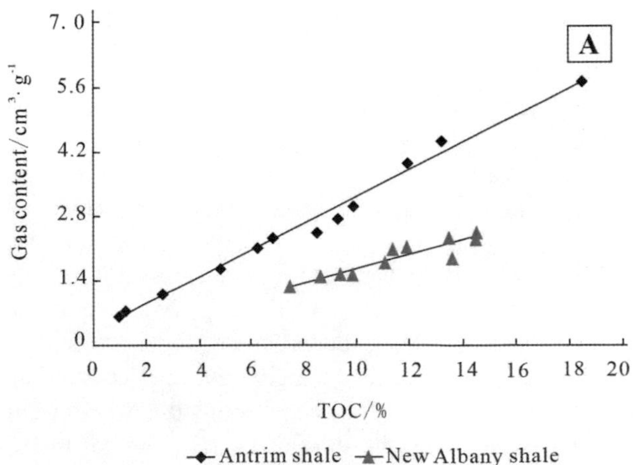

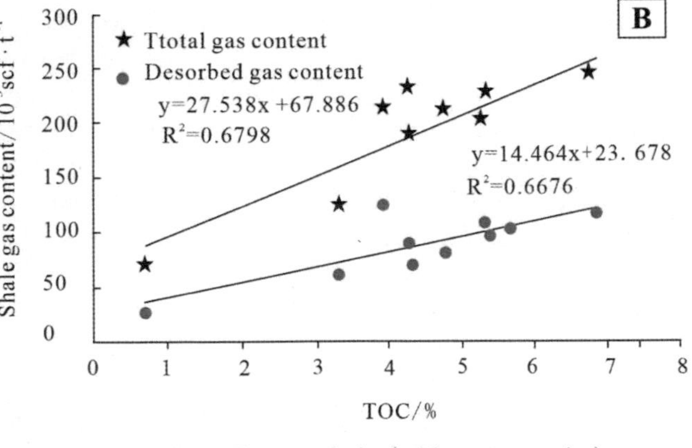

Fig. 2.1 Relationship between TOC and gas content of North American shale gas (after Chen et al. 2011a, b; Wang et al. 2010)

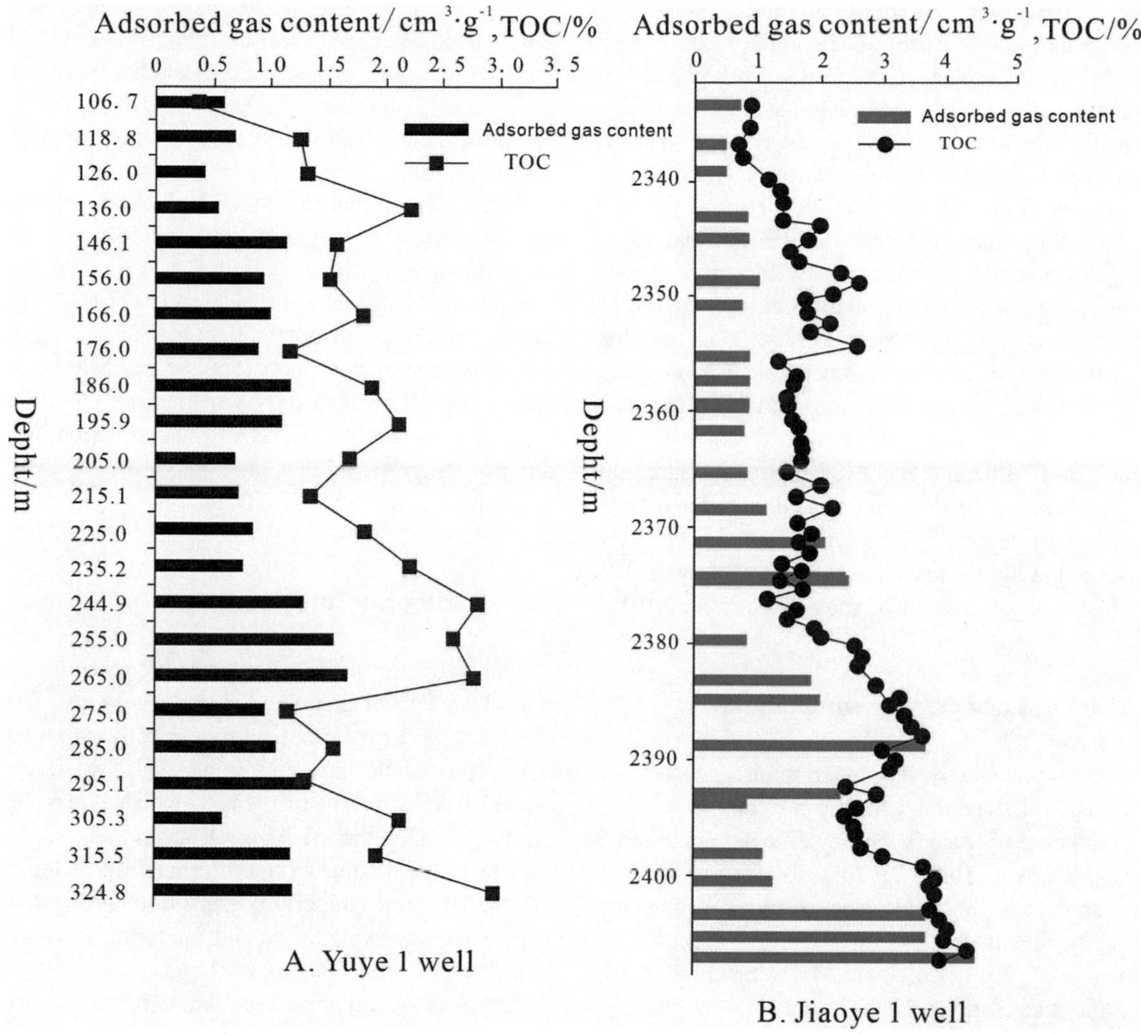

Fig. 2.2 Relationship between TOC and gas content of black rocks in Longmaxi Formation of Southern Sichuan Basin and its periphery (A. YY1 well, after Han et al. 2013a, b; B. JY1 well, after Guo et al. 2013)

and accordingly the increases of adsorbed shale gas content (Xu et al. 2011). As adsorbed gas is the main type of shale gas, organic matter content influences the shale gas content the most. Studies have shown that once the organic-rich mudstone/shale must have sufficient organic matter, generally TOC > 2%, the shale gas accumulated as shale gas play (Mou et al. 2016).

2.3.2 Types and Maturity of Organic Matter

Organic matter type is also one of the important indicators to evaluate the gas generation quality of organic-rich shale, which plays a decisive role in the gas generation potential and nature of shale (Wang et al. 2013a, b, c). Because the chemical composition and structural characteristics of different kerogens are significantly different, the gas production rate varies greatly. Under the experimental conditions, the gas generation of organic matter at different heating rates is basically the same, but the Ro values corresponding to the main gas generation period (70–80% of the total gas generation) are different for different kerogen types. Ro of type I kerogen is 1.2–2.3%, type II kerogen is 1.1–2.6%, type III kerogen is 0.7–2.0%, and marine petroleum cracking into gas is 1.5–3.5% (Zhao et al. 1996).

The degree of thermal evolution of organic matter can affect the hydrocarbon generation potential of shale (Han et al. 2013b), and the degree of thermal evolution (or maturity) is a key indicator to determine the hydrocarbon generation from organic matter (Li et al. 2013a, b). Wu et al. (2013a, b) found that the maturity of organic matter has a certain negative correlation with micropore volume and mesopore volume and has no correlation with macropore volume when Ro < 2.0%. However, it has a certain positive correlation with macropore volume when Ro > 2.0%. Both Ro > 2.0% and high maturity organic carbon content were positively correlated with macropore volume, which may be related to the increase of macropore volume caused by the development of nanoscale microcracks in high-maturity organic matter. It can be concluded that the maturity of organic matter affects the reservoir space type of shale and consequently affects the gas content of shale. The higher the thermal maturity of gas-bearing shale is, the more gas exists in shale, indicating that the higher the thermal maturity is, the more gas generated. As the maturity increases, the formation pressure caused by hydrocarbon gas generation can also improve the adsorption capacity of shale gas. Therefore, thermal maturity is an important geochemical parameter to evaluate the potential of high shale gas production (Bai et al. 2011).

In conclusion, different types of organic matter have different hydrocarbon generation capacities under different thermal evolution degrees. Therefore, organic matter types not only affect the hydrocarbon generation capacity of shale, but also affect the gas content of shale (Zou et al. 2010).

2.3.3 Thickness of Gas-Bearing Shale

It is well known that widely distributed mudstone/shale is an important basis for shale gas generation (Yang et al. 2009). A certain thickness of shale is the fundamental condition for shale gas enrichment. Shale thickness is also an important influence factor for the shale gas resources abundance, which directly affects the size of shale gas resources (Li 2009; Liang et al. 2011; Lu et al. 2012). Bai et al. (2011) proposed that the commercial accumulation of shale gas requires sufficient thickness and certain burial depth of shale. The deposition thickness is the prerequisite to ensure sufficient organic matter and reservoir space. Therefore, the thickness of organic shale is positively related to the enrichment degree of shale gas; organic shale is the carrier for the generation and occurrence of shale gas and an important condition for ensuring sufficient storage and permeability space (Tu et al. 2014).

In addition, the shale thickness and roof and floor conditions have controlled the storage condition (Li et al. 2014). Mudstone/shale itself has sealing capability and can be used as shale cap rock of gas reservoir. Especially for larger thickness of mudstone/shale, when thickness is greater than the maximum distance of hydrocarbon expulsion during the climax period of hydrocarbon generation, the gas will effectively trapped in mudstone/shale (Bai et al. 2011; Hu et al. 2014). Therefore, if the mudstone/shale itself has a certain thickness, it can self-seal and obtain a certain amount of shale gas (but not enough for commercial exploitation) (Hu et al. 2014). In order to form a large-scale shale gas reservoir, the thickness of shale must be greater than the effective hydrocarbon expulsion thickness, normally more than 30 m. In addition, under the same deep burial conditions, from the test analysis data, the Lower Paleozoic marine mudstone in Sichuan Basin has very low permeability and is self-sealing (Hu et al. 2014), which may also be one of the factors for it to become a high-quality shale gas reservoir.

It can be seen that the effective thickness of deposition is the prerequisite to ensure sufficient organic matter and reservoir space. The thicker the shale is, the stronger the sealing ability of shale has, which is benefit to gas preservation and shale gas accumulation (Yang et al. 2009). For example, continental lacustrine basin, Marine basin and slope area are all regions where mudstone/shale developed widely, with large thickness and wide distribution area, and are also the preferred and favorable prospect areas for shale gas exploration (Xu et al. 2011).

2.3.4 Mineral Composition

As the matrix permeability of shale gas reservoir is generally nano-Darcy level and the lithology is dense, fracturing is needed to generate fracture network to improve the seepage capacity of shale gas. Therefore, shale gas reservoir itself should have a certain brittleness, which is easy to produce fractures under the fracking (Wang et al. 2013b). Brittle mineral content controls the reformability of shale (Li et al. 2014). Therefore, mineral composition and content of shale often affect the exploitation and fracturing effect of shale gas reservoirs (Li et al. 2013a, b).

There is an obvious correlation between rock mineral composition and organic matter content in black shale of Silurian Longmaxi Formation in Southern Sichuan and its periphery (Wang et al. 2015a), which further affects the gas content of shale. There is a slightly negative correlation between clay content and organic carbon content. Therefore, Nie et al. (2011) believes that with the increase of clay content, the adsorbed gas content of shale shows a slightly downward trend. In addition, there are micropores in clay minerals, illite, illite/montmorillonite and chlorite with certain specific surface areas, which can be used as adsorption media for organic matter and one of the main adsorption media for adsorbing gas (Xue et al. 2013). It can be seen that the type and content of clay minerals not only affect the organic matter content, but also affect the adsorption and adsorption gas volume of shale. In addition, the change of rock mineral composition affects the rock mechanical properties and pore structure of shale. Compared with quartz and carbonate, clay minerals have more micropores and larger surface area and have stronger adsorption capacity for shale gas. As seen from the three-dimensional relationship diagram of quartz, organic matter content and adsorbed gas content, the organic matter content of rock increases with the

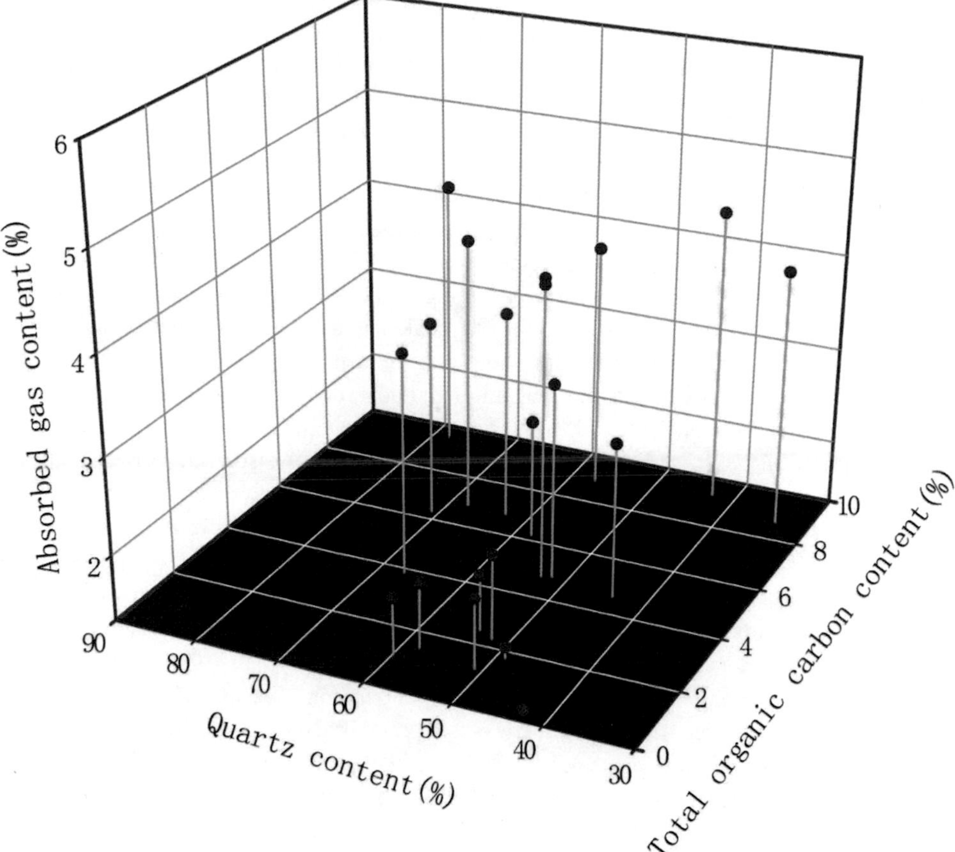

Fig. 2.3 Three-dimensional diagram of quartz, TOC and absorbed gas content in black rocks (according to the internal data)

increase of quartz content, and the adsorbed gas content of shale also increases with the increase of quartz content (Fig. 2.3, according to enterprise internal report). The content of clay minerals and quartz is positively correlated with adsorbed gas, while quartz content is negatively correlated with clay mineral content. When studying the relationship between mineral composition and shale gas content, it is necessary to find a favorable range between clay minerals, quartz and carbonate content.

2.3.5 Reservoir Characteristics

The volume and aperture of micropores and fractures in shale are the main storage space of free gas in shale. The distribution and volume of the pores and fractures can significantly affect the occurrence form of shale gas and control the content of free gas in shale. Organic pores and intergranular pores of clay minerals are the two pore types with most wide development in shale, which are of great significance for gas adsorption and storage. Microfractures are not only the storage space of free gas, but also the main channel of gas seepage (Yang et al. 2013). Chalmers et al. (2012) believe that porosity is positively correlated with the total shale gas content; thus, the total shale gas content increases with the increase of shale porosity. Ross et al. (2009) found that when the porosity increased from 0.5 to 4.2%, the content of free gas increased from 5 to 50%.

Shale is a low porosity and ultralow permeability reservoir, and its permeability is generally less than 0.01×10^{-3} μm^2. However, with the development of microfracture system, the permeability of rock is increased significantly, and the relative accumulation of free shale gas is larger. Barnett shale gas in North America is related to the development of microfracture system, and its free gas content accounts for about 55–75%. This is mainly because the natural microfractures in Barnett shale developed. Although most of them are cemented by calcite, the fractures in the rock can be effectively increased after fracturing, thus increasing the permeability of the rock.

2.3.6 Burial Depth and Formation Pressure

Lin et al. (2012) tested isothermal adsorption of 21 core samples with different TOC content of the Longmaxi shale in the southeast of Chongqing region. The results revealed the relationship among shale buried depth, gas content and the TOC under the condition of the organic matter maturity. With the same depth, the higher the shale content of TOC value is, the higher adsorbed gas content is. If TOC value is constant, the shale adsorbed gas content increases gradually with the increase of burial depth and formation pressure, and the increase rate decreases when it reaches about 1200 m. And the adsorbed gas content gradually stabilizes and tends to a constant value. In order to achieve the same gas content, the shallower the burial depth is, the higher TOC content is required, that is, the shale burial depth and adsorbed gas content have a mutual compensation relationship.

Formation pressure is also a factor affecting shale gas production. The amount of gas absorbed in shale is also affected by formation pressure. Studies show that there is a positive correlation between formation pressure and adsorbed gas capacity, and the higher formation pressure is, the greater the adsorption capacity is and the higher the adsorbed gas content is (Wang et al. 2010; Bai et al. 2011). Free gas content also increases with the increase of pressure, and the two basically show a linear relationship (Wang et al. 2010). Hu et al. (2014) believes the shale gas accumulation is endogenous. As hydrocarbon generation increases shale pore pressure and leads abnormal high pressure in the hydrocarbon source rocks, by action of abnormal pressure and hydrocarbon concentration difference, hydrocarbon always migrates outside, if the sealing ability for shale gas is poor, shale gas discharge greatly and pressure reduced quickly too, and form low pressure shale formation. Otherwise, a good sealing ability will lead a higher formation pressure and shale gas will be maintained. Therefore, formation pressure coefficient is a good indicator of shale gas preservation conditions. Pressure coefficient is a comprehensive discriminant index of preservation conditions, and there is a positive correlation between pressure coefficient and shale gas production (Hu et al. 2014). When the local pressure rises to a certain degree, the microfractures is also a good storage space for shale gas (Bai et al. 2011). The total gas content of organic-rich shale increases with the increase of pressure. Adsorbed gas increases rapidly under low pressure. When the pressure reaches a certain level, the increased rate slows down significantly, while free gas still increases significantly and becomes the main body of shale gas (Li et al. 2011). If shale reservoir pressure is abnormally high, it means that a large amount of oil and gas has been generated in shale, and there may be no large-scale migration or loss in geological history. High shale reservoir pressure also means high shale gas content and initial production (Wang et al. 2013b).

Li et al. (2011) proposed that pressure is directly related to burial depth. For shale gas play, in stable structure areas, the higher the burial depth is, the higher the formation pressure is. This also coincides with the proposition of Shirley (2001), Pu et al. (2010) and Li et al. (2011). They proposed that organic carbon content and formation pressure are the most important factors affecting shale adsorption capacity.

2.3.7 Storage Conditions

Compared with conventional reservoir gas, shale gas accumulation belongs to the natural gas without secondary migration or with very short distance secondary migration. It accumulated independently without the help of trap and preservation conditions in the conventional sense (Zhang et al. 2011a, b, c). Shale gas plays are typical in situ accumulation. In the long process of hydrocarbon generation, the natural gas generated firstly meets the adsorption needs of organic matter and clay mineral particles on the surface of rock, and also meets the needs of matrix pores and various reservoir spaces. When the adsorbed gas and dissolved gas reached saturation, the enriched natural gas migrated and dispersed in free or dissolved phases, forming conventional gas reservoirs when conditions are suitable (Nie et al. 2009b; Wang et al. 2010).

Most sedimentary basins in China have experienced superimposed reconstruction of multiphase tectonics in geological history. For a large number of complex structural faults, the structural framework of the original sedimentary basin and the integrity of the original sedimentary strata of organic-rich mudstone/shale, it is very difficult to understand the preservation conditions of shale gas (Nie et al. 2011; Guo et al. 2012). So, the preservation conditions, as an important research content in geological theory, cannot be ignored in China's shale gas exploration and development. Compared with the North American stable tectonic background of sedimentary basin, the shale gas preservation condition is especially content and unique proposition in China's shale gas geology theory study (Zeng et al. 2011; Li et al. 2013a, b). The key to shale gas exploration in China is to find favorable areas with relatively stable structure, continuous distribution of organic-rich shale and good preservation conditions based on the restoration of prototype basins (Zou et al. 2011). Factors influencing and characterizing the preservation conditions of shale gas include tectonic movement, development degree of faults and microfractures, development characteristics of roof and bottom strata, magmatic thermal events, hydrogeological conditions, current shale pressure, etc. Preservation conditions of shale gas should be considered comprehensively (Li et al. 2009a, b; Nie et al. 2012). However, shale gas reservoir has the characteristics of "self-generation, self-storage and self-capping", and due to its adsorption properties, it consequently has relatively low requirements for preservation conditions (Chen et al. 2011a, b; Wang et al. 2012a, b). Guo and Liu (2013) analyzed the shale gas reservoir with its characteristics of gas adsorption, physical property, continuous gas accumulation and so on, and combining with the actual shale gas breakthrough of JY1, they believe that shale gas reservoir has relatively weak influences on preservation conditions compared to conventional gas reservoir from the aspects of oil and gas migration channel including porosity, unconformity, fault and so on. Compared with carbonate and sandstone, mudstone/shale usually has characteristics of stronger plasticity and lower permeability; thus, the shale has certain ability to resist tectonic deformation. But when tectonic movement is too strong initiating uplift, denudation, fold, fracture, surface water infiltration or pressure system damage, the sealing and preservation conditions of mudstone/shale become poor (Hu et al. 2014). Therefore, the study of shale gas preservation conditions became one of the important contents for shale gas exploration and development in China. And the research should be based on a recognition using the basin evolution, structure characterization, cap and floor rock development, burial depth and pressure conditions to determine the shale gas preservation condition (Li et al. 2013a, b).

2.3.8 Comprehensive Analysis of Influencing Factors

Organic matter, as hydrocarbon generating material, controls the existence of shale gas reservoirs. In the process of hydrocarbon generation of organic matter, not only natural gas is generated, but also the organic pore is generated as the main storage space, which controls the adsorption potential and gas volume of shale gas. There is a certain correlation between mineral composition and organic matter content, which not only affects the physical properties of shale, but also affects the gas content of shale. Shale storage space and physical properties are mainly affected by organic carbon content, mineral composition and thermal maturity of organic matter and tectonic movement. It can be concluded that shale reservoir characteristics directly affect the gas content. The organic pores and intergranular pores are the main storage spaces. These pores are formed by the extensive transformation of clay minerals and hydrocarbon generation of organic matter when the organic-rich shale enters the middle-diagenetic stage A (Wang et al. 2015b). Therefore, the effective shale reservoir is controlled by organic matter content and comprehensive evolution of organic and inorganic diagenesis. Chen et al. (2013) proposed that the influence of clay minerals on pore formation is far less than TOC and brittle mineral content. And TOC content is the most critical and significant factor influencing pore formation in mudstone/shale. Wu et al. (2015) analyzed the pores in shale samples of Longmaxi Formation and Xujiahe Formation in Southeast Chongqing. The mathematical statistics of the test results showed that pore type was not the main controlling factor of gas content, while TOC was the most essential factor of shale gas reservoirs. It is suggested that the shale gas content is the result of the comprehensive influence of many factors, but the most direct and fundamental factor is the organic matter characteristics, and the content of organic matter is the most important factor.

However, the characteristics of organic matter (like organic carbon content, organic matter type) and the thickness of organic-rich shale are controlled by sedimentary environment or sedimentary facies. The characteristics of mineral composition are also controlled by sedimentary environment. The correlation between mineral composition and organic carbon content is caused by the joint influence of sedimentary environment on them. Therefore, sedimentary environment is the fundamental factor determining the shale gas enrichment. It does not only control the thickness, distribution area, organic matter content and other characteristics of mudstone/shale, but also seriously affect the sedimentary rock types and rock mineral composition. And the differences in rock types and mineral composition determine the characteristics of reservoir physical property development, thus affecting the accumulation of shale gas (Wang et al. 2013a, b, c). Shale gas in Jiaoshiba area has been made a breakthrough, and it is believed that the development of organic-rich shale provides a rich material basis for the generation and storage of shale gas (Guo et al. 2014).

If organic matter has good sealing ability in the process of hydrocarbon generation and expulsion, good shale gas play can be formed, which is characterized by high-pressure or abnormal high-pressure distribution area. In stable tectonic zone, formation pressure is related to burial depth. Therefore, under certain tectonic background, the main geological factors affecting formation pressure are organic carbon content, thermal maturity of organic matter and burial depth. The shale gas play formed under certain tectonic setting and sedimentation. The fine-grained sediments rich in organic matter developed in anoxic, reduction environment; after the diagenetic evolution of water–rock reaction and organic matter, hydrocarbon expulsion, without strongly fracturing damage (kept good seal), the shale gas play formed (Fig. 2.4). Shale gas must be "residual gas" staying in rich organic matter in mudstone/shale. The size of the shale gas play is controlled by the thickness and distribution of fine-grained sedimentary rocks rich in organic matter. And it is showed by the characteristics of the very short migration distance and self-generation and self-storage. And the fine-grained sedimentary rocks are controlled by the sedimentary facies. Therefore, sedimentary facies (lithofacies paleogeography) influence the development of shale gas reservoirs.

Therefore, to conduct comprehensive shale gas evaluation and favorable area selection, it is necessary to analyze the sedimentary facies, diagenesis and diagenetic evolution with detailed tectonic background understanding, to determine the key factors affecting the enrichment of shale gas reservoirs and to select appropriate parameter criteria. Lithofacies paleogeography, as a comprehensive reflection of facies and

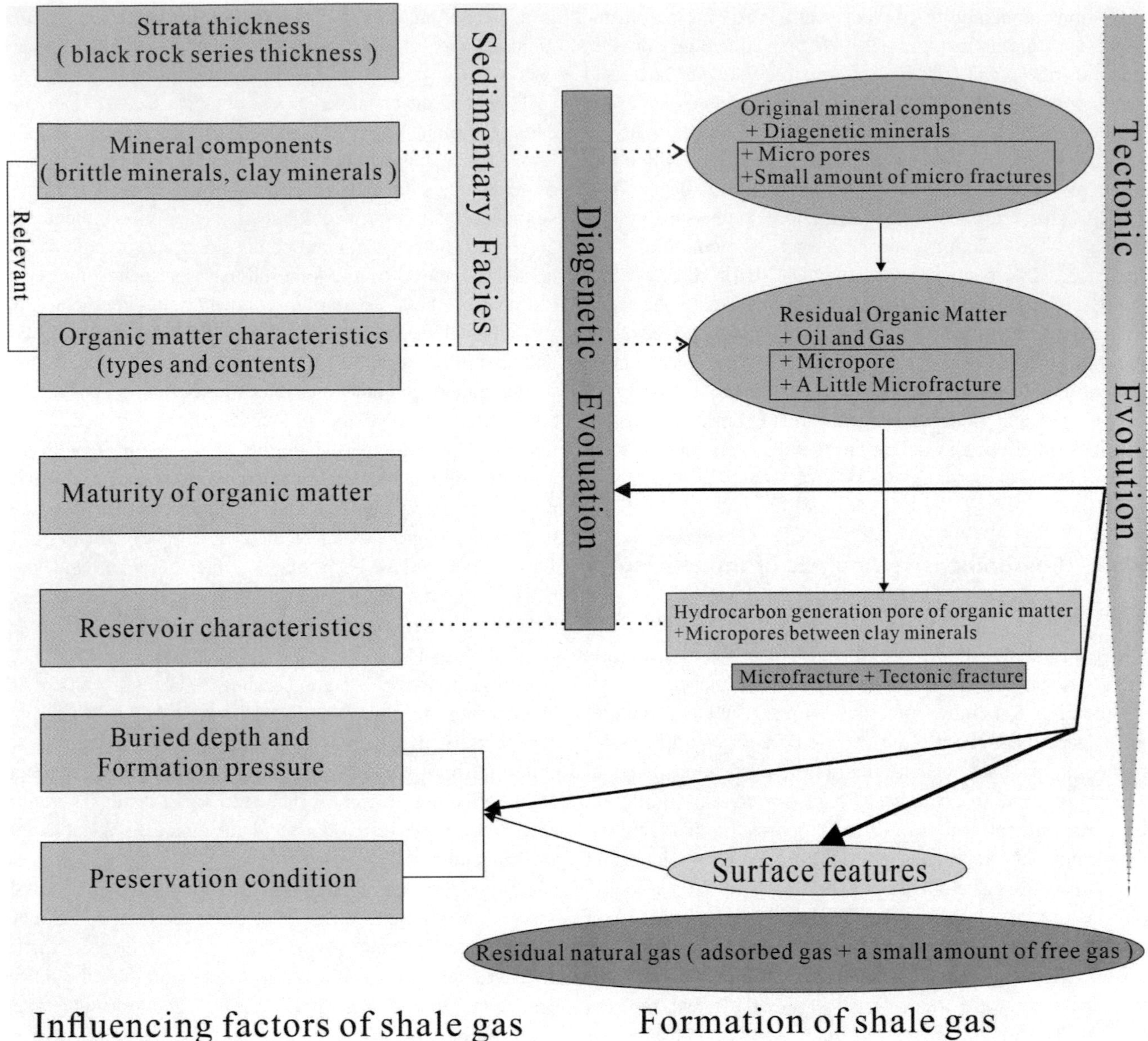

Fig. 2.4 Related schematic diagram of the factors affecting the shale gas enrichment

paleosedimentary environment, is the main controlling factor and basic element affecting the development of organic-rich shale and should be the basis and key of shale gas geological survey. This study, from analysis of the main geological parameters of shale gas, based on the actual data, choose to sedimentary facies (lithofacies paleogeography) map as basic layout, considered organic carbon content contour map, the rich organic shale thickness contour map and vitrinite reflectance contour map as main control consideration. And take the mineral components (clay minerals, brittle minerals) and the buried depth map of the study area as further restriction, and use the gas content data to correct the conclusion. See Chap. 4 for detailed steps and methods.

References

Bai ZH, Shi BH, Zuo XM (2011) Study on shale gas and its aggregation mechanism. Nat Gas Oil 29(3):54–57

Bowker KA (2007) Barnett shale gas production Fort Worth Basin: issues and discussion. AAPG Bull 91(4):523–533

Chalmers GR, Bustin RM (2012) Light volatile liquid and gas shale reservoir potential of the cretaceous Shaftesbury formation in Northeastern British Columbia, Canada. AAPG Bull 96(7):1333–1367

Chen GS, Huang YZ (2011) Research progress of shale gas geology in China. Petroleum Industry Press, Beijing

Chen B, Guan XQ, Ma J (2011a) Early Paleozoic shale gas in Upper Yangtze region and North American shale gas potential Barnett contrast. J Oil Gas Technol (J Jianghan Petrol Inst) 33(12):23–27

Chen X, Wang M, Yan YX et al (2011b) Accumulation conditions for continental shale oil and gas in the Biyang depression. Oil Gas Geol 32(4):568–576

Chen SB, Xiaohong X, Yong Q et al (2013) Classification of pore structures in shale gas reservoir at the Longmaxi formation in the south of Sichuan Basin. J China Coal Soc 38(5):760–765

Curtis JB (2002) Fractured shale–gas systems. AAPG Bull 86 (11):1921–1938

Daniel MJ, Ronald JH, Tim ER et al (2007) Unconventional shale gas systems: the Mississippian Barnett shale of north–central Texas as one model for thermogenic shale–gas assessment. AAPG Bull 91 (4):475–499

Gault B, Stotts G (2007) Improve shale gas production forecasts. E&P 80(3):85–87

Guo TL, Liu RB (2013) Implications from Marine Shale gas exploration breakthrough in complicated structural area at high thermal stage: taking Longmaxi formation in well JY1 as an example. Nat Gas Geosci 24(4):643–651

Guo XS, Guo TL, Wei ZH et al (2012) Thoughts on shale gas exploration in Southern China. Strat Stud CAE 14(6):101–105

Guo XS, Hu DF, Wen ZD et al (2014) Major factors controlling the accumulation and high productivity in marine shale gas in the Lower Paleozoic of Sichuan Basin and its periphery: a case study of the Wufeng-Longmaxi Formation of Jiaoshiba area. Geol China 41 (3):893–901

Han SB, Zhang JC, Brian H et al (2013a) Pore types and characteristics of shale gas reservoir: a case study of Lower Paleozoicshale in Southeast Chong qing. Earth Sci Front 20(3):247–253

Han SB, Zhang JC, Li YX et al (2013b) The optimum selecting of shale gas well location for geological investigation in Niutitang formation in lower Cambrian, Northern Guizhou area. Nat Gas Geosci 24 (1):182–187

Hu DF, Zhang HR, Ni K et al (2014) Main controlling fac- tors for gas preservation conditions of marine shales in Southeastern margins of the Sichuan Basin. Nat Gas Ind 34(6):17–23

Jarvie DM, Hill RJ, Pollastro RM (2005) Assessment of the gas potential and yields from shales: the Barnett shale model. In: Cardott BJ (ed) Unconventional energy resources in the Southern Midcontinent. Norman Oklahoma University Press, pp 37–50

Jia CZ, Zheng M, Zhang YF (2014) Four important theoretical issues of unconventional petroleum geology. Acta Petrolei Sinica 35(1):10

Jiang ZX (2003) Sedimentology. Petroleum Industry Press, Beijing

Jiang YQ, Dong DZ, Qi L et al (2010) Basic features and evaluation of shale gas reservoirs. Nat Gas Ind 30(10):7–12

Li YL (2009) Calculation methods of shale gas reserves. Nat Gas Geosci 20(3):466–470

Li DH, Li JZ, Wang SJ et al (2009a) Analysis of controls on gas shale reservoirs. Nat Gas Ind 29(5):22–27

Li XJ, Lv ZG, Dong DZ et al (2009b) Geologic controls on accumulation of shale gas in North America. Nat Gas Ind 29(5):27–32

Li YX, Qiao DW, Jiang WL et al (2011) Gas content of gas–bearing shale and its geological evaluation summary. Geol Bull China 30(2–3):308–317

Li SZ, Jiang WL, Wang Q et al (2013a) Research status and currently existent problems of shale gas geological survey and evaluation in China. Geol Bull China 32(9):1440–1446

Li XQ, Zhao P, Sun J et al (2013b) Study on the accumulation conditions of shale gas from the Lower Paleozoic in the south region of Sichuan Basin. J China Coal Soc 38(5):864–869

Li JQ, Gao YQ, Hua CX et al (2014) Marine shale gas evaluation system of regional selection in South China: enlight enment from North American exploration experience. PGRE 21(4):23–27

Liang X, Ye X, Zhang JH et al (2011) Reservoir forming conditions and favorable exploration zones of shale gas in the Weixin Sag, Dianqianbei depression. Pet Explor Dev 38(6):693–699

Lin LM, Zhang JC, Liu JX et al (2012) Favorable depth zone selection for shale gas prospecting. Earth Sci Front 19(3):259–263

Lu SF, Huang WB, Chen FW et al (2012) Classification and evaluation criteria of shale oil and gas resources: discussion and application. Pet Explor Dev 39(2):249–250

Martineau DF (2007) History of the Newark East field and the Barnett Shale as a gas reservoir. AAPG Bull 91(4):399–403

Martini AM, Walter LM, Budai JM et al (1998) Genetic and temporal relations between formation waters and biogenic methane: Upper Devonian Antrim Shale, Michigan Basin, USA. Geochim Cosmochim Acta 62(10):1699–1720

Mou CL, Wang QY, Wang XP, et al (2016) A study of lithofacies-palaeogeography as a guide to geological survey of shale gas [J]. Geol Bull China 35(1):10–19

Nie HK, Tang X, Bian RK (2009a) Controlling factors for shale gas accumulation and prediction of potential development area in shale gas reservoir of South China. Acta Petrolei Sinica 30(4):484–491

Nie HK, Zhang JC, Zhang PX et al (2009b) Shale gas reservoir characteristics of Barnett Shale gas reservoir in Fort Worth Basin. Bull Geol Sci Technol 28(2):87–93

Nie HK, He FQ, Bao SJ (2011) Peculiar geological characteristics of shale gas in China and its exploration countermeasures. Nat Gas Ind 31(11):111–116

Nie HK, Bao SJ, Gao B et al (2012) A study of shale gas preservation conditions for the Lower Paleozoic in Sichuan Basin and its periphery. Earth Sci Front 19(3):280–294

Nie YS, Leng JG, Han JH et al (2013) Exploration potential of shale gas in Cen'gong block, Southeastern Guizhou province. Oil Gas Geol 34(2):274–280

Pang XQ, Li SM, Jin ZJ et al (2004) Geochemical evidences of hydrocarbon expulsion threshold and its application. Earth Sci J China Univ Geosci 29(4):384–390

Pu BL, Jiang YL, Wang Y et al (2010) Reservoir–forming conditions and favorable exploration zones of shale gas in Lower Silurian Longmaxi formation of Sichuan Basin. Acta Petrolei Sinica 3(12):225–230

Qiu XS, Hu MY, Hu ZG (2014) Lithofacies palaeogeographic characteristics and reservoir–forming conditions of shale gas of lower Cambrian in middle Yangtze region. J Central South Univ (Sci Technol) 9:3174–3185

Ross JK, Bustin RM (2007) Shale gas potential of the Lower Jurassic Gordondale Member, Northeastern British Columbia, Canada. Bull Can Pet Geol 55(1):51–75

Ross DJK, Bustin MR (2008) Characterizing the shale gas resource potential of Devonian-Mississippian strata in the Western Canada sedimentary basin: application of an integrated formation evaluation. AAPG 92(1):87–125

Ross DJ, Bustin RM (2009) The important of shale gas composition and pore structure upon gas storage potential of shale gas reservoirs. Mar Pet Geol 26:916–927

Shirley K (2001) Shale gas exciting again. AAPG Explorer 22(3):24–25

Tu Y, Zou HY, Meng HP (2014) Evaluation criteria and classification of shale gas reservoirs. Oil Gas Geol 35(01):153–158

Wang SQ (2013) Shale gas exploration and appraisal in China: problems and discussion. Nat Gas Ind 33(12):13–29

Wang X, Liu YH, Zhang M et al (2010) Conditions of formation and accumulation for shale gas. Nat Gas Geosci 21(2):350–356

Wang PW, Chan ZL, He XY et al (2012b) Shale gas accumulation conditions and play evaluation of the Devonian in Guizhong depression. Oil Gas Geol 33(3):353–363

Wang SQ, Wang SY, Man L et al (2013a) Evaluation method and key parameters of shale gas selection area. J Chengdu Univ Technol Nat Sci Edn 40(6):609–620

Wang ZH, Xiao ZH, Yang RF et al (2013b) A study on carboniferous ceshui formation shale gas generation material basis in Central Hunan. Coal Geol China 25(5):19–21

Wang LB, Jiu K, Zeng WT et al (2013c) Characteristics of Lower Cambrian marine black shales and evaluation of shale gas prospective area in Qianbei area, Upper Yangtze region. Acta Petrol Sinica 29(9):3263–3278

Wang M, Li JF, Ye JL (2012a) Shale gas knowledge reader. Science Press, Beijing

Wang XP, Mou CL, Ge XY, et al (2015a) Mineral component characteristics and evaluation of black rock series of Longmaxi Formation in Southern Sichuan and its periphery[J]. Acta Petrolei Sinica, 36(2):150–162

Wang XP, Mou CL, Ge XY, et al (2015b) Diagenesis of black shale in Longmaxi Formation, southern Sichuan Basin and its periphery[J]. Acta Petrolei Sinica 36(9):1035–1047

Wu JS, Yu BS, Zhang JC et al (2013a) Pore characteristics and controlling factors in the organic–rich shale of the Low Silurian Longmaxi Formation revealed by samples from a well in Southeastern Chongqing. Earth Sci Front 20(3):260–269

Wu X, Ren ZY, Wang Y et al (2013b) Situation of world shale gas exploration and development. Resour Indust 15(5):61–67

Wu YY, Cao HH, Ding AX et al (2015) Pore characteristics of a shale gas reservoir and its effect on gas content. Pet Geol Exp 37(2):231–236

Xu GS, Xu ZX, Duan L et al (2011) Status and development tendency of shale gas research. J Chengdu Univ Technol (Sci Technol Edn) 38(6):603–610

Xu ZY, Jiang S, Xiong SY et al (2015) Characteristics and depositional model of the lower paleozoic organic rich shale in the Yangtze continental block. Acta Sedimentol Sin 33(1):21–34

Xue HQ, Wang HY, Liu HL et al (2013) Adsorption capability and aperture distribution characteristics of shales: taking the Longmaxi formation shale of Sichuan basin as an example. Acta Petrolei Sinica 34(5):826–832

Yang ZH, Li ZM, Shen BJ et al (2009) shale gas accumulation conditions and exploration prospect in Southern Guizhou depression. China Petrol Explor 14(3):24–28

Yang RD, Cheng W, Zhou RX (2012) Characteristics of organic–rich shale and exploration area of shale gas in Guizhou province. Nat Gas Geosci 23(2):340–347

Yang F, Ning ZF, Hu CP et al (2013) Characterization of microscopic pore structures in shale reservoirs. Acta Petrolei Sinica 34(2):301–311

Zeng XL, Liu SG, Huang WM et al (2011) Comparison of Silurian Longmaxi formation shale of Sichuan Basin in China and Carboniferous Barnett formation shale of Fort Worth Basin in United States. Geol Bull China 30(2–3):372–384

Zhang JC, Xue H, Zhang DM et al (2003) Shale gas and its accumulation mechanism. Geoscience 17(4):466

Zhang JC, Jin ZJ, Yuan MS (2004) Reservoiring mechanism of shale gas and its distribution. Nat Gas Ind 24(7):15–18

Zhang JC, Nie HK, Xu B et al (2008a) Geological condition of shale gas accumulation in Sichuan Basin. Nat Gas Ind 28(2):151–156

Zhang JC, Wang ZY, Nie HK et al (2008b) Shale gas and its significance for exploration. Strat Stud CAE 22(4):640–646

Zhang JC, Bian RK, Jing Tieya et al (2011) Fundamental significance of gas shale theoretical research[J]. Geol Bull China 30(2–3):318–323

Zhang JC, Bian RK, Jing TY et al (2011a) Fundamental significance of gas shale theoretical research. Geol Bull China 30(2–3):318–323

Zhang LY, Li YX, Li JH et al (2011b) Accumulation conditions for shale gas and it's future exploration of Silurian in the central–Upper Yangtze region. Geol Sci Technol Inform 30(6):90–93

Zhang WD, Guo M, Jiang ZX (2011c) Parametersand method for shale gas reservoir evaluation. Nat Gas Geosci 22(6):1093–1099

Zhao WZ, Zou CN, Song Y et al (1996) Advances in petroleum geology theory and methods. Petroleum Industry Press, Beijing

Zou CN, Dong DZ, Wang SJ et al (2010) Geological characteristics, formation mechanism and resource potential of shale gas in China. Pet Explor Dev 37(6):641–653

Zou CN, Tao SZ, Hou LH et al (2011) Unconventional petroleum geology, 2nd edn. Geological Publishing House, Beijing

Shale Gas Geological Survey

3

Abstract

It is clearly proposed that "lithofacies paleogeography can be used as a guide for shale gas geological survey work". Lithofacies paleogeography is not only the theoretical guidance of shale gas geological survey work, but also its related mapping technology method is the key method to "find" shale gas.

Keywords

Constituency evaluation • Working methods • Shale gas • Geological survey • Lithofacies paleogeography

3.1 The Task of Shale Gas Geological Survey

In recent years, China's consumption and production of conventional energy fuels have reached a record high level. Oil and natural gas still dominate the energy consumption structure. The sharp rise in domestic unconventional oil and gas production has basically balanced domestic oil and gas supply and demand (Jia et al. 2014). However, in the production of unconventional oil and gas resources, relatively cleaner shale gas resources have little contribution to the production, which is inconsistent with China's huge potential status of shale gas. Compared with the unconventional resources such as tight gas and coalbed methane that have achieved large-scale industrial production, the large-scale production of shale gas has made relatively slow progress, so it is necessary to seek further to realize the large-scale and effective development of shale gas resources.

Shale gas, as retained gas in source rocks (Li et al. 2012) and residual gas (Mou et al. 2016), is the accumulation of oil and gas in source rocks. In China, there are huge differences in the research basis and degree between petroliferous basins and sedimentary areas outside of petroliferous basins or where the degree of oil and gas exploration is very low. The former may have a certain working basis for basic data such as gas-bearing shale development series (including shale distribution and thickness), organic matter content and type, thermal evolution degree, gas display, rock characteristics and mineral components of gas-bearing shale.

However, for the latter, the above basic information is relatively lacking. In view of the research status in China, such as large differences in the basic work related to source rock research, weak foundation, no system, no detail and late start of research as shale gas, it is necessary to clarify its main tasks before the large-scale geological survey of shale gas.

Based on the author's own actual experience in shale gas geological survey, combined with the analysis of domestic shale gas research status, existing problems and research trends, it is concluded that China's shale gas geological survey has three basic and main tasks at present:

(1) Clarify the basic geological characteristics of source rocks, including lithologic characteristics, sedimentary environment, sedimentary microfacies (rock) types and characteristics, organic matter types and contents, mineral composition, etc. Mainly through core and field observation and relevant sampling work, carry out the analysis of lithologic characteristics, depositional environment and TOC and Ro of shale of source rock and clarify the types and content of organic matter in source rocks. Further, analyze the types and contents of conventional minerals and clay minerals in source rocks, summarize the conventional mineral composition of source rocks and the variation law of clay mineral composition in the section and comprehensively study the basic geological characteristics of source rocks.
(2) Clarify the temporal and spatial distribution law of source rocks, including their thickness, buried depth, fine distribution, area, etc. Mainly through the field, combined with seismic and drilling data, the distribution range and distribution law of source rocks are

C. Mou et al., *Lithofacies Paleogeography and Geological Survey of Shale Gas*, The China Geological Survey Series, https://doi.org/10.1007/978-981-19-8861-5_3

studied, and their thickness variation law, buried depth and variation trend are analyzed.

(3) Optimize the prospective area and favorable area of shale gas reservoir. After completing tasks (1) and (2), we have the basic conditions to carry out the optimization of shale gas scenic spots and even favorable areas. On the basis of clarifying the scope of shale gas scenic spots, we can further carry out the research on the gas-bearing property of source rocks, reservoir capacity and resource potential analysis and select appropriate parameters in combination with geochemical indicators to further optimize the favorable areas or even dessert areas of shale gas.

3.2 Methods of Shale Gas Geological Survey

At present, China's shale gas geological research is in the stage of rapid development. Local exploration units represented by China Geological Survey are carrying out large-scale shale gas geological surveys, moving from theoretical learning to scientific practice, from key area survey to multi-regional and multi-type evaluation and from local test to large-scale investment. However, the geological conditions of shale gas in China are extremely complex, the theoretical research of shale gas geology is extensive, there are many problems, and the theoretical research is relatively backward. Combined with the basic purpose and task analysis of shale gas geological survey, the author believes that in addition to the guidance of basic geological theory, the key problem is to select appropriate technical methods.

In oil and gas exploration, although different disciplines have played different roles in identifying or making oil and gas breakthroughs, in the early stage of oil and gas exploration, lithofacies paleogeography research is indispensable and one of the key foundations for making breakthroughs in oil and gas resources. It is also a method of oil and gas exploration (Mou et al. 2010, 2011), and the same is true for shale gas (Mou et al. 2016).

At present, the mainstream view on the formation of oil and gas (including shale gas) is still the theory of organic genesis (Wang et al. 2003). Therefore, the formation of oil and gas is closely related to lithofacies paleogeographic environment from the generation and development of primitive organic life to the later death, burial, decomposition and migration and from the material of oil and gas reservoir to the formation of its preservation body. For shale gas, the study and reconstruction of lithofacies paleogeography can help clarify the scientific problems related to its occurrence carrier—organic-rich shale. Faced with shale, a restricted area of oil and gas exploration in the past, it is difficult to evaluate the target area of shale gas exploration, and the exploration risk and cost are high. Therefore, the understanding of lithofacies and paleogeography of shale gas target area has become more important, which should be paid attention to by geological investigators. However, oil and gas geologists and sedimentary geologists who conduct geological investigation and research on shale gas pay more attention to the geological characteristics of shale gas carriers (organic-rich shale), such as TOC, Ro, mineral composition. Secondly, it pays attention to the structural preservation conditions similar to conventional oil and gas reservoirs; however, the basic research on the fine sedimentary environment for the development of organic-rich shale, the carrier of shale gas and the research on the optimization of prospective area, favorable area and target area of shale gas as a key method is relatively few. A large number of studies show that the sedimentary environment not only controls the thickness, distribution area and organic carbon content of organic-rich shale, but also determines the sedimentary rocks type and mineral composition of the rocks, and the difference between rocks type and mineral composition determines the characteristics of physical property development of source rocks reservoir and then affects the accumulation of shale gas. Therefore, the sedimentary environment is the fundamental factor determining shale gas accumulation. Therefore, based on the study of regional sedimentary facies, the temporal and spatial distribution of favorable facies zones of source rocks can be clarified through lithofacies paleogeography mapping, so as to provide the basis and direction for shale gas exploration (Mou et al. 2016). For the basic geological survey of shale gas, the research on the sedimentology and lithofacies paleogeography of shale gas source rocks and reservoir and taking this as a method to delineate the scenic spots, favorable areas and target areas should be the eternal theme throughout the whole geological survey of shale gas and the process of further exploration and development. In fact, the fundamental goal of a shale gas geological survey is to find prospective areas and favorable areas of shale gas, so as to provide a scientific basis for shale gas exploration and development (Mou et al. 2016). Therefore, the theory and method of "using superposition method to evaluate the prospective area, favorable area and target area of shale gas based on the mapping research and mapping of lithofacies paleogeography" should be the key technical method of large-scale shale gas geological survey. Research and practice have proved that the research and mapping of lithofacies paleogeography can be used as the basic method and key technology and can guide the geological survey of shale gas.

3.2.1 Sedimentary Basin and Shale Gas

A large number of studies show that sedimentary basins play the most basic and important role in controlling the occurrence of a variety of sedimentary minerals and oil and gas (Wang and Li 2003), and the formation of sedimentary basins is controlled by tectonic activities and evolution. Zhang et al. (1991) pointed out that the three Chinese mainland plates were affected by China's three tectonic cycles during the geological history, including the splitting, drifting, collision and convergence of China's continental tectonic cycle. These tectonic activities control the evolution of petroliferous basins and lithofacies paleogeography in China and the evolution of the types of petroliferous basins in China from generation to generation and then control the distribution of oil and gas. The idea of activity theory and stage theory runs through the book lithofacies paleogeography and oil and gas in China's petroliferous basins. Academician Tian (1997) believes that paleogeography and its changes are mainly controlled by the degree of crustal differentiation and plate tectonic evolution, and the types of sedimentary basins are controlled by tectonic action (Tian and Zhang 1997), which further controls the distribution of paleogeography and the generation of oil and gas. In fact, with the advent of the theory of plate tectonics, people gradually realized that the formation of sedimentary basins is closely related to mantle material activities and is controlled by the tectonic system (movement); structure plays an important role in controlling the formation, evolution and sedimentary sequence of sedimentary basins and then affects the distribution of oil and gas (Wang et al. 1985; Liu and Xu 1994; Mou et al. 1999, 2010, 2011; Wang and Li 2003). For example, Mou et al. (2010, 2011) analyzed from the structural point of view and believed that today's South China could be divided into the Yangtze Block and Cathaysian Block. From the Cryogenian (Nanhua period), due to the influence of basement properties and tectonic activities, the sedimentary evolution of South China has been differentiated. There are different sedimentary environments and sedimentary filling sequences in the Yangtze area and Cathaysian Block. The main type of sedimentary basin in South China belongs to the Craton Basin. Most of its periods are epicontinental sea and limited shallow sea environment, which constructs stable carbonate platform sedimentation and widely distributed black shale sedimentation. The Cathaysian Block is in an active tectonic environment with strong volcanic activity in the early stage. The sedimentary basin type belongs to the rift basin. There is no unified carbonate platform deposition, and the main body is terrigenous clastic deposition and filling. The different tectonic activities determine different types of sedimentary basins, further determine their different sedimentary environments and sequences and finally determine the different basic geological conditions of oil and gas in the two regions: The Yangtze region has rich source rocks, good reservoirs and sealing layers, while the Cathaysian Block has bad source rocks and reservoirs. Some scholars also believe that tectonic activity and tectonic pattern directly control the lithofacies paleogeographic environment and sedimentary system, determine the distribution of sedimentary areas and provenance areas, sedimentation, changes in sedimentary environment and paleogeography and often affect the changes in climate environment (Jing et al. 2012). In fact, the above laws can be summarized as that the structure controls the formation of the basin. The basin is the accumulation space of materials. Therefore, the structure determines the sedimentary filling sequence and sedimentary style of the basin and finally determines the original basic geological conditions and combinations of oil and gas, that is, "basic geological conditions of structure controlling basin, basin controlling facies and facies controlling oil and gas" (Mou et al. 2010). In the field of petroleum geology, many scholars have always advocated the epistemological view that "structure is dominant, sedimentation is the foundation, oil generation is the key, and transportation, accumulation, and preservation are the conditions". Therefore, the analysis of prototype sedimentary basin types and the study of detailed sedimentary characteristics and evolution sequences under their control is the basis for analyzing basic geological conditions of oil and gas. Shale gas is no exception.

For shale gas reservoirs, although they are different from conventional oil and gas reservoirs and do not need structural traps, shale gas may be formed as long as there is a place where organic shale is developed (with space for accumulation of organic shale). That is, to some extent, shale gas may be formed in any sedimentary basin as long as there is organic-rich shale (Wu et al. 2013). Based on the author's understanding and definition of shale gas, organic-rich shale, as the filling sequence and sedimentary style (or one) of these sedimentary basins, it is also the most primitive and basic material basis for the formation of shale gas. Therefore, in the geological survey of shale gas, the first thing to do is to find the places where these primitive and basic material bases are accumulated, find favorable structural units and judge whether it is a "complete shale basin" conducive to the development of organic shale, to analyze the prototype sedimentary basin type of organic shale, restore the complete shale basin rich in organic shale as much as possible and, on this basis, screen the structural sedimentary units conducive to the development of shale gas reservoirs, such as the stable Craton Basins (generally conducive to the development of biogas) and the Foreland Basin (generally, the basin with relatively developed conventional oil and gas is also conducive to the development of shale gas reservoir of thermogenic gas type). Among those "fragmented" sedimentary basins, such as some inter-arc basins, pre-arc basins and back-arc basins with volcanic rocks under the background of tectonic compression, although mud shale will also develop and be source rocks and full of "residual gas" to form shale gas, but due to the influence of factors such as the area, thickness and capping layer of shale gas source rock (they are the influencing factors of shale gas reservoir enrichment), shale gas reservoirs with industrial development value are often not formed.

Theoretically, from the time scale, the sedimentary basin should be a syngenetic sedimentary basin. However, it is undeniable that due to the control of tectonic, the deformation of sedimentary basins during and after deposition and the superposition of different types of structures in the later stage, there may be a great controversy on the understanding of basin types in the original organic shale-rich sedimentary period; from the perspective of sedimentary basin analysis, the formation, development and extinction of a specific basin are a quasi-continuous evolution process, and the superimposed basin has nothing to do with the basin formed in the specific period and specific dynamic mechanism studied (i.e., the prototype basin in the development and sedimentation period of organic-rich shale). The prototype sedimentary basin can be the basis or basement of the later basin and even determine various elements of the later basin to varying degrees; the late basin has no effect on the formation and evolution of the early basin. It plays a role in the transformation of the early basin, and the fundamental reason for this transformation is the transformation of the dynamic mechanism. The research shows that no basin has crossed the whole Phanerozoic and no region, and the basin type has not changed since ancient times. Therefore, the practice of dividing all sedimentary strata in a certain area into one basin and taking successive basins formed under different dynamic mechanisms as different stages of a basin is contrary to objective facts and geological principles (Wang and Li 2003).

Gong et al. (2012) analyzed and compared the reservoir forming conditions of shale gas reservoirs in the USA, especially the control of structures and sedimentary basin types under their control of shale gas reservoir forming conditions and considered that the Foreland Basins are favorable places for shale gas formation and accumulation. The reason is that the lower strata of the Foreland Basin are usually stable craton shale deposits rich in organic matter, which provides sufficient material basis for the formation of shale gas. The upper strata of the Foreland Basin experienced multi-stage thrust folds, and the resulting tectonic thermal events provided thermal and dynamic conditions for the maturity of the lower source rocks and the generation of natural fractures, making the Foreland Basin an ideal place for shale gas accumulation. Li et al. (2013) also pointed out that the research about the control effect of structure on the distribution law of gas-bearing shale should be strengthened by analyzing the research status and problems faced by domestic shale gas. In other words, it is also the research on the type of prototype sedimentary basin. Shale gas reservoirs have the basic characteristics of integration of source and reservoir. Research shows that relatively high-quality reservoirs are often formed in a specific period with a small span, and the thickness may be only a few meters. For example, the most high-quality source rocks and shale gas development interval of the Fuling shale gas field of the Wufeng Formation-Longmaxi Formations at its bottom (Guo 2016), which is the first large shale gas field in China, and its material composition and lithofacies environment are significantly different from the similar lithology underlying and overlying it. What makes this obvious difference must be that there are some mechanism working. If we can accurately restore the types of sedimentary basins which controlled by structural evolution, clarify the distribution law of sediment filling process and lithofacies environment development under its control and master the internal law between them, it is possible to characterize the fine organic-rich shale section of shale gas development. This process is actually the analysis of sedimentary basins. Of course, it also has the analysis of fine sedimentary facies under its control, which will be discussed in detail below. In addition, in the process of analyzing the type of prototype sedimentary basin, we will have a very in-depth understanding and understanding of the geological background of the development of source rocks (such as the development degree of existing faults and microfractures, development characteristics of caprock, hydrogeological conditions, current pressure conditions), which will help to solve the research problems of shale gas preservation conditions. This is a great significance for the research on the important scientific problem of shale gas exploration and development, but it is not a high status in the research of shale gas preservation conditions in China.

The ultimate goal of the shale gas geological survey is to find shale gas reservoirs with industrial exploration and development value. Research and practice show that shale gas reservoirs with exploration and development value often exist in (or develop in) complete sedimentary basins. Most of those that have realized shale gas development or have great shale gas potential are Foreland Basins and the Craton Basins. Taking the USA as an example, nearly 20 shale gas blocks found in its 48 states widely develop gas-bearing black shale, most of which are developed in the Craton Basins and the Foreland Basins, such as the currently exploited shale gas reservoirs in the Foreland Basin on the west side of Appalachian-Quachita in the Eastern USA and the back-uplift basin group. At the same time, its potential shale gas blocks are also mainly distributed in the Foreland Basins and back-uplift basin on the east side of Rocky

Mountain (USGS 2013). The tectonic evolution of the main development stages of these two sedimentary basins types and the paleogeographic conditions under their control, such as an anoxic reduction environment, suitable hydrodynamic conditions and less supply of terrigenous debris in most layers, control the formation of high-quality source rocks and shale gas with different genesis and maturity (Li et al. 2009). In China, taking the marine shale gas in Southern China as an example, the research shows a very large shale gas prospect in Sichuan Basin. The main development horizons are the Cambrian Niutitang Formation developed in the Craton Basin and the Late Ordovician Wufeng Formation-Early Silurian Longmaxi Formation in the back-uplift basin (Mou et al. 2010, 2011; Ge et al. 2014; Liang et al., 2014; Zhou et al. 2015) (Fig. 3.1). Or most researchers believe that the type of sedimentary basin is a Foreland Basin in the Late Ordovician-Early Silurian period formed on the eastern margin of the Yangtze Block due to the convergence of the Yangtze Block and the Cathaysian Block (Wang 1985; Liu et al. 1993; Xu et al. 1996; Yin et al. 2001, 2002; Su et al. 2007; Wang et al. 2012a, b, c, d; Wang et al. 2012a, b, c, d). In the Dabashan area on the northern edge of the Yangtze plate (the transition between the south side and the Sichuan Basin), the fracturing yield of shale gas in the Jurassic Zhenzhuchong section of well Yuanba 9 on Northeast Sichuan exceeds 1×10^4 m^3, Jiannan 111 well. After fracturing, the daily natural gas production in the Dongyuemiao section is 3000 m^3. The geological survey and exploration and development of shale gas in organic shale-rich strata have made a preliminary breakthrough (He and Zhu 2012). Further analysis also shows that the maximum thickness area of organic-rich shale in the Lower Jurassic Ziliujing Formation is distributed in the front of the Dabashan orogenic belt and controlled by the tectonic activity and tectonic load of the Dabashan orogenic belt, while the tectonic activity of Longmenshan orogenic belt is relatively weak and has little impact on its sedimentary and sedimentation. They all have a complete orogenic belt Foreland Basins system. At the western edge of the North China plate, in April 2011, the gas was produced by shale gas fracturing test of Chang 7 section of Liuping 177 well. The depth of the vertical well is 1000–2000 m, and the test output of shale gas is generally 1500–2000 m^3 [Li et al. 2015 (internal meeting data)], which is the first continental shale gas well in China. The study shows that the western margin of the Ordos Basin was a Foreland Basin in the Late Triassic, and the basement has the nature of a Craton Basin. These relatively stable tectonic backgrounds and the types of favorable sedimentary basins under their control have laid a favorable sedimentary environment for the development of source rocks and are the decisive factor for the enrichment of shale gas reservoirs.

In addition, the sedimentary pattern of the South China Basin is very different from the Mesozoic, and the sedimentary mechanism of the South China Basin is very different. Moreover, the sedimentary mechanism of the South

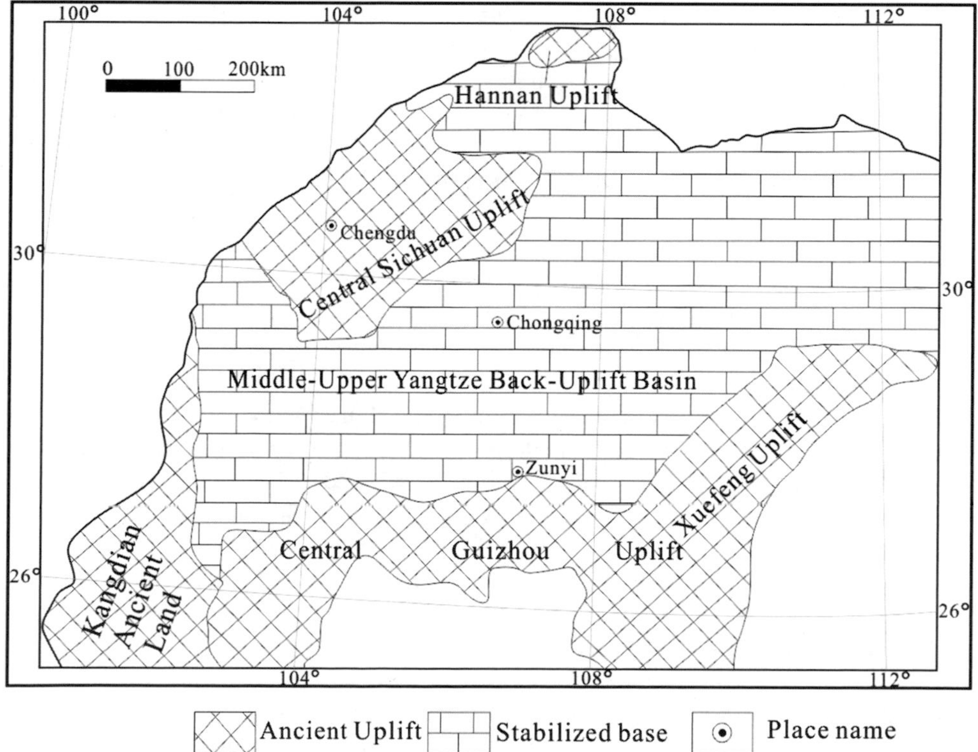

Fig. 3.1 Distribution of Late Ordovician-Early Silurian tectonic framework and sedimentary basin types in the Middle-Upper Yangtze region (modified by Mou et al. 2010, 2016)

China Basin is very different from the Mesozoic, and there are many tectonic changes. In the Mesozoic and even the Paleozoic, the basin experienced great tectonic deformation. The South China Basin has experienced tectonic uplift and Paleozoic, and the tectonic stability is very different. The sedimentary environment, tectonic setting and preservation conditions are extremely complex (Mou et al. 2010, 2011, 2016). The above factors have brought great difficulties to find relatively stable and well-preserved shale zones, which is the basic task of shale gas geological survey and the key to exploration and development in China. From the perspective of theory and practice, whether there is a complete shale gas basin is very important for shale gas development. Therefore, in the process of actual geological survey and exploration, development and production, it is very necessary to reconstruct a complete and fine prototype sedimentary basin of shale gas development in combination with the structural background. It is necessary to study the nature of the sedimentary basin, the size, shape, scale, etc. This work affects the work deployment of shale gas geological survey, and the calculation of shale gas resources and reserves, and further affects the exploration and development of shale gas. In the process of shale gas geological survey, even in the face of the basin that has been strongly transformed, on the one hand, we should study its evolution history and trace the prototype of the basin from the perspective of geological history evolution; on the other hand, from the perspective of "tectonic basin control", we must study the plate tectonic background of the basin and the dynamic mechanism determined by it, so as to objectively and historically identify the basin type, so as to restore the complete sedimentary basin type in the development period of organic shale-rich shale conducive to the formation of shale gas.

3.2.2 Sedimentary Facies (Environment) and Shale Gas

The history of shale gas research and exploration practice in the USA for hundreds of years shows that one of the reasons why organic-rich shale in North America can make a major breakthrough in shale gas is that favorable sedimentary facies zones are generally developed. Although the mining area is small, it is in a favorable deep-water sedimentary environment. In addition, the favorable anoxic and stagnant sub-environment, which can be explained by the "Black Sea" detention quiet sea model and coastal upwelling model, has basically laid the foundation for the rich organic shale in its mining area and the high abundance of organic matter. Thicker thickness is a good material basis for shale gas development (Montgomery et al. 2005; Martineau 2007; Loucks and Ruppel 2007; Zhang et al. 2008a, b; Zou et al. 2010, 2013; Guo et al. 2012; Ma et al. 2012; Wang et al. 2012a, b, c, d; Li et al. 2013; Wu et al. 2013; Mou et al. 2016). They are the main factors for shale gas accumulation.

It is generally believed that in the deep-water slope and basin environment: first, organisms are relatively prosperous, which provides a rich material basis for oil and gas production and is conducive to the formation of a large amount of organic matter; second, except for the short-term strong hydrodynamic conditions in some areas during the turbidity current excitation period, the hydrodynamic conditions in most other sedimentary areas are generally weak, which are relatively quiet semi-deep water to deep-water low-energy environment, which is conducive to the development of source rocks and the preservation and transformation of organic matter. For example, in the development of black shale (source rocks) in Sichuan Basin and its surrounding, the Datangpo Formation is obviously controlled by the evolution of the rift basin: black rock series developed in areas with strong rifting or depression; from the center to the edge, the thickness of source rocks of Datangpo Formation gradually becomes thinner, and the content of organic matter gradually decreases (Fig. 3.2). Therefore, in the deep-water slope and basin environment, it is a favorable oil and gas generation area (Feng et al. 1997), which is also a favorable shale gas development area. It is concluded that the development and accumulation of shale gas are mainly controlled by the sedimentary environment.

In fact, unlike conventional oil and gas accumulation and distribution mainly controlled by the comprehensive control of "source, reservoir, cap, circle, transportation and protection", shale gas is a key reservoir forming system element such as source rock, reservoir and even caprock. In the same organic-rich shale rock series, the lithofacies type, development thickness, distribution area, organic matter type and content, maturity and other geochemical parameters, porosity and other physical parameters of its source rock influence the shale gas-rich integrated reservoirs. In terms of the industrial accumulation and exploration and development conditions of shale gas, the first thing we need is a rich gas source material basis, that is, high-quality source rock. The hydrocarbon generation conditions are required to meet certain standards, such as the large thickness of dark shale, good type of organic matter, high content of organic carbon, appropriate maturity and content of mineral components. This has been discussed earlier. Combined with the re-understanding of the definition of shale gas and the understanding of its development and distribution characteristics in this book, it is considered that the basic characteristics of the development of these source rocks, that is, the control factors affecting the rich accumulation of shale gas, are obviously controlled by sedimentary facies belts and their changes in plane and space (vertical and horizontal direction), and the development of organic-rich shale is

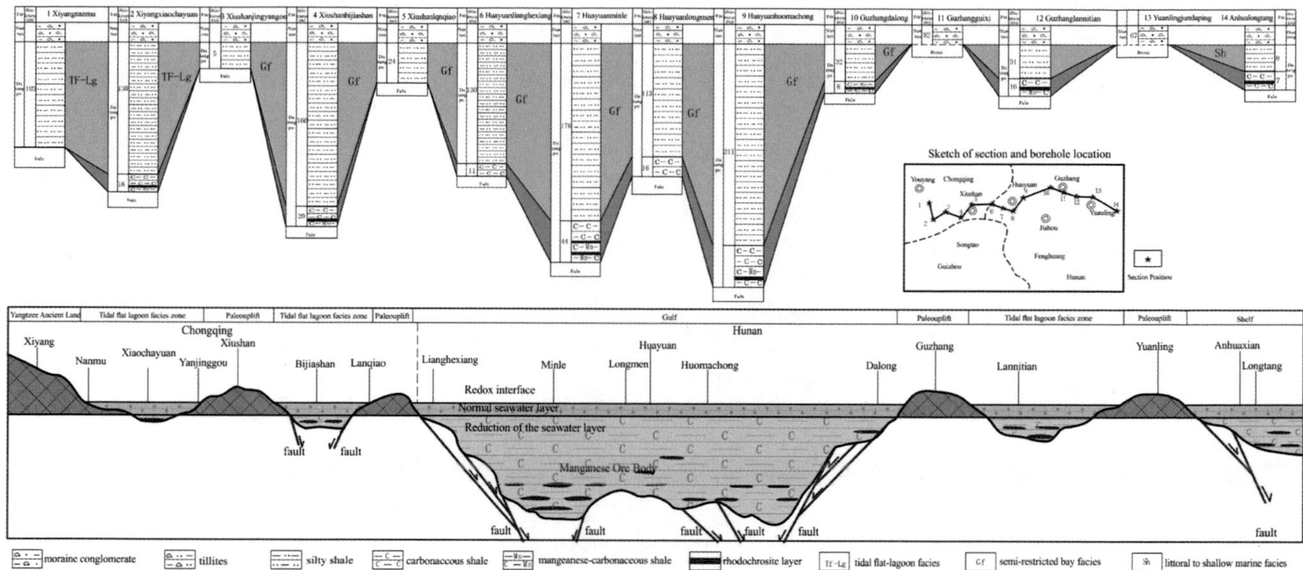

Fig. 3.2 Development model of black rock series of Datangpo formation of Nanhua system in Sichuan Basin and its periphery (modified according to Zou et al. 2020)

different in different facies belts, that is, the formation of sedimentary relatively organic-rich shale has an internal control role (Zheng et al. 2013; Mou et al. 2016).

Firstly, the sedimentary environment controls source rocks' thickness, distribution area and distribution law. It is of great significance to study the formation environment of source rocks and their control over shale gas development (Ma et al. 2014). For the geological survey, exploration and development of shale gas, source rocks is the basic material basis. Today's research and exploration practice show that the main basis for the generation and occurrence of shale gas is the thickness of source rock, and the hydrocarbon generation potential of source rock formed in shallow sea shelf (including deep-water shelf area), slope, semi-deep sea and other environments is greater. If the thickness of the source rock is greater than 30 m, the gas remaining in shale is more conducive to enrichment and accumulation due to the influence of its own sealing layer; its shale gas development conditions are better (Li et al. 2008; Xu et al. 2011; Wang et al. 2013). At the same time, the main distribution law of shale gas is also directly reflected in the distribution area and fine temporal and spatial distribution of this set of source rocks. This objectively reflects the main (fundamental) factor position of sedimentation and sedimentary environment on shale gas development, which controls the thickness, distribution area and temporal and spatial distribution law of high-quality source rocks (Wang et al. 2015; Mou et al. 2016). The preliminary study of shale gas potential strata such as the Niutitang Formation and Longmaxi Formation in the whole South China shows that they are developed in a favorable deep-water sedimentary environment similar to the shale gas development strata in North America, and the "congenital conditions" are better; further sedimentary facies, microfacies and even fine lithofacies types, such as (calcium-containing) carbonaceous (siliceous) mudstone and silty (calcium-containing) carbonaceous mudstone in deep-water shelf environment, have better shale gas development conditions. In addition, factors such as underwater hypoxia retention sub-environment and slow deposition rate basically control the development and distribution of high-quality source rocks of Niutitang Formation and Longmaxi Formation in this area, so that the source rocks in whole area have the characteristics of large thickness and wide distribution area and are the main target strata for shale gas geological survey and further exploration and development (Zhang et al. 2012; Zhang 2015).

The actual exploration and development of the Longmaxi Formation shale gas reservoir in Jiaoshiba area further show that the source rocks section realizing industrial gas flow is mainly concentrated in the lower part of its black rock series (Guo 2016), which puts forward higher requirements for the identification of fine sedimentary sequences of source rocks. However, it is well known that the sedimentary environment plays a decisive role in controlling the transgression and regression sequences of source rocks during the sedimentary period and then affects the development of their sedimentary sequences. Therefore, it further reflects the control of sedimentary facies and sedimentary environment on shale gas development. In addition, some studies show that some specific sedimentary environments also affect the transformation and development of source rocks after deposition. For example, statistical analysis shows that the favorable sedimentary environment makes the favorable lithofacies types formed in the early stage of the reservoir space of source rocks which are not easy to be transformed by later diagenesis.

To sum up, as the carrier of shale gas, the source rock itself has a very close relationship with the sedimentary environment. The latter controls the formation of source rocks and the geometric shape and spatial distribution of source rocks. Through the final paleogeographic analysis, it is helpful to recognize the shape of source rocks and predict the fine spatial distribution law of source rocks, especially the fine spatial distribution law of shale gas favorable intervals in Jiaoshiba area.

Secondly, the enrichment of organic matter is the main controlling factor and basic element conducive to the development of shale gas (reservoir). It affected the amount of hydrocarbon generation and controlled the adsorption capacity of shale gas. It is not only the key evaluation factor of shale gas reservoir formation, but also the main research object of shale gas geological survey and exploration and development (Fu et al. 2008; Li et al. 2008). Studies have shown that the types and distribution of sediments of high-quality source rocks, such as the types and abundance of hydrocarbon-generating parent materials and kerogen type of source rocks, were not only controlled by sedimentary facies and sedimentary environment, but also provided favorable conditions for biological activities, reproduction and prosperity of hydrocarbon-generating parent materials, including high original productivity, limited retention anoxic reduction environment, appropriate deposition rate, rising ocean current submarine hydrothermal activity, ice age, etc. It is concluded that the sedimentary facies (environment) determine the conditions for the enrichment (degree) and preservation of organic matter in organic-rich shale, so as to control the organic matter content and type of organic-rich shale as a whole.

During the deposition period of the organic-rich shale development section of the Early Silurian Longmaxi Formation in Southern China, the paleoclimate warmed rapidly, resulting in the rapid melting of the Late Ordovician Hirnantian glaciers and the rapid rise of sea level in the initial stage. Combined with the tectonic pattern at that time, the area rapidly evolved into an occluded "bay-type" sedimentary environment (Li et al. 2008; Zhang et al. 2012; Mou et al. 2010, 2014; Ge et al. 2013a, b, 2014). Under the rapid transgressive environment, not only the intake of terrigenous clastic materials unfavorable to the development of organic-rich shale is greatly reduced (Hickey and Henk 2007), but also the formation of the layered sedimentary water body is exacerbated. Firstly, the rapid warming of paleoclimate makes the surface water of sedimentary water rich in oxygen due to direct solar radiation, the bottom water cannot get solar radiation due to rapid transgression, and it continues to maintain the ancient water temperature during the ice age for a long time, so it lacks oxygen. The former is suitable for the growth, life and reproduction of a large number of organisms. Together with the abundant planktonic fossils dominated by graptolites in the bottom water body, it creates high productivity conditions, provides rich sources of organic matter for sediments and is conducive to the enrichment of organic matter; in the latter, anoxic reduction conditions hinder the decomposition of organic matter and are conducive to the preservation of organic matter (Clavert 1987; Chen et al. 1987, 2000; Curtis 2002; Chen et al. 2011; Chen et al. 2006; Li et al. 2008; Fu et al. 2011; Cheng et al. 2013; Wang 2015; Mou et al. 2016). At the same time, due to the deep-water shelf environment and the low temperature of the bottom water body near the center of the basin, the carbonate minerals in the water body are difficult to be saturated for chemical precipitation, while the settlement of silicon-rich organisms in the upper water body promotes the formation of siliceous shale, which is conducive to the hydraulic fracturing of shale gas. In addition, geological practice and simulation experiments show that when the deposition rate is slow, it is not conducive to the preservation of organic matter; when the deposition rate is fast, the content of organic matter per unit volume weight is significantly diluted and reduced. Therefore, an appropriate deposition rate is also a favorable condition for the deposition of source rocks and the enrichment of organic matter (Chen et al. 2006; Zhang et al. 2013; Wang et al. 2015; Mou et al. 2016). Due to the lowest loss of rock porosity, it is conducive to the formation of good shale gas reservoirs (He et al. 2010). Feng et al. (2008) believe that the early deposition thickness of Longmaxi Formation in Southern China is 20–200 m. Its deposition rate is about 6–60 m/Ma, which is very conducive to the development of source rocks and the enrichment and preservation of organic matter, which is closely related to the favorable sedimentary environment in the early stage of Longmaxi Formation and the conditions such as hypoxia and high biological productivity in this environment. In sharp contrast, the late deposition of Longmaxi Formation, although the overall sedimentary environment is still in a relatively stagnant environment, is already a relatively shallow sedimentary environment (Zhang et al. 2012). In this environment, it not only has a rapid deposition rate, but the large addition of terrigenous clastic sediments increases the dilution of organic matter, reduces the abundance of organic matter in source rocks and is not conducive to the development and accumulation of shale gas.

Hill et al. (2007) also found through the detailed study of Barnett shale that under the deep-water sedimentary environment, the biomarkers of this set of shale show favorable conditions such as hypoxia, normal salinity and strong upwelling. Under this specific sedimentary environment (deep water), in addition to the anoxic conditions similar to the sedimentary period of organic-rich shale in the Early Longmaxi Formation in Southern China, the biological activities in the sedimentary environment of Barnett shale are also very active due to the active upwelling current. Barnett shale is often symbiotic with phosphorus-rich minerals, which ultimately determines that this set of source rocks has high organic matter content, good organic matter type and good enrichment degree.

In fact, the enrichment and preservation process of organic matter in Barnett organic-rich shale in the USA and the early organic-rich shale in Longmaxi Formation in Southern China, as well as the types and abundances determined by this process, is closely related to favorable factors such as biological original productivity, ancient water depth, ancient climate, limited retention hypoxia, rising ocean current and appropriate deposition rate, which affect the development of hydrocarbon-generating parent materials of source rocks; when these factors reach the optimal configuration, it is conducive to the development of higher-quality source rocks and the enrichment and preservation of organic matter. These factors are restricted by the sedimentary environment, which controls the development of relevant conditions. They are the relevant parameters of sedimentary facies. In a specific period and sedimentary environment, one or two factors can affect or change other factors, which has become the most critical factor affecting the development of high-quality source rocks and the enrichment and preservation of organic matter. Therefore, the sedimentary environment plays a very important role in the development of source rocks with high organic matter abundance (Fu et al. 2008; Li et al. 2008; Liang et al. 2011; Wang et al. 2015; Zhang et al. 2015). Sedimentary facies are the basis and main controlling factor for the development of source rocks and then control the type and abundance of organic matter in source rocks.

Moreover, as the key factor of shale gas content (i.e., resource evaluation and favorable block optimization) (Yang et al. 2012), the thickness, distribution area, organic carbon content and types of organic shale are not only controlled by the sedimentary environment, but also as the key factor of shale gas "dessert" area, whether shale gas can be economically developed and how the output is; reservoir characteristics such as rock types, mineral composition, brittleness, porosity (including fractures) and permeability of source rocks are also controlled by sedimentary environment (Kinley et al. 2008; Loucks et al. 2009; Yang et al. 2012; Wang et al. 2014, 2015; Guo et al. 2015; Mou et al. 2016).

The sedimentary environment controls the development of source rocks, and it affects the microlithofacies types and mineral components of source rocks. There are many types of lithofacies of source rock, the carrier of shale gas, including siliceous shale, calcareous shale, (calcium-containing) carbonaceous (siliceous) mudstone, silty sand (calcium-containing) carbonaceous mudstone, carbonaceous (calcium-containing) silty mudstone, carbonaceous mudstone and carbonaceous argillaceous limestone, which are formed in different sedimentary environments or in different sub-environments or microenvironments of the same environment, sedimentary system, paleowater depth; many environmental factors such as paleoclimate directly affect their development characteristics and determine their vertical and horizontal changes and migration; it is the development of lithofacies types, and mineral components of source rocks in different sedimentary facies belts, sedimentary surfaces belts and microfacies belts are different. This difference and related environmental conditions not only determined the development of reservoir physical properties such as porosity and pore structure of source rocks itself, but also directly affected the enrichment and accumulation of shale gas as an internal controlled factors, but also affected the development and formation of fractures in source rocks and then affected the later development effect of shale gas as an external controlled factor, which were also the key factors affecting the brittle content of shale (Curtis 2002; Jarvie et al. 2007; Wang et al. 2012a, b, c, d; Wang et al. 2013, 2014, 2015; Mou et al. 2016).

The sedimentary environment determines the characteristics of rock facies types and mineral components of source rocks. Lithofacies types and mineral components are also correlated with organic matter abundance. Wang et al. (2014, 2015) analyzed the fine lithofacies types of source rocks in the shale gas development interval of Longmaxi Formation in Southern Sichuan and its periphery and found that the (calcium-containing) carbonaceous (siliceous) mudstone and silty (calcium-containing) carbonaceous mudstone developed in the deep-water shelf environment have high organic carbon content, and most of them were developed in the early stage of transgression. If the sedimentary water body was deep, the carbonate mineral content was low and the siliceous content was high; most of them were siliceous shale, which was conducive to the enrichment and accumulation of shale gas and the later exploration and development. Carbonaceous (calcium-containing) silty mudstone, carbonaceous mudstone and carbonaceous argillaceous limestone were mainly developed in shallow shelf environments. Although their organic carbon content was also relatively high, the organic carbon content gradually decreases with the increase of the content of clastic particles and carbonate minerals. The correlation between the former and organic carbon content was more sensitive, indicating that terrigenous clasts provide a large amount of oxygen-rich sedimentary environment, which was not conducive to the preservation of organic matter. The increase of carbonate minerals may only dilute the abundance of organic matter, which was consistent with the previous view. The flat tidal environment with shallow water body, high oxygen content and rapid warming is mainly a set of silty mudstone and argillaceous siltstone, which are not conducive to the preservation of organic matter, and the content of terrigenous debris and carbonate minerals was very high, so it is difficult to form a high abundance of organic carbon. Wu et al. (2016) divided and studied the shale facies of the Wufeng Formation-Longmaxi Formation in the Fuling gas field and founded that there were mainly eight kinds of lithofacies: siliceous shale facies, mixed siliceous shale facies, clay-containing siliceous shale facies, ash/silicon-containing mixed shale facies, clay/silicon-containing mixed shale facies, mixed shale facies, silicon-containing clayey shale facies and clay/ash-containing mixed shale facies. Different types of lithofacies indicate different sedimentary environments. It is determined that the mixed siliceous shale facies and clay-containing siliceous shale facies in the study area are class I dominant facies, and the clay/silica-containing mixed shale facies are class II dominant facies. The above research results comprehensively reflected that not only the development of organic matter in source rocks was controlled by the sedimentary environment, but also the uncompensated and anoxic deep-water environment with low carbonate mineral content was also conducive to the preservation of organic matter (Zhang et al. 2012; Cheng et al. 2013), but also the development characteristics of fine lithofacies types and mineral components of source rocks are also controlled by the sedimentary environment. It is precisely because the related sedimentation in this environment jointly controls the organic matter, lithofacies types and mineral components of source rocks, resulting in the correlation between lithofacies types, mineral components and organic matter abundance.

From the control factors of the development characteristics of source rocks reservoir, the organic carbon content, mineral composition and organic matter maturity of source rocks are the three most important factors for the development of source rocks reservoir (Curtis 2002; Jarvie et al. 2005, 2007). It is obvious that these three important factors are controlled by the sedimentary environment. The higher the TOC content, the appropriate mineral component content and the gas content of the source rocks, the easier it is to form a favorable source rocks reservoir and the more conducive it is to the development and accumulation of shale gas. To a certain extent, it can be directly summarized that the sedimentary environment controls the development and distribution of the source rocks and reservoir (Bowker 2007). For example, Guo et al. (2015) took the source rocks of the Shanxi Formation in the north of Ordos Basin as the research object. They studied the control of different sedimentary (micro) relative source rocks reservoirs of the Shanxi Formation. Firstly, seven types of lithofacies were identified through the study of sedimentary microfacies of source rocks of the Shanxi Formation. Further, through the study of the organic matter abundance, type characteristics and mineral composition characteristics of these lithofacies types, it was pointed out that the organic matter abundance of the four lithofacies types ((gray) black carbonaceous shale, (gray) black carbonaceous shale, (gray) black carbonaceous (silty) sandy shale and (dark) gray carbonaceous shale) controlled by the plant-rich swamp microfacies and the plant poor swamp microfacies was higher; the content of mineral components was more conducive to the development of shale gas; there is a better correlation between these microlithofacies types and their own mineral component content and organic matter abundance, and there is an obvious positive correlation between organic matter content (TOC) of favorable lithofacies types and adsorbed gas content. This is consistent with Bowker (2007). Considering the symbiosis of Shanxi Formation source rocks reservoir and coal seam and the symbiosis of plant-rich swamp microfacies and plant poor swamp microfacies with coal forming peat swamp microfacies and peat flat microfacies, it is a more favorable source rocks development environment and indirectly reflects the important influence and control of sedimentary environment on the development and distribution of source rock reservoir.

3.2.3 Lithofacies Paleogeography and Shale Gas

In the geological survey of shale gas, the study of regional sedimentary facies and sedimentary environment can not only clarify the basic geological characteristics of source rocks, but also understand the basic geological characteristics of shale gas development; on this basis, the temporal and spatial distribution of favorable facies zones of source rocks can be clarified through the mapping method and technology of lithofacies paleogeography; finally, on the distribution map of favorable sedimentary microfacies and lithofacies paleogeography of shale gas, further comprehensive research will be carried out to determine the prospective area, favorable area and target area of shale gas development, so as to provide a scientific basis for shale gas exploration.

From the perspective of the actual effect of China's shale gas geological survey and exploration and development, although the preliminary shale gas survey shows that the early Paleozoic marine source rocks strata in China, especially in Southern China, have great shale gas resource potential, through nearly a decade of research and geological survey, only the organic shale rich in Ordovician Wufeng Formation-Silurian Longmaxi Formation in some areas of Sichuan Basin has realized industrial development. It is reasonable to say that the study of these sets of source rocks strata in Southern China also has a certain history, and it should have a comprehensive and clear grasp of the basic geological conditions of oil and gas. However, the reality is that it not only has a shale gas reservoir, but its basic resource background also has not been clarified, there is still a long way to go, and there are still many studies to be strengthened. Moreover, even as a source rocks of conventional oil and gas reservoirs, its specific distribution law and other related research are still slightly insufficient.

First of all, there are a lack of comprehensive and profound understanding of shale gas research. The understanding of history refers to the understanding from the sedimentary pattern and characteristics of various geological ages; it is from the pattern and characteristics of paleogeography and paleostructure and the law of historical evolution. The overall understanding and research are to understand the comprehensive and systematic geological evolution law of “the ancient structure controls the development of sedimentary basins during the development period of source rocks, the types of sedimentary basins controls the sedimentary facies of source rocks, and the sedimentary environment determines the development of various geological factors of shale gas”. Deep understanding refers to the fine understanding and research on the petrology, geochemistry and reservoir space of source rocks segments and the qualitative leap from qualitative to quantitative.

Secondly, taking the Sichuan Basin in Southern China as an example, in the process of sedimentary evolution of the basin, there has been a sedimentary environment which is very conducive to the formation of source rocks and the development of favorable factors for the formation of shale gas. However, the research and evaluation of shale gas in this area, as well as the optimization of prospective areas, favorable areas and target areas, still need to be further explored (Mou et al. 2011, 2012, 2016). Therefore, on the basis of reasonable theory, using appropriate working means to carry out the basic geological survey is the key to shale gas exploration in the Sichuan Basin with superior structural and sedimentary conditions. That is, we should start with the most fundamental geological situation research and start with the geological background of hydrocarbon source rocks development, that is, the basic paleogeographic research, and then follow the sequence of shale gas research, shale gas development conditions research and shale gas exploration and development without taking shortcuts.

Predecessors have carried out at least three rounds of systematic studies on sedimentary facies and lithofacies paleogeography with different scales and fine degrees in Sichuan Basin (Mou et al., 1992, 2010, 2011, 2016; Xu et al. 1993; Liu and Xu 1994; Ma and Chu 2008), preliminarily showing the Early Cambrian Qiognzhusi period (Niutitang Formation) during the deposition of high-quality source rocks in the Late Ordovician Wufeng period (the Wufeng Formation) and the Early Silurian long period (the Longmaxi Formation); these sets of source rocks are widely distributed and stably distributed regionally, which are potential shale gas development strata. The research shows that the basic geological research and determination of source rocks, especially the determination of high-quality source rocks, can play a very important role and significance in scientifically evaluating the oil and gas resource potential of the basin and deeply revealing the laws of oil and gas generation, migration, accumulation and enrichment (Lu et al. 2006; Hou et al. 2008). Among them, for the shale gas reservoir integrating source and reservoir, the fine research on its material basis and control factors is particularly important. Therefore, systematic fine sedimentary microfacies and large-scale lithofacies paleogeography research should be carried out with the source rocks development area as the center of focus and the paleogeographic sedimentary environment for shale gas development as the center or premise, so as to determine the source rock characteristics and spatial distribution of shale gas development intervals, and take lithofacies paleogeography mapping technology as the key method and technology, based on the favorable microfacies belt or sedimentary environment for shale gas development. Conduct research on shale gas evaluation parameters such as source rocks thickness, organic matter content and mineral composition, and finally select appropriate parameters for superposition research, so as to find and delineate the distribution locations and zones with good coupling, such as the favorable sedimentary microfacies belt of source rocks, high organic matter content and appropriate mineral composition content of source rocks and take them as the preferred target and preferred area for shale gas exploration and development (Mou et al. 2016).

In fact, the research and practice of China's energy resources show that the research and mapping technology related to lithofacies paleogeography can actively and effectively serve the prediction and exploration of various minerals (Feng et al. 1997, 1999, 2005; Mou et al. 2010, 2011, 2106). Nowadays, the scope of prospecting work guided by lithofacies paleogeography research and related mapping technology has been extended to almost all mineral resources, especially in the field of oil and gas. It has a more prominent effect in guiding oil and gas geological survey, prediction, exploration and development. It not only serves as basic geological science research for the geological survey of oil and gas resources. Moreover, it is often used as a key method and technical means to serve the prediction and exploration of oil and gas resources (Mou et al. 2016). It should be said that this is one of the main characteristics of lithofacies paleogeography in China, and it is also an important reason for the prosperity of lithofacies paleogeography.

Of course, the oil and gas resources mentioned above often refer to conventional oil and gas resources. As for the development of shale gas reservoirs, our views and suggestions are still the research of lithofacies paleogeography and its related mapping methods and technologies, when we choose what theories and related methods and technologies to serve the geological survey or prediction, exploration and development of shale gas, can accurately and well complete the relevant tasks (Mou et al. 2016). Based on the study of regional sedimentary facies and the mapping method of lithofacies paleogeography, the temporal and spatial distribution of favorable facies zones of source rocks can be clarified, so as to provide the basis and direction for shale gas exploration. Therefore, taking lithofacies paleogeography research and mapping as the basic method and key technology can also provide guidance for the geological survey of shale gas (Mou et al. 2016).

In short, sedimentary facies control the basic geological conditions of shale gas development and its related properties and spatial distribution (such as the thickness, distribution area, organic matter content and type, mineral composition) for the geological survey of shale gas under stable structural units, and the detailed and systematic study of sedimentary facies and lithofacies paleogeography is an important basis and premise for the optimization of favorable areas for shale gas. It is also theoretical guidance and key methods and technologies (Mou et al. 2016). To carry out the geological survey of shale gas, we must first carry out comprehensive, overall and fine research and mapping on the paleogeographic environment of source rocks, especially high-quality source rocks. Then, on this basis, focusing on the sedimentary microfacies belt or sedimentary environment conducive to the development of shale gas, comprehensively study the conditions or parameters of shale gas development or evaluation, select appropriate shale gas influence parameters and carry out the coupling superposition of relevant contour maps on the fine lithofacies paleogeographic map, so as to carry out the optimization of shale gas scenic spots, favorable areas and target areas. After such a systematic step, it is possible to grasp the key and key of shale gas geological investigation and research, so as to provide a more scientific basis for the final exploration and development of shale gas and realize the breakthrough of shale gas.

As mentioned above, Xie (1947) put forward the forward-looking view of paleogeography as a guide for prospecting as early as several decades ago. The research and mapping of sedimentary facies, lithofacies and paleogeography have also been widely proved to play a basic and key guiding role in the success of oil and gas exploration industry. As Mou et al. (2011, 2012) said, when conducting a geological survey of oil and gas resources and further exploration and development in an area, the different disciplines will put forward corresponding suggestions and understanding from different angles. However, sedimentary geology, as one of the most basic disciplines, should be the basis and necessary content of research and work, but also a method. Its core is sedimentology and lithofacies paleogeography. The geological survey of shale gas is no exception. For the basic geological survey of shale gas in China, which is already in the stage of extensive development, the research and mapping of sedimentary facies and lithofacies paleogeography are indispensable. It can be used as the key basic geological theory and method to provide guidance for the geological survey of shale gas, which is the first step of the basic work.

Theoretically, because the information of lithofacies paleogeography reconstruction comes from geological records, any information on the sedimentary environment in the sedimentary period of source rocks will be directly or indirectly branded and preserved in the geological stratigraphic records. However, due to the limited technical means and cognitive level of human beings, as well as the destruction and change of geological records, some congenital deficiencies or even sedimentary discontinuities are caused by later structural reasons. It brings great difficulties for researchers to obtain, interpret and retrieve these paleo-environmental information, resulting in great asymmetry in quantity, authenticity and reliability between the extremely rich and complex information in geological records and the parts available to researchers at present (Wang et al. 2003), which hinders the reconstruction research of lithofacies paleogeography and the geological investigation of shale gas resources to a certain extent. Therefore, on the basis of paying attention to the research and mapping of sedimentology and lithofacies paleogeography, we should also pay attention to the research of other disciplines or theoretical technology, such as structural geology and the selection of geochemical analysis and testing methods, so as to more effectively carry out and complete the task and goal of shale gas geological survey. However, the understanding of shale gas geological survey with sedimentology and paleogeography as the core theory and key technical methods cannot be deviated. Extensive and comprehensive lithofacies paleogeography can be used as a guide for shale gas geological survey, and relevant lithofacies paleogeography mapping technology can be used as a key technical method to realize the task of shale gas geological survey, so as to provide a solid foundation and scientific basis for further exploration and development of shale gas reservoir (Mou et al. 2016). The means of other disciplines (such as geochemistry) can only be an auxiliary means. The main purpose is to evaluate the geological conditions of shale gas development and then provide a reference basis for the optimization of shale gas favorable areas based on the study of lithofacies paleogeography and mapping. This consensus and methodological understanding should be the product of the high combination of paleogeography and shale gas exploration and development practice (Mou et al. 2016). It should be emphasized that in the process of shale gas geological survey, sedimentology and lithofacies paleogeography are not only a regional, multi-information and multi-disciplinary comprehensive basic research, but also more importantly, they are a method of shale gas geological survey and a method of "looking for" shale gas.

3.2.4 Specific Methods and Steps

1. Restoration of prototype sedimentary basin

Organic-rich shale can be developed in rift basins (also divided into the intercontinental rift and intracontinental rift), passive continental margin basins, ocean basins, Craton Basins, trench-arc basin, system basins (such as intra-arc basins), residual basins, back-arc Foreland Basins, peripheral Foreland Basins and other qualitative basins. They represent the types of sedimentary basins under different plate tectonic backgrounds and dynamic mechanisms. It reflects the different evolution geological history from the Cratonic Basins extension stage, decline stage, residual stage and suture orogeny stage. However, most of today's sedimentary basins have been transformed by different tectonic activities in the later stage and show the type of superimposed sedimentary basin. Importantly, theory and practice have proved that a complete shale basin is very important for the development of shale gas. At the same time, those complete favorable structural units not only have high-quality source rocks required for shale gas development, but also have good caprock and other conditions which will be the key objects and areas for specific shale gas geological survey in the next step. It affects the specific deployment of shale gas geological survey.

Therefore, the first step of a shale gas geological survey is to restore the original properties of places rich in organic shale.

Firstly, with the idea of "tectonic controlled the development of basins ", it is not only necessary to study the plate tectonic background and the dynamic mechanism of the sedimentary basin during the development of organic shale in combination with the analysis of the relevant tectonic background. Secondly, we should study the evolutionary history of the sedimentary basin in this period, judge its evolution stage and trace the original nature of the sedimentary basin in this period from the perspective of geological history. Finally, on the basis of previous research results and practical basis, comprehensively and objectively identify and restore the types of sedimentary basins in the development period of organic-rich shale, judge their complete types of sedimentary basins and further screen or optimize the types of sedimentary basins that are more conducive to the development of high-quality source rocks for shale gas accumulation, such as the Craton Basin and the Foreland Basins. So as to restore their scope, the size, shape and scale lay the foundation for the following specific shale gas geological survey and provide a measurement reference and scientific basis.

2. Lithofacies paleogeography research and mapping

(1) Research on point

The research on point includes the sedimentary facies, sedimentary microfacies, rock types and characteristics, geochemical characteristics, mineral composition characteristics and vertical variation law of source rocks, in order to clarify the relevant basic geological characteristics of source rocks.

The specific approach is, through a large number of detailed field work, first to carry out the research on the fine lithology characteristics of source rocks from specific points, establish the backbone section or even "iron pillar" of the petrological characteristics of source rocks and then carry out the fine research on the types and characteristics of sedimentary microfacies (rock). If conditions permit, the observation and research of cores can also be strengthened. On this basis, through a certain degree of fine sampling, TOC, Ro and other parameters of source rocks reflecting shale gas development conditions are analyzed, various indoor identification, analysis, laboratory test and comprehensive research work are carried out, the variation law on the profile is summarized, and different gas-bearing source rock segments on a single point are divided. Through geochemical analysis, the types and contents of conventional minerals and clay minerals in different gas-bearing source rocks sections are analyzed, and their variation laws in profile and plane are summarized. Combined with the field and indoor comprehensive research results, the formation environment of different gas-bearing source rocks segments is further determined. Finally, establish a comprehensive histogram of the development characteristics of source rocks on the point, comprehensively clarify the various characteristics of the development of different gas-bearing source rocks segments from the point and complete the first goal of the shale gas geological survey.

(2) Research from point to surface

The research from point to surface is mainly to conduct vertical sedimentary sequence and horizontal comparison, clarify the spatial distribution law of source rocks and the vertical and horizontal changes of shale gas development conditions and prepare single-factor maps of shale gas development conditions.

After comprehensively clarifying the development characteristics of different gas-bearing source rocks segments at each point, through the field profile connection and drilling well-connection profile, comprehensively analyze the thickness, TOC, Ro, mineral composition, occurrence and distribution of different gas-bearing source rocks segments from point to surface and then analyze their vertical and horizontal variation laws. If possible, it is also necessary to study the buried depth through seismic and detailed drilling data to further clarify the temporal and spatial distribution law of source rocks and their characteristic conditions, such as area, and prepare various single-factor maps of shale gas development conditions.

(3) Lithofacies paleogeography research and mapping

With the mapping idea of "tectonic controlled the development of basins and sedimentary basins controlled the development of sedimentary facies", on the basis of the above two studies, through the research and mapping technology of lithofacies paleogeography, it shows the variation law of the thickness and range of source rock development, the evolution law of favorable sedimentary (micro)facies, the variation trend of buried depth, etc. and then comprehensively compiles the favorable sedimentary (micro)facies map and lithofacies paleogeography map of shale gas development.

Through the research of the above steps, we can clarify the temporal and spatial distribution law of source rocks, that is, the second task of the shale gas geological survey.

3.3 Optimization of Prospective Area, Favorable Area and Target Area

(1) Optimization of prospective area

After completing the basic research content of point 2, the shale gas geological survey work area has the basic conditions for the optimization of shale gas scenic spots. Only according to the actual situation of different areas of shale gas geological survey work, different evaluation factors for shale gas development should be selected, such as rock and mineral data of gas-bearing shale or 1–3 parameters such as thickness and distribution range comprehensively determined according to TOC and Ro of source rock; on the basis of lithofacies paleogeographic map, the scope of shale gas prospect can be basically defined by stacking relevant parameters and their isoline map with superposition method.

(2) Optimization of favorable area

Firstly, on the basis of detailed basic geological research such as point, line and plane and lithofacies paleogeography mapping, further carry out the research on fine favorable sedimentary microfacies and even fine favorable lithofacies of shale gas development, select more favorable gas-bearing shale intervals and prepare more elaborate lithofacies paleogeography map. Secondly, different evaluation parameters affecting shale gas development are comprehensively studied and selected based on more detailed lithofacies paleogeographic map and the actual situation of different zoning of shale gas geological survey. Generally, TOC (which can be divided into different standards such as >1.0% and >2.0%) and Ro are the main geochemical indicators of source rocks, combined with the research results of the mineral composition of source rocks. In the favorable sedimentary microfacies belt where the source rocks are developed, the superposition method is further used to stack the relevant parameters and their isoline map to couple the favorable blocks where the source rocks are developed. Finally, combined with the data obtained from the previous study on the gas content of source rocks and the preliminary data of resource potential, the areas with better comprehensive conditions are further selected in the coupled favorable blocks for the development of favorable source rocks as the preferred areas for the development of shale gas.

(3) Optimization of target area

Similarly, based on the lithofacies paleogeographic map and step (2), within the scope of the selected favorable area, the research results and maps of the reservoir capacity of source rocks (including the map of reservoir space type, reservoir physical properties, reservoir fractures and their change law, reservoir porosity, permeability and other data and index change map), the research data and results in map of diagenesis, the buried depth and occurrence of source rocks and other maps can be further superimposed and improve the geochemical index parameters reflecting the development degree of shale gas (e.g., increase the data of TOC contour map to more than 3.0%). Finally, according to the specific situation of shale gas geological survey zoning, we can comprehensively optimize the hydrocarbon source rocks development which are suitable for commercial development, that is, the shale gas development target area, considering the preservation conditions such as structural faults and other factors that need to be considered in industrial exploration and development.

Through the step-by-step method in Figs. 3.3 and 3.4, the third task of the shale gas geological survey, such as the

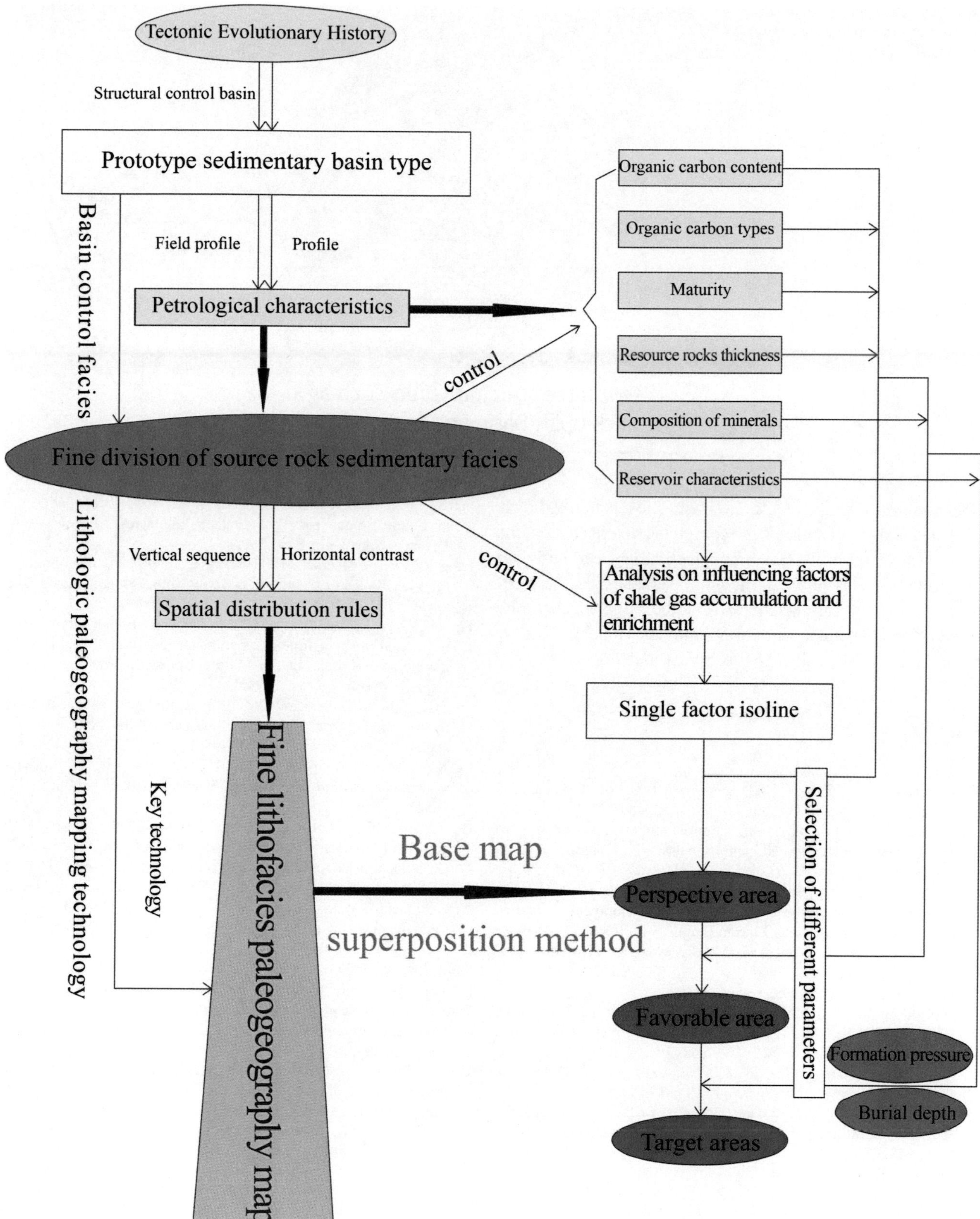

Fig. 3.3 Working methods (technical) steps of shale gas geological survey

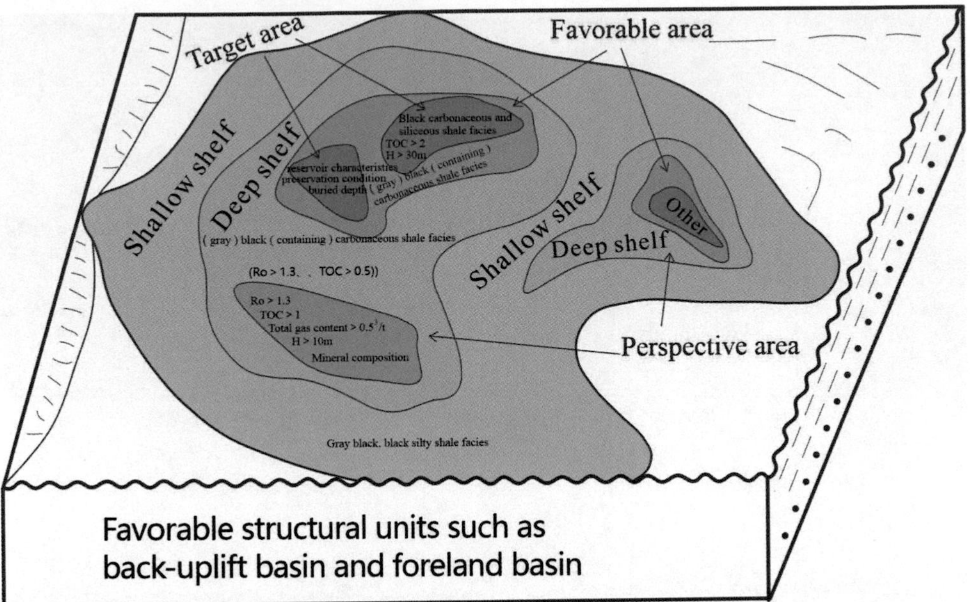

Fig. 3.4 Schematic diagram of the working method of shale gas geological survey

optimization of shale gas development prospect area, favorable area and target area, can be completed. In conclusion, the basic task of the shale gas geological survey can be better completed. In this process, the author emphasizes the theoretical basis and guiding role of lithofacies paleogeography research and the understanding of lithofacies paleogeography mapping technology as a key technical method.

References

Bowker KA (2007) Barnett Shale gas production Fort Worth Basin: issues and discussion. AAPG Bull 91(4):523–533

Calvert SE (1987) Oceanographic controls on the accumulation of organic matter in marine sediments. Geol Soc London Spec Publ 26 (1):137–151

Chen JF, Zhang SC, Bao ZD et al (2006) Main sedimentary environments and influencing factors for development of marine organic–rich source rocks. Marine Origin Petrol Geol 11(3):49–54

Chen S, Zhu Y, Wang H et al (2011) Shale gas reservoir characteristisation: a typical case in the Southern Sichuan Basin of China. Fuel Energy Abst 36(11):6609–6616

Chen X, Xiao CX, Chen HY (1987) Wufengian (Ashgillian) Graptolite Faunal Differentiation and anoxic environment in South China. Acta Palae Ontologica Sinica 26(3):106–118

Chen X, Rong JY, Zhang YD et al (2000) Acommentary on Ordovician Chronostratigraphy. J Stratigr 24(1):18–26

Cheng LX, Wang YJ, Chen HD et al (2013) Sedimentary and burial environment of black shales of Sinian to Early Palaeozoic in Upper Yangtze region. Acta Petrologica Sinica 29(8):2906–2912

Curtis JB (2002) Fractured shale–gas systems. AAPG Bull 86 (11):1921–1938

Feng ZZ (2008) Formation, development, characteristics and prospects of paleogeography in China. In: National conference on palaeogeography and sedimentology

Feng ZZ, Yang YQ, Jin ZK et al (1999) Potential of oil and gas of the carboniferous in South China from the viewpoint of lithofacies palaeogeography. J Palaeogeogr 1(4):86–92

Feng ZZ (2005) Discussion on petroleum exploration of marine strata in South China from quantitative lithofacies palaeogeography. J Palaeogeogr (Chin Edn) 79(1):1–11

Feng ZZ, Bao ZD et al (1997) Why an important breakthrough in Marine Oil—Gas exploration in South China So far having not been made?. MOPG 2(4):4–7

Fu XD, Qin JZ, Tenger et al (2011) Mineral components of source rocks and their petroleum significance: a case from Paleozoic marine source rocks in the Middle–Upper Yangtze region. Petrol Explor Develop 38(6):671–684

Fu XD, Qing JZ, Tenger (2008) Evaluation on excellent marine hydrocarbon source layers in Southeast area of the Sichuan Basin: an example from well D-1. Petrol Geol Experim 30(6):621–628, 642

Ge XY, Mou CL, Zhou KK et al (2013a) Characteristics of Ordovician sedimentary evolution in Hunan Province. Geol China 40(6):1829–1841

Ge XY, Mou CL, Zhou KK et al (2013b) Sedimentary characteristics and depositional model in the sandbian–early katian ages of late Ordovician in Hunan area. J Palaeogeogr 15(1):59–68

Ge XY, Mou CL, Zhou KK, et al (2014) Sedimentary Facies and Lithofacies Palaeogeography in the Late Katian-Hirnantian of Late Ordovician in Hunan Province [J]. Acta Sedimentary Sinica 32(1):8–18

Gong JM, Wang SJ et al (2011) Foreland basin: a favorable. place for shale gas accumulation. Marine Geol Front 28(12):25–29

Guo TL (2016) Key geological issues and main controls on accumulation and enrichment of Chinese shale gas. Petrol Explor Dev 43 (3):317–326

Guo W, Liu HL, Xue HQ et al (2015) Depositional facies of Permian Shanxi formation gas shale in the Northern Ordos Basin and Its impact on shale reservoir. Acta Geol Sin 89(5):931–941

Guo XS, Guo TL, Wei ZH et al (2012) Thoughts on shale gas exploration in Southern China. Strat Stud CAE 14(6):101–105

He FQ, Zhu T (2012) Favorable targets of breakthrough and built up of shale gas in continental facies in Lower Jurassic, Sichuan Basin. Pet Geol Exp 34(3):246–251

He XH, Liu Z, Liang QS et al (2010) The influence of burial history on mudstone compaction. Earth Sci Front 17(4):167–173

Hickey JJ, Henk B (2007) Lithofacies summary of the Mississippian Barnett Shale, Mitchell 2 T. P. Sims well, Wise County, Texas. AAPG Bull 91(4):437–443

Hill RJ, Zhang E, Katz BJ et al (2007) Modeling of gas generation from the Barnett Shale, Fort Worth Basin, Texas. AAPG Bull 91(4):501–521

Hou YJ, Zhang SW, Xiao JX et al (2008) The excellent source rocks and accumulation of stratigraphic and lithologic traps in the Jiyang depression, Bohaibay Basin, China. Earth Sci Front 15(2):137–146

Jarvie DM, Hill RJ, Ruble TE et al (2007) Conventional shale–gas system: the Mississippian Barnett Shale of North–Central Texas as one model for thermogenic shale–gas assessment. AAPG Bull 91 (4):75–499

Jarvie DM, Hill RJ, Pollastro RM (2005) Assessment of the gas potential and yields from shales: the Barnett shale model. In: Cardott BJ (ed) Unconventional energy resources in the Southern Midcontinent. Norman Oklahoma University Press, pp 37–50

Jia CZ, Zheng M, Zhang YF (2014) Four important theoretical issues of unconventional petroleum geology. Acta Petrolei Sinica 35(1):1

Jing L, Pan JP, Xu GS et al (2012) Lithofacies paleogeography characteristics of the marine shale series of strata in the Xiangzhong depression, Hunan, China. J Chengdu Univ Technol (Sci Technol Edn) 39(2):215–222

Kinley TJ, Cook LW, Breyer JA et al (2008) Hydrocarbon potential of the Barnett Shale (Mississippian), Delaware Basin, West Texas and Southeastern New Mexico. AAPG Bull 92(8):967–991

Li SJ, Xiao KH, Wo YJ et al (2008) REE geochemical characteristics and their geological signification in Silurian, West of Hunan Province and North of Guizhou Province. Strat Stud CAE 22(2):273–282

Li SZ, Jiang WL, Wang Q et al (2013) Research status and currently existent problems of shale gas geological survey and evaluation in China. Geol Bull China 32(9):1440–1446

Li XJ, Lv ZG, Dong DZ et al (2009) Geologic controls on accumulation of shale gas in North America. Nat Gas Ind 29(5):27–32

Li YX, Zhang JC, Jiang SL et al (2012) Geologic evaluation and targets optimization of shale gas. Earth Sci Front 19(5):332–338

Liang C, Jiang ZX, Guo L et al (2011) Characteristics of black shale, sedimentary evolution and shale gas exploration prospect of shelf face taking Weng'an Yonghe profile Niutitang group as an example. J Northeast Petrol Univ 35(6):11–22

Liang W, Mou CL, Zhou KK et al (2014) Lithofacies palaeogeography of the Cambrian in central and Southern Hunan province. J Palaeogeogr 16(1):41–54

Liu BJ, Xu XS, Pan XN et al (1993) Crustal evolution and mineralization of the ancient continental deposits in South China. Science Press, Beijing

Liu BJ, Xu XS (1994) Atlas of the palaeogeography of South China. Science Press, Beijing

Loucks RG, Reed RM, Ruppel SC et al (2009) Morphology, genesis, and distribution of nanometerscale pores in siliceous mudstones of the Mississippian Barnett Shale. J Sediment Res 79:848–861

Loucks RG, Ruppel SC (2007) Mississippian Barnett Shale: lithofacies and depositional setting of a deep–water shale–gas succession in the Fort Worth Basin, Texas. AAPG Bull 91(4):523–533

Lu JC, Li YH, Wei XX et al (2006) Research on the depositional environment and resources potential of the Oil Shale in the Chang 7 member, Triassic Yanchang formation in the Ordos Basin. J Jilin Univ (Earth Sci Edn) 36(6):928–932

Ma LY, Li JH, Wang HH et al (2014) Global Jurassic source rocks distribution and deposition environment: lithofacies paleogeographic mapping research. Acta Geol Sin 88(10):1982–1991

Ma YS, Chu ZH (2008) Building–up process of carbonate platform and high–resolution sequence stratigraphy of reservoirs of reef and oolitic shoal facies in Puguang gas field. Oil Gas Geol 29(5):548–556

Ma YS, Feng JH, Mou ZH et al (2012) The potential and exploring progress of unconventional hydrocarbon resources in SINOPEC. Strat Stud CAE 14(6):101–105

Martineau DF (2007) History of the Newark East field and the Barnett Shale as a gas reservoir. AAPG Bull 91(4):399–403

Montgomery SL, Jarvie DM, Bowker KA et al (2005) Mississippian Barnett Shale, Fort Worth Basin, North–Central Texas: gas–shale play with multi-trillion cubic foot potential. AAPG Bull 89(2):155–175

Mou CL, Xu XS (2010) Sedimentary evolution and petroleum geology in south China during the early palaeozoic. Sediment Geol Tethyan Geol 30(3):24–29

Mou CL, Ge XY, Xu XS et al (2014) Lithofacies palaeogeography of the late ordovician and its petroleum geological significance in middle-upper Yangtze region. J Palaeogeogr (Chin Edn) 16(4):427–440

Mou CL, Wang J, Yu Q et al (1999) The evolution of the sedimentary Basin in landing area during Mesozoic–cenozoic. Min Petrol 19 (3):30–36

Mou CL, Wang QY, Wang XP, et al (2016) A study of lithofacies–palaeogeography as a guide to geological survey of shale gas. Geol Bull China 35(1):10–19

Mou CL, Xu XS, Lin M (1992) Sequence stratigraphy and compilation of Lithofacies and palaeogeographic maps—an example from the Devonian Strata in Southern China. Sediment Facies Palaeogeogr 4 (1):1–9

Mou CL, Zhou KK, Liang W, et al (2011) Early Paleozoic sedimentary environment of hydrocarbon sourcerocks in the Middle-Upper Yangtze region and petro-leum and gas exploration [J]. Acta Geol Sinica 85(4):526–532. (in Chinese)

Mou CL, Liang W, Zhou KK, et al (2012) Sedimentary facies and palaeogeography of the middleupper Yangtze area during the Early Cambrian (Terreneuvian-Series 2) [J]. Sediment Geol Tethyan Geol 32(3):41–53

Su WB, Li ZM, Ettensohn FR et al (2007) Distribution of black shale in the Wufeng-Longmaxi formations (Ordovician–Silurian), South China: major controlling factors and implications. Earth Sci J China Univ Geosci 32(6):819–827

Tian ZY, Zhang QC (1997) Lithofacies paleogeography and oil and gas of petroliferous basins in China. Geological Publishing House, Beijing

Tian ZY, Wan LK (1993) Lithofacies palaeogeography and petroliferous prospect, Jurassic, China. Xinjiang Petrol Geol 14(2):101–115

USGS (2013) World petroleum assessment [EB/OL]. http://pubs.usgs.gov/dds/dds-060

Wang ZG (2015) Preakthrough of Fuling shale gas exploration and development and its inspiration[J]. Oil Gas Geol 36(1):1–6

Wang CS, Li XH (2003) Principles and methods of sedimentary Basin analysis. Higher Education Press, Beijing

Wang DY, Zheng XM, Li FJ (2003) Some problems on lithofacies paleogeography in oil and gas area. Acta Sedimentol Sin 21(1):133–136

Wang HZ (1985) Atlas of ancient geography of China. Geological Publishing House, Beijing

Wang J, Duan TZ, Xie Y et al (2012a) The tectonic evolution and its oil and gas prospect of Southeast margin of Yangtze Block. Geol Bull China 31(11):1739–1749

Wang M, Li JF, Ye JL (2012b). Science Press, Beijing, pp 1–69

Wang SY, Liu SG, Sun W, et al (2012) Features of the shales from upper Ordovician–lower Silurian in the north of middle Guizhou-uplift, China. J Chengdu Univ Technol Sci Technol Edn 39(6):599–605

Wang Y, Chen J, Hu L, et al (2013) Sedimentary environment control on shale gas reservoir: A case study of Lower Cambrian Qiongzhusi Formation in the Middle Lower Yangtze area [J]. J China Coal Soc 38(5):845–850

Wang XP, Mou CL, Ge XY et al (2014) Study on clay minerals in the Lower Silurian Longmaxi formation in Southern Sichuan Basin and Its Periphery. Nat Gas Geosci 25(11):1781–1794

Wang XP, Mou CL, Ge XY, et al (2015) Mineral component characteristics and evaluation of black rock series of Longmaxi Formation in Southern Sichuan and its periphery[J]. Acta Petrolei Sinica 36(2):150–162

Wang YM, Dong DZ, Li JZ et al (2012c) Reservoir characteristics of shale gas in Longmaxi formation of the Lower Silurian, Southern Sichuan. Acta Petrolei Sinica 33(4):551–561

Wang ZJ, Xie Y, Yang PR et al (2012d) Marine basin evolution and oiland gas geology of Sinian-Early Paleozoic period on the Western-side of the Xuefeng Mountain LJ. Geol Bull China 31(11):1795–1811

Wu LY, Hu DF, Lu YC et al (2016) Advantageous shale lithofacies of Wufeng formation–Longmaxi formation in fulling gas field of Sichuan Basin, SW China. Pet Explor Dev 43(02):189–197

Wu X, Ren ZY, Wang Y et al (2013) Situation of world shale gas exploration and development. Resour Indus 15(5):61–67

Xie JR (1947) The geology and mineral resources of the new coal field and the great Southern Basin. Geol Rev 12(5):317–348

Xu GS, Xu ZX, Duan L et al (2011) Research status and development trends of shale gas. J Chengdu Univ Technol (Sci Technol Edn) 38 (6):603–610

Xu XS, Mou CL, Lin M (1993) Characteristics and depositional model of the lower paleozoic organic rich shale in the Yangtze Continental Block. Chengdu University of Science and Technology Press, Chengdu

Xu XS, Xu Q, Pan GT et al (1996) Comparison of the evolution of south China continent and global paleogeography. Geological Publishing House, Beijing

Yang RD, Cheng W, Zhou RX (2012) Characteristics oforganic–rich shale and exploration area of shale gas in Guizhouprovince. Nat Gas Geosci 23(2):340–347

Yin FG, Xu XS, Wan F et al (2001) The sedimentary response to the evolutionary process of Caledonian Foreland Basin system in South China. Acta Geosci Sin 22(5):425–428

Yin FG, Xu XS, Wan F et al (2002) characteristic of sequence and stratigraphical division in evolution of Upper Yangtze region during caledonian. J Stratigr 26(4):315–319

Zhang CM, Zhang WS, Yinghai G (2012) Sedimentary environment and its effect on hydrocarbon source rocks of Longmaxi formation in Southeast Sichuan and Northern Guizhou. Earth Sci Front 19 (3):136–145

Zhang JC, Nie HK, Xu B et al (2008a) Geologicalcondition of shale gas accumulation in Sichuan Basin. Nat Gas Ind 28(2):151–156

Zhang JC, Wang ZY, Nie HK et al (2008b) Shale gas and its significance for exploration. Strat Stud CAE 22(4):640–646

Zhang K (1991) on the disintegration, displacement, collision and convergence of Pan—China plate and evolution of RTS oil and gas bearing basins. Xinjiang Petrol Geol 12(2):91–106

Zhang WS (2015) The effect of ancient sedimentary environment on source rocks: an example of the Longmaxi formation in Southeast Sichuan. Inner Mongolia Petrochem Indus 18(1):141–143

Zhang XL, Li YF, Lv HG et al (2013) Relationship between organic matter characteristics and depositional environment in the Silurian Longmaxi formation in Sichuan Basin. J China Coal Soc 38 (5):851–856

Zheng HR, Gao B, Peng YM et al (2013) Sedimentary evolution and shale gas exploration direction of the Lower Silurian in Middle–Upper Yangtze area. J Palaeogeogr (Chin Edn) 15(5):645–656

Zhou KK (2015) Middle Ordovician–Early Silurian lithofacies paleogeography of Middle–Upper Yangtze and its Southeastern margin. Chinese Academy of Geological Sciences, Beijing

Zou CN, Dong DZ, Wang SJ et al (2010) Geological characteristics, formation mechanism and resource potential of shale gas in China. Pet Explor Dev 37(6):641–653

Zou CN, Yang Z, Cui JW et al (2013) Formation mechanism, geological characteristics and development strategy of nonmarine shale oil in China. Pet Explor Dev 40(1):14–22

4 Examples—Taking the Ordovician Wufeng Formation-Silurian Longmaxi Formation in Southern Sichuan and Its Periphery as an Example

Abstract

Taking the Ordovician Wufeng Formation-Silurian Longmaxi Formation as an example of shale gas development in the Southern Sichuan Basin and its periphery. According to the guiding of "Lithofacies paleogeographical research and mapping compilation can be a guide for shale gas geological survey" and methods of area selection evaluation, lithofacies paleogeography and basic geological conditions of shale gas of the Ordovician Wufeng Formation-Silurian Longmaxi Formation in this area have been carefully studied, and potential shale gas areas have been selected.

Keywords

Ordovician Wufeng Formation—Silurian Longmaxi Formation • Lithofacies paleogeography • Shale gas geological survey • Shale gas evaluation • Southern Sichuan Basin and its periphery

4.1 Regional Geological Profile

4.1.1 Location of the Study Area

The Sichuan Basin belongs to a secondary structural unit of the Yangtze quasi-platform. It's a superimposed basin developed on the basis of the Upper Yangtze Craton Basin. It's a Mesozoic and Cenozoic basin. The Himalayan Movement was fully folded and finally formed into a tectonic basin with a diamond-shaped boundary (Bai 2012). The Sichuan Basin is located in the west of the Yangtze quasi-platform, with the Micangshan uplift-Dabashan fold belt in the north, the Emeishan-Liangshan thrust belt in the south, the Longmenshan orogenic belt in the west, and the Hunan-Guizhou-Hubei thrust belt in the east (Su et al. 2007; Zeng et al. 2011) (Fig. 4.1). The study area is located in the Southern part of the Sichuan Basin, including Southeastern Sichuan, Western Chongqing, Northern Guizhou and Northeastern Yunnan. The Huaying line is mainly the area bounded by the Emeishan-Liangshan thrust belt and the Hunan-Guizhou-Hubei thrust belt (Fig. 4.1).

4.1.2 Regional Geological Background

1. Regional structure

In the long tectonic and sedimentary evolution history, the Sichuan Basin and the Yangtze region have experienced multi-period and multi-directional marginal deep fault activities, which are characterized by multiple cycles. According to the regional tectonic features, current tectonic traces and previous research results (Sichuan Provincial Bureau of Geology and Mineral Resources 1991), the basin is divided into three structural regions bounded by two deep and large faults, Huaying Mountain and Longquan Mountain, from Northwest to Southeast. They are the depression area (Northwest Sichuan), the uplift area (central Sichuan) and the depression area (Southeast Sichuan); they are further divided into six sub-level tectonic zones, from Northeast to Southwest, and they are the high-steep fault-fold belt in Eastern Sichuan, Low and steep fault-fold belts in Southern Sichuan, low and gentle fault-fold belts in SouthWestern Sichuan, gentle fault-fold belts in central Sichuan, gentle fault-fold belts in Northern Sichuan, and low-slow fault-fold belts in Western Sichuan. The study area mainly includes the SouthWestern Sichuan paleo-uplift structural belt, the Southern Sichuan low and gentle anticline structural belt, the central Sichuan low-flat structural belt, the Western Sichuan low uplift structural belt and the Southern part of the eastern Sichuan high-steep fault-fold belt, and the Wuling fold belt around the basin is properly considered. The Western part of the Northeastern Yunnan thrust fold belt and the Northern part of the broad and gentle fold belt in the NorthWestern

C. Mou et al., *Lithofacies Paleogeography and Geological Survey of Shale Gas*, The China Geological Survey Series, https://doi.org/10.1007/978-981-19-8861-5_4

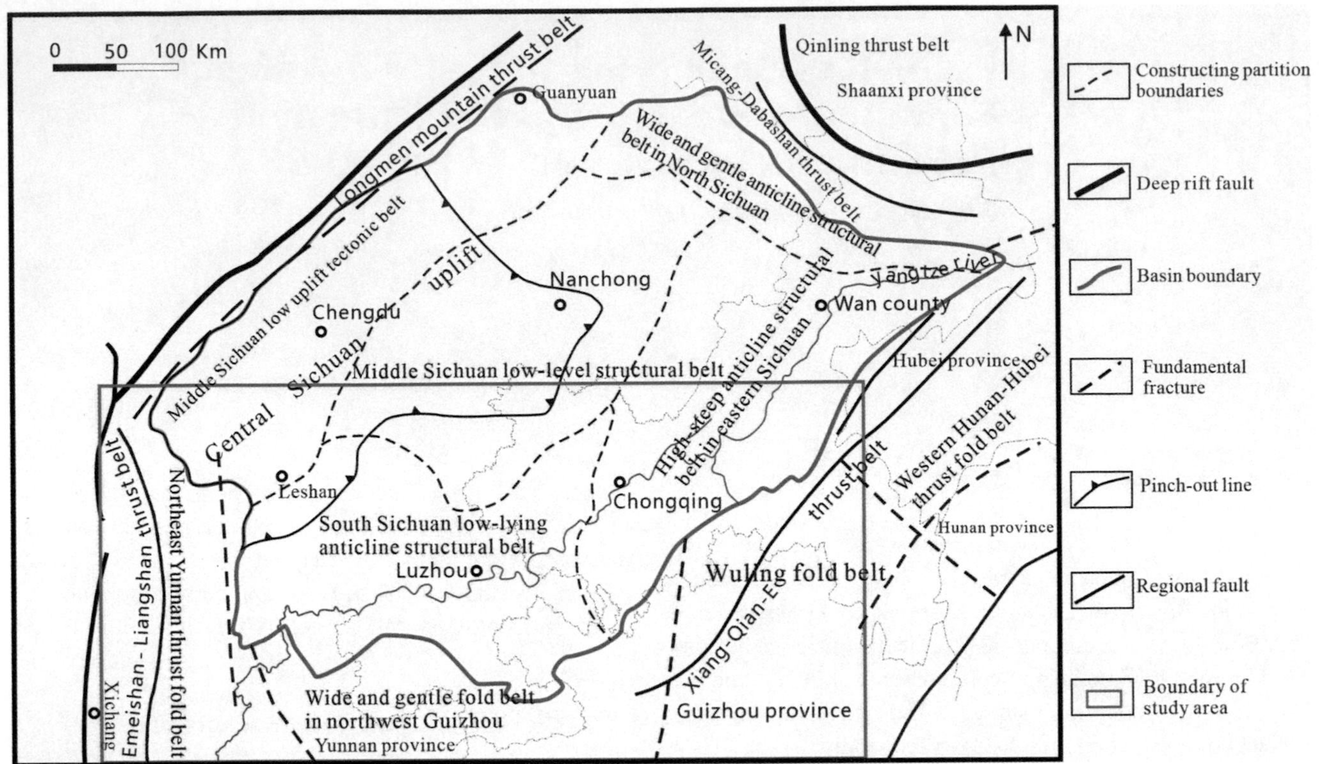

Fig. 4.1 Location of sampling sites and the study area (modified from Huang et al. 2012c)

Guizhou (Fig. 4.1). This area is one of the areas with the best shale gas exploration potential in the Upper Yangtze region. At present, shale gas breakthroughs have been achieved in the lower assemblage of Weiyuan, Changning and its adjacent Zhaotong, Pengshui, Qianjiang and Fuling blocks. The potential of rock gas resources is good.

2. Sedimentary tectonic evolution

The geological pattern of the main body of Southern Sichuan and its periphery is closely related to the tectonic evolution of the entire Sichuan Basin and even the middle and Upper Yangtze region. In the long tectonic and sedimentary evolution history, the Sichuan Basin and the Yangtze region have experienced multi-period and multi-directional marginal deep fault activities, which are characterized by multiple cycles.

(1) Yangtze tectonic cycle

Including the Jinning Movement and the Chengjiang Movement, the Jinning Movement is the most important, which made the Pre-Sinian geotrough folds return, and the Yangtze quasi-platform was generally consolidated into a unified base. The late Upper Proterozoic was the initial stage of the sedimentary caprock of the Yangtze quasi-platform. On the paleotopographic background of crystalline basement and folded basement, in the Early Sinian, the ancient land was located in the west of the Qianjiang River and Duyun line, and the sediments were dominated by rifted terrigenous clastic rocks along the Dongrongjiang River and Huaihua (Nantuo Period) was the Great Ice Age, including large areas to the east of Yibin and Chongqing covered by glacial moraine conglomerates. In the late Sinian, the initial characteristics of relatively stable passive continental margins began to appear. The Doushantuo period was the transgressive sedimentation after the melting of glaciers during the Great Ice Age. The uneven seafloor topography was characterized by relatively occluded, anoxic and open environments. The sedimentary black shale is mainly composed of two sets of different combinations of argillaceous dolomite and phosphorite, and purple-red, yellow-gray sandy mudstone, and argillaceous dolomite with phosphorite. During the Dengying Period in the Late Sinian, the characteristics of the sea basins were high in the west and low in the east, shallow in the west and deep in the east.

(2) Caledonian sedimentary tectonic cycle

The Zhijin Movement (or Tongwan Movement) at the end of Sinian was characterized by large-scale uplift, and the upper part of the Dengying Formation was extensively denuded, especially in Southern Sichuan and Northern Guizhou. The Lower Cambrian Niutitang Formation was overlying different layers of the Dengying Formation. Above, there is an uneven erosion surface. The Early Cambrian continued to develop on the basis of inheriting the paleogeomorphic pattern of the Late Sinian, and the sea basin was still clearly characterized by a continental surface sea that was high in the west and low in the east and shallow in the west and deep in the east. Due to the deepening of the water body and the occurrence of anoxic events in the early Cambrian, the black carbonaceous shale and silt shale of the Niutitang Formation were deposited, with a thickness of 50–200 m, which was the main source layer in this area, and gradually transitioned to it contains sandy shale, interbedded sand and mudstone, and limestone assemblages in the late stage. In the Middle and Late Cambrian, a semi-limited to lagoon facies depositional environment appeared in a large area, and sandy shale and interbedded sand mudstone were deposited. The Yunnan-Guizhou Movement at the end of the Cambrian period led to the occurrence of NE-trending underwater uplifts in the early Ordovician in the central and Northern Guizhou regions south of Southeastern Sichuan, and a set of limestone assemblages were deposited.

In the middle and late Ordovician–Silurian, due to the Westward subduction of the ancient Pacific plate, the passive continental margin, which was relatively stable, began to move, and the Southern part of the Upper Yangtze Platform was uplifted as a whole. The Wan Formation has been denuded in a large area, and some areas have been denuded to the Honghuayuan Formation. The Leshan-Longnvsi uplift in Sichuan, the central Guizhou-Eastern Yunnan uplift and the Southern Xuefeng uplift in Guizhou all uplifted rapidly at this time, during which the Sichuan-Guizhou depression appeared in the Southeastern Sichuan region, and the depositional Wufeng-Longmaxi period was relatively closed. The stagnant and anoxic intraplatform basin facies strata are another set of high-quality source rocks.

The Guangxi Movement at the end of the Silurian period led to the splicing of South China and the Yangtze Platform to form a new South China Block, also known as the South China Plate. The differential uplifting and uplifting in the plate formed the uplifting pattern, and the uplift of the NE-trending South China fold belt and the Western platform area announced the end of the Caledonian tectonic cycle. At the same time, it was the Hercynian tectonic development. The Southern continental margin rift laid the foundation.

(3) Hercynian cycle

After the Guangxi Movement, it was not until the Late Devonian that the Devonian-Carboniferous sediments with small thickness were "filled and filled" in the Northern and Eastern regions of Southeastern Sichuan. Since most of the Southeastern Sichuan region has been in the paleocontinental environment, and the Yunnan Movement at the end of the Carboniferous was uplifted and denuded again, the Devonian-Carboniferous deposits were missing.

In the Early Permian, due to the rise of global sea level, seawater swept and submerged the entire Yangtze paleocontinental area from south to north, depositing thicker shallow-water carbonate rocks; influenced by the role, a regional Dongwu movement emerged.

In the Late Permian, the water body deepened from west to east. In the early period (Wujiaping period), the west was high in the west and the east was low, the west was sand mudstone of alluvial plain facies, and the east was coastal plain facies coal-bearing sand mudstone; in the late period (Changxing period), the water body deepens, forming littoral-neritic sediments.

The Early and Middle Triassic basically inherited the paleogeographic features of the end of the Late Permian. At the beginning of the Early Triassic, the sedimentary environment in Southeastern Sichuan was dominated by the coastal plain, and the main sedimentary strata were shale and siltstone; in the Triassic, the degree of sealing of the sea basin was increased, and the sediments of dolomite and argillaceous dolomite were the main ones. Vertically, the water regression cycle combination of sandstone, limestone, dolomite, and gypsum is formed.

(4) Hercynian cycle

Indosinian cycle (Foreland uplift evolution stage)

At the end of the Middle Triassic, due to the rapid closure of the Paleo-Tethys Ocean, with the Western and Northern margins of the Yangtze block and the Qiangtang block, respectively, from Northwest to Southeast, while the North China block occurred in two directions from north to south. High-angle collision caused strong compression, deflection and subsidence of the lithosphere in the Songpan-Ganzi area on the west side of the Yangtze block and the Southern Qinling area on the Northern margin to form a "peripheral foreland basin" (Zheng et al. 2012). During the Late Indosinian Movement in the Late Triassic (Nori-Reti period), due to the comprehensive return of the Ganzi-Aba trough area, the Longmen Mountains in the Northwest of the Sichuan Basin were strongly depressed and entered the evolutionary stage of the Western Sichuan-like foreland basins. The subsidence sloping from east to west forced the foreland uplift to migrate rapidly to the Southeast, thus causing the Southeastern Sichuan region to enter the foreland slope belt gently sloping to the Northwest, and then the Late Triassic Xujiahe Formation occurred from Northwest to Southeast

The extended sedimentary overlay, characterized by the rapid increase in the thickness of the deposits from east to west, formed a large-scale front of the coastal- lake-fluvial coal-bearing clastic lithofacies and the Eastern lacustrine-fluvial coal-bearing clastic facies from west to east. The continental-continental lake basin has since ended the marine sedimentary history of the whole region and ended the 2300 Ma marine sedimentary history in Southern China from the Lvliang Period to the Late Triassic.

(5) Yanshan cycle (evolutionary stage of preland uplift and post-depression)

Since the Yanshan Movement that began in the Jurassic, due to the Westward oblique subduction of the Pacific plate, the tectonic activity belt has shifted from the Longmen Mountains in the Northwest of the Indosinian Basin to the Southeast of the basin. Thrust and nappe, intracontinental depression appeared on the Northwest side of the ancient land and was strongly connected with the Southern side of the Dabashan Ancient Land, forming the depression in Wanxian, Nanchuan and Gulin, and the relative uplift in the west of Lzhou and Chishui. The basin area extends to the Jiangnan Ancient Land in the east, the Longmen Mountain Ancient Land in the west, and the Daba Mountain Ancient Land in the north. In the sedimentary background of the Western Sichuan foreland basin, the depression is more of a post-uplift depression after the foreland uplift (Luzhou-Dachuan), and this pattern continued until the end of the Jurassic.

(6) Late Yanshan-Himalayan cycle (foreland shrinking, decaying evolution stage)

From the end of the Jurassic to the late Early Cretaceous, due to the oblique subduction of the Pacific plate, further lateral compression of the Yangtze plate was formed, and the main episode of the Yanshan movement appeared, leading to the shrinking of the depositional range of the foreland basin. At the end of the Early Tertiary and the middle Eocene, there was a strong fold fault movement in Eastern Sichuan and a strong fold fault and nappe movement in Western Sichuan between the New and Old Tertiary collectively referred to as the Sichuan Movement, which was the first episode of the Himalayan Movement. The time is early in the east and late in the west; that is, the early period is the middle and late Eocene of the Tertiary, and the late period is between the new and old Tertiary. The Sichuan Movement ended the depositional history of large continental lake basins, and it was an important period for the formation of the current tectonic pattern of the Sichuan Basin's compressional folds. More folds in Southeastern Sichuan were formed at this stage, and then the Himalayan Movement Phase II and Himalayan Movement Phase III uplifted. The transformation is further shaped.

The Sichuan Basin belongs to the Middle and Upper Yangtze Craton Basin. The Yangtze Craton Basin completed cratonization in the late Paleoproterozoic (about 1800–1600 Ma) and is one of the oldest continental block basins in China. Its Western boundary is the Honghe fault-Longmenshan fault, the Northern boundary is the Micang-Daba fault-Chengkou-Xiangfan fault-Jiujiang-Huai'an fault, and the Southeast is "loosely connected" with the Cathaysian Basin in the territory of Zhejiang, Jiangxi, Hunan and Guangxi. Huang et al. (2011) divided the Sinian-Middle Triassic craton evolution process of the Middle and Upper Yangtze into four stages based on the differences in the nature and characteristics of the basins in different periods: Stage I, Sinian-Early Ordovician, in an extensional environment, it has the characteristics of evolving from an early rift basin to a rift basin, the sedimentary formation is dominated by carbonate rocks, and the profile structure is composed of two large cycles with gradually increasing carbonate rocks (Sinian and Cambrian); the second stage, the late Early Ordovician–Silurian, was in a compression stress environment, the nature of the basin was an inherited compression depression type basin within the craton, and the craton margin was generally compressed and uplifted. It is a basin structure bounded by uplift divisions. The sedimentary formations are dominated by clastic rocks and mixed types, and the profile structure has a trend of decreasing carbonate rocks and increasing clastic rocks; in the third stage, Devonian-Carboniferous or early Permian, under the background of the overall uplift of the Middle and Upper Yangtze after the Caledonian Movement, the craton margin was in an extensional environment. The nature of the basin was dominated by craton margin rift basins, and the sedimentary construction was controlled by regional structural characteristics, mainly carbonate rock type and mixed type, and the profile structure has an upward trend of increasing carbonate rock. Stage IV, Permian-Middle Triassic, the stress environment changed from the extensional environment of the Dongwu Movement to the compressional environment of the Indosinian period. The nature and evolution of the basin were obviously differentiated, and the formation of carbonate rocks was dominant; under the influence of tributary movement, the seawater exited the Middle and Upper Yangtze region and was in a regional compression environment. The nature and evolution of the basin have undergone strong changes, and it has the properties of foreland basins and its sedimentary filling characteristics, thus opening a new history of basin evolution in the Middle and Upper Yangtze region.

4.1.3 Stratigraphic Division and Comparison

4.1.3.1 Regional Stratigraphic Characteristics

In the Upper Yangtze region, including Southern Sichuan and its periphery, the basement is the Upper Archean-Mesoproterozoic crystalline basement and the Sibao-Jinning folded basement, except for some areas where the Ordovician, Silurian and Devonian are missing. In addition to the Carboniferous strata, the Sinian-Jurassic strata are developed (Table 4.1). Controlled by structural conditions and sedimentary systems, during the Early Cambrian Niutitang Period, the Late Ordovician-Early Silurian Longmaxi Period, the Late Permian Longtan Period, the Late Triassic Xujiahe Period, and the Early Jurassic Artesian Period Abundant organic-rich shale is deposited. The Lower Paleozoic organic-rich shale in Southern Sichuan is mainly developed in the Niutitang Formation, Wufeng Formation and Longmaxi Formation.

1. Nanhua

The Nanhua System is the filling and filling sediments on the basis of the Neoproterozoic rift, mainly depositing the Nantuo Formation continental moraine, which transitions to marine moraine along the Southeast direction.

2. Sinian

A set of cap dolomite is generally developed at the bottom of the Sinian Doushantuo Formation, which transitions to mud shale upwards, the bottom "ice cap" dolomite is an important comparison mark in the region, with a general thickness of 50–80 m. The fourth member of the Doushantuo Formation is an organic-rich lithological member, which is in pseudo-conformity contact with the underlying Nantuo Formation. The Dengying Formation can be divided into four members, the first member of Dengying Formation is algae-poor dolomite, the second member of Dengying Formation is algae-rich dolomite, crust-like dolomite, algal reef dolomite, the top dissolved pores are developed, and the storage capacity is good; It is gray feldspar quartz sandstone, gray-brown, purple-brown, light gray to dark gray mud shale, 10–50 m dark gray mud shale developed in the central Sichuan area and has certain hydrocarbon generation potential; the Deng 4 Member is algal laminar dolomite, siliceous dolomite; integrated contact with the underlying Doushantuo Formation.

3. Cambrian

The lower Cambrian area in SouthWestern Sichuan is the Maidiping Formation. The lithology is mainly phosphorus-bearing dolomite and siliceous dolomite. It is characterized by the appearance of small shell fossils. A set of regional organic-rich shale intervals are deposited in the Qiongzhusi Formation; the lower part is gray to gray-black mud shale, and the upper part is gray to dark gray silty mud shale; pseudo-integrated with the underlying Maidiping Formation Contact, some areas can be respectively contacted with the fourth segment of the lamp, the third segment of the lamp, and the second segment of the lamp. The Canglangpu Formation is a neritic clastic rock deposit that integrates with the underlying Qiongzhusi Formation. The Longwangmiao Formation is mainly composed of a set of carbonate rock deposits in arid environment. The lithology is mainly composed of oolitic limestone, granular dolomite intercalated with argillaceous dolomite and gypsum-salt rock, with good storage capacity. Group integration contacts. At the bottom of the Niutitang Formation in Northern Guizhou, black carbonaceous mudstone and siliceous rock are interbedded rhythmically with phosphorus nodules; a set of black carbonaceous mudstone is generally developed in the lower part, rich in pyrite, 40–120 m thick, and the distribution is stable. It is a strip of dark gray, gray-green mud shale intercalated with siltstone, about 30–50 m thick. Mingxinsi Formation and Jindingshan Formation are demarcated by gravel-bearing quartz sandstone and quartz sandstone in Northern Guizhou area. The lithological assemblage is close, mainly gray-green mudstone and siltstone interbedded with limestone, and the siltstone gradually increases upwards. The lithology of the Douposi Formation in the Cambrian Middle and Upper Series in the Western Sichuan-Southern Sichuan area is argillaceous quartz siltstone intercalated with argillaceous dolomite, which is in contact with the underlying Longwangmiao Formation. The carbonate deposition in the evaporative environment of the Xixiangchi Formation is composed of micrite microcrystalline dolomite and granular dolomite intercalated with gypsum-salt rock, with well-developed dissolved pores and good storage capacity, which is in integrated contact with the underlying Douposi Formation. The Cambrian Middle-Upper System in the Northern area of Guizhou is a large set of carbonate formations, and carbonate platform facies are deposited. The lithology of the Gaotai Formation is gray sand, argillaceous dolomite intercalated with mica-quartz sandstone and calcareous shale, and gypsum or gypsum-bearing sediments are developed in local intervals. The main body of Loushanguan Formation is gray medium-thick layer—mass dolomite, dolomitic limestone and limestone, with local development of sandy limestone and oolitic limestone.

4. Ordovician

The lower Ordovician is the Tongzi Formation and the Honghuayuan Formation. The lithology is mainly light gray

Table 4.1 Lower Paleozoic strata division of Sichuan Basin and its periphery (modified from Qian et al. 2012)

Erathem	System	Series	Formation	Stratum code	Brief lithology	Remarks
Paleozoic erathem	Silurian System	Middle upper series			Absence	Local distribution
		Lower series	Hanjiadian Fm	S_1h	Gray-green, gray-yellow shale, siltstone sandwiching siltstone, biotite lenticular	Local organic-rich mud shale
			Shiniulan Fm	S_1s	Dark gray, dark gray marl, siltstone with thin bioclastic tuff, muddy siltstone, sandy marl, tuberous marl and calcareous mudstone	
			Longmaxi Fm	S_1l	The upper part is dark gray mudstone interspersed with chalky mud shale, and the lower part is black shale rich in penstones	
	Ordovician System	Upper series	Kuanyinchiao Fm	O_3g	Gray, dark gray biogenic tuffs, mudstones and ferromanganese mudstones with Hernandezite	
			Wufeng Fm	O_3w	Black silica-bearing graywacke shale, topped by common dark gray marl	
			Jiancaogou Fm	O_3j	Gray and light gray verrucose marl	
Paleozoic erathem	Ordovician System	Middle series	Baota Fm	O_2b	Light gray, gray bioclastic-bearing horseshoe tuff	
			Shizipu Fm	O_2sh	Gray, dark gray bioclastic-bearing tuffs and muddy tuffs occasionally interspersed with shales	
		Lower series	Meitan Fm	O_1m	Upper part gray and gray-green shale with siltstone shale interbedded with tuff. Middle part yellow-green siltstone interbedded with dark gray mud-bearing tuff. Lower yellow-green shale, siltstone shale interbedded with biotite tuff lenses	
			Honghuayuan Fm	O_1h	Gray and dark gray bioclastic tuffs interspersed with small amounts of shale, dolomitic tuffs and sandy tuffs, commonly containing siliceous bands (nodules)	
			Tongzi Fm	O_1t	Upper gray and gray-yellow shale, dark gray bioclastic tuff and oolitic tuff, lower light gray and gray dolomitic tuff, gray dolomite, muddy dolomite interspersed with shale bioclastic tuff, sandy tuff and oolitic tuff	
	Cambrian	Upper series	Loushanguan Gr	ϵ_3ls	Dolomite with fine-grained quartz sandstone underlain by cloudy mudstone	
		Middle series	Shilengshui Fm	ϵ_2s	Dolomite, mud-bearing dolomite and chert with gypsum	
			Douposi Fm	ϵ_2d	Mud-bearing quartz siltstone	
		Lower series	Qingxudong Fm	ϵ_1q	The lower section is dominated by tuffs, and the upper section is dolomite interspersed with muddy claystones	
			Jindingshan Fm	ϵ_1j	Mud shale, muddy siltstone and siltstone interbedded with chert	
			Mingxinsi Fm	ϵ_1m	Mud and sandstone dominate, with more chert in the lower part	
			Niutitang Fm	ϵ_1n	The mudstone and silt-bearing mudstone are dominant, interspersed with siltstone. Pseudo-integrated contact with the Meishucun Formation	
Proterozoic erathem	Sinian System	Upper series	Dengying Fm	Z_2d	Dolomite with siliceous rocks	
		Lower series	Doushantuo Fm	Z_1ds	Mudstone is dominated by dolomite at the bottom and sandy mudstone with colloidal phosphorite nodules at the top	

to gray granular limestone, intercalated with mud shale, and gradually transitions to dolomite in the Northern Guizhou area. It is better to integrate and contact with the underlying Cambrian system. The Meitan Formation is mainly composed of marine clastic rocks, the upper part is gray and gray-green shale, and the silty shale is intercalated with limestone; the middle part is interbedded with yellow-green siltstone and dark gray argillaceous limestone; the lower part is yellow-green shale. It is composed of silty shale intercalated with bioclastic limestone lenses; in the Eastern Guizhou and Western Hunan and Hubei areas, it transitions to the purple-red nodular marl deposits of the Dawan Formation, which integrates and contacts with the underlying Honghuayuan Formation. Shizipu Formation, Baota Formation and Linxiang Formation are mainly gray to dark gray bioclastic limestone, bioclastic tortoise-cracked limestone, mixed with purple-red bioclastic limestone, unstable in distribution, and missing uplift in the middle of Guizhou; they are integrated and contacted with each other, and also integrated contact with the underlying Meitan Formation. The lithology of the Wufeng Formation is gray-black to black mud shale and siliceous mud shale, with abundant graptolites, stable distribution and good hydrocarbon generation potential; it is in integration and contact with the underlying Linxiang Formation. The lithology of Guanyinqiao Formation is dark gray bioclastic limestone and bioclastic marl, radiant coral, Hirnantianbei, and Dalmania are the "fashionable" biological assemblages in this formation. Besides, crinoids stem fragments can also be seen. The Northeastern part of the Sichuan Basin is missing, and it is an important marker layer for shale gas exploration; it is in contact with the underlying Wufeng Formation.

5. Silurian

The lower part of the Silurian is the Longmaxi Formation, which is an important interval for shale gas exploration; the lower part is dark gray to gray-black mud shale, rich in graptolites, 30–120 m thick, and the upper part is gray to dark gray silty mud shale mixed with siltstone or banded and lenticular limestone, the distribution is unstable; most areas are in contact with the underlying Guanyinqiao Formation, and the uplift near the middle of Guizhou is gradually in pseudo-conformity contact. The lithology of the Shiniulan Formation is dark gray, dark gray mud shale, silty sandy mudstone intercalated with thin bioclastic limestone, argillaceous siltstone, sandy marl, nodular marl and calcareous mudstone. Banded coral reef limestones are developed in the Southern Sichuan and Northern Guizhou areas. Fossils such as corals, sponges, nucleoids, brachiopods and trilobites are found, which are in contact with the underlying Longmaxi Formation. The Hanjiadian Formation is composed of gray-green, purple-red mud shale, silty mud shale and calcareous siltstone, with well-developed tidal bedding, prosperous benthic organisms such as brachiopods, gills and trilobites, with layered remains and burrows. Drilling holes and wave marks are abundant. It is in conformity with the underlying Shiniulan Formation, and in contact with the overlying Permian pseudo-conformity, and most areas lack the Middle and Upper Silurian deposits.

4.1.3.2 Stratigraphic Characteristics of the Ordovician Wufeng Formation-Silurian Longmaxi Formation

In the actual research process of shale gas development, we usually think that the black shale of the Wufeng Formation and the black shale of the Longmaxi Formation are the same. There is usually a shell layer "Guanyinqiao Formation" between the Longmaxi Formation and the Guanyinqiao Formation. The Guanyinqiao Formation is deposited between two sets of black shale, which are generally crustal marl, dolomitic limestone, claystone, etc., and its biological assemblage has obvious environmental indications. Small Dalmania (trilobites), Hirnantianbei fauna (brachiopods), corals and crinoids are commonly developed. The sedimentary thickness is usually only ten to tens of centimeters. Hirnantianbei fauna is in an ecological location. It is equivalent to benthic assemblage BA2-3, mainly BA3, and the water depth does not exceed 60 m (Fang et al. 1993; Rong 1984; Rong and Chen 1987). The lithofacies assemblages also indicate that the Guanyinqiao Formation is shallower than the deep-water shelf at the bottom of the Wufeng Formation and Longmaxi Formation (Chen 2018; Li et al. 2009a; Liu et al. 2012c; Mu et al. 2014; Wang 2016; Wang et al. 2015b, c; Su 2017). It is confirmed from the side that although the Wufeng Formation and the Longmaxi Formation are both black shale, their black shale formation mechanisms are different, Yan et al. (2009a, b, 2010a), Zhang et al. (2000), Zhou et al.(2015) and Zou et al. (2018) five peaks- Corresponding geochemical analysis of the surface profile samples of the Longmaxi Formation shows that the sedimentary water bodies in South China experienced a change process of anoxic-oxygen-enriched-anoxic from the end of the Late Ordovician to the beginning of the Early Silurian. The black pages of the Wufeng Formation and the Longmaxi Formation although the rocks were deposited in an anoxic water environment, their genetic mechanisms are different.

(1) Biostratigraphy

The black shale of Wufeng Formation-Longmaxi Formation spanned the Ordovician and Silurian ages. It spans the Ordovician Kaidi Stage, Hirnantian Stage, Ludan Stage,

Elonian Stage and Trechi Stage, and there are 13 corresponding graptolite belts in total (Chen et al. 2006c; Fan et al. 2012; Chen et al. 2015, 2017, 2018).

Among them, the global standard stratotype profile and point (GSSP) of the Henantian at the top of the Ordovician fall on the Wangjiawan profile in Yichang, Hubei-Nankinolithus band, Dicellograptus complexus band and Paraorthograptus pacificus band. The Paraorthograptus pacificus belt is further divided into three sub-zones, namely Lower Subzone, Tangyagraptus typicus and Diceragraptus mirus. The Hirnantianian contains two graptolithic belts. The bottom is the Normalograptus extraordinarius graptolithic belt, which corresponds to the top strata of the Wufeng Formation. The upper part is the Persculptograptus persculptus graptolithic belt corresponds to the black shale at the bottom of the Longmaxi Formation. The Rudanian contains four graptolite belts, from bottom to top, the Akidograptus ascensus belt, the Parakidograptus acuminatus belt, the Cystograptus vesiculosus belt and the Coronograptus cyphus belt. The Elonian includes three graptolite belts, from bottom to top, the Demirastrites triangulates belt, the Lituigraptus convolutes belt and the Stimulograptus sedgwickii belt. The Trechianian contains two graptolite belts, from the bottom to the top, that are the *Spirograptus guerichi* Zone and the *Spirograptus turriculatus* belt. According to the current data, in the black graptolite shale of the Longmaxi Formation in the Yangtze Platform, only the Spirograptus guerichi belt, the bottommost graptolite belt of the Trechian stage, has been found. In order to facilitate the division and comparison of the Ordovician and Silurian black shale underground formations, Academician Chen et al. (2015) used WF and LM to refer to the corresponding graptolitic belts and then used to divide and compare the two sets of the Wufeng Formation and the Longmaxi Formation. For black shale, for example, WF3 is called the "Wusan member", that is, the third graptolite zone of the Wufeng Formation; LM2 is called "Longer member", the second graptolite zone of Longmaxi Formation.

(2) Lithostratigraphy

From the end of the Late Ordovician to the beginning of the Early Silurian, there are three target layers in total. From bottom to top, they are Wufeng Formation, Guanyinqiao Formation and Longmaxi Formation. The characteristics of each group from bottom to top are briefly described as follows.

(1) Ordovician Wufeng Formation

The Wufeng Formation, which evolved from the Wufeng Shale named by Sun (1931), is characterized by black graptolite shale and siliceous rocks and is widely distributed in the Yangtze region. In Eastern Sichuan, its bottom boundary and the Linxiang Formation or the nodular limestone of Jiancaogou Formation are in conformity with contact, and the top boundary is in conformity with Guanyinqiao Formation, with a thickness of 3–30 m. It mainly develops graptolites and a few brachiopods, and radiolarians or siliceous sponge spicules develop in the siliceous shale. There are gray-white and brown-yellow bentonite layers distributed among them, with a single layer thickness ranging from 0.1 to 40 cm (Sichuan Provincial Bureau of Geology and Mineral Resources 1991).

(2) Ordovician Guanyinqiao Formation

The Guanyinqiao Formation was proposed by Sheng (1958) in the article "Ordovician strata on the border of Sichuan and Guizhou". In 1974, it was classified into the Silurian system and was renamed the Guanyinqiao Formation in the 1974 Handbook of Paleontological Fossils in Southwest China (edited by Nanjing Institute of Geology and Palaeontology) (Sichuan Provincial Bureau of Geology and Mineral Resources 1991). The Guanyinqiao Formation is a set of shell facies sediments, and its lithology is mainly composed of biological limestone, calcareous siltstone, mudstone or calcareous dolomite. The Guanyinqiao Formation is characterized by the abundance of Hirnantia-Dalmanitina *Fauna* (Rong and Chen 1987; Zhou et al. 1993; Zhan et al. 2010a), and the sedimentary lithology includes biological limestone, marl rock, sandy limestone, mudstone, etc.; the sedimentary thickness is relatively small, most of which are in the range of ten to several tens of centimeters, and the largest is only more than ten meters.

(3) Silurian Longmaxi Formation

The Longmaxi Formation evolved from the Longmaxi shale named by predecessors in Longmaxi, Zigui Xintan, Hubei Province. The lower part of the formation is in contact with the Guanyinqiao Formation and is in contact with the underlying Ordovician Guanyinqiao Formation. The overlying strata are the Silurian Luoreping Formation (formed with neritic sand and marl near the coast, mainly distributed in Northern and Northeastern Sichuan)/Shiniulan Formation (mainly limestone, mainly found in SouthWestern Sichuan) or Xiaoheba Formation (it is a neritic marl deposit, more common in Eastern Sichuan and Western Sichuan Basin). The Permian Liangshan Formation in the central Sichuan area overlies the Ordovician and Cambrian strata, respectively, and the Lower Silurian Wengxiang Group in the central Guizhou area directly overlies the Lower Ordovician Dawan Formation. According to its lithological and sedimentary characteristics, the Longmaxi Formation can be divided into upper and lower members, of which the lower member is basically the same as the Wufeng Formation. In the early stage (*Glyptograptus perscuptus was* brought to *Pristiograptus Leei* belt), it inherited the depositional characteristics of the five peaks, mainly black carbonaceous, siliceous shale and gray-black calcareous mudstone assemblages. The deposition rate was slow, and the deposition time exceeded the Longmaxi period. In the late Longmaxi (Demirastrites *triangulates* brought to the *Monograptus sedgwickii* belt), it is mainly gray-green in the upper Longmaxi Formation, yellow-green mudstone, silty mudstone and siltstone assemblages, usually sandwiched with marl lenses, are non-black rock series (Zhang et al. 2012). The color gradually deepens from top to bottom, the sandy quality decreases, and the organic matter content increases. The shale generally contains sandy quality, and silty quartz is locally enriched. The black organic-rich mud shale in the lower section is thick, usually ranging from 20 to 200 m, which is an important set of source rocks and a key interval for shale gas exploration in this area.

4.1.4 Geological Background of Ordovician Wufeng Formation-Silurian Longmaxi Formation in the Study Area

From the Sinian to the Early Ordovician, the entire Middle and Upper Yangtze region was in the background of extensional rifting, and a stable craton basin was formed inside the continental block (Huang et al. 2011; Huang et al. 2012c). Mu et al. (2011) pointed out that the Middle and Upper Yangtze region has generally experienced the evolution from a craton marine basin (Cambrian-Middle Ordovician) to a craton-based post-uplift basin (Ordovician-Early Silurian). During the Longmaxi Period of the Early Silurian, the margins of the Middle and Upper Yangtze Continent were in the process of compressional fold orogeny, the Western Sichuan-Central Yunnan Ancient Land and the Hannan Ancient Land expanded, the Central Sichuan Uplift continued to expand, and the Central Guizhou Uplift, Wuling Uplift, the Xuefeng uplift and the Miaoling uplift are basically connected to form the Yunnan-Guizhou-Guangxi uplift belt, forming a tectonic pattern of multiple uplifts and one bend, which transformed the sea area with the characteristics of broad sea in the early and middle Ordovician into a limited sea area bounded by the uplift. Continental shelf sedimentary system has the evolution process of upward and shallower (Su et al. 2007; Zeng et al. 2011; Liu et al. 2011; Huang et al. 2011), resulting in a large area of low-energy, undercompensated, anoxic depositional environment (Mu et al. 2011). The study area is located in the southern part of the Sichuan Basin, close to the Xuefeng Mountain-Central Guizhou paleo-uplift, and the Central Sichuan Uplift in the north, which is part of the post-uplift basin of the Yangtze Foreland Basin (Fig. 4.2).

The distribution and sedimentary evolution of the Wufeng Formation-Longmaxi Formation sedimentary system in the Sichuan Basin are mainly affected by the shifting and uplifting of the foreland uplifts including the Central Guizhou Palaeohigh (Guo et al. 2004). During the Late Ordovician-Early Silurian, a strong global sea-level rise occurred (Loydell 1998), and a large-scale transgression occurred, resulting in the formation of a large number of black organic-rich shale in North Africa and Arabia (Lüning et al., 2005). The Early Silurian Longmaxi period was the most intense period of compression in Southern China (Guo et al. 2004; Zhang et al. 2011), and at the same time affected by global transgression (Rong 1984), the Longmaxi Formation formed a set of distribution ranges wide and thick fine-grained clastic rocks dominated by black shale constitute a regional source rocks series (Guo et al. 2004; Zhang et al. 2013b). The Middle Ordovician–Silurian was in the main active period of the Caledonian movement, and the early ancient land and the paleo-uplift formed due to the Caledonian movement were the supply areas for terrigenous clastic provenance (Zeng et al. 2011), it is inferred that the terrigenous clastic materials mainly come from the foreland uplift belt including the Central Guizhou paleo-uplift and the Xuefeng paleo-uplift, and the Central Sichuan paleo-uplift may only provide dissolved materials (Guo et al. 2004), so the Early Silurian sedimentary filling has many mixed continental shelf sedimentary properties of the source (Guo et al. 2004; Zhang et al. 2013b).

4.1.5 Research Ideas and Methods

This chapter is guided by sedimentology, petrology, oil and gas geology and reservoir geology, on the basis of fully absorbing domestic and foreign research results, through detailed observation and description of outcrop profiles in the whole area, making full use of various macroscopic and microscopic analysis and geochemical methods for comprehensive research. Based on the results of various analyses and tests, the histogram, cross-section and plane distribution map of the organic matter content (TOC), thickness and maturity of organic matter (*R*o) and brittle mineral content in the study area were compiled. On this basis, the lithofacies and paleogeographic maps of the study area are compiled, and the distribution characteristics of lithofacies, mineral components and organic matter in the study area are clarified. Combined with the analysis of the reservoir space, the types and characteristics of diagenesis in the study area are given. According to the classification criteria, the

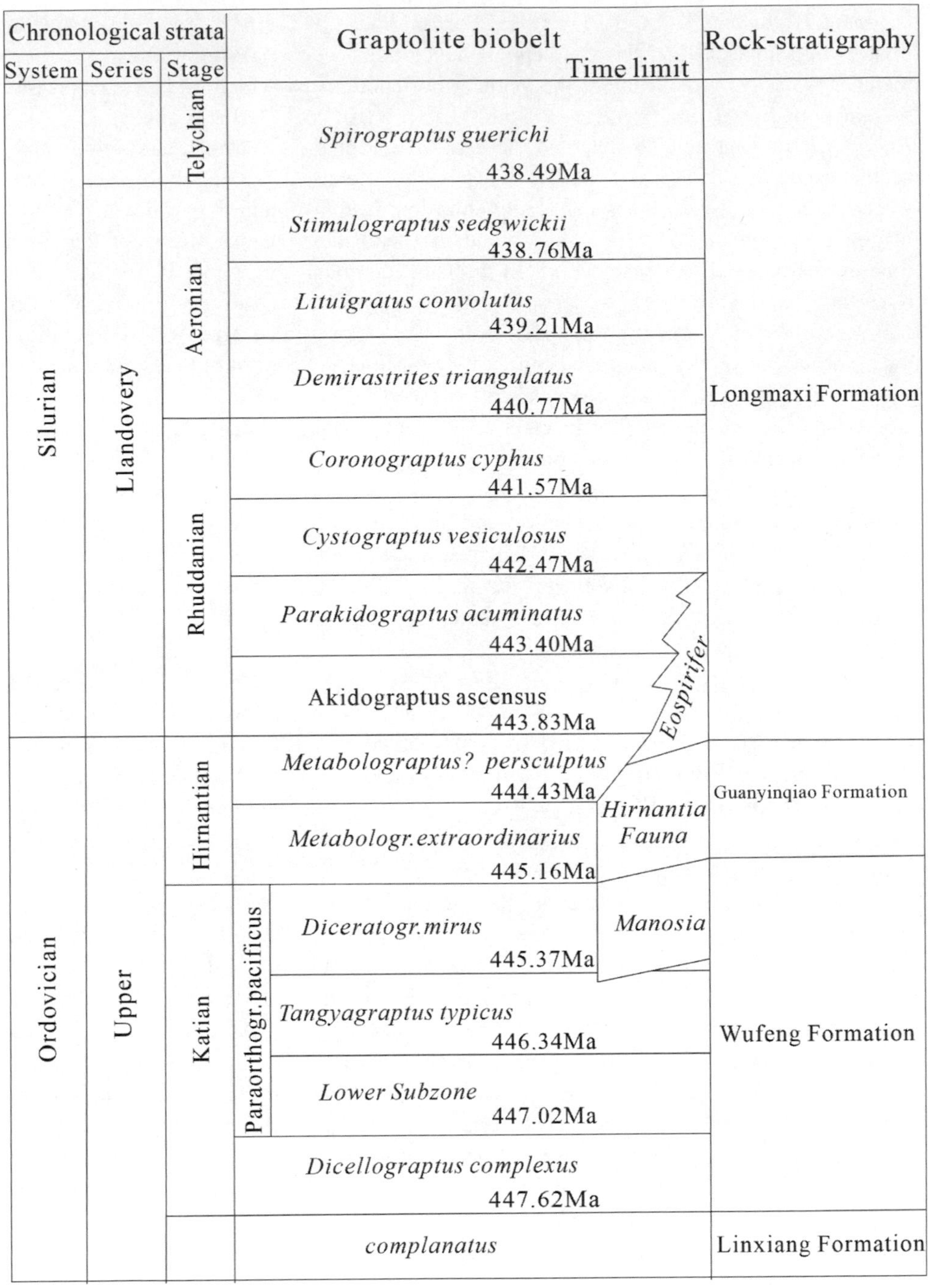

Fig. 4.2 Upper Ordovician-Lower Silurian graptolite zone (according to Chen et al. 2015)

Chronological strata: System	Series	Stage		Graptolite biobelt	Time limit		Rock-stratigraphy
Silurian		Telychian		*Spirograptus guerichi*	438.49Ma		Longmaxi Formation
Silurian	Llandovery	Aeronian		*Stimulograptus sedgwickii*	438.76Ma		Longmaxi Formation
Silurian	Llandovery	Aeronian		*Lituigratus convolutus*	439.21Ma		Longmaxi Formation
Silurian	Llandovery	Aeronian		*Demirastrites triangulatus*	440.77Ma		Longmaxi Formation
Silurian	Llandovery	Rhuddanian		*Coronograptus cyphus*	441.57Ma		Longmaxi Formation
Silurian	Llandovery	Rhuddanian		*Cystograptus vesiculosus*	442.47Ma		Longmaxi Formation
Silurian	Llandovery	Rhuddanian		*Parakidograptus acuminatus*	443.40Ma	*Eospirifer*	Longmaxi Formation
Silurian	Llandovery	Rhuddanian		Akidograptus ascensus	443.83Ma	*Eospirifer*	Longmaxi Formation
Ordovician	Upper	Hirnantian		*Metabolograptus? persculptus*	444.43Ma	*Eospirifer*	Guanyinqiao Formation
Ordovician	Upper	Hirnantian		*Metabologr.extraordinarius*	445.16Ma	*Hirnantia Fauna*	Guanyinqiao Formation
Ordovician	Upper	Katian	*Paraorthogr.pacificus*	*Diceratogr.mirus*	445.37Ma	*Manosia*	Wufeng Formation
Ordovician	Upper	Katian	*Paraorthogr.pacificus*	*Tangyagraptus typicus*	446.34Ma		Wufeng Formation
Ordovician	Upper	Katian	*Paraorthogr.pacificus*	*Lower Subzone*	447.02Ma		Wufeng Formation
Ordovician	Upper	Katian		*Dicellograptus complexus*	447.62Ma		Wufeng Formation
Ordovician	Upper			*complanatus*			Linxiang Formation

characteristics of diagenesis and pore evolution are determined. On the basis of the tectonic background, and finally all the above factors are integrated, the favorable distribution areas of shale gas in the study area are divided according to reasonable evaluation criteria. The technical route of the specific implementation is shown in Fig. 4.3.

The main research object in this book is the black organic-rich shale developed in the Wufeng Formation of the Upper Ordovician-Lower Silurian Longmaxi Formation, which is distributed in the Wufeng Formation and the lower member of the Longmaxi Formation. Based on the distribution and development of the Ordovician Wufeng Formation and the Silurian Longmaxi Formation in Southern Sichuan and adjacent areas, a total of 33 outcrop profiles covering the whole area were selected for observation (Fig. 4.4). Data from Wells Changxin 1, Yuye 1, Jiaoye 1 and some old wells and literature data are used to study the sedimentation-diagenesis of the black rock series of the Longmaxi Formation and their controlled on shale gas. The samples used were collected from 33 outcrop profiles and

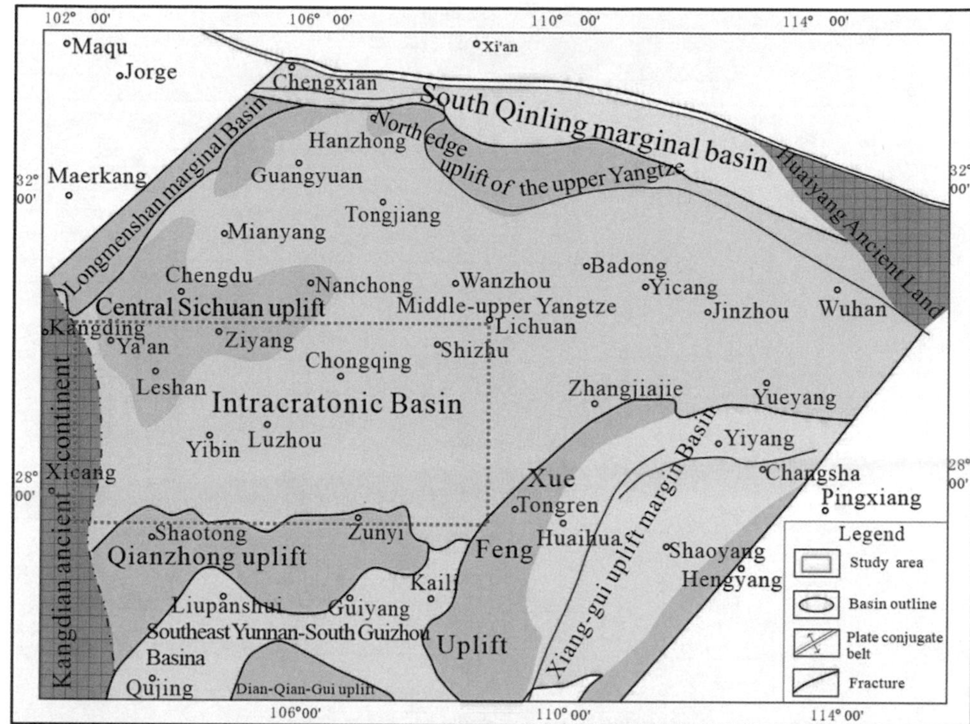

Fig. 4.3 Tectonic outline and basin distribution of Early Ordovician–Silurian in Southern Sichuan Basin and its periphery [modified from Xu et al. 2011 (internal data)]

Well Daoye 1 covering almost the whole area. The sampling density was high. The observation profiles were sampled at 1 m intervals, and the 7 measured profiles were at 0.5 m intervals. A total of 845 samples were collected. In order to meet the requirements of various analysis, the samples sampled in the field weigh about 1 kg. Select fresh block samples to grind thin slices and probe pieces, respectively, for petrological and mineralogy identification; and select fresh samples to grind to a particle size of less than 0.2 mm for the analysis of total organic carbon; under pollution-free conditions, the fresh samples were crushed into 300 mesh powder for X-ray diffraction analysis.

The identification of common flakes was done with Zeiss Axio Scope A1; Electron backscatter diffraction and scanning electron microscope analysis were done on Hitachi S-4800 scanning electron microscope equipped with electron backscatter diffraction system (EBSD), and the test standard was GB/T 17359–1998 "General Principles for Quantitative Analysis of Electron Probe and Scanning Electron Microscope X-ray Spectroscopy", the laboratory temperature is 20 °C and the humidity is 40% RH; the electron probe analysis is using Shimadzu EMPA-1600 electron probe, with GB/T17359–1998 is the test standard. The laboratory temperature is 20 °C, and the humidity is 58% RH. The above analysis was completed in the Key Laboratory of Sedimentary Basin and Oil and Gas Resources of the Ministry of Land and Resources (Chengdu, China).

The argon ion profile-scanning electron microscope analysis of the samples was completed in the Experimental Research Center of Wuxi Petroleum Geology Research Institute, Sinopec Petroleum Exploration and Development Research Institute and SY/T 5162–1997 (2005) "Analysis Method of Rock Samples by Scanning Electron Microscope", the experimental instrument is XL 30 scanning electron microscope, the experimental temperature is 25 °C, and the humidity is 50% RH. The X-ray diffraction analysis results of the whole-rock components and clay mineral components of the samples were obtained by using a D8 DISCOVER X-ray diffractometer at a laboratory temperature of 22 °C and a humidity of 30% RH. The mineral components were quantified. The detection basis of the analysis is SY/T 5163–2010 "X-ray Diffraction Analysis Method of Clay Minerals and Common Non-Clay Minerals in Sedimentary Rocks"; organic matter analysis includes determination of total organic carbon content, kerogen microscopy analysis, vitrinite reflectance determination, Kerogen carbon isotope analysis and rock pyrolysis analysis; the determination of total organic carbon was obtained by the American Leco CS-200 carbon and sulfur analyzer, and the laboratory temperature was 22 °C; the humidity was 30% RH.GB/T19145–2003 "Determination of Total Organic Carbon in Sedimentary Rocks" standard; kerogen microscopic analysis is carried out by biological microscope (Axioskop 2 plus), and the determination standard is SY/T 5125–1996 "Transmitted Light-Fluorescence Kerogen Microscopy" Component identification and type classification method"; vitrinite reflectance according to SY/T 5124–1995 "Determination of vitrinite emissivity in sedimentary rocks", using a microphotometer (MPV-SP), the indoor temperature is 22 °C, under the condition of humidity of

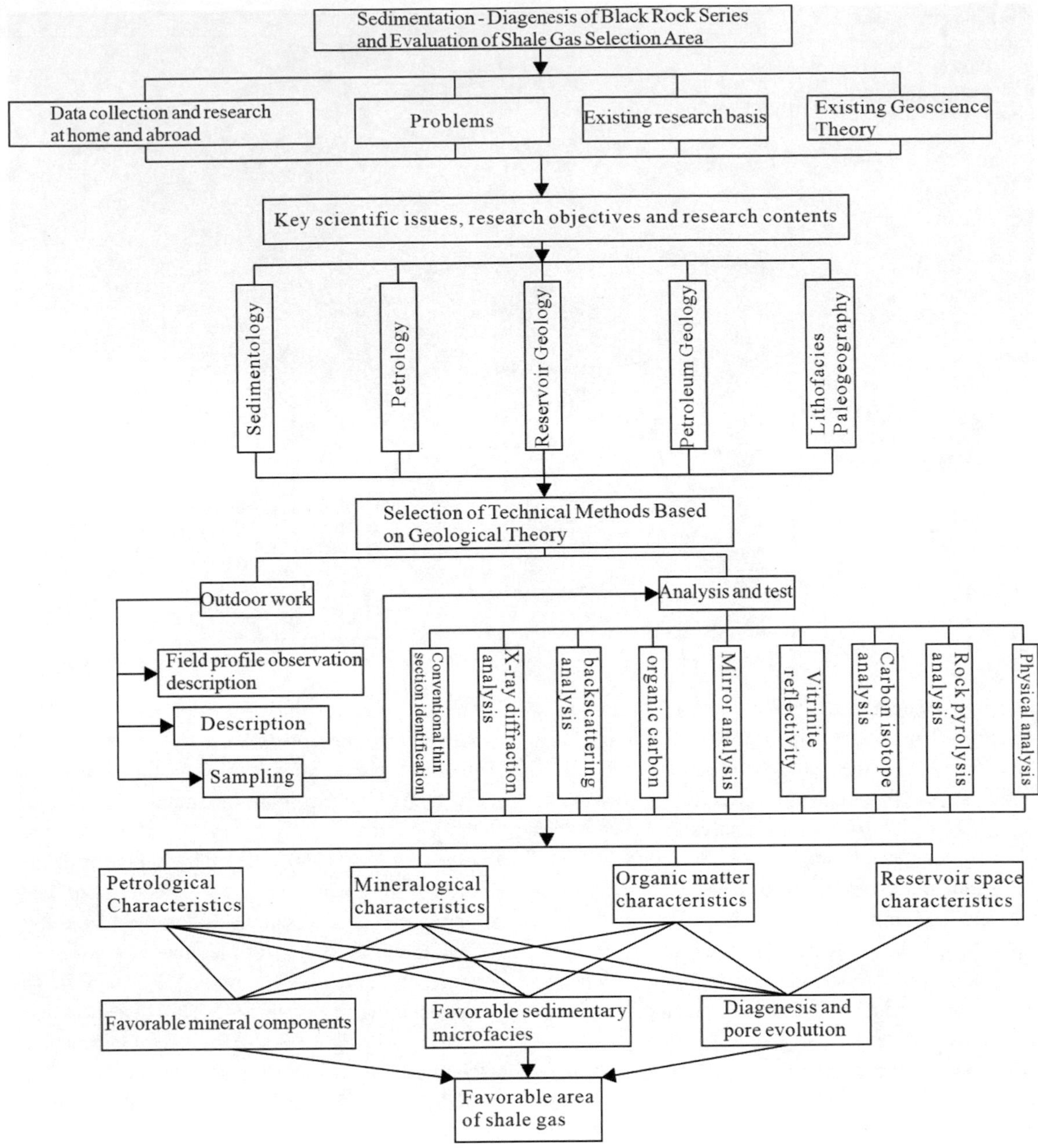

Fig. 4.4 Technical route of this book

45% RH; the carbon isotope of kerogen uses Finngan MAT-252 isotope mass spectrometer, based on SY/T5238–2008 "Analysis Method of Carbon and Oxygen Isotopes of Organic Matter and Carbonate", at 22 °C, 45 completed in the experimental environment of 45% RH; rock pyrolysis analysis according to GB/T 18602–2001 "Rock pyrolysis analysis" standard, under the condition of indoor temperature of 24 °C, using oil and gas display evaluation instrument; physical property analysis includes mercury intrusion analysis and core. For routine physical property analysis, the detection conditions of mercury intrusion analysis are that the maximum pressure is 50 MPa, and the detection equipment is 9510-IV type mercury intrusion instrument. The room temperature is 10 °C, and the humidity is under the conditions of 22% RH and atmospheric pressure of 1037 HPa; it was completed according to the standard of SY/T 5346–2005 "Determination of Capillary Pressure Curve by Mercury Intrusion Method"; the conventional physical property analysis of the core was based on SY/T 5336–2006 "Core Analysis Method", and the indoor environment was

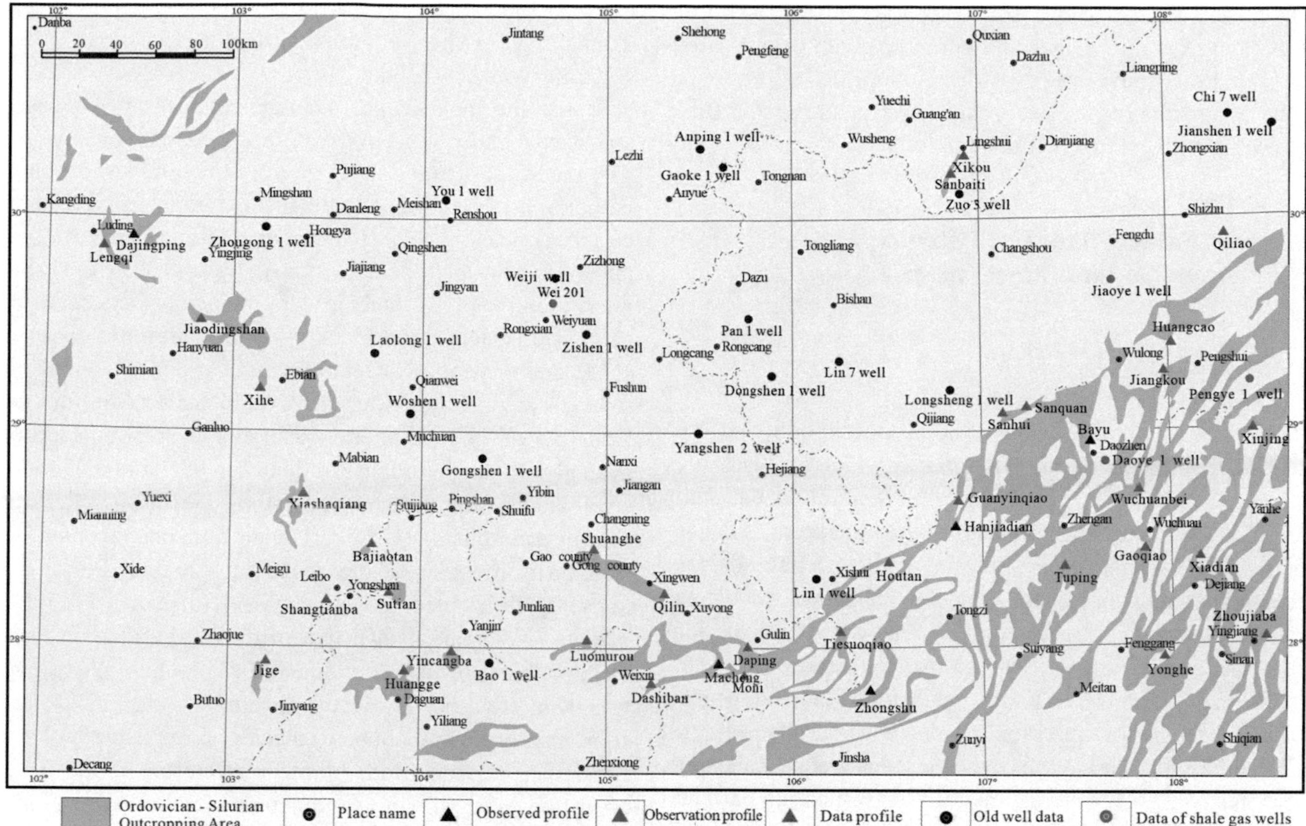

Fig. 4.5 Location of outcrop and wells sites of Wufeng-Longmaxi Formation in Southern Sichuan Basin and its periphery. Sanbaiti of Huaying, Sichuan (SBTP); Xikou of Huaying, Sichuan (XKP); Daping of Gulin, Sichuan (GDP); Dajingping of Tianquan, Sichuan (DJP); Jiading Mountain of Hanyuan, Sichuan (JDSP); Xiashaqiang of Mabian, Sichuan (XSQP); Xihe of E'bian, Sichuan (XHP); Luomurou of Junlian, Sichuan (LMRP); Suoqiao of Gulin, Sichuan (GTP); Qilin of Xingwen, Sichuan (XQP); Shuanghe of Changning, Sichuan (CSP); Lengji of Luding, Sichuan (LQP); Moni of Xuyong, Sichuan (MNP); Macheng of Xuyong, Sichuan (XMP); Shangtianba of Leibo, Sichuan (STBP); Yonghe Dongkala of Fenggang, Guizhou (FDP); Zhongshu of Renhuai, Guizhou (RZP); Hanjiadian of Tongzi, Guizhou (HJDP); Bayu of Daozhen, Guizhou (DZP); Houtan of Xishuiliangcun, Guizhou (XLP); Tuping of Zhengan, Guizhou (ZTP); Jige of Jinyang, Yunnan (JGP); Sutian of Yongshan, Yunnan (STP); Yinchangba of Yanjin, Yunnan (YYP); Huanggexi of Daguan, Yunnan (DHP); Dashipan of Weixin, Yunnan (WDP); Bajiaotian of Yongshan, Yunnan (YBP); Xinjing of Yanhe, Chongqiang (XJP); Qiliao of Shizhu, Chongqing (QLP); Guanyinqiao of Qijiang, Chongqing (QGP); Huangcao of Wulong, Chongqing (HCP); Jiangkou of Wulong, Chongqing (JKP); North Wuchuan, Chongqing (CWP); Gaoqiao of Wuchuan, Chongqing (GQP); Sanquan of Dejiang, Chongqing (SQP); Xiadian of Dejiang, Chongqing (XDP); Zhoujiaba of Yinjiang, Chongqing (YZP); Zhougong 1 Wwell (ZG1); You 1 Wwell (Y1) Weiji Well (WJ) Zishen 1 well (ZS1) Laolong 1 well (LL1); Woshen 1 well (WS1) Gongshen 1 well (GS1) Yangshen 2 well (YS2) Anping 1 well (AP1) Gaoke 1 well (GK1) Pan 1 well (P1) Zuo 3 well (Z3) Lin 7 well (L7) Dongshen 1 well (DS1) Lin 1 well (L1) Chi 7 well (CH1) Daoye 1 well (DY1) Jianshen 1 well (JS1) Peng Page 1 Wwell (PY1); Longsheng 1 well (ZG1) Bao 1 well (B1) Wei 201 well (W201) Jiaoye 1 well (JY1)

20 °C, 40% RH and 1025 HPa atmospheric pressure, the detection conditions are porosity measurement reference chamber pressure is 0.07 Mpa, analysis and testing instruments are ULTRAPOE-200A helium porosimeter and ULTRA-PERMTM200 permeability analyzer; mineral composition analysis, organic matter analysis and physical property analysis were carried out in the Sedimentation Laboratory of PetroChina Huabei Oilfield Exploration and Development Research Institute (Fig. 4.5).

4.2 Sedimentary Characteristics and Lithofacies Paleogeography

Since "shale" in shale gas is often used as a geological formation term, a non-lithologic term (Bust et al. 2013), shale gas formations are actually mainly composed of fine-grained sedimentary rocks, rather than "shale" in the traditional sense. "shale". Swanson (1961) pointed out that black shale (Black Shale) is an organic-rich black mudstone containing organic matter and silt–clay-grade debris. However, for the convenience of description, according to the current research habits, fine-grained sedimentary rocks (mud shale) are still briefly described as "shale".

4.2.1 Characteristics of Petrology and Sedimentary Facies

1. Ordovician Wufeng Formation

According to the field observation description and thin section identification, the main rock types of the black rock series of the Ordovician Wufeng Formation in the study areas are carbonaceous mudstone, carbonaceous siliceous mudstone, carbonaceous siltstone, carbonaceous siliceous calcareous mudstone, and siliceous mudstone. Mudstone, calcium-bearing carbonaceous silty mudstone, carbonaceous silty mudstone, carbonaceous mudstone, carbonaceous silty mudstone, carbonaceous silty mudstone, calcium-bearing carbonaceous silty mudstone, calcareous argillaceous silt sandstone, silt-bearing carbonaceous calcareous mudstone, carbonaceous silty-bearing micrite, calcic-bearing carbonaceous argillaceous siltstone, carbonaceous calcareous mudstone, cloud-bearing silt-bearing carbonaceous mudstone, cloud-bearing siliceous mudstone, dolomitic mudstone and marl. Affected by sedimentary differentiation, there is a certain distribution regularity in vertical and plane. Carbonaceous mudstone and siliceous mudstone are mostly distributed in Luding-Hanyuan in Western Sichuan, Yibin-Luzhou in Southern Sichuan and Wulong-Shizhu in Chongqing. Correspondingly, with the increase of carbonate and terrigenous clastic minerals, carbon-bearing silt-bearing calcareous mudstone and calcium-bearing carbonaceous silty mudstone are mostly distributed in the margins of Central Sichuan Uplift, Western Sichuan Uplift-Central Yunnan Uplift-Central Guizhou Uplift which are E'bian-Yanjin-Weixin-Renhuai-Zheng'an-Tongzi areas.

Under the microscope, the mudstone has two types of sedimentary structures: massive and laminar, mostly with silty muddy structure or silty muddy structure, and its clastic particles are floating in the mud; the amount of terrigenous clastic particles mostly 10–50%, including quartz, feldspar, mica, etc., the particle size is usually less than 0.05 mm; interstitial materials include clay minerals, siliceous and carbonate cements, and black organic matter in a dispersed state, the content of which is usually > 50%, authigenic minerals are mainly pyrite, quartz particles account for more than 75% of the total debris, clean and bright, good sorting, average rounding, particle size between 0.02 and 0.05 mm, a small amount can reach 0.1 mm, the edges of the grains are mostly metasomatized by carbonate minerals, and the dissolution is irregular in the shape of a harbor. Feldspar is unevenly distributed, its content is less than 8%, and its particle size is larger than that of quartz. In addition to the visible albite polycrystal twins and microcline feldspar lattice twins, feldspar is mostly covered by clay minerals or some carbonate minerals. Replacing, irregular dissolution edges can be seen. Both biotite and muscovite are developed, the latter are more numerous, in the shape of thin strips, with certain compaction deformation, biotite often corrodes into chlorite and exhibits a certain water swelling along the cleavage crack. Pyrite is an authigenic mineral formed in the early stage of sediment formation with low content, usually less than 5%, but it is ubiquitous, often with euhedral or semi-eumorphic crystals, or strawberrylike aggregates, and its intercrystalline pores are developed.

By analyzing the microscopic characteristics of Wufeng Formation lithology in each section and drilling, the sedimentary facies of Wufeng Formation are divided into several microfacies according to different lithologic combinations (Table 4.2).

Table 4.2 Sedimentary lithofacies division of Wufeng Formation organic-rich shale in Southern Sichuan and adjacent areas

Target layer system	Sedimentary microfacies	Sedimentary subfacies	Sedimentary facies
Wufeng Fm	Olomitic claystone + Marl	Intertidal zone	Tidal
	Carbonaceous silty micrite limestone + Carbonaceous silty micrite limestone		
Wufeng Fm	Silt-bearing Carbonaceous Mudstone + Carbon-bearing siltstone	Shallow shelf	Neritic shelf
	Silt-bearing Calcareous Mudstone + Carbon-bearing Silty Mudstone		
	Carbargilite, Siliceous claystone	Deep shelf	
	Calcareous carbonaceous silty mudstone + Silty carbonaceous calcareous mudstone		
	Carbonaceous siliceous mudstone + Dolomitic siliceous mudstone		

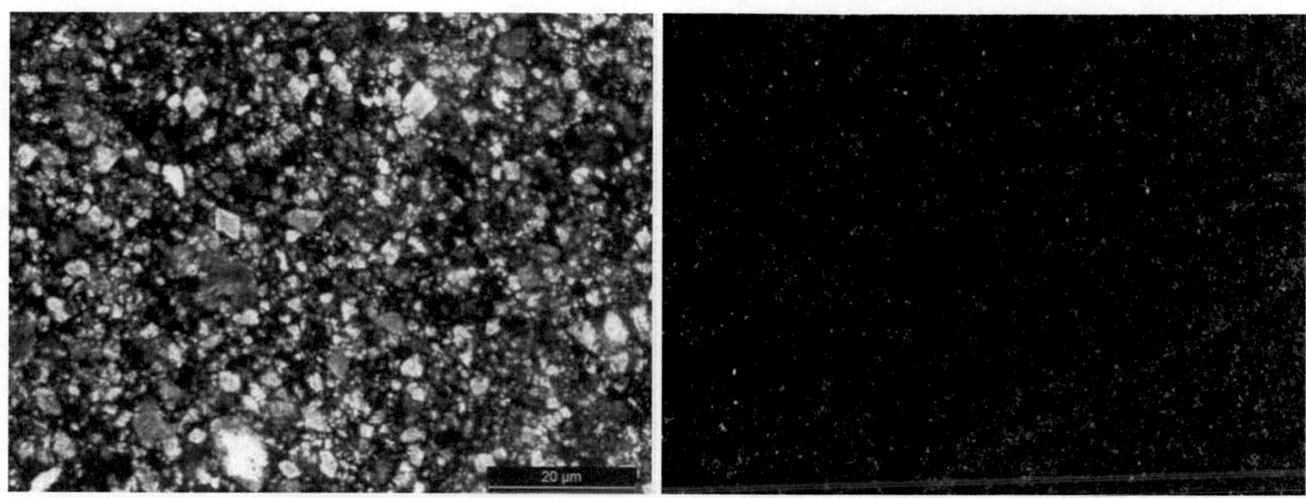

Fig. 4.6 Carbon-bearing silt-bearing micrite (A, 10 × 10(+)) and calcium-bearing silt mudstone (B, 10 × 10(+)) in each Wufeng Formation in Jige of Jinyang

(1) Tidal flat facies

The tidal flat facies are developed in the limited or gentle coastal zone of the nearshore zone composed of fine clastic materials (clay, silt) and carbonate rocks, which are dominated by tidal action without strong wave action, with extremely gentle slope. The tidal flat facies of the Ordovician Wufeng Formation are divided into dolomitic mudstone + marl and carbonaceous silt-bearing micrite + calcareous siltstone microfacies according to different sedimentary lithologies.

(i) Dolomite mudstone + marl

It is mainly located below the average low tide level and has been subjected to hydrodynamic effects such as tides for a long time. It is a mixed deposition of terrigenous fine clastic rocks and carbonate rocks. Often it developed around the Central Sichuan Uplift, dolomitic mudstone is mainly composed of mudstone with a particle size of less than 0.005 mm, mainly carbonate cementation, and the cement is mainly dolomite, with a diamond-shaped structure, and is filled between muddy grains; Marlstone mainly composed of mud-powder crystal calcite, and the crystal is brighter, mainly in other forms or semi-hedrons, and it can be seen that local recrystallization into fine-grained calcite particles.

(ii) Carbon-bearing silt-bearing micrite + calcareous siltstone

Mainly developed in the Western Sichuan-Central Yunnan Uplift and the Northern margin of the Central Guizhou Uplift, it is characterized by the mixed deposition of terrigenous fine clastic rocks and carbonate rocks or calcium-bearing silty mudstone, etc. (Fig. 4.6), carbon-bearing silt-bearing micrite limestone is mainly composed of micrite calcite. About 75%, quartz particle size is 0.01–0.03 mm, sub-angular-sub-circular, first-class gray-white interference color, contains a small amount of feldspar, the size of feldspar particles is roughly the same as that of quartz particles, and the feldspar is mainly polycrystalline twin crystals The plagioclase feldspar and the microplagioclase with lattice twin crystals are mostly replaced by clay minerals or calcite at the edge of the feldspar, and the organic carbon content is mostly between 0.5 and 1. The calcareous siltstone rocks are mainly composed of calcite-cemented clastic particles, and the overall cementation is basal; a large number of silt clasts and a small amount of mica clasts are distributed in it, mainly quartz silt, which are well-sorted and rounded poor, sub-angular-sub-round, the siliceous part is cryptocrystalline-granular, and part is accompanied by mud; forming a siliceous muddy mass contains a small amount of feldspar, with plagioclase and polycrystalline twin crystals. There are argillized potassium feldspar, etc., and trace mica fragments are needle-like and leaf-like, with local directional arrangement; calcareous cementation, distributed in irregular contiguous sheets, mostly metasomatic quartz, feldspar and other debris; with dissolution fill the cracks with calcareous filling, see a small amount of asphaltene filling the primary and dissolution pores unevenly, and the organic carbon content is < 0.5 or none.

(2) Shelf facies

The shelf facies range from sea level to the slope break of the continental slope with a water depth of about 200 m. It is widely distributed in the study area. It is characterized by abundant sunlight and prosperous organisms. The sediments are mainly terrigenous fine clastic materials and biochemical sedimentary materials, rich in biological remains; generally dark mud shale mixed with silt fine sandstone, with a small amount of marl in part, and often mixed with iron colloid deposits; the sedimentary structure has oblique bedding and erosion, bioclasts, etc. Traces of vigorous movement of seawater, and periodic and variable sedimentary layers. According to the sedimentary filling sequence and sedimentary structure characteristics, which can be divided into shallow shelf and deep-water shelf. This study takes the storm wave base as the boundary.

(i) Shallow shelf facies

The shallow shelf facies in the study area refer to the shallow-water area below the low tide surface that borders the tidal flat on the continental side and above the storm wave base and locally developed storm deposits. During the depositional period of the Wufeng Formation, the shallow shelf facies were mainly composed of dark gray-gray-black silt-bearing calcareous mudstone, carbon-bearing silty sandy mudstone, cloud-bearing silt-bearing carbonaceous mudstone, and carbon-bearing siltstone.

a. Cloud-bearing silt carbonaceous mudstone + carbon-bearing siltstone

The cloud-bearing silt-bearing carbonaceous mudstone + carbon- bearing siltstone is located in the shallow shelf area close to the tidal flat side and is intermittently affected by waves and wave backflow, and the hydrodynamic force is relatively strong. The cloud-bearing silt-bearing carbonaceous mudstone has a silt-bearing argillaceous structure, mainly composed of clay minerals and quartz particles. The clay minerals are mostly microscopic scale-like directional distribution; about 0.01–0.05 mm, good sorting, rounded and sub-angular-sub-circular; with a small amount of feldspar, mostly plagioclase, see polyplate twin crystal structure, some feldspar surfaces are metasomatized into clay minerals; mostly calcareous cementation, mainly dolomite, diamond-shaped, a small amount of calcite particles, calcite particles are mostly distributed in clay minerals or between the particles, the organic carbon content is high, more than 2%. The carbonaceous siltstone rock has a silt structure, and the clastic components are mainly composed of quartz and feldspar. The quartz particle size is about 0.01–0.05 mm, and the largest particle is about 0.2 mm. It is well-sorted and is sub-angular-sub-circular. There is a small amount of feldspar, and the feldspar is mostly plagioclase; the mica content is less, mostly needle-like, and the arrangement is directional, and the pyrite particles are black quadrangular or semi-self-shaped distributed among the particles; the cements are mainly calcite and dolomite. The calcite cements are mostly calcite microcrystalline particles, most of which are less than 0.01 mm in size, and some calcite particles replace quartz particles. Part of the dolomite has a good degree of self-shape, most of which are rhombic, and the particle size is mostly less than 0.05 mm; the organic carbon content is high, mostly 0.5–2%, and it is filled between clay minerals.

b. Carbon-bearing silt-bearing calcareous mudstone + carbon-bearing silty sandy mudstone

Carbon-bearing silt-bearing calcareous mudstone + carbon-bearing silty sandy mudstone are mainly distributed in the Southeastern margin of the Sichuan Basin and the north side of the Central Guizhou Uplift (Fig. 4.7). The carbonaceous silt-bearing calcareous mudstone has a silt-bearing and argillaceous structure as a whole, mainly composed of clay

Fig. 4.7 Carbon-bearing silty sandy mudstone in the Wufeng Formation of Xiadian, Dejiang

minerals. The microscopic scale-like clay minerals are arranged in a directional arrangement to form a directional structure, and the carbonaceous powder aggregates are rendered black along the layers strip-shaped; clastic particles are mainly composed of quartz and feldspar, the size of quartz particles is about 0.01–0.05 mm, with silt structure, it is in other shape, sub-angular-sub-circular; contains a small amount of feldspar, with agglomerated double crystalline plagioclase and argillized potassium feldspar, etc., are sub-angular-sub-circular, and trace mica fragments are arranged in needle-like and leaf-like directions; calcareous cementation, the calcium content is mainly calcite, and a small amount of dolomite, the particle size is from microcrystalline to fine crystal, some are continuous or intermittent stripes, some are agglomerates, some are single crystal scattered, mostly metasomatic feldspar and other debris, the organic carbon content is 0.5–2%, mostly filled in microcracks or between particles. The carbonaceous silty sandy mudstone has a silty muddy structure as a whole, mainly composed of clay minerals with a particle size of < 0.005 mm and silt-grade debris. The texture is rendered as black strips along the layer, and the content is between 0.5 and 2%; the clastic particles are mainly composed of quartz and feldspar, and the quartz particle size is about 0.01–0.05 mm, with a silt structure, and the content is about 15%. It's in other shape, sub-angular-sub-circular; the feldspar content accounts for about 4%, its particle size is roughly the same as that of quartz particles, with a silt structure, including plagioclase with polylamellar twin crystals and argillized potassium feldspar, etc., which are sub-angular-sub-circular; The cement is cemented with calcium mainly, the calcium component is mainly dolomite, and the dolomite is filled between clay minerals or quartz particles in the form of euhedral microfine crystals.

(ii) Deep-water shelf facies

The deep-water shelf facies is located in the deep-water area outside the shallow shelf (below the storm wave base), where the wave action is reduced and still water deposition is dominant. The depositional period of the Wufeng Formation is divided into three microfacies according to the different lithological and sedimentary assemblages, namely calcic-carbonaceous silty mudstone + silt-bearing carbonaceous calcareous mudstone shed, carbonaceous mudstone + siliceous mudstone shed and carbon-siliceous mudstone + cloud-bearing siliceous mudstone shed.

a. Calcium-bearing carbonaceous silty mudstone + silt-bearing carbonaceous calcareous mudstone

Calcium-bearing carbonaceous silt mudstone + silt-bearing carbonaceous calcareous mudstone are located in the sea area with low body energy, the hydrodynamic conditions are relatively weak and are basically not affected by ocean currents and storms. Their depositional characteristics are shown in Fig. 4.8. The calcium-bearing carbonaceous silty mudstone has a silt-muddy structure as a whole, the clastic particles are mainly composed of quartz and feldspar, the quartz particle size is about 0.01–0.05 mm, and the content is about 26%, sub-angular–sub-circular; contains a small amount of feldspar, including plagioclase with polylamellar twin crystals and argillized potassium feldspar, etc., which are sub-angular-sub-circular, and trace mica fragments are needle-shaped and leaf-shaped oriented arrangement; calcareous cements are calcite and dolomite, the particle size is from microcrystalline to fine grain, some are continuous or intermittent stripes, some are lumps, some are single crystal scattered, mostly metasomatic feldspar and other debris; microscopic scale-like clay

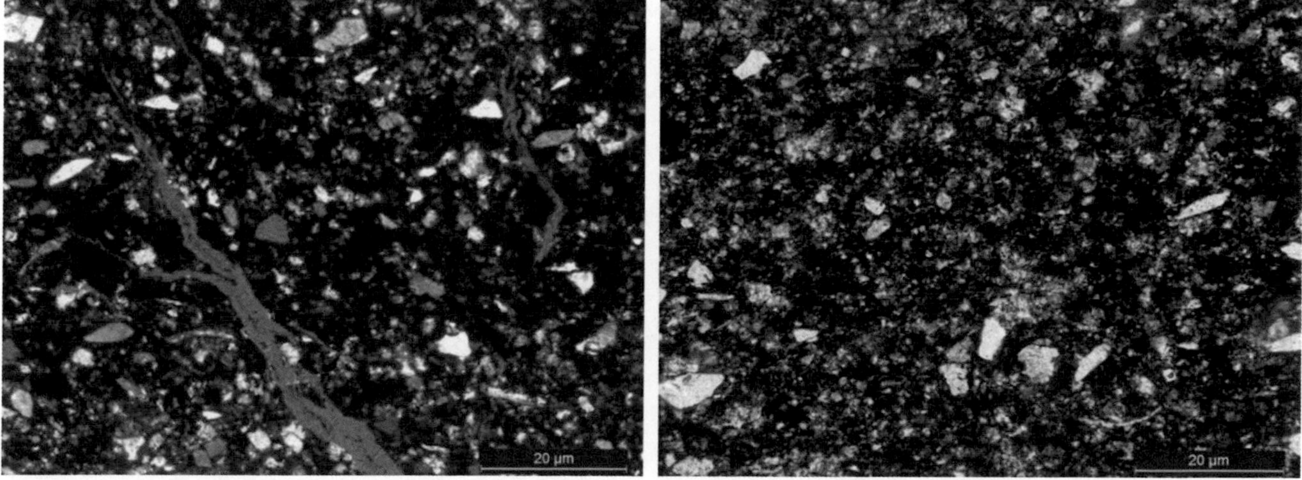

Fig. 4.8 Silt-bearing carbonaceous calcareous mudstone (left) and calcareous carbonaceous silty mudstone (right) in the Wufeng Formation of Luomu, Junlian

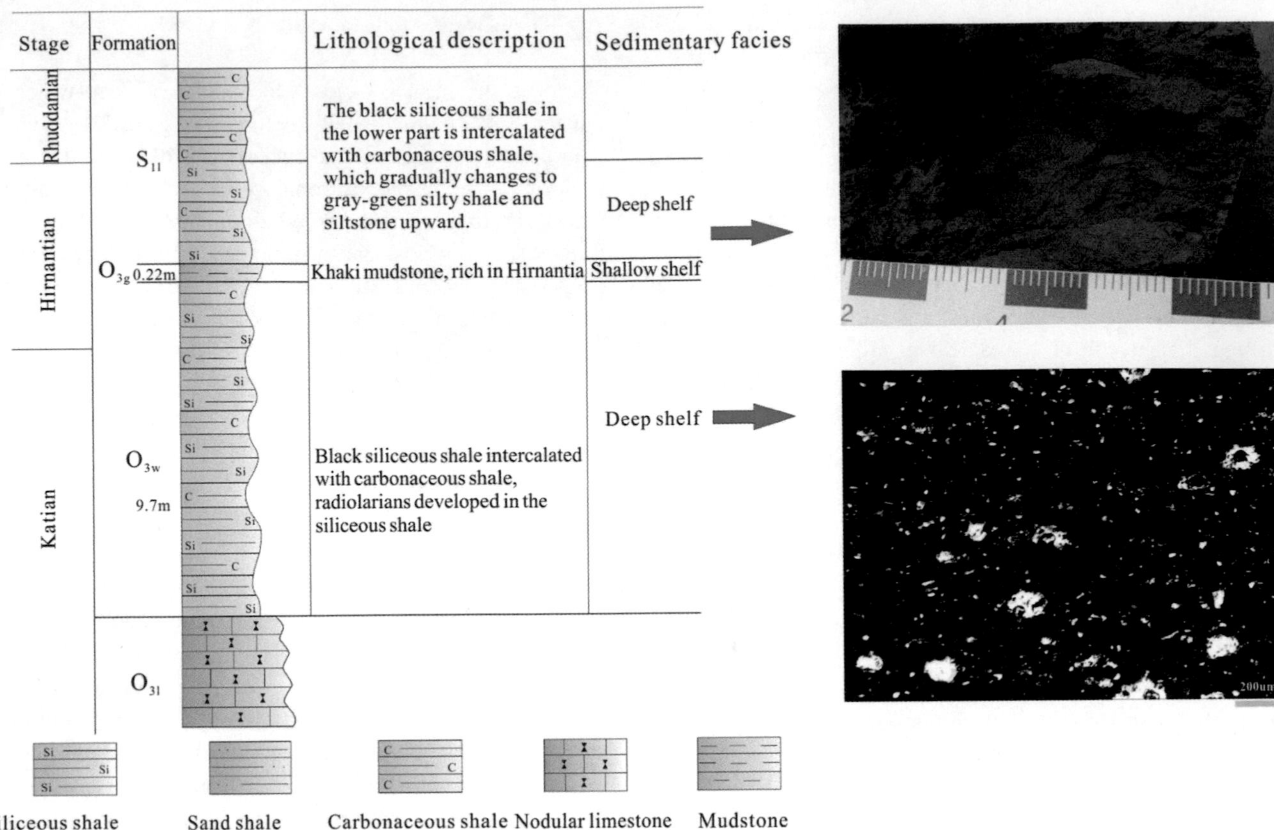

Fig. 4.9 Carbonaceous and siliceous mudstone of Wufeng Formation in Huangcao Wulong

minerals are arranged in a directional arrangement to form a directional structure, and the carbonaceous powder aggregates are rendered in black stripes along the layer, and their organic carbonaceous content is greater than 2%. The silt-bearing carbonaceous calcareous mudstone rock has a silt-bearing argillaceous structure as a whole, mainly composed of clay minerals. The microscopic scale-like clay minerals are arranged in a directional arrangement to form a directional structure, and the powdery carbonaceous aggregates are black stripes. Filled in microcracks, the content is more than 2%; the clastic particles are mainly composed of quartz and feldspar, the size of quartz particles is about 0.01–0.05 mm, with a silt structure, in other shapes, sub-angular-sub-circular contains a small amount feldspar, including plagioclase with polycrystalline twin crystals and argillized potassium feldspar, etc., which are sub-angular-sub-circular, and trace mica fragments are arranged in needle-like and leaf-like orientations. Calcium cementation, calcite, is mainly composed of calcite, a small amount of dolomite, particle size microcrystalline-fine-grained, more metasomatic feldspar and other particles.

Carbonaceous mudstone + siliceous mudstone are developed in sea areas with weak hydrodynamic conditions and are basically unaffected by currents and storms, and nodular and infested pyrites are more developed. Its deposition characteristics are shown in Fig. 4.9.

The carbonaceous mudstone is mainly composed of microscopic scale-like hydromica clay minerals and organic matter. The organic carbon content is greater than 2%. A small amount of silt debris and trace mica debris are distributed in it, mainly quartz silt, which is cryptocrystalline granular, and part of it's accompanied by mud to form siliceous muddy agglomerates and contains a small amount of feldspar; there are plagioclase with polylamellar twin crystals and argillized potassium feldspar, etc., which are sub-angular-sub-circular, and trace mica fragments are arranged in needle-like and leaf-like orientation; a small amount of carbonate cementation is calcite and dolomite, and the particle size is from microcrystalline to fine crystal, mainly in the scattered distribution of single crystal, and mostly metasomatic feldspar and other debris. Fractures are relatively developed, have certain directionality, and are mostly filled with siliceous and organic matter. The siliceous mudstone has an argillaceous structure as a whole and contains a small amount of quartz and feldspar, and the quartz has a silt structure. The mica fragments are arranged in needle-like and leaf-like directions; siliceous radiolarians are developed in the rocks which are elliptical-circular, with a content of about 20%, and some siliceous radiolarians are dissolved and filled with organic matter; the calcareous cement is mainly calcite. It is composed of dolomite, mainly

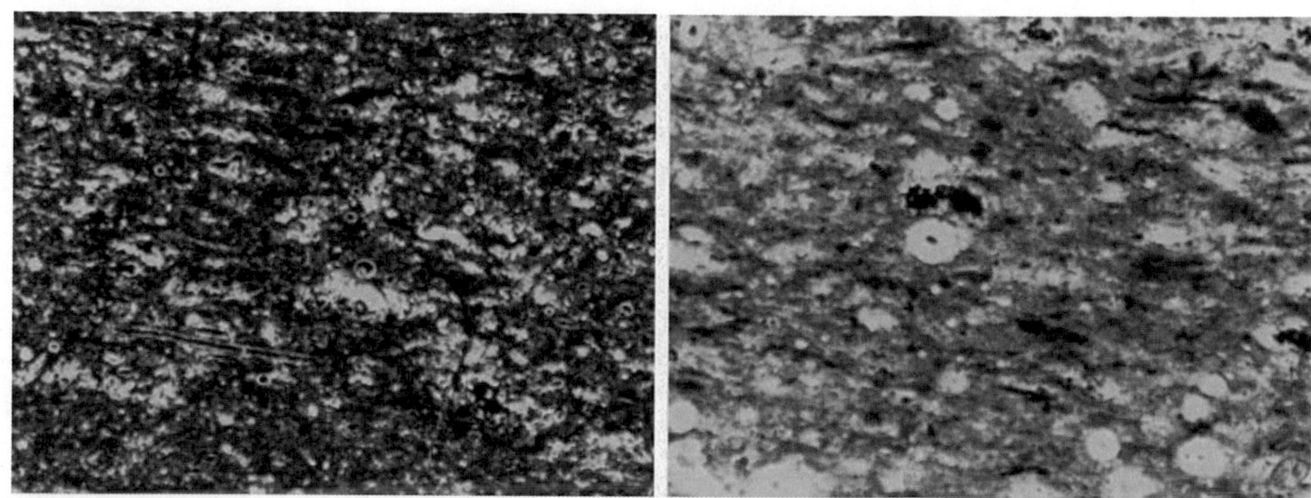

Fig. 4.10 Dolomitic siliceous mudstone (A, 10 × 10(−)) and carbonaceous siliceous mudstone (B, 10 × 10 (−)) in the Wufeng Formation of Dajingping, Tianquan

calcite, and the calcite is filled between the edge of the grain and its pores in a microcrystalline-fine crystal form; the microscopic scale-like clay minerals are arranged in a directional arrangement to form a directional structure.

c. Carbonaceous siliceous mudstone + cloudy siliceous mudstone

Carbonaceous siliceous mudstone + cloud-bearing siliceous mudstone are mainly developed in the areas of Luding and Tianquan in the Western Sichuan Basin. The distinctive feature is that siliceous sponge spicules are developed in the carbonaceous siliceous mudstone, and the siliceous sponge spicules are concentric; it's distributed in layers, and some bone spicules are filled with organic carbon after dissolution; the organic carbon content is more than 2%. In addition to the siliceous sponge spicules developed inside the siliceous mudstone, the carbonate cement is mainly dolomite, and the rhombic crystals of dolomite are well-developed. Its sedimentary characteristics are shown in Fig. 4.10.

2. Ordovician Guanyinqiao Formation

(1) Tidal flat facies

The tidal flat facies of the Guanyinqiao Formation are mainly distributed in the periphery of the Central Guizhou Uplift, and the lithology is dominated by silty limestone, silty argillaceous limestone and bioclastic mudstone. It is mainly composed of carbonate deposits mixed with terrigenous detrital particles. The silt content in the near-shore area near the Central Guizhou Uplift is relatively high, and the silt content gradually decreases in the offshore direction, while the argillaceous content increases, showing silty argillaceous limestone. The thickness is thin, and the sedimentary structure is basically not developed. In the production of brachiopods, a small amount of trilobites and crinoids, bryozoans, gastropods, bivalves and calcium algae were also seen (Li et al. 2005b; Li et al. 2008c) (Fig. 4.11).

(2) Shoal facies

The shoal facies here mainly refer to the intertidal zone within the tidal flat facies in the sediments of the Guanyinqiao Formation. Due to frequent changes in water level and relatively high energy in local areas, they gradually developed into shoal deposits. The Guanyinqiao Formation in here and other places is dominated by granular limestone, including sprite sandstone, bioclastic limestone and oolitic limestone. The bioclasts are mostly brachiopods, crinoids and other organisms. Dongkala section in Fenggang is the place with the thickest deposition of the Guanyinqiao Formation, and a sandy beach dominated by bright crystal sandy limestone develops. In the Renhuai area, taking the central section as an example, the lower part of the area is composed of calcareous mudstone with limestone nodules (Zhan et al. 2010a), which is the main output part of the Henantebe fauna; the upper part is the habitat with less mud content. Clastic limestone, sandy clastic limestone, mainly produces coral, a small amount of trilobites and non-Hirnantiane shells brachiopods and bryozoans (Fig. 4.12).

Fig. 4.11 Sedimentary filling sequence of tidal flat facies of Guanyinqiao Formation (Fenghuang profile, Youyang county, Chongqing City, left; Xiadian profile, Dejiang county, Guizhou Province, right)

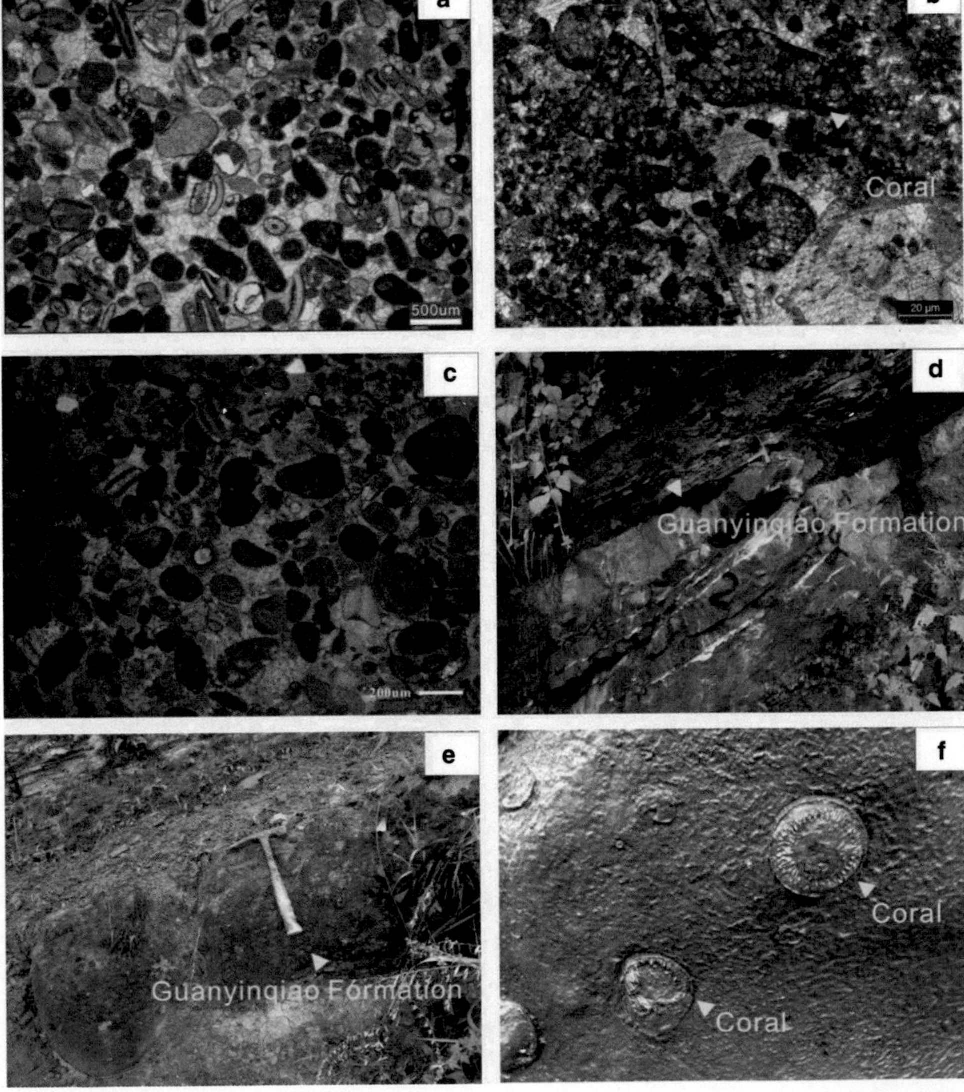

Fig. 4.12 Typical characteristics of shoal facies of Guanyinqiao Formation **a** the sandy limestone of Guanyinqiao Formation, Dongkala in Fenggang; **b** coral organisms in the biogenic limestone of the Guanyinqiao Formation in the central part of Renhuai; **c** the sandy limestone of Yonghe, Fenggang; **d** macro of the boundary of the Wufeng-Guanyinqiao-Longmaxi Formation, Yonghe, Fenggang; **e** macro of the Guanyinqiao Formation in the central Guanyinqiao in Renhuai; **f** coral organisms seen on the limestone surface of the Guanyinqiao Formation in the central Guanyinqiao in Renhuai

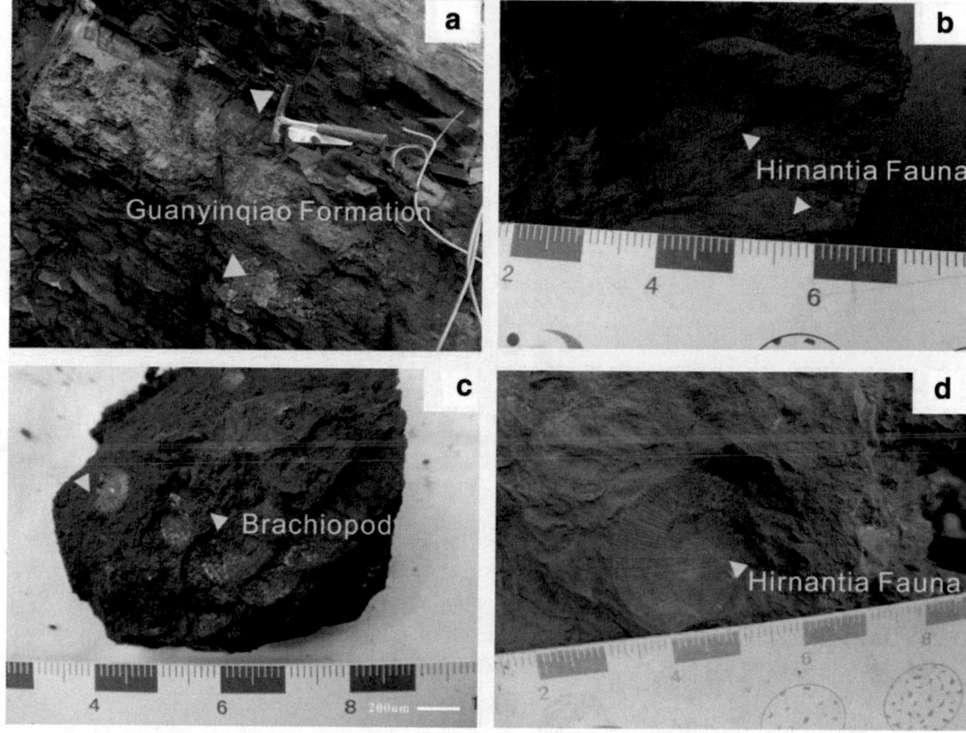

Fig. 4.13 Characteristics of shallow shelf facies of the Ordovician Guanyinqiao Formation in Southern Sichuan and its periphery **a** Biological mudstone of Guanyinqiao Formation, Wulong Huangcao profile; **b** Hirnantian brachiopod developed in biogenical mudstone of Guanyinqiao Formation, Wulong Huangcao profile; **c** Hirnantian brachiopod developed in biogenic marl of Guanyinqiao Formation, Xishui liangcun profile; **d** Hirnantian brachiopod developed in the biomarl-bearing marl of Guanyinqiao Formation, Changning Shuanghe profile

(3) Shallow shelf facies

The sedimentary lithology of shallow shelf facies during the depositional period of Guanyinqiao Formation is mainly dark gray bioclastic marl and biogenic limy mudstone. Yellow and yellowish-brown, these sedimentary facies are distributed in most areas of the Sichuan Basin and is dominated by brachiopods (especially Hirnantianbei, Fig. 4.13), trilobites, crinoids, gastropods and hornworts.

3. Silurian Longmaxi Formation

(1) Lithofacies types

Through the field name and microscopic analysis of the black organic-rich shale of the Longmaxi Formation, combined with the whole-rock mineral analysis of X-ray diffraction, seven main lithofacies types are distinguished in the lower member of the Longmaxi Formation in the study area.

(i) (Calcium-containing) carbonaceous (siliceous) shale

Carbonaceous shale is the most developed (Fig. 4.14a), and the mineral components are mainly clay minerals and quartz; the former is more than 35%, and the latter is more than 40%. It is relatively uniform black under single polarized light, and silt-grade quartz and other detrital particles are distributed in star-like shape, with a content of less than 5%; the content of organic carbon is relatively high; pyrite grain aggregates are often developed, and the content is mostly 1–5%. 5%, which individual samples can reach more than 10%. Carbonaceous siliceous shale is mostly distributed at the bottom of the Longmaxi Formation, and the siliceous content is high, more than 70%, and the highest is more than 85%. The siliceous particles are mainly radiolarians, and the siliceous particles are unevenly distributed in the black mud among them; the body cavity is mostly filled with organic matter (Fig. 4.14b); some areas contain a small amount of calcium, mainly distributed in Dajingping in Tianquan and Jiaodingshan in Hanyuan in Western Sichuan, and the content of organic carbon is also high. Such rock microfacies are mainly developed in the relatively deep-water area of the study area and are distributed in the Yibin-Changning-Luzhou area and the Western Sichuan area. The bottom is dominated by carbonaceous siliceous mudstone, and silt and calcareous components can be seen upwards, resulting in a rapid decrease in organic carbon content.

(ii) Carbonaceous shale containing silt (containing calcium)

The silty-bearing carbonaceous shale (Fig. 4.14c) is relatively developed in the study area. The mineral components are mainly quartz + feldspar, more than 60%, and clay minerals are mostly 30–60%. The clastic content is less than 15%, mostly 10% to 15%, mainly quartz, with the characteristics of overall dispersion and local enrichment. The content of carbonate minerals is less than 10%, mainly

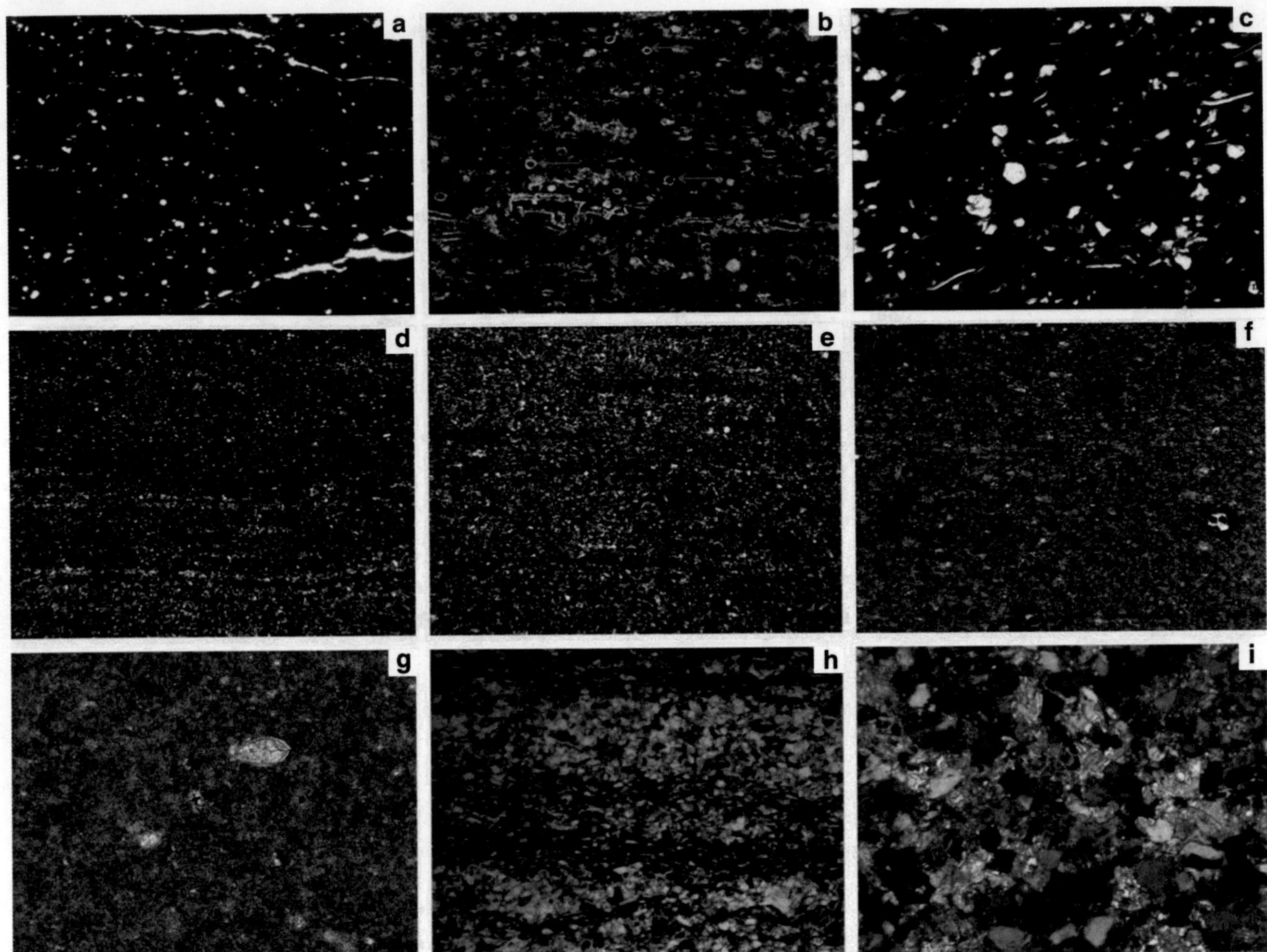

Fig. 4.14 Lithofacies characteristics of black rocks of Longmaxi Formation in Southern Sichuan Basin and its periphery **a** carbonaceous shale, Qijiang Guanyinqiao, single polarized light; **b** carbonaceous siliceous shale (siliceous radiolarian indicated by the arrow), Tianquan Dajingping, single polarized light; **c** silt-bearing carbonaceous shale, Xishui Liangcun, single polarized light; **d** carbon-bearing and calcium-bearing silt shale, Gulin Tiesuobridge, single polarized light; **e** carbonaceous siliceous carbonate shale, Renhuai Zhongshu, single polarized light; **f** carbon-bearing shale, Huaying Sanbaiti, single polarization; **g** carbonaceous argillaceous limestone, Daguan Huanggexi, single polarization; **h** silt shale, Yanhe Xinjing, single polarization; **i**. calcareous siltstone, Dejiang Xiadian, Orthogonal Polarization

calcite, and the content of carbonate minerals in local areas can reach more than 25%, only be founded in Xuyong area, Sichuan; pyrite is founded in most samples, and its content is less than 5%. The distinctive feature of this kind of rock microfacies is that it generally contains clastic silt, and its content has a significant effect on the organic carbon content. A small amount of calcium components can be seen in local areas, mainly distributed on the side near the uplift area of Qianzhong and Xuefeng Mountain.

(iii) Carbon (calcium-containing) silty shale

The carbon- and calcium-bearing silty shale is the most developed, and the mineral components are mainly quartz + feldspar, the content of which is mostly 40–70%, and the content of clay minerals is mainly between 30 and 50%; calcite is mainly composed of 10–20%; the content of detrital quartz is relatively high, which is mostly distributed in the dark basement, and has the characteristics of local enrichment and banded distribution. The latter is microscopic with the dark mud. Laminated (Fig. 4.14d), the content of clasts is mostly 25–40%, and pyrite is distributed in some areas; as a whole, the content of organic carbon decreases with the increase of clastic content. The distinctive features of this type of rock microfacies are the high content of silt with high compositional maturity and low structural maturity, as well as horizontal bedding and a small amount of wavy bedding. Carbonaceous (silt/silty) carbonate shale is developed in local areas (Fig. 4.14e), and its carbonate mineral content is relatively high, mostly between 25 and 40%, including calcite and dolomite, The content of the two is not much different, mainly in the form of cement coexisting with organic matter and argillaceous matter, which is only observed in the Dajingping section of Tianquan in Western Sichuan.

(iv) Carbon-bearing shale

The microfacies of this kind of rock are mainly argillaceous, with a content of > 85%, brown and gray-black under single polarized light (Fig. 4.14f). The mineral components are mainly quartz, feldspar and clay minerals, in which the content of clay minerals is more than 35%; the content of clastic particles and carbonate minerals is less than 10%, and they are unevenly distributed in the dark mud, and a small amount of pyrite are also seen. Mineral grains low organic carbon content. It is mainly distributed in the Northern part of the study area and the Eastern side of the Central Sichuan Uplift, indicating that the supply of terrigenous debris is less, and the sedimentary water body is relatively limited.

(v) Carbonaceous argillaceous limestone

Carbonaceous argillaceous limestone (Fig. 4.14g) is less distributed in the study, only distributed in Daguan Huanggexi profile, Jinyang Jige profile and other areas, and these types are also shown in individual samples from Dajingping profile in Tianquan. Its calcite content is very high, mainly ranging from 40 to 75%, mainly in micrite structure, mostly recrystallized, and contains a small amount of dolomite with a good degree of eumorphism, about 10%; it rarely contains clastic particles, a small amount of brachiopod bioclasts were seen in individual samples; the brown and gray-black mud was irregular mass, mainly ranging from 15 to 30%; the organic matter coexisted with terrigenous detrital mud, and the content was low. The carbonaceous argillaceous limestone specimens are black and grayish black, and the presence of organic matter indicates that they were formed in quieter water bodies and are mostly symbiotic with massive calcareous mudstones.

(vi) `Silt/calcareous shale

The silty shale is generally composed of ordinary mud shale intercalated with siltstone and silty mudstone, with a microlaminated structure (Fig. 4.14h). The clastic particles arc mainly quartz silt with trace needles, plate-like, leaf-like mica, which are mainly calcareous cementation, continuous or intermittent stripes, agglomerates or scattered single crystals, and mostly metasomatic feldspar and other debris. A small amount of iron minerals were founded, which in the form of particles or microtetragonal single crystals are sporadically distributed, and powder-like aggregates are unevenly rendered on the surface of hydromica, showing yellow–brown or brown-black intermittent fine stripes distribution. Clay content is high, mainly 15–40%, with illite as the main, contains a small amount of organic matter and is symbiotic with mud in thin strips. The carbonate mineral content of calcareous shale is relatively high, mainly 35–50%, mainly calcite, and contains a certain amount of dolomite, microcrystalline fine crystal. The silty mudstone has microlaminar features, indicating that its sedimentary water fluctuates frequently, and the calcareous mudstone is mixed as a whole gray, gray-black.

(vii) (Calcium/calcareous) argillaceous siltstone, siltstone

The siltstone is mainly distributed in the intertidal zone which is close to the provenance uplift and has strong hydrodynamic force. It is composed of clastic grains and calcareous cements or argillaceous matrix (Fig. 4.14i). The clastic grains are composed of quartz. Mainly, clean and bright, the particle size is 0.01–0.05 mm, the sorting is good, the rounding is poor, the sub-angular sub-round shape, and the base-porous cementation. A small amount of feldspar and mica, and the particle size of feldspar is relatively small compared to quartz. Large, weakly clayed or partially replaced by carbonate minerals, with many irregular dissolution edges; both biotite and muscovite are developed, the latter are more numerous, in the shape of thin strips, with certain compaction deformation, biotite is often altered to chlorite. It exhibits a certain water swelling along the cleavage crack. The calcareous cement is the product of early diagenesis, while the calcareous metasomatism is the product of late diagenesis, which is the product of late diagenesis. It has the characteristics of high compositional maturity and low structural maturity. Such rock microfacies are mainly distributed on the edge of the uplift in central Guizhou and central Sichuan and which are mostly wavy bedding and small sand-grained bedding, which are tidal flat deposits.

On the basis of previous data and research results, combined with field data and laboratory analysis results, according to the analysis of lithofacies development characteristics, sedimentary structure, profile sequence,

Table 4.3 Sedimentary facies of Lower Longmaxi Formation in Southern Sichuan Basin and its periphery

Depositional system	Sedimentary facies	Sedimentary subfacies	Distribution areas
Barrier coast system	Tidal	Intertidal zone, subtidal zone	Northern edge of the Qianzhong Uplift, Southern edge of the Sichuan-Central Uplift
Continental shelf system	Neritic shelf	Shallow shelf, deep shelf	South Sichuan, Southeast Sichuan

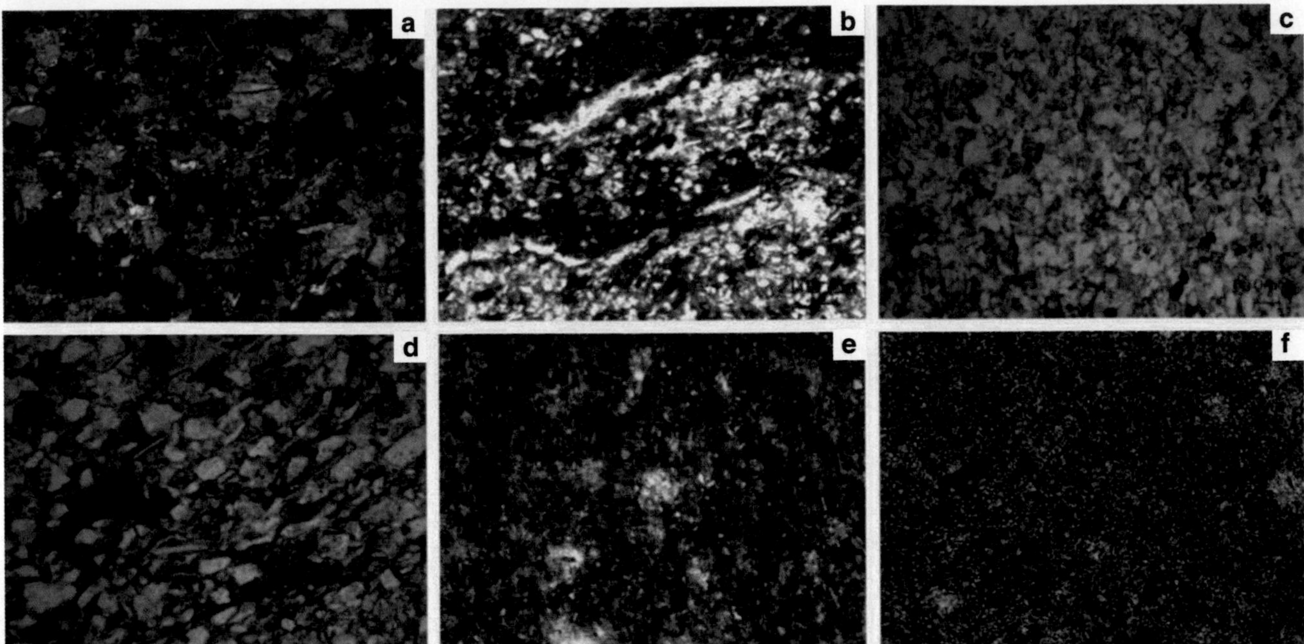

Fig. 4.15 Lithofacies of tidal flat facies of Lower Longmaxi Formation in Southern Sichuan Basin and its periphery. **a** calcareous siltstone, cross-polarized light, Dejiang Xiadian; **b** carbon-bearing silt shale, single polarized light, Dejiang Xiadian; **c** calcareous siltstone, single polarized light, Wuchuan Gaoqiao; **d** calcareous powder Sandstone, single polarization, Yinjiang Zhoujiaba; **e** calcareous shale, single polarization, Jinyang Jige; **f** micritic limestone, single polarization, Jinyang Jige

biological assemblage, etc., the black rock series of the Silurian Longmaxi Formation in Southern Sichuan and adjacent areas developed the layers (mainly the lower Longmaxi Formation) can be divided into two sedimentary facies (Table 4.3).

(2) Tidal flat facies

In the study area, the tidal flat facies of the Longmaxi Formation are mainly distributed on the edge of the uplift, and the area bounded by the Central Guizhou Uplift and the Xuefengshan Uplift (i.e., the area where the Wuchuan-Dejiang-Shiqian spreads from north to south). Mainly silt shale, silty shale and calcareous/dolomitic siltstone (Fig. 4.15a–d), with high content of detrital quartz and mostly carbonate cement, dark color. The argillaceous content is low, and the hand specimens are mainly yellow-green and gray. The Jinyang Jige profile is dominated by calcareous shale and micrite (Fig. 4.15e, f), and the carbonate mineral content is relatively high. High and relatively low detrital content, indicating that this area has a different provenance supply from the former. Horizontal bedding, vein bedding, wavy bedding, lenticular bedding, and small-scale flowing sand-grained bedding are developed (Fig. 4.16), small cross-bedding development can be seen in the local interval.

Vertically, compared with the Ordovician Wufeng Formation, the Longmaxi Formation is characterized by a relatively shallow-water body. For example, the Longmaxi Formation in Gaoqiao, Wuchuan and Dejiang Xiadian in Chongqing are also believed to have developed a set of relatively shallow tidal flat calcareous siltstone deposits. The Gaoqiao area in Wuchuan is mainly characterized by a combination of yellow-green, red-brown, and gray-green thin-layered silty mudstone and calcareous siltstone, with limestone interlayers, and yellow-green thin-layered mud shale deposits in the lower part. The content of siltstone increased significantly (Fig. 4.16). The lower part of the Dejiang Xiadian area in the South of Wuchuan is mainly a set of red-brown, gray-black, gray-green.

Mud shale, silty mud shale and siltstone assemblages upwards are siltstone, argillaceous siltstone and argillaceous limestone assemblages; silty shale and siltstone develop corrugated bedding and lenticular layers in which the small limestone lens body was founded. The limestone lens body has an increasing trend upwards. The upper marl sees argillaceous strips, in the form of veins.

(3) Shelf facies

(i) Shallow shelf subfacies

The shallow shelf subfacies of the Longmaxi Formation in Southern Sichuan and its periphery are relatively developed, and the overall characteristics are as follows: (a) It is mainly

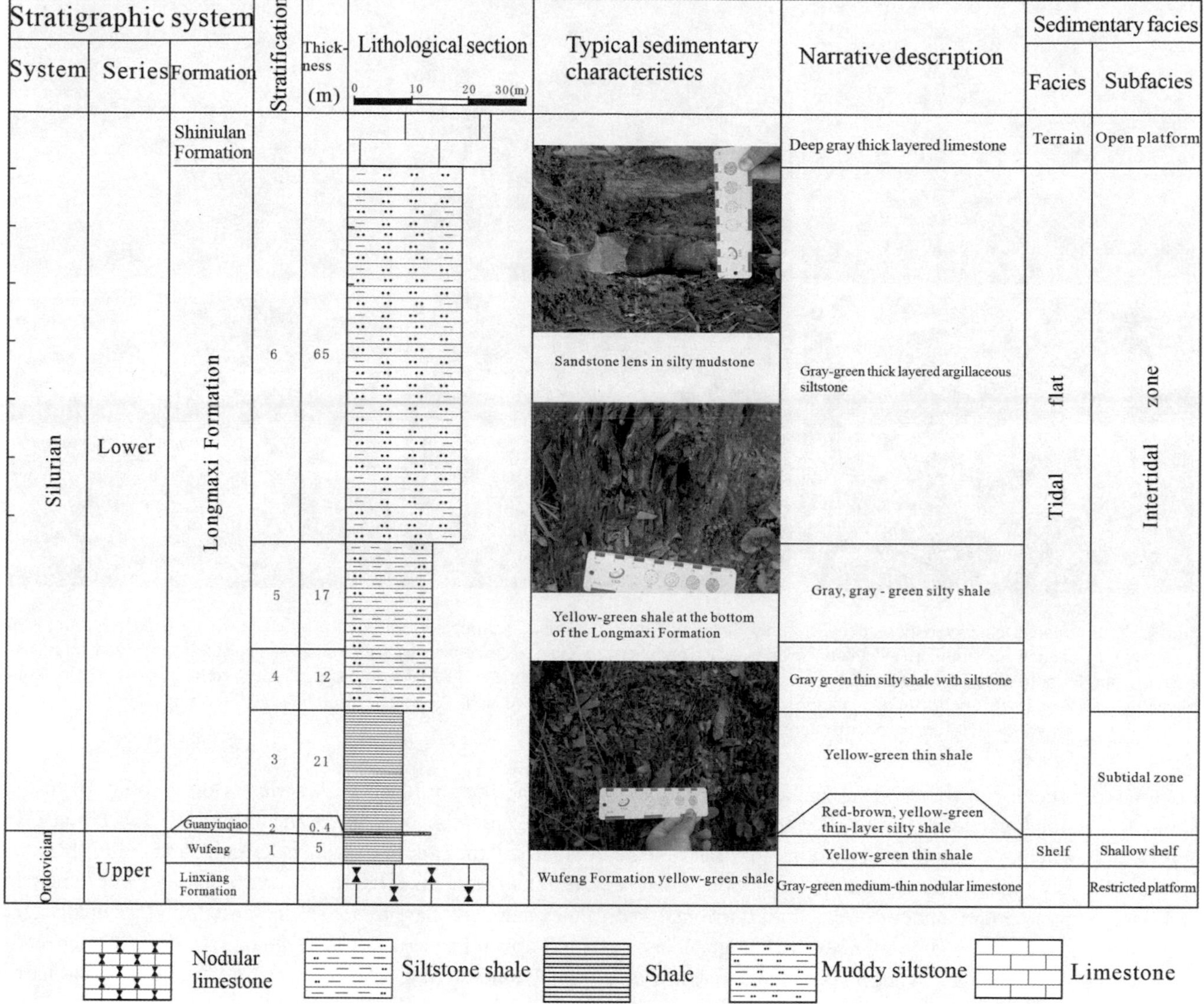

Fig. 4.16 Sedimentary characteristics of tidal flat facies of Longmaxi Formation in Wuchuan Gaoqiao, Chongqing

distributed in the post-uplift basins limited by the Kangdian-Central Guizhou Uplift and the Central Sichuan Uplift. Outside the tidal flat facies, there is a vast sea area that is generally distributed in an east–west direction; (b) The rock types are more complex, with carbonaceous/carbonaceous shale, siliceous calcium-bearing carbonaceous shale, (calcium/calcareous) silty carbonaceous shale is dominant (Fig. 4.17), followed by (carbon-containing) argillaceous siltstone and carbonaceous argillaceous limestone. Due to the relatively shallow-water body, the water body is turbulent and is often intermittently affected by storms, due to the influence of tidal currents and ocean currents, the sediments were transformed. According to the characteristics of the sedimentary water bodies, the study area mainly developed sandy muddy (Figs. 4.18 and 4.19) and stucco shallow shelf deposits; (c) The colors of sediments are mainly dark gray, gray, black, gray-black, etc.; (d) The structures of sediments such as horizontal bedding, sand-grained bedding, sandy agglomerates and calcareous nodules are developed. There are many different shapes. The individual size of graptolites becomes longer, and the types and numbers decrease. Secondly, a small number of brachiopods, corals and other biocides can be seen.

The Silurian Longmaxi Formation profile in the Northern Wuchuan is located near the north side of Gaoqiao (Fig. 4.18). Mainly sandy shale, the sedimentary environment in the Gaoqiao area of Wuchuan is deeper, horizontal bedding is seen, and graptolites are developed; upward clastic silt and carbonate minerals gradually increase, and calcareous argillaceous siltstone is mainly developed, indicating that the body of water gradually becomes shallower. The Yanhe Xinjing profile near the Xuefeng Uplift, the Silurian

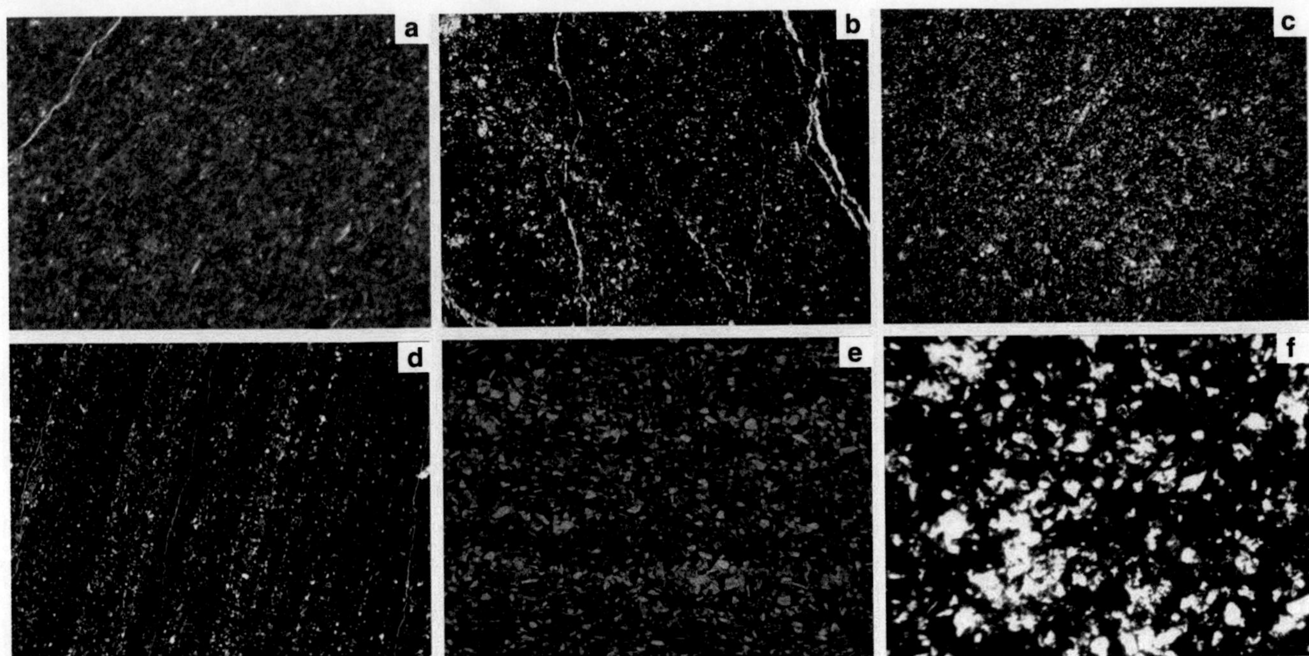

Fig. 4.17 Lithofacies characteristics of shallow shelf facies of Lower Longmaxi Formation in Southern Sichuan Basin and its periphery **a** Marl, single polarization, Daguan Huanggexi; **b** carbon-bearing calcareous shale, single polarization, Yanjin Yinchangba; **c** calcium-bearing silty shale, single polarization, Yongshan Sutian; **d** cloud-bearing silty carbonaceous shale, single polarized light, E'bian Xihe; **e** carbonaceous silty sandy shale, single polarized light, Gulin Tiesuobridge; **f** calcium-bearing silty carbonaceous shale, Single polarized light, Yanhe Xinjing

Longmaxi Formation is about 240 m thick, and the bottom is about 100 m thick calcium-bearing silty carbonaceous shale (Fig. 4.17f), calcium-bearing carbonaceous silty sandy shale, with a relatively high clastic content, mainly about 35–40%, and containing a certain amount of calcium; horizontal bedding is developed, and a large number of orthographites are seen. It gradually changes to argillaceous siltstone upward, and the clastic particles are dominated by quartz and contain a large amount of powder-fine short columnar mica fragments; higher debris content indicates that it is strongly influenced by the Xuefeng Uplift. Located in the Tongzi Hanjiadian profile on the north side of the Central Guizhou Uplift, the lower member of the Longmaxi Formation develops black silt-bearing-calcium-bearing carbonaceous shale, cloud-bearing carbonaceous siliceous shale, which is enriched in pyrite and organic matter. And a large number of graptolites were developed. In the later stage of the Longmaxi Formation, with the intensified tectonic compression, the range of the Central Guizhou Uplift continued to expand to the north, and the depth of the water body decreased. Shale, argillaceous siltstone and marl, and carbonate rocks gradually increase upward, and the top is bounded by the Shiniulan Formation by the appearance of gray-green and thin-layered nodular limestone. Further the south, it is closer to the Renhuai central profile of the Central Guizhou Uplift area (Fig. 4.19), where the Longmaxi Formation sedimentary stratum is relatively thin, with a total thickness of about 26 m, and the bottom black rock series is only about 10 m thick about, the carbonate (calcite and dolomite) silty carbonaceous shale and the carbon-bearing calcium-bearing silty shale have different contents relative to Tongzi Hanjiadian clastic particles and carbonate minerals. Increased, and gradually changed upward to argillaceous siltstone, quartz siltstone with a small amount of bioclastic limestone. In the long and narrow area of Western Sichuan, which is bounded by the ancient land of Kangdian Uplift, Central Guizhou Uplift and Central Sichuan Uplift, taking the Sutian profile of Yongshan as an example, the sedimentary thickness of the Longmaxi Formation is relatively increased, with a total thickness of about 230 m. The thickness is up to 85 m, and it is mainly composed of calcium-bearing silt-bearing carbonaceous shale. The clast content is relatively small, mainly ranging from 10 to 25%. The pyrite and organic matter are developed, and a large amount of travertine and curvilinear are enriched, indicating that the sedimentary water body is deep.

In the Huaying area, which is located in the Northeast of the study area, the west side is close to the Central Sichuan Uplift and extends to the Northeast-East direction. The overall dark shale of the Longmaxi Formation in Xikou and Sanbaiti profiles are not very developed, while the yellow-green shale is not well-developed. The shale is relatively thick, and the organic matter content is low. Among them, the sedimentary thickness of the Longmaxi Formation in the Sanbaiti profile is larger, with a total thickness of 350

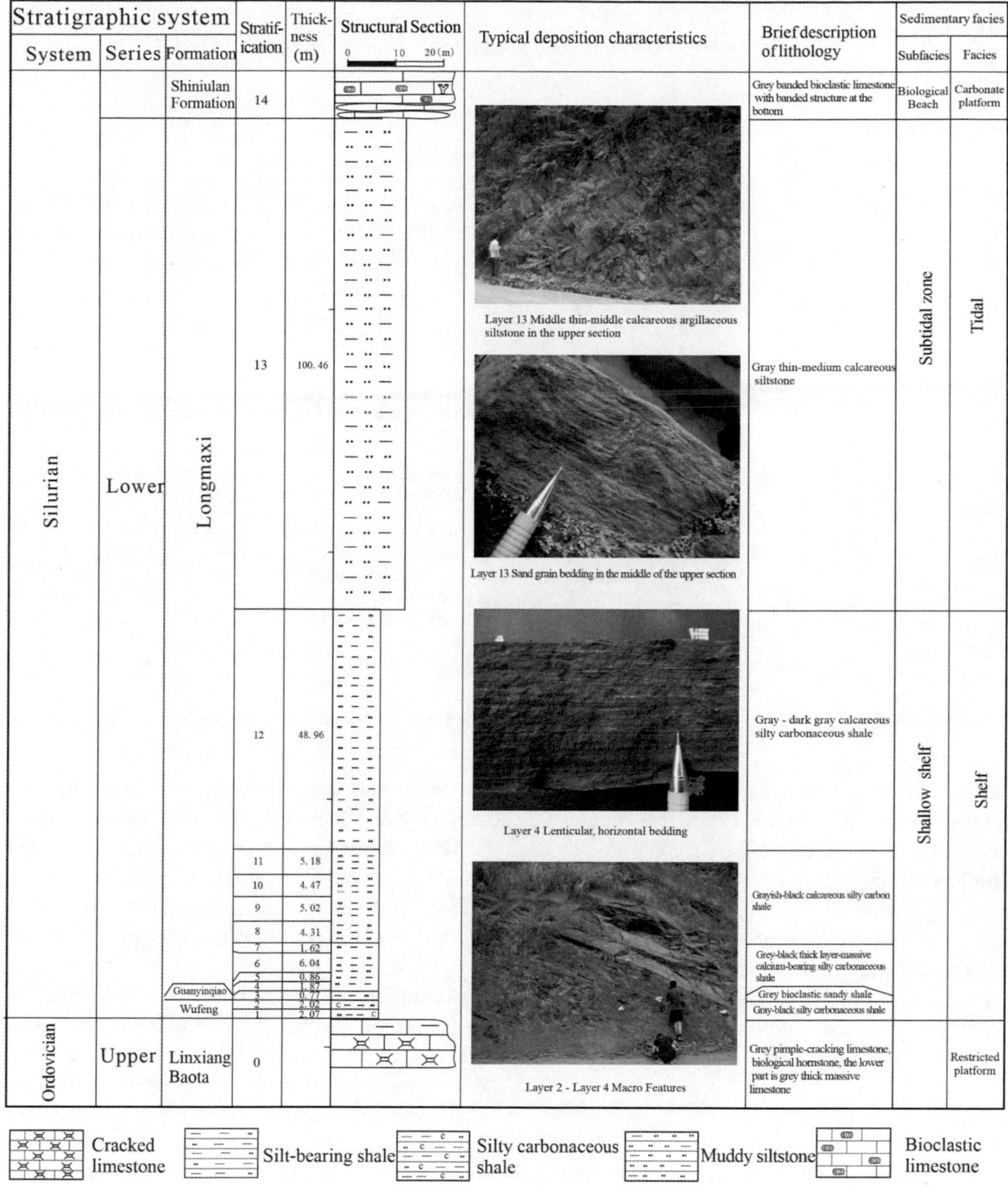

Fig. 4.18 Sedimentary characteristics of shallow shelf facies of Longmaxi Formation in Northern Wuchuan county, Chongqing

m. The black rock series is only distributed at the bottom, with a thickness of about 35 m. The black rock series is mainly composed of carbon-bearing shale, which may be gradually expanded by the Uplift of the Central Sichuan, and the sedimentary water body is shallow. From the development lithology, it can be seen that the above areas are all characterized by sandy-muddy shallow-water continental shelf deposits. The Weixin-Daguan-Zhaojue areas, which are close to the Kangdian Uplift and the Central Guizhou Uplift, have a high content of carbonate minerals in the Silurian Longmaxi Formation, such as the Daguan Huanggexi profile and the Yanjin Yinchangba profile. The Longmaxi Formation is about 110 m thick, and the black rock series is about 13 m thick, mainly carbonaceous argillaceous limestone (Fig. 4.17a), above which are about 23 m of black shale and gray-black marl interbedded. The latter Longmaxi Formation is about 160 m thick, and the black shale development interval is about 35% thick, mainly composed of

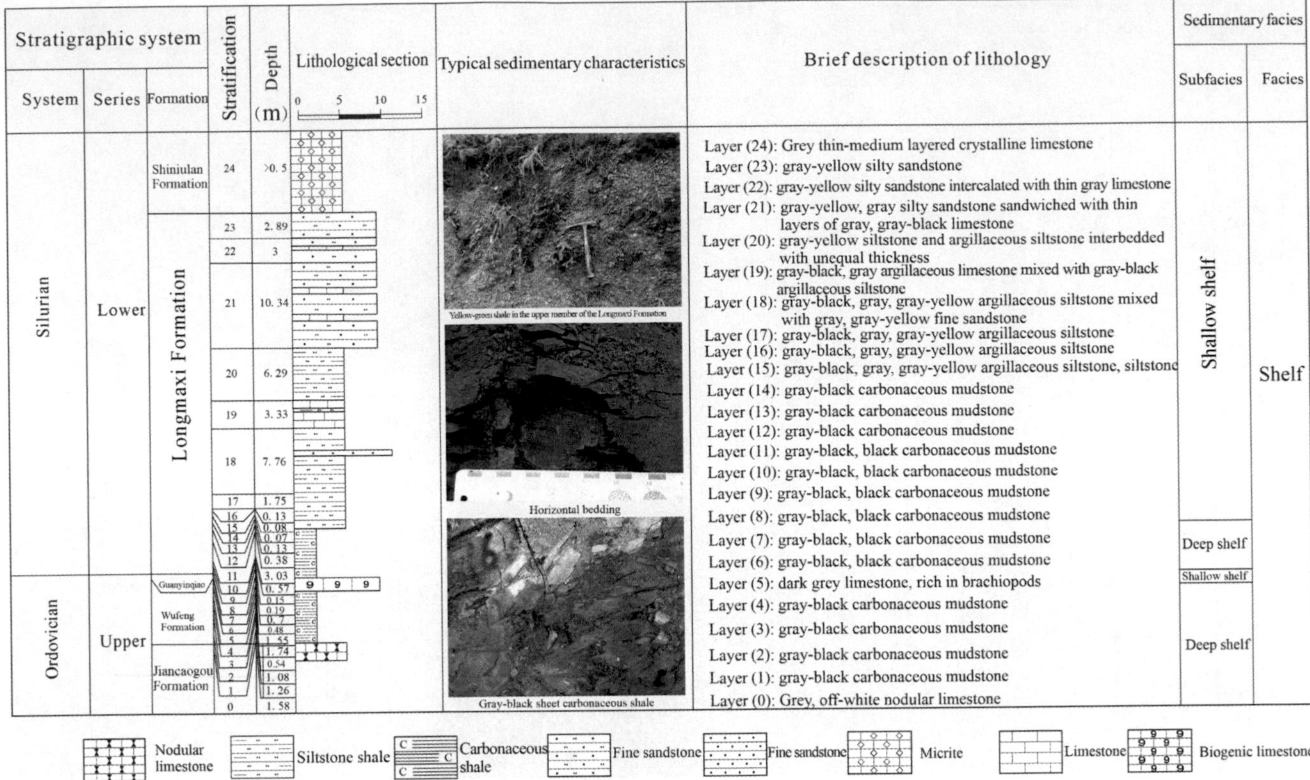

Fig. 4.19 Sedimentary characteristics of shallow shelf facies of Longmaxi Formation in Renhuai Zhongshu profile, Sichuan Province

carbon-bearing calcium/calcareous shale (Fig. 4.17b), and the carbonate minerals gradually increase upwards. Mud-bearing/silt-bearing marl or argillaceous limestone is mainly developed (Fig. 4.17b). It can be seen that this area is characterized by stucco shallow shelf sedimentary and can be matched with the tidal flat deposits on its outer side—the carbonate-rich mineral characteristics of the Longmaxi Formation in each profile of Jinyang Jige, indicating that it may be affected by the same influence of provenance supply and other depositional factors other than geomorphology.

(ii) Deep-water shelf subfacies

The deep-water shelf subfacies are mainly black carbonaceous siliceous shale and carbonaceous shale (Fig. 4.20), including silty carbonaceous shale and siliceous shale. In the entire deep-water shelf facies depositional area, a large number of horizontal bedding or discontinuous horizontal bedding develops (Figs. 4.21 and 4.22); graptolite species are abundant and numerous (Figs. 4.21 and 4.22), a very small amount of coral, brachiopod and other bioclasts can be seen in some areas. Pyrite grains are distributed in carbonaceous shale in the form of dispersion and strip. In a word, the deep-water shelf subfacies in the study area were calm, with little or no disturbance of benthic animals, and the deposition rate was relatively small, which belonged to relatively undercompensated deep-water sedimentary.

Deep-water shelf subfacies are located in the center of the basin or relatively deep water in the basin and are bounded by the shallow shelf facies. Similar to the shallow shelf facies and tidal flat facies, they are affected by the sedimentary water body and provenance area which have distinct lithological differences. In the deep-water shelf facies, black shale is relatively developed which located in the Shizhu-Wulong areas on the Northeast edge of the study area, the Silurian Longmaxi Formation are rich in silty/silty carbonaceous shale and silty-bearing shale. Sandy calcium-bearing carbonaceous shale (Fig. 4.20a, b) is dominant, such as the Wulong Huangcao profile (Fig. 4.21). The Longmaxi Formation is about 200 m thick, the black rock series is about 67 m thick, and clastic particles which the content is relatively low and gradually increases from the bottom to the top. The thickness of the single layer is relatively uniform, mainly ranging from 5 to 15 cm. The pyrite and organic matter are relatively developed, the graphite is very rich, and a small amount of horizontal bedding was founded which indicates that the sedimentary water body is relatively deep and stable. The Qijiang-Xishui areas on the west side of the Shizhu-Pengshui areas began to be bounded by a relatively obvious north–south uplift, and its silt content increased

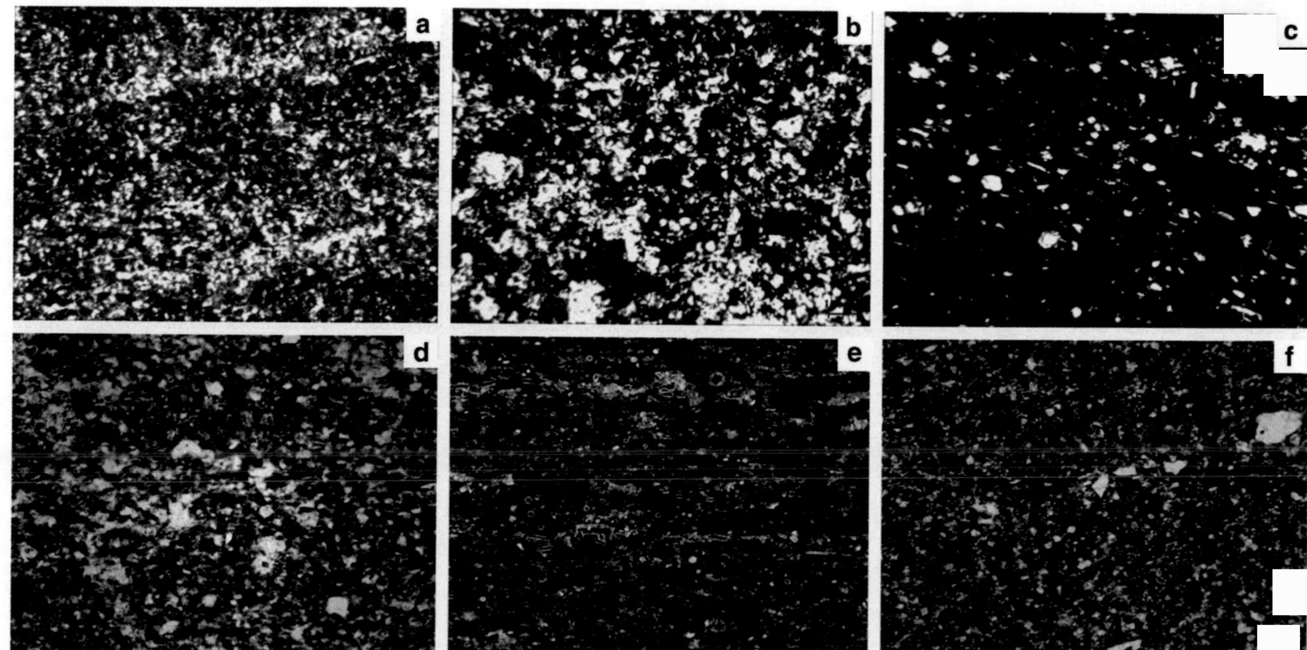

Fig. 4.20 Lithofacies of deep-water shelf facies of Lower Longmaxi Formation in Southern Sichuan Basin and its periphery **a** silt-bearing carbonaceous shale, single polarized light, Shizhu Qiliao; **b** silt carbonaceous shale, single polarized light, Wulong Huangcao; **c** carbonaceous shale, single polarized light, Qijiang Guanyinqiao; **d** carbonaceous siliceous shale, single polarization, Xingwen Qilin; **e** carbonaceous siliceous shale, single polarization, Tianquan Dajingping; **f** silt-bearing carbonaceous shale, single polarization, Hanyuan Jiaodingshan

relatively, mainly silt-bearing carbonaceous shale, and the Daozhen Bayu profile. Take the profile as an example, after the Hirnantian period, with the occurrence of global transgressive events due to the melting of glaciers, a set of thick black silt-bearing (calcium-bearing) carbonaceous shale was deposited in the lower member of the Longmaxi Formation. The lithological color of the upper Longmaxi Formation becomes lighter, the corresponding calcareous components and clastic particles also gradually increase, and the sedimentary gray-blue-gray calcareous shale with marl interlayers, with the increase of carbonate rocks at the top, gray calcareous mudstone and biological limestone interbedded with the Shiniulan Formation. The Qijiang Guanyinqiao profile is located near the center of the basin. The sedimentary thickness of the Longmaxi Formation is large, up to 340 m, and the black rock series can reach 100 m. It is mainly composed of silt-bearing carbonaceous shale. Pyrite grains are developed; graptolite is very rich, and organic matter content is high. Upwards gradually transitions to silt-bearing carbonaceous shale intercalated with a thin layer of carbonate rock. From bottom to top, carbonate minerals increase significantly, and the top is the argillaceous limestone and micrite limestone are predominant, but the content of detrital particles has not increased significantly, which indicates that the Qijiang area is still far away from the terrigenous area in the later stage of the Longmaxi Formation deposition, but the sedimentary water body becomes shallow.

In the limited area bounded by the Central Sichuan Uplift, the Central Guizhou Uplift and the Kangdian Ancient Land, the water body is relatively deep. Because it is located in the basin, there are few surface data. Near the Xingwen Qilin profile on the south side, the Longmaxi Formation is generally black to gray-black, with a thin deposition thickness of about 35 m, which shows the characteristics of starvation deposition; The black shale in the lower section is relatively thick, accounting for more than 70% of the total thickness, and is composed of siliceous carbonaceous shale (Fig. 4.20d) Mainly, it shows that the Xingwen Qilin profile, as the area at the center of deposition, indicates a typical deep-water, stagnant depositional environment. In the Luding-Hanyuan area located in the Western Sichuan area, the black shale of the Longmaxi Formation is thicker, and the bottom is dominated by siliceous carbonaceous shale, and siliceous radiolarians are developed (Fig. 4.21). The boundary between the uplift and the ancient land of Kangdian is represented by deep-water continental shelf deposits, such as the Dajingping profile in Tianquan (Fig. 4.22), the Longmaxi Formation is about 220 m thick, and the mineral components are mainly siliceous, with an average content of > 45%; there are not only a certain amount of quartz grains, but also a large number of siliceous radiolarians and some siliceous cements, which indicates that this area is bounded by the nearby Central Sichuan Uplift and the ancient land of Kangdian, showing as the deeper

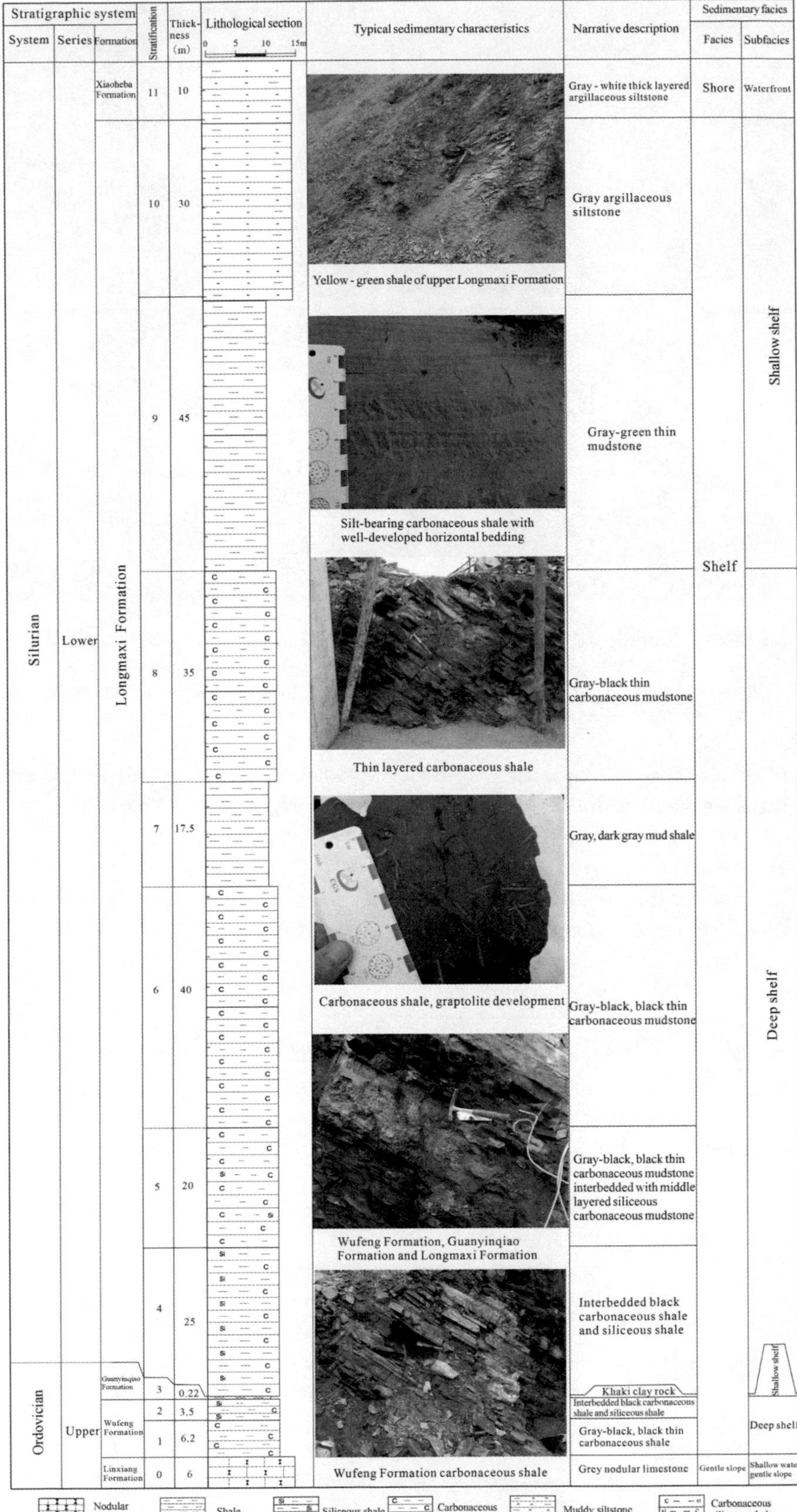

System	Series	Formation	Stratification	Thickness (m)
Silurian	Lower	Xiaoheba Formation	11	10
		Longmaxi Formation	10	30
			9	45
			8	35
			7	17.5
			6	40
			5	20
			4	25
Ordovician	Upper	Guanyinqiao Formation	3	0.22
		Wufeng Formation	2	3.5
			1	6.2
		Linxiang Formation	0	6

Fig. 4.21 Sedimentary characteristics of deep-water shelf facies of Longmaxi Formation in Wulong Huangcao profile, Chongqing

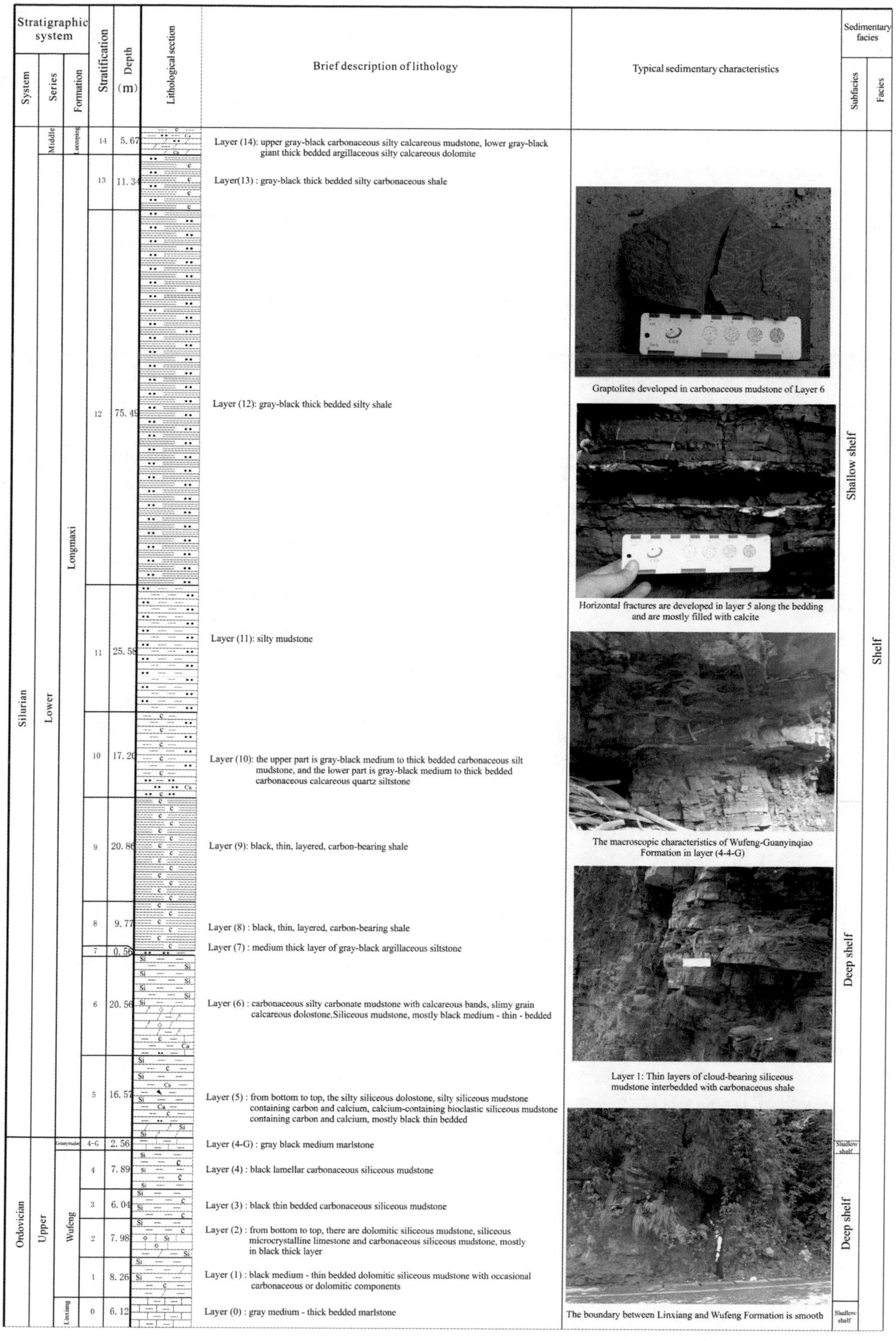

System	Series	Formation	Stratification	Depth (m)	Brief description of lithology	Subfacies	Facies
Silurian	Middle	Luoreping	14	5.67	Layer (14): upper gray-black carbonaceous silty calcareous mudstone, lower gray-black giant thick bedded argillaceous silty calcareous dolomite	Shallow shelf	Shelf
	Lower	Longmaxi	13	11.34	Layer(13) : gray-black thick bedded silty carbonaceous shale		
			12	75.49	Layer (12): gray-black thick bedded silty shale		
			11	25.58	Layer (11): silty mudstone		
			10	17.20	Layer (10): the upper part is gray-black medium to thick bedded carbonaceous silt mudstone, and the lower part is gray-black medium to thick bedded carbonaceous calcareous quartz siltstone		
			9	20.86	Layer (9): black, thin, layered, carbon-bearing shale	Deep shelf	
			8	9.77	Layer (8) : black, thin, layered, carbon-bearing shale		
			7	0.56	Layer (7) : medium thick layer of gray-black argillaceous siltstone		
			6	20.56	Layer (6) : carbonaceous silty carbonate mudstone with calcareous bands, slimy grain calcareous dolostone,Siliceous mudstone, mostly black medium - thin - bedded		
			5	16.57	Layer (5) : from bottom to top, the silty siliceous dolostone, silty siliceous mudstone containing carbon and calcium, calcium-containing bioclastic siliceous mudstone containing carbon and calcium, mostly black thin bedded		
Ordovician	Upper	Guanyinqiao	4-G	2.56	Layer (4-G) : gray black medium marlstone	Shallow shelf	
		Wufeng	4	7.89	Layer (4) : black lamellar carbonaceous siliceous mudstone	Deep shelf	
			3	6.04	Layer (3) : black thin bedded carbonaceous siliceous mudstone		
			2	7.98	Layer (2) : from bottom to top, there are dolomitic siliceous mudstone, siliceous microcrystalline limestone and carbonaceous siliceous mudstone, mostly in black thick layer		
			1	8.26	Layer (1) : black medium - thin bedded dolomitic siliceous mudstone with occasional carbonaceous or dolomitic components		
		Linxiang	0	6.12	Layer (0) : gray medium - thick bedded marlstone	Shallow shelf	

Typical sedimentary characteristics:

Graptolites developed in carbonaceous mudstone of Layer 6

Horizontal fractures are developed in layer 5 along the bedding and are mostly filled with calcite

The macroscopic characteristics of Wufeng-Guanyinqiao Formation in layer (4-4-G)

Layer 1: Thin layers of cloud-bearing siliceous mudstone interbedded with carbonaceous shale

The boundary between Linxiang and Wufeng Formation is smooth

Fig. 4.22 Sedimentary characteristics of deep-water shelf facies of Longmaxi Formation in Tianquan Dajingping profile, Sichuan Province

sedimentary water body has the sedimentary characteristics of near provenance, and which is supplied by nutrients from the Northwest side of the sea, showing complex components. Among them, the development of siliceous radiolarians may be related to upwelling, and it is likely to be inherited. Liu et al. (2010b) pointed out that during the depositional period of the Wufeng Formation, this area had good connectivity with the open sea and was deposited in the deep-water area of semi-restricted shallow seas. In general, from west to east, the content of detrital particles in deep-water shelf sediments gradually decreased, which may be due to the eastward direction of the study area; the Northeast is connected to the vast open sea, and the supply of terrigenous clasts is less related. For example, the Youyang Heishui profile in the Southeast of Pengshui County (Liang et al. 2012a), black shale with a thickness of about 40 m which developed at the bottom, and powder can be founded. Sandy shale and argillaceous siltstone of turbidite origin are interbedded, developed slump deformation structure and channel model structure, which are rich in graptolite and pyrite and are sediments in a deep-water reducing environment. It indicates the sedimentary characteristics of deep-water shelf facies.

4.2.2 Vertical Depositional Sequence of Typical Profiles

At the end of the Middle Ordovician, the nature of the Sichuan Basin changed from a craton basin to a post-uplift basin which developed on the craton and surrounded by various uplifts (Mu et al., 2010, 2011). The salt limestone facies changed into black shale facies of Wufeng Formation. The black shale of Wufeng Formation mainly develops graptolites, and its types are mainly double graptolites. At the end of the depositional stage of Wufeng Formation (mid-Henante period), continental glacial activity erupted, and the formation of glaciers caused a significant drop in sea level. On the black shale, the Guanyinqiao Formation with a thickness of only tens of centimeters to a few meters and a maximum of about ten meters began to be deposited. The area is developed with high-energy oolitic limestones, and in these limestones and mudstones, cool-water crustaceans, the Hirnantianbei fauna, are generally developed and often accompanied by trilobites dominated by Dalmanitina. By the end of the glacial period at the end of the Hirnantian period, with the large-scale melting of glaciers and the gradual warming of the climate, the relative sea level rose rapidly, and large-scale transgression occurred. In the early Silurian Ludan period, black carbonaceous shale at the bottom of the Longmaxi Formation began to be deposited. This set of black rock series has a wide distribution range and a large thickness, and the organisms are mainly graptolites. In addition to the main body of the Wufeng Formation, the graptolithic types include the double graptolithic family, and the two-shaped graptolithic family and the single graptolithic family are also more common. Late Rudan—In the early Elonian period, tectonic compression from the Southeast Cathaysian block became the dominant factor in paleogeographic evolution, followed by global sea-level decline. The upper part of the Longmaxi Formation begins to deposit gray-green, gray-yellow sandy shale and siltstone. The color of the sedimentary lithology becomes lighter than that of the lower part, the terrigenous clastic components increase, and the grain size becomes relatively coarse. The Longmaxi Formation is upward from bottom to top. Shallow sedimentary sequence, in the late Elonian period, with the continuous intensification of compressional tectonic action, the Xuefeng Uplift and the land in the central and Southern Hubei area were uplifted and expanded, and the marginal facies (clastic coast, delta) also continued to advance, which is a prograde type. The development of clastic sand bodies laid the foundation. Eastern Sichuan, Southeastern Chongqing, and Western Hunan mainly deposit medium-thick layered gray-green siltstone and shale in the Xiaoheba Formation, with local intercalation of bioclastic limestone. The biogenic limestone is rich in corals, brachiopods, crinoids and other organisms. The Western Sichuan-Central Yunnan-Central Guizhou Uplift did not continue to expand, but it showed tectonic subsidence, and its periphery also became a gentler shallow-water coastal area. This phenomenon obviously inhibited the supply of terrigenous detrital materials. In addition, the global warm and humid climate at that time and other factors led to the recovery of the carbonate depositional environment. Limestone, marl, the upper section is bioclastic limestone, biogenic limestone and dolomite, and the whole is a carbonate rock platform deposit.

Taking the Hanjiadian profile of Tongzi in Southeastern Sichuan as an example, in the early Late Ordovician, the nodular micrite limestone of Jiancaogou Formation was deposited, and various types of brachiopods, trilobites and other organisms developed in the limestone, which was a shallow-water gentle slope depositional environment. However, with the occurrence of Caledonian tectonic movement and relative sea-level rise, the entire Upper Yangtze block, except for the central Sichuan, Central Guizhou and Xuefeng Uplifts, shifted from carbonate rock gentle slope environment to limited deep-water shelf facies and deposited black carbonaceous mudstones inWufeng Formation. The dark gray marl of the Guanyinqiao Formation was deposited at the end of the Late Ordovician (Hirnantian), which is a shallow shelf depositional environment. In the early stage of the Longmaxi Formation, black carbonaceous mud shale, cloudy carbonaceous silt mudstone, which is rich in pyrite and organic matter were deposited in the lower part of the Longmaxi Formation. It's a deep-water shelf environment

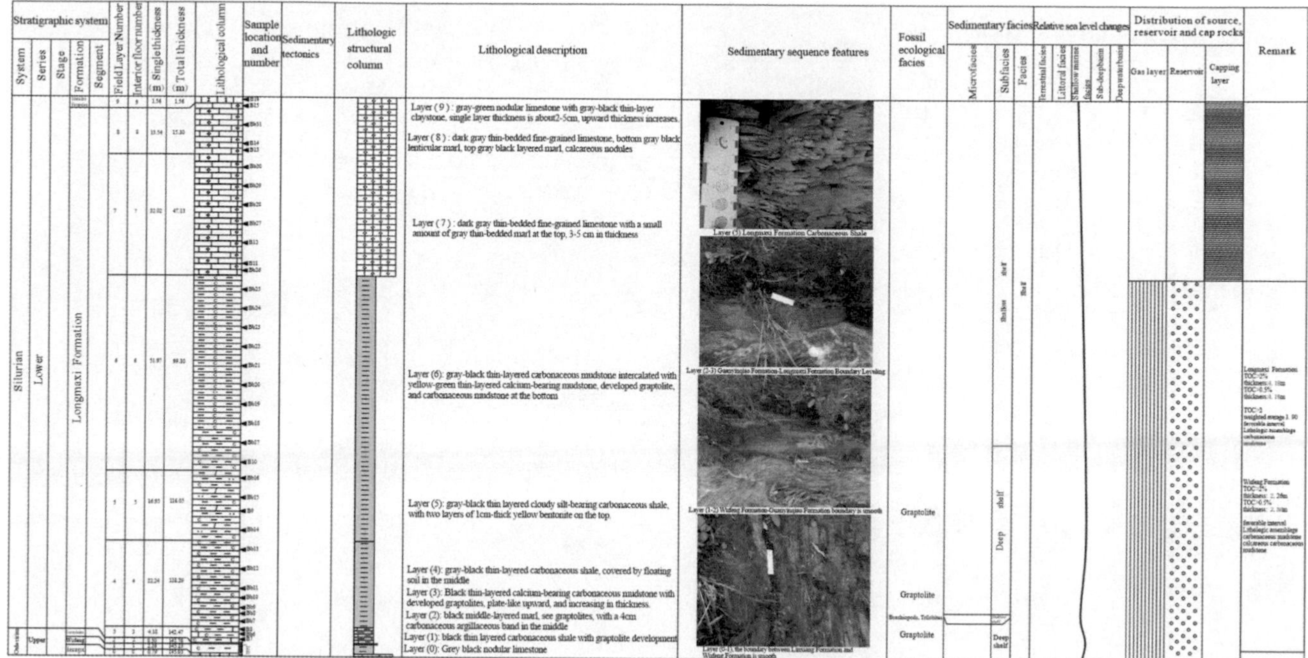

Fig. 4.23 Sedimentary filling sequence of Wufeng-Longmaxi Formation in Hanjiadian, Tongzi

with low energy and strong reduction. In the later stage of deposition, with the intensification of tectonic compression, the range of the Uplift continued to expand and began to expand to the Upper Yangtze region, and the depth of the surrounding water body decreased. Upwards, the carbonate rocks gradually increase, and the top is bounded by the Shiniulan Formation by the appearance of gray-green thin-layered nodular limestone (Fig. 4.23).

The Jiancaogou Formation in the Daozhen Bayu profile is composed of thick gray marl and calcareous mudstone, while the Wufeng Formation is mainly composed of carbonaceous silt mudstone, and the sedimentary structure is dominated by horizontal bedding. The water body of the Formation became deeper, and it was a deep-water shelf facies deposition. After the depositional period of the Wufeng Formation, the silt-bearing micrite limestone of the Guanyinqiao Formation began to be deposited. The limestone developed individual complete Hirnantian brachiopods. After the Guanyinqiao period, with the glaciers the occurrence of ablation global transgression events, black carbonaceous mudstone and carbonaceous silt mudstone were deposited in the lower member of the Longmaxi Formation in Daozhen Bayu, with a large number of graptolithic organisms development, which is a deep-water shelf environment, and the lithological color of the upper member of the Longmaxi Formation becomes shallow shelf; it is a shallow shelf facies deposition, and the corresponding calcareous components also gradually increase, starting to deposit gray-blue-gray calcareous mudstone, and gradually there are marl interlayers. Mudstone and biolimestone interbeds are bounded by the Shiniulan Formation (Fig. 4.24).

4.2.3 Relative Depositional Comparison

In order to analyze the regional distribution characteristics of Wufeng Formation-Longmaxi Formation more clearly, there are seven profiles, Leibo Bajiaotan-Junlian Luomurou-Xuyong Macheng-Lin 1 well-Dingshan 1 well-Wulong Huangcao-Pengye 1 well from west to east and were selected in order for regional comparison (Fig. 4.25). The lack of carbonaceous mudstone, silty mudstone and calcareous mudstone is deposited in the Wufeng Formation of Leibo Bajiaotan profile on the west side, with a thickness of about 36 m, which developed in deep-water sedimentary environment. To the east, the thickness of the Wufeng Formation in the Luomurou area of Junlian was reduced to 12 m, and the lithology changed to black carbonaceous calcareous siltstone, which increased in silt content compared to the Bajiaotan profile. The thickness of the Wufeng Formation in Macheng, Xuyong continues to decrease to 9.24 m. The lithology of the Wufeng Formation in these areas is calcium-bearing carbonaceous siltstone. The lithology of the Wufeng Formation in Well Lin 1 and Well Dingshan 1 is black carbonaceous mudstone, silty mudstone, and argillaceous siltstone, with a large difference in sedimentary thickness, 16 m and 4 m, respectively: eastward to Wulong and Pengshui in Southeastern Chongqing. The Wufeng Formation in the Huangcao profile has a black

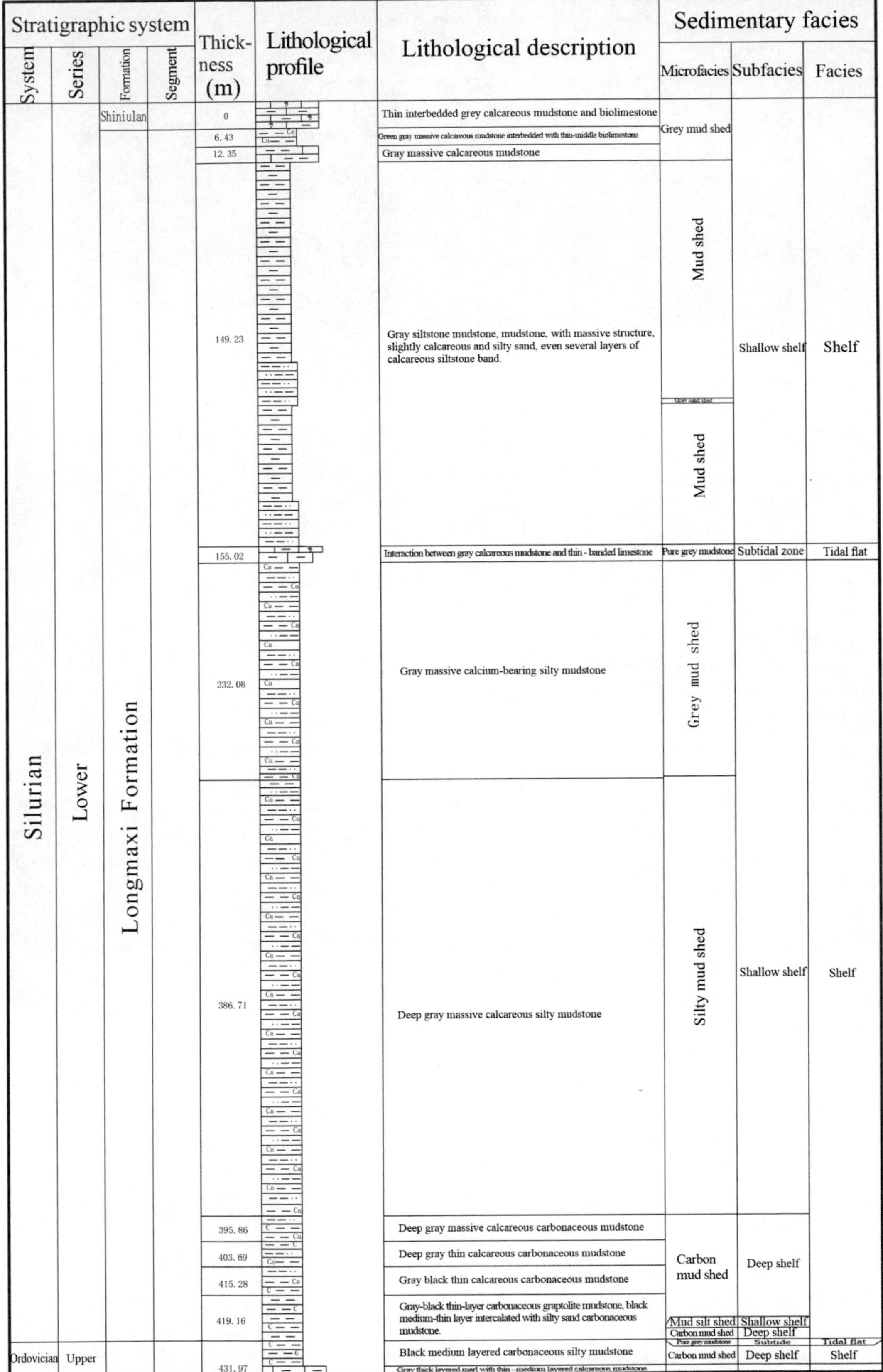

Fig. 4.24 Sedimentary filling sequence of Wufeng-Longmaxi Formation, Daozhen Bayu

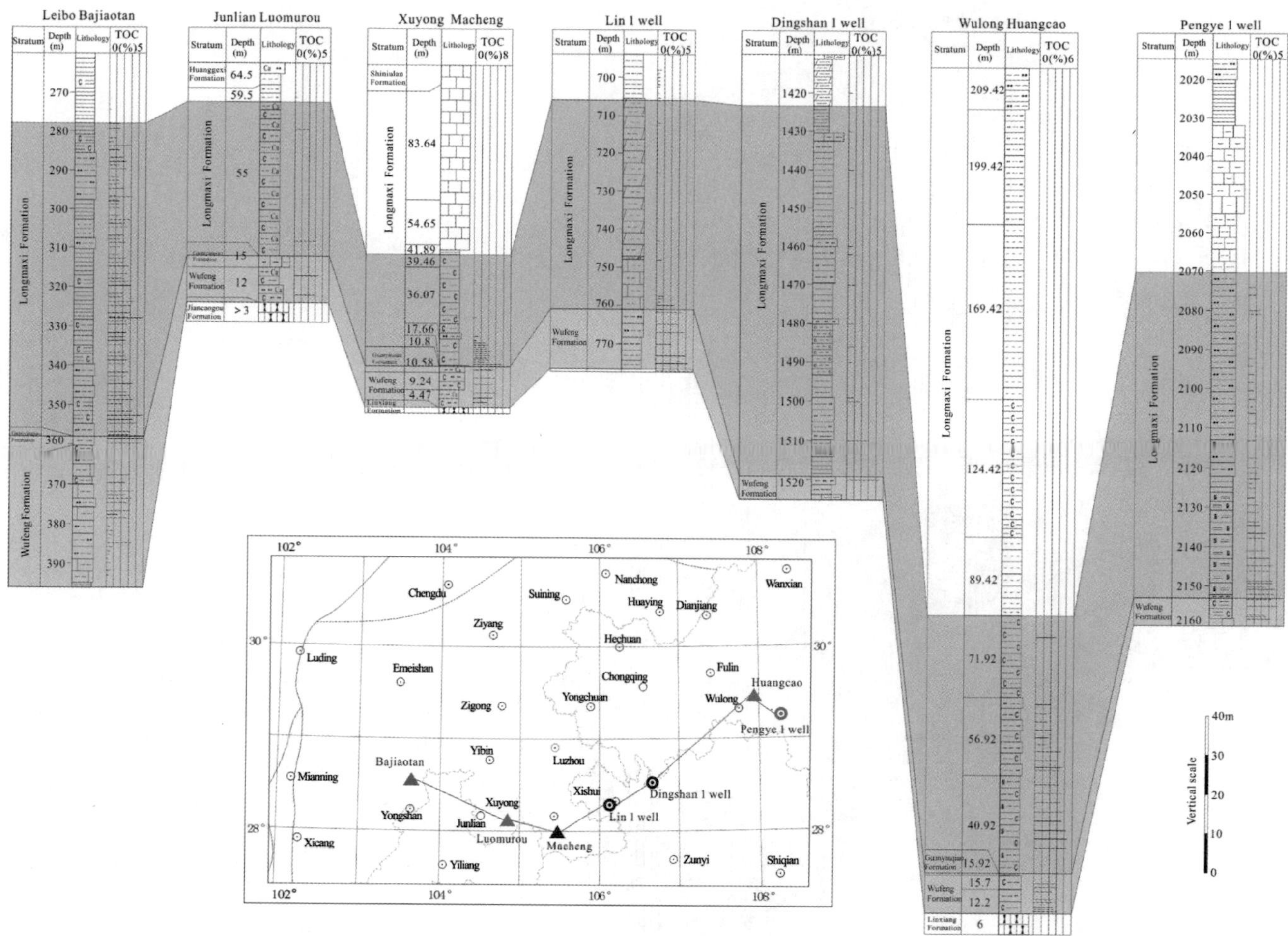

Fig. 4.25 Comparison of east–west sedimentary and shale gas characteristics of Wufeng-Longmaxi Formation

carbonaceous siliceous mudstone with a thickness of 9.7 m. The Wufeng Formation in Well Pengye 1 is carbonaceous shale with a thickness of 8 m. From west to east, the sedimentary thickness of the Wufeng Formation gradually decreases, and the thickness of the Wufeng Formation near Well Dingshan 1 (Qijiang) decreases to the thinnest.

During the deposition of the Longmaxi Formation, the black lithological section of the lower Longmaxi Formation in the Leibo Bajiaotan profile on the west side mainly black carbonaceous shaled, silty shale and mudstone is developed, with a favorable thickness of about 64 m, extending eastward to the Luomurou area. The thickness of the black calcareous carbonaceous mudstone section of the Longmaxi Formation is reduced to 40 m, and the carbonate content is higher than that of Bajiaotan. The black shale section of the Longmaxi Formation in Macheng, Xuyong continues to thin to 6.36 m, and the lower section of the Longmaxi Formation in this area mainly deposits black carbonaceous mudstone. Calcareous mudstone, dolomitic mudstone, mudstone and silty mudstone, mudstone, carbonaceous mudstone, and argillaceous siltstone are developed in the lower member of Longmaxi Formation in Well Lin 1 and Well Dingshan 1, respectively, with a deposition thickness of 63 m and 100 m, respectively. To the east, the lower part of the Longmaxi Formation in the Huangcao area of Wulong, Southeast Chongqing, black carbonaceous mudstone and siliceous mudstone with a thickness of 56 m are deposited. The lower black shale section of the Longmaxi Formation in Well Pengye 1 is siliceous shale and silty mudstone with a thickness of 82 m. From west to east, the thickness of the black shale section of the Longmaxi Formation gradually decreases, and the thickness decreases to the thinnest in the Macheng area of Xuyong. To the east, the sedimentary thickness of Well Lin 1, Well Dingshan 1 and Well Pengye 1 begins to thicken again.

In the north–south direction, five profiles of stone pillars in Southeastern Sichuan are selected for comparison, including Shizhu Qiliao, Wulong Huangcao, Yanhe Xinjing, Zheng'an Tuping and Fenggang Dongkala (Fig. 4.26). The Wufeng Formation deposits carbonaceous shale and siliceous shale with a thickness of 6 m; the black shale in the lower Longmaxi Formation is a deep-water shelf facies

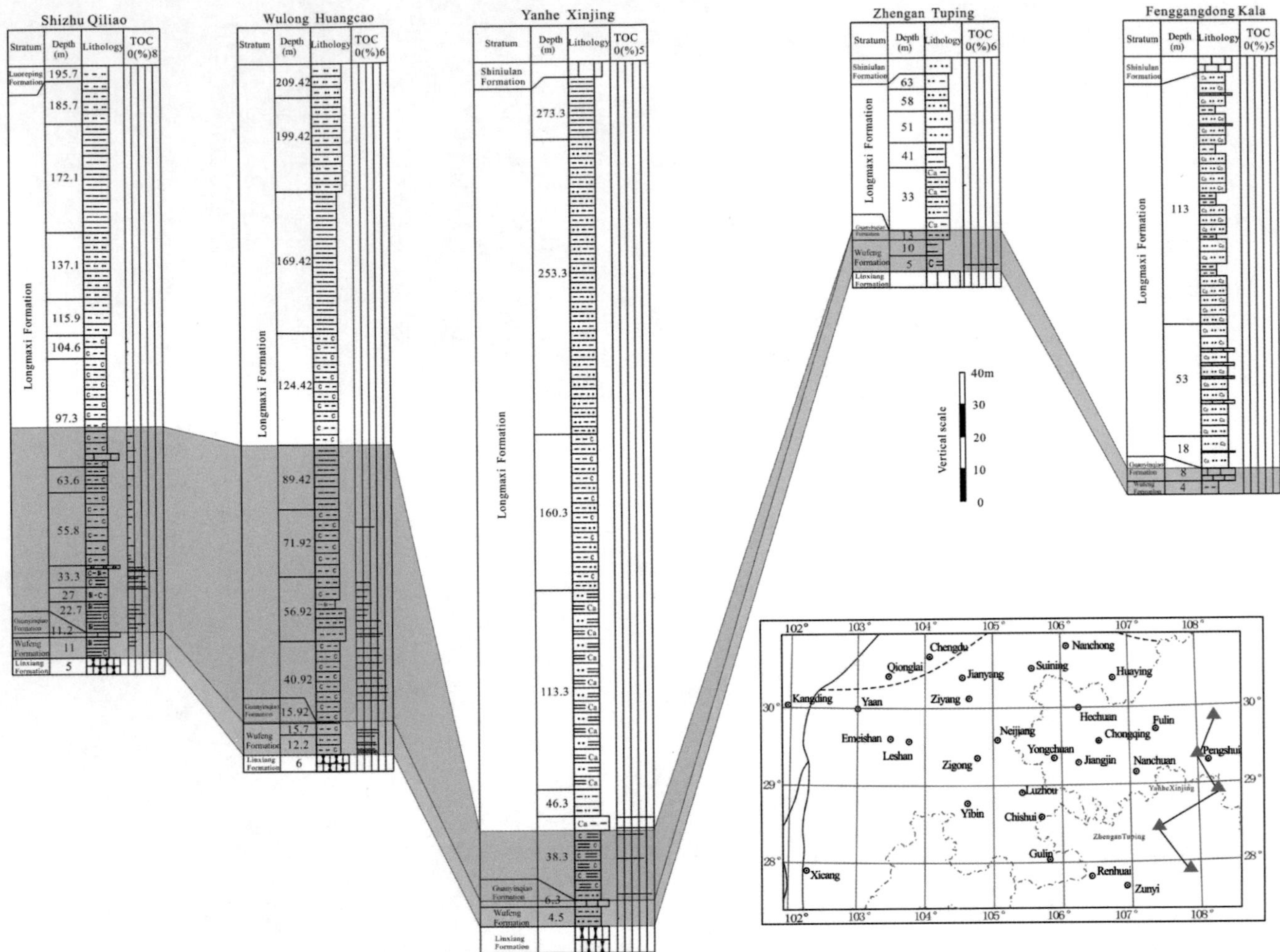

Fig. 4.26 Comparison of the characteristics of the Wufeng-Longmaxi Formation NS-trending sediments and shale gas

deposit. To the south, the Huangcao area, black carbonaceous silty mudstone with a thickness of 9.7 m is deposited in the Wufeng Formation, and black carbonaceous mudstone and siliceous mudstone with a thickness of 56 m are deposited in the lower part of the Maxi Formation. The Wufeng Formation in Yanhe Xinjing profile is only 4.5 m thick carbonaceous silt mudstone, the lower Longmaxi Formation is mainly 32 m thick carbonaceous shale, and Wufeng Formation is mud shale and carbonaceous shale in the Zhengan Tuping profile. The Longmaxi Formation has no carbonaceous or carbonaceous mud shale since the lower member and has obvious shallow-water characteristics. The Dongkala profile of Fenggang on the Southernmost side is also the black carbonaceous mud shale with relatively thin sedimentary thickness in the Wufeng Formation, but the Longmaxi Formation has no black carbonaceous or carbonaceous shale sections since the beginning of deposition and is a gray calcareous siltstone deposit.

4.2.4 Lithofacies Paleogeography of Favorable Intervals for Shale Gas

1. Lithofacies paleogeography of the Ordovician Wufeng Formation and its roof and floor depositional period

Due to the Caledonian movement at the end of the Middle Ordovician, plate tectonic compression in South China continued to intensify (Zhou et al. 1993; Liu et al. 1993; Liu and Xu 1994; Xu et al. 1996, 2004; Shu et al. 2008; Shu 2012; Yan et al. 2010a; Charvet et al. 2010; Faure et al. 2009). The Central Sichuan Uplift and the Southern Sichuan-Central Yunnan-Central Guizhou Uplift continue to expand. In other areas, due to the continuous enclosure of each uplift, the relative sea-level rises, and the bordered carbonate platform is submerged (Zhou et al. 1993; Li et al. 1997; Huang et al. 2011; Yu et al. 2011b; Ge 2012), a carbonate gentle slope

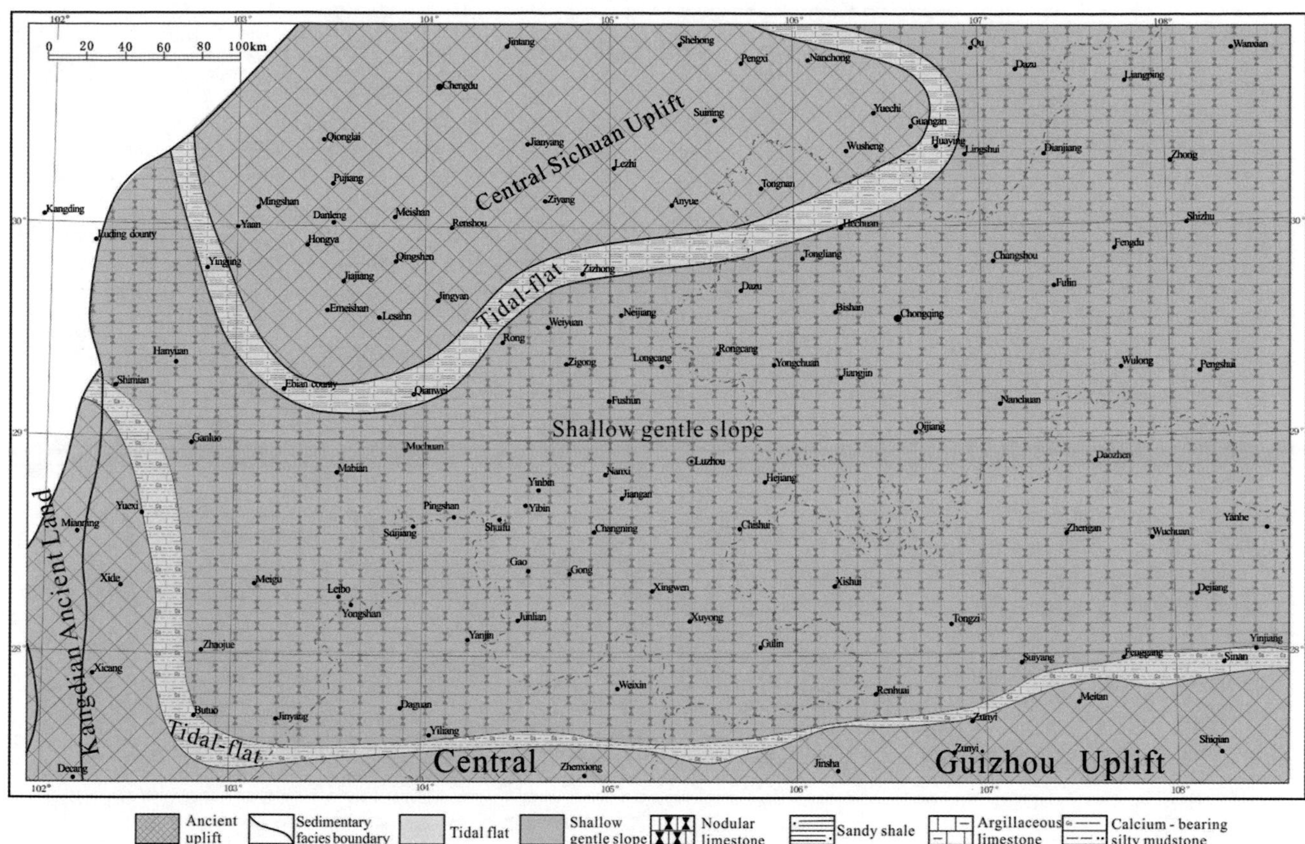

Fig. 4.27 Lithofacies paleogeographic map of the Linxiang Formation in Southern Sichuan and its periphery during the depositional period

characterized by tortoise-cracked limestone-nodular limestone in the Baota Formation and Linxiang Formation were deposited (Fig. 4.27), the carbonate gentle slope occupies most of the area around the Sichuan Basin, roughly located in Southeast of the line of Tianquan-Yingjing-E'bian-Qianwei-Weiyuan-Zizhong–Huaying. In the north of the line of Xichang-Jinyang-Daguan-Renhuai-Suiyang-Fenggang-Yinjiang, the Central Sichuan Uplift roughly surrounds the line of Baoxing-Lushan-Ya'an-Wusheng-Guangan-Nanchong; the tidal flat facies deposits between it and the line of Tianquan-Yingjing-Ebian-Qianwei-Weiyuan-Zizhong-Huaying. Sandy shale and argillaceous limestone are the main properties, and the sedimentary structure is dominated by vein-like bedding and small sand-grained bedding. The uplift range in the Southern part of Central Guizhou is roughly along the line of Butuo-Yiliang-Jinsha-Zunyi-Shiqian, and the tidal flat facies deposits between it and the line of Xichang-Jinyang-Daguan-Renhuai-Suiyang-Fenggang-Yinjiang. The lithology is mainly calcium-bearing silty mudstone and argillaceous limestone.

In the late Ordovician, with the continuous collision and extrusion of the Yangtze and Cathaysia landmasses (Zhou et al. 1993), the areas of the Central Guizhou Uplift, Xuefeng Uplift and Central Sichuan Uplift continued to increase. The Yangtze Continent has changed from the original craton basin to a post-uplift basin which is developed above the craton and bounded by various marginal uplifts. The sedimentary lithology has also changed from pure carbonate rocks to black carbonaceous and siliceous shale in the Wufeng Formation. The Wufeng Formation in the E'bian and Qianwei areas around the Central Sichuan Uplift is mainly composed of dolomitic shale and marl, while the Zhaojue, Weixin and Renhuai areas deposits in the Northeast of the Western Sichuan-Central Yunnan-Central Guizhou Uplift are mainly composed of calcareous siltstone and calcareous siltstone. The carbon-bearing silt-bearing micrite limestone and the supply of terrigenous clasts near the uplift margin are also relatively abundant, mainly composed of mixed tidal flat deposits.

The shallow shelf facies are located outside the tidal flat facies, and the lithology is dominated by carbon-bearing silt-bearing calcareous mudstone, carbon-bearing silt-bearing mudstone, carbon-bearing siltstone, cloud-bearing silt-bearing carbonaceous mudstone, which located in E'bian-Mabian-Ganluo in Western Sichuan. These areas mainly develop cloud-bearing silt-bearing carbonaceous mudstone and carbon-bearing siltstone, and the pyrite lens develops in the carbon-bearing siltstone. The Southern Yanjin-Weixin and Southeastern Wuchuan-Dejiang area mainly develops carbonaceous mudstone compared with the

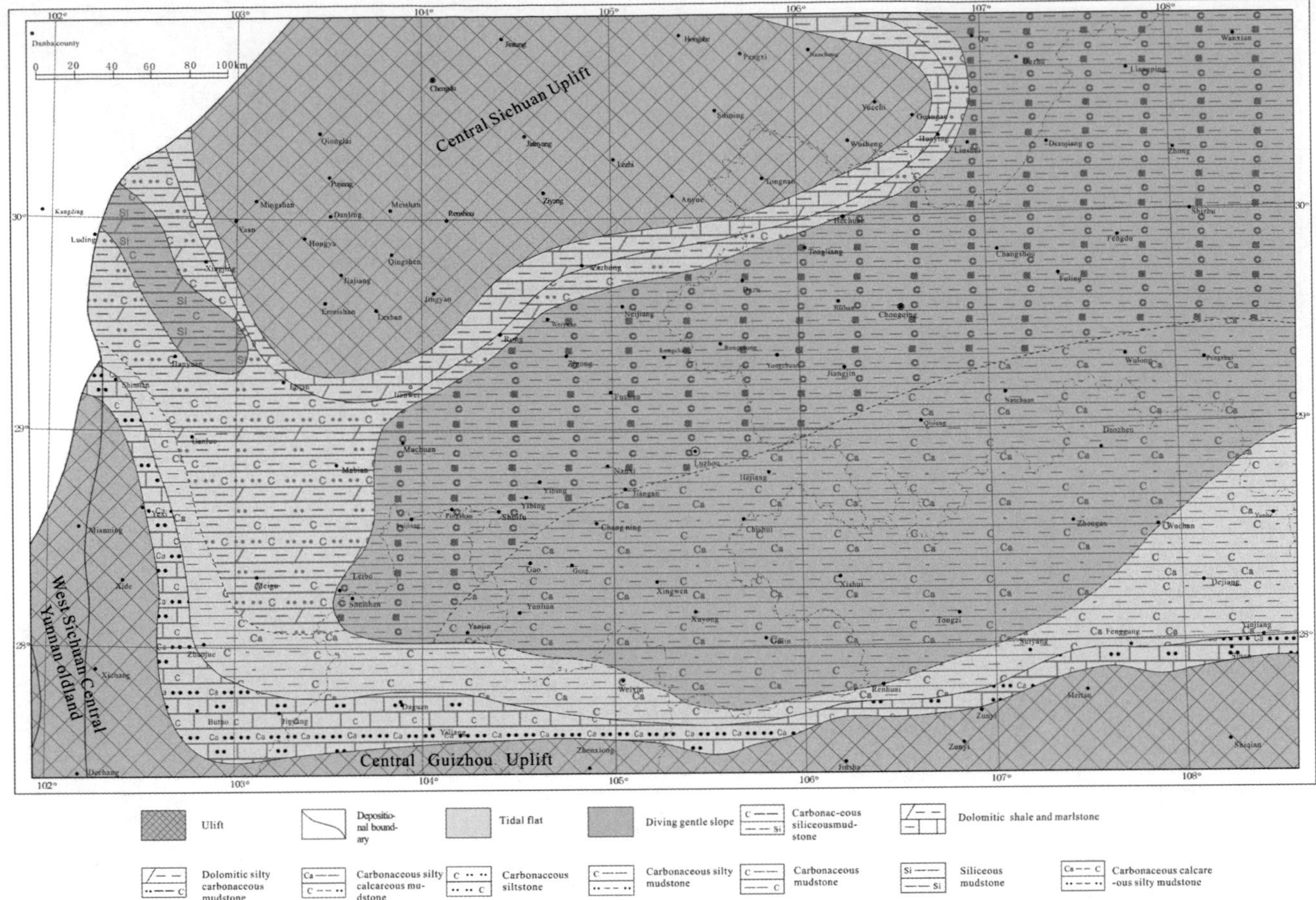

Fig. 4.28 Lithofacies paleogeographic map of the organic-rich mud shale member of the Wufeng Formation in Southern Sichuan and its periphery during the depositional period

west, the silt-bearing calcareous mudstone and the carbon-bearing silty sandy mudstone are gradually more calcareous, but they are still shallow shelf deposits.

Deep-water shelf facies mainly develop in the Northeast of the line of Yanhe-Zheng'an-Tongzi-Weixin-Yanjin-Leibo-Muchuan, which are divided into two types of microfacies according to different sedimentary lithologies. The Southeastern side of Pengshui mainly deposits calcic-carbonaceous silt mudstone and silt-bearing carbonaceous calcareous mudstone. Compared with shallow shelf facies, the sedimentary facies have more carbonaceous components, generally more than 2%, and carbonate components such as calcite or dolomite also occupy a large proportion in the rock. The main rock types in the Northwest of the line of Changning-Hejiang-Qijiang-Nanchuan-Pengshui are mainly carbonaceous mudstone and siliceous mudstone depositions. The rocks in these sedimentary facies contain almost no carbonate components or only a small amount, mainly composed of siliceous minerals. In addition to graptolite organisms, siliceous radiolarians are also quite developed in this set of rocks. There are also small-scale deep-water shelf facies in the Tianquan-Hanyuan areas in the Western Sichuan (Fig. 4.28) Basin. The sedimentary lithology is dominated by dolomitic siliceous mudstone and carbonaceous siliceous mudstone. The main feature of this microfacies is the lower part of the Wufeng Formation. The deposited black mudstone not only contains abundant siliceous minerals (mostly siliceous sponge spicules), but also contains more dolomite minerals, and the rhombic crystal form of dolomite is completed. As far as the Wufeng Formation is concerned, black shale spreads throughout the entire strata. Generally speaking, the organic carbon content in the deep-water shelf facies is generally greater than 2%, which is the most favorable distribution area of black shale. It followed by the shallow shelf facies area, and the organic carbon content in this facies area is mostly between 0.5 and 2%, which is a sub-favorable area.

In the mid-Hirnantian period, continental glaciers erupted around the world, causing a significant drop in sea level. The black graptolite shale facies deposited in the Wufeng Formation was gradually replaced by the shallow shell facies of the Guanyinqiao Formation. The Guanyinqiao Formation, which is several tens of centimeters to several meters, with a maximum of about ten meters, during the depositional

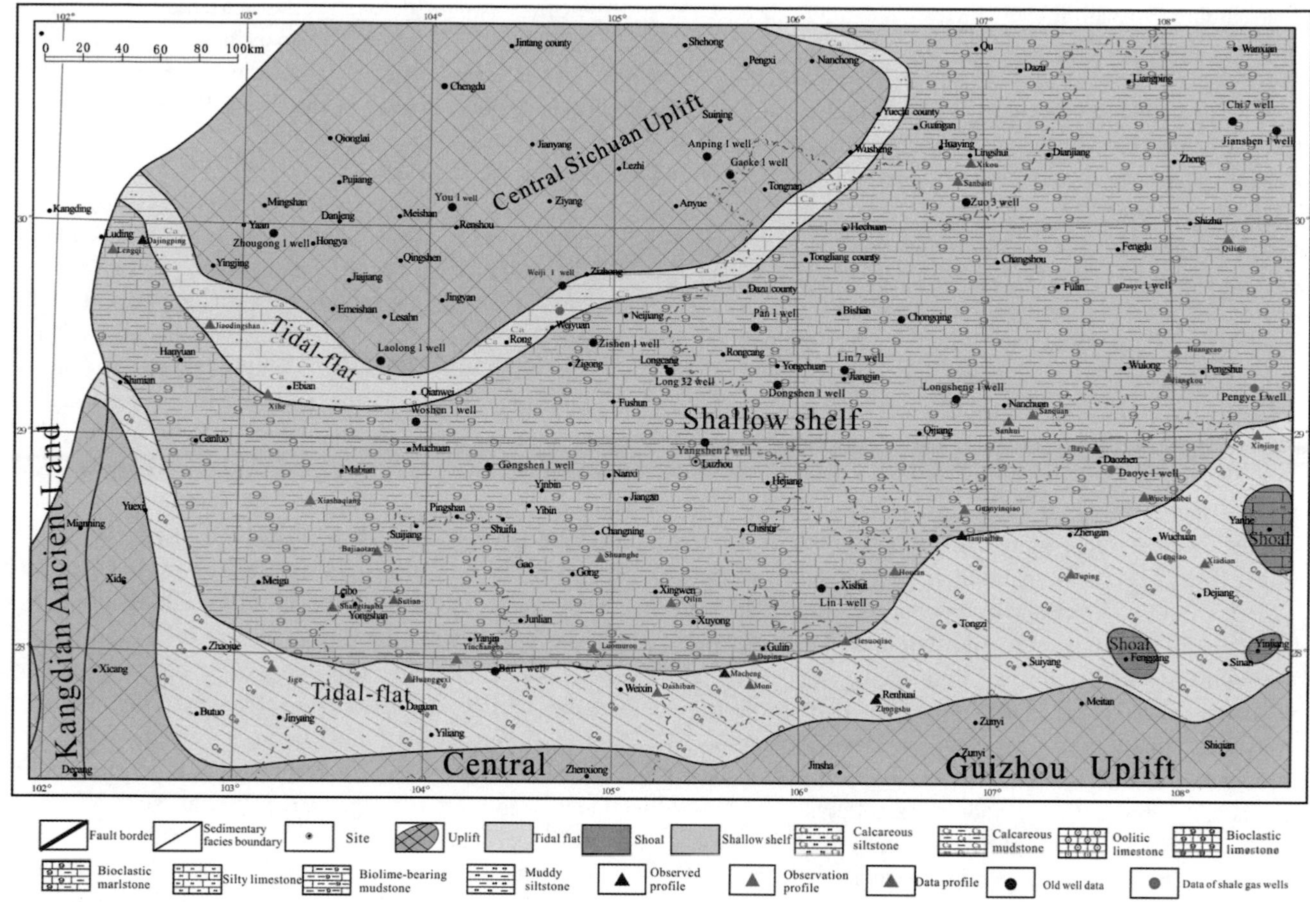

Fig. 4.29 Lithofacies paleogeographic map of the Guanyinqiao Formation in Southern Sichuan and its periphery during the depositional period

period of the Guanyinqiao Formation, the area of the Western Sichuan-Central Yunnan-Central Guizhou Uplift in the Southern part of the Central Sichuan Uplift are also increased accordingly, and the edge of the uplift is still tidal flat facies. The cross-sectional data of Xihe and Hanyuan-Jiaodingshan profiles show that the tidal flat facies calcareous siltstone and limestone are mainly developed around the Central Sichuan Uplift, and their range is roughly located in between the line of Baoxing-Ya'an-Laolong 1 well-Zizhong-Guang'an-Nanchong and the line of Dajingping-Hanyuan-Ganluo-Qianwei-Rongxian-Weiyuan-Neijiang-Dazu-Tongliang-Hechuan-Huaying. There are large-scale deposits of shallow shelf facies on the outside. Detrital marl and biogenic calcareous mudstone are predominant, and the Hirnantianbei fauna generally develops in this set of rocks. The brachiopod assemblage on the bottom of the sea (Chen et al. 2000), often accompanied by trilobites dominated by Dalmanitina, indicating a shallow-water continental shelf environment with a water depth of 20–60 m, and the facies belt extends Southward to the line of Jinyang-Daguan-Weixin-Tiesuoqiao-Hanjiadian-Zheng'an-Yanhe Xinjing; The periphery of the Western Sichuan-Central Yunnan-Central Guizhou Uplift in the south is a ring-shaped clastic tidal flat facies, sedimentary calcareous mudstone and silty mudstone, with small sand-grained bedding and vein bedding are dominant; And high-energy beach facies deposits are founded in the tidal flats on the Southeast side of the tidal flat, such as Fenggang Dongkala profile, Yanhe Xinjing profile, Yinjiang Zhoujiaba profile, etc. The Dongkala area develops bright crystal oolites limestone (Fig. 4.29), and the Xinjing and Zhoujiaba areas are dominated by the sedimentary biogenic limestone.

2. Lithofacies paleogeography during the deposition of the Silurian Longmaxi Formation

With the end of the ice age at the end of the Hirnantian period, the glaciers melted on a large scale, the climate gradually warmed up, large-scale transgression occurred, and the relative sea level gradually rose. In addition, the strong compressional uplift and crustal deflection became deeper since the late Ordovician. The neritic environment after intracontinental uplift after the tectonic transformation in the middle and late Ordovician created a unique development environment for black carbonaceous shale in the early (Lower) Longmaxi Formation, except for the

Guanyinqiao Formation shell facies sandwiched between it and the Wufeng Formation Marl. Dolomitic limestone, claystone and other shallow facies (the sedimentary water body is relatively shallower) were deposits. It can be said that the black carbon-bearing and carbonaceous mud shale of shelf deposition since the Ordovician Wufeng Formation has been developed in succession. This set of black rock series has a wide distribution range and is thicker than Wufeng Formation. The organisms are mainly graptolites. In addition to the main body of the Wufeng Formation, the graptolite types include the double graptolite family, the two-shaped graptolite family and the single graptolite family. It is more common, and its differentiation and abundance are increasing, which may be related to the melting of glaciers and the changes in the living environment caused by the first sea-level rise in the Silurian period. Judging from the development of the entire study area, the lithology of this profile is the most favorable interval for the development of shale gas, but this set of black carbonaceous mud shale is not developed in the study area. The shelf facies are a favorable development area for black carbonaceous mud shale gas, while the tidal flat facies developed in this period are an unfavorable environment for shale gas development, but its distribution range is relatively limited. The characteristics of the ancient maps of this period are as follows.

The periphery of the uplift is a narrow tidal flat zone of detritus intercalated with carbonate rocks, which quickly enters the shallow shelf. It indicates that the terrain slope is relatively large and the distribution range is relatively limited (Fig. 4.30). Generally, it's dominated by mudstone, siltstone and fine sandstone. Bioclastic limestone and marl are developed in the sandstone and mudstone in some areas. The Northern margin is represented by the Mabian Xiashacai profile, where a set of gray, gray-green sandy shale and light purple-red dolomitic siltstone intercalated with thin layers of purple-red calcareous silt mudstone is deposited. Small-scale cross-bedding can be seen in local sections of the stratum, and a large number of light purple-red siltstone and dolomitic siltstone lenses can be seen in the gray-green sandy shale and purple-red silty mudstone layers. Its distribution range is roughly along the line of Tianquan-Mabian-Hechuan-Guangyuan along the outer edge of the Central Sichuan Uplift; it's distributed in a ring shape, and it's in the shape of a dustpan in some parts of Mabian. The tidal flat facies deposits in the Southern margin show two characteristics: First, mixed tidal flat deposits developed in Butuo profile and other profiles in the SouthWestern margin, and mudstone, sandstone and marl deposits developed, which may reflect the large uplift area in the Southern margin, and the uplift area is characterized by rich source supply types. Second, on the Southeast side of the study area, combined with field and indoor thin section research, it was found that there was a set of siltstone deposits in Gaoqiao and Xiadian areas of Wuchuan, and local layers were calcareous siltstone.

The shallow shelf is mainly developed in the Ganluo-Meigu-E'bian areas in Western Sichuan and the Zheng'an-Yanhe areas in Southeastern Sichuan (Fig. 4.31). While in the Zheng'an Tuping and Yanhe Xinjing areas, it is mainly a set of silty carbonaceous mud shale, calcareous silty carbonaceous mud shale. Along the Jinyangjige-Weixin and other places in the Southwest corner of Sichuan Basin, there are tidal flat facies. These lithofacies depositional areas may be affected by the Kangdian Ancient Land and the Central Guizhou Uplift, and the marl developed in this area and the argillaceous limestone interlayer developed in the tidal flat facies are presumed to have the same provenance. The shale gas development conditions are general, and it's an area with poor shale gas development conditions in the shallow shelf facies depositional environment.

The deep-water shelf environment occupies most of the study area. The east side is roughly located in the north and west of the line of Huaying-Tongliang-Zhong-Rongxian-Leibo-Yanjin-Junlian-Weixin-Tongzi-Daozhen, and the west is Tianquan-Hanyuan. The source is a small-scale deep-water continental shelf environment with central deposition. There are mainly four lithofacies assemblages in the area: a. calcareous mudstone + carbonaceous argillaceous limestone; b. calcium-bearing silty carbonaceous mudstone + siltstone with calcium-bearing Carbonaceous mudstone; c. Carbonaceous mudstone + silty mudstone; d. Calcium-siliceous carbonaceous shale + calcium-bearing siliceous carbonaceous shale. In the "Tianquan-Hanyuan" deep-water shelf deposition center, calcium-bearing siliceous carbonaceous mud shale + calcium-bearing siliceous carbonaceous mud shale are mainly developed, with the Tianquan Dajingping profile as a typical representative. Its

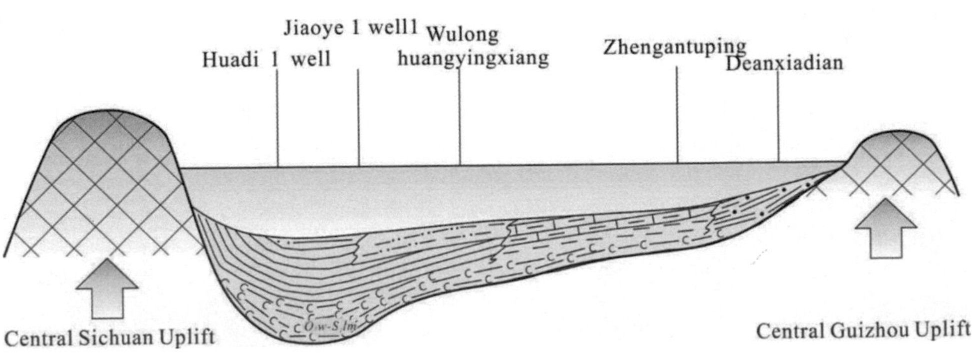

Fig. 4.30 Sedimentary model of the Silurian Longmaxi Formation in Southern Sichuan and its periphery

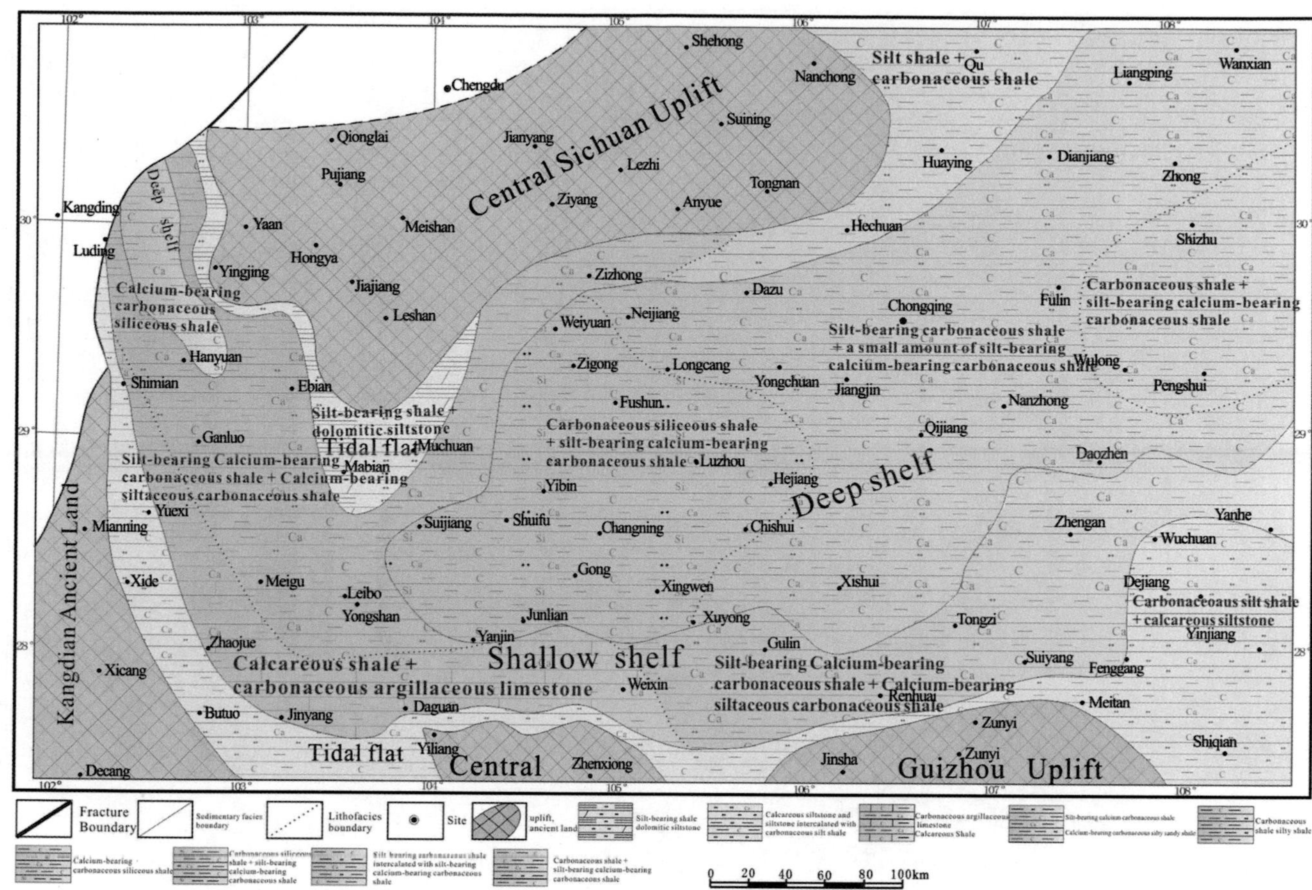

Fig. 4.31 Lithofacies paleogeographic map of the upper Longmaxi Formation in Southern Sichuan and its periphery

influence range is mainly in Tianquan-Hanyuan areas, Northwest-Southeast which is a set of black siliceous carbonaceous mud shale develops in the deep-water shelf sedimentary center of "Yibin-Zigong-Longchang-Luzhou-Nanxi" in a narrow strip in this direction which can extend Xingwen, Gongxian and other places to the south, north to Fushun, Luxian and Yongchuan areas. The "Wulong-Pengshui" deep-water shelf deposition center mainly develops black carbonaceous mud shale, and local intervals are calcium-bearing silty sandy carbonaceous mud shale.

At the beginning of the late Ludan period, the tectonic compression from the Southeast Cathaysian block became the dominant factor in the evolution of paleogeography, followed by global sea-level decline. With the continuous conduction of the extrusion, the whole area showed an overall uplift and regression. The upper part of the Longmaxi Formation began to deposit gray-green, gray-yellow sandy shale and siltstone, and the upper part of the Longmaxi Formation near the Northern part of the Central Guizhou Uplift is mostly gray-green sandy shale, siltstone and argillaceous limestone. The color of sedimentary lithology becomes lighter than that of the lower member, and the terrigenous debris increases and the grain size becomes coarser. The Longmaxi Formation is a sedimentary sequence that becomes shallower from bottom to top. At the late stage of Longmaxi Formation, with the continuous intensification of compressional tectonic action, the Xuefeng Uplift and the land uplift and expansion in central and Southern Hubei, and the marginal facies (clastic coast, delta) also continued to advance, laying the foundation for the development of progressive clastic sand bodies. In short, the entire Longmaxi Formation, except for the tidal flat facies still developed around the Ancient continents and Uplifts, which was basically in a shallow shelf environment during the lithologic depositional period of the middle and upper members, occupying most of the study area (Fig. 4.32).

4.3 Organic Geochemical Characteristics

4.3.1 Types of Organic Matter

The type of organic matter is one of the most important factors determining the source rock quality, but it does not affect the amount of gas produced by source rock, but only affects the adsorption rate and diffusion rate of natural gas. The total

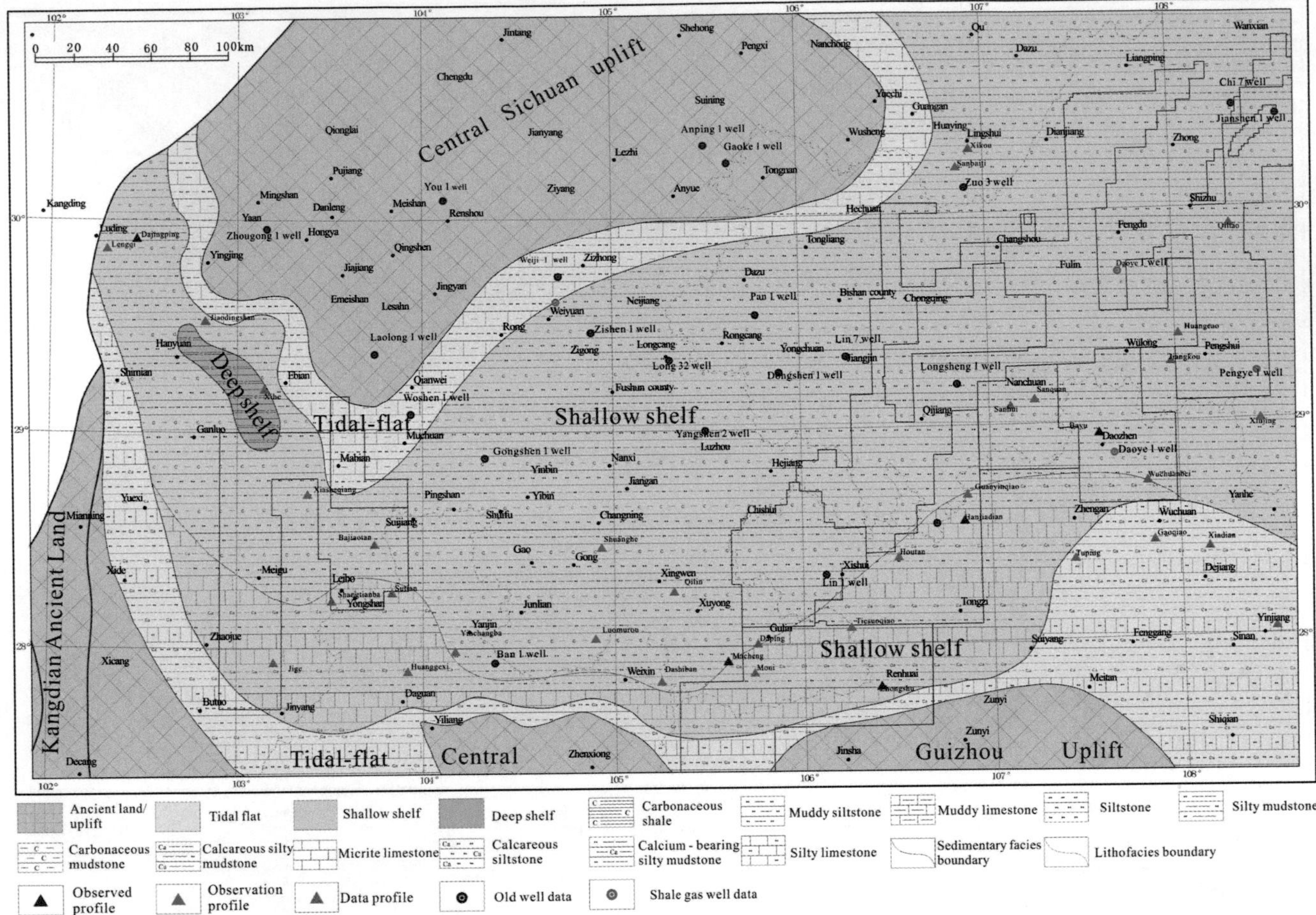

Fig. 4.32 Lithofacies paleogeographic map of the Guanyinqiao Formation in Southern Sichuan and its periphery during the depositional period

amount and thermal maturity of organic matter are the important variables controlling the gas-producing capacity of source rock (Bai 2012). The type of organic matter is classified by the input organisms and the physicochemical and biochemical environment of deposition and initial burial. Organic geochemistry and organic petrology can be used to evaluate organic matter types. For high and overmature source rocks, most of the conventional evaluation parameters that can be applied have failed, but the organic petrological characteristics of kerogen composition and carbon isotope ($\delta^{13}C$) of kerogen in organic geochemistry are still reliable.

1. Kerogen composition

(1) Ordovician Wufeng Formation

The outcrop sample's result shows that kerogen macerals of the Ordovician Wufeng Formation are mainly composed of sapropelic amorphous and inertinite. The relative abundance of sapropelic amorphous is 30–80%, with an average of 67.5%. The relative abundance of inertinite ranges from 1 to 72%, with an average of 23.8%. A few samples contained vitrinite, with occasional traces of algal bodies. Kerogen's color is brown-dark brown. Statistics showed that the kerogen type index was 10–64, and type II_1 was the main type of kerogen, followed by type II_2 (Table 4.4). The carbon isotope values of black shale kerogen in Wufeng Formation range from −31.0 to −27.4% indicating that the organic matter is dominated by type I and type II_1 kerogen, and the sapropelic formation content varies greatly. The content of asphalt is high, and it mostly filled in the cracks and micropores of mud and shale. Organic matter is gray-dark brown and predominantly brown, with irregular clumpy aggregates (Fig. 4.33).

(2) Silurian Longmaxi Formation

The black shale of the lower part of Longmaxi Formation in the study area has reached the overmature stage of maturity, and the organic matter has lost its fluorescence and becomes completely opaque, so the maceral analysis can only be carried out by reflectance observation. The kerogen macerals of the black shale in the lower part of Longmaxi Formation are mainly sapropelic amorphous, inertinite and vitrinite

Table 4.4 Microscopic components and types of carbonaceous mudstone caseous roots of the Wufeng Formation in Southern Sichuan and its periphery

Layer	Original number	Lithology	Sapropelite %			Humic amorphous %	Vitrinite %	Inertinite %	Sapropel group color	Type index	Type
			Amorphous	phycoplast	Total						
O_3w	DHP-S3	Mud-bearing limestone	80		80			20	Tan	60.00	II1
O_3w	DJP-B4	Carbonaceous siliceous shale	69		69			31	Tan	38.00	II2
O_3w	DJP-B6	Siliceous shale	60		60			40	Tan	20.00	II2
O_3w	DJP-B9	Carbonaceous siliceous shale	63		63			37	Tan	26.00	II2
O_3w	DJP-B12	Carbonaceous siliceous shale	55		55			45	Tan	10.00	II2
O_3w	DJP-B14	Carbonaceous shale	60		60			40	Tan	20.00	II2
O_3w	DJP-B16	Carbonaceous siliceous shale	62		62			38	Tan	24.00	II2
O_3w	DJP-B30	Calcium-bearing carbonaceous shale	65		65			35	Tan	30.00	II2
O_3w	HCP-B1	Silica-bearing carbonaceous shale	58		58			42	Tan	16.00	II2
O_3w	HCP-B3	Carbonaceous shale	60		60			40	Tan	20.00	II2
O_3w	HJDP-B2	Carbonaceous shale	40		40		25	35	Tan	−13.75	III
O_3w	HJDP-B4	Carbonaceous shale	30		30		15	55	Tan	−36.25	III
O_3w	JDSP-B10	Dolomitic shale	68		68			32	Tan	36.00	II1
O_3w	JDSP-B13	Dolomitic shale	65		65			35	Tan	30.00	II2
O_3w	JGP-B6	Carbonaceous mudstone	76		76			29	Brown	47.00	II1
O_3w	JKP-B2	Carbonaceous shale	58		58			42	Tan	16.00	II2
O_3w	XHP-B1	Argillaceous siltstone	71		71			29	Dark Brown	42.00	II1
O_3w	XHP-B3	Carbonaceous mudstone	78		78			22	Dark Brown	56.00	II1
O_3w	LQP-B1	Carbonaceous shale	82		82			18	Tan	64.00	II1
O_3w	QGP-S2	Carbonaceous shale	56		56			44	Tan	12.00	II2
O_3w	QGP-S3	Carbonaceous shale	60		60			40	Tan	20.00	II2
O_3w	RZP-S1	Carbonaceous calcareous mudstone	78		78			22	Brown	56.00	II1
O_3w	RZP-S2	Carbonaceous calcareous mudstone	72		72			28	Brown	44.00	II1
O_3w	RZP-S3	Carbonaceous mudstone	68		68		6	26	Brown	37.50	II2
O_3w	RZP-S4	Carbonaceous mudstone	65		65		5	30	Tan	31.25	II2
O_3w	RZP-S5	Carbonaceous mudstone	66		66		12	22	Tan	35.00	II2
O_3w	RZP-S6	Carbonaceous mudstone	62		62		10	28	Tan	26.50	II2
O_3w	STBP-B5	Carbonaceous mud shale	75		75			25	Brown	50	II1
O_3w	XSP-S3	Black shale	62		62			28	Tan	34	II2
O_3w	XQP-S4	Carbonaceous mudstone	82		82			18	Tan	64	II1
O_3w	YYP-S2	Carbonaceous calcareous mudstone	73		73			27	Tan	46	II1

Note See Fig. 4.5 for section code

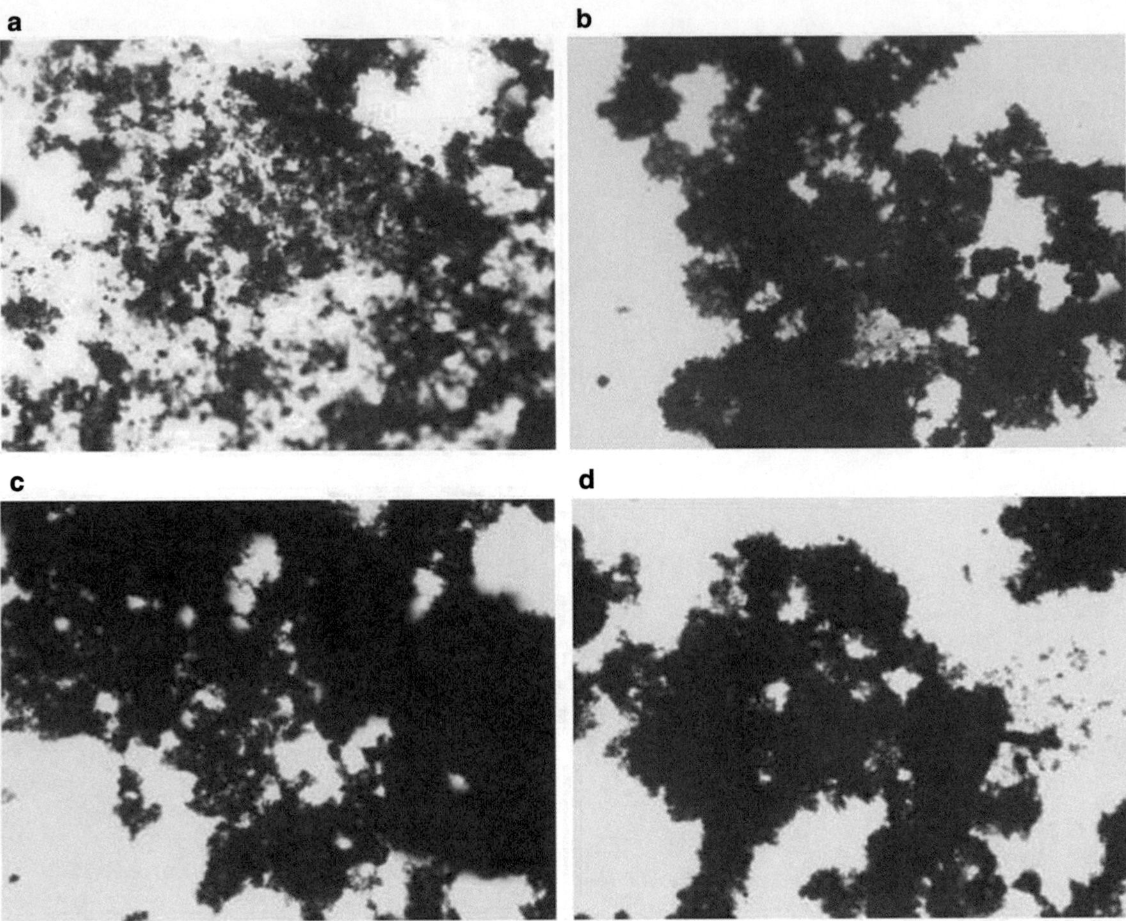

Fig. 4.33 Microscopic characteristics of Kerogen of Wufeng Formation in Southern Sichuan and its periphery **a** gray-brown organic matter (Wulong Jiangkou); **b** brown-black organic matter (E'bian Xihe); **c** Brown-black organic matter (Xuyong Macheng); **d** brown-black organic matter (Nanchuan Sanquan)

(Table 4.5). The sapropelic content varies greatly, and the organic matter is gray-dark brown and black-brown, showing an irregular mass aggregation (Fig. 4.34). The content of sapropelic amorphous in the black shale from outcrop samples is 38–89%, with an average of 70%. The content of inertinite ranges from 11 to 45%, with an average of 28.2%. Unstructured vitrinite was found in some samples with an average of 8.7%. The kerogen-type index ranges from 10 to 78, with an average of 40.4. Type II_1 is the dominant kerogen type, followed by type II_2. The sapropelic content in well Daoye 1 ranges from 94 to 99%, with an average of 96.9%, and inertinite and vitrinite contents are both < 5%. The kerogen-type index ranges from 89 to 98%, with an average of 94%, and is typical of type I kerogen (Table 4.6).

Generally, in the Western region and southwestern region of Sichuan, the sapropelic amorphous content is 66–89%, with an average of 76.2%, inertinite content is 11–34%, with an average of 23.8%, excluding vitrinite. The content of sapropelic amorphous in Southeast Chongqing and Northern Guizhou is 28–70%, and inertinite content is 17–39%, with an average of 30.4%. A small amount of vitrinite was detected in Eastern Sichuan, showing type II kerogen. The content of vitrinite in some samples from Guanyinqiao profile in Chongqing is as high as 45%, type III kerogen, which may be affected by the local enrichment of vitrinite. Therefore, the types of kerogen in different regions are different, which may be caused by the influence of palaeogeographic environment. However, there is little difference in the types of kerogen among different sedimentary facies. In outcrop profile, the sapropelic formation in South Sichuan near the center of the basin has the highest content of 82%, which may be related to the deep sedimentary water.

2. Carbon isotopes of kerogen

The criteria for carbon isotope classification of kerogen organic matter are as follows: $\delta^{13}C \leq -29\%$ for type I, −29 to −26% for type II_1, $\delta^{13}C$ −26 to −25% for type II_2, $\delta^{13}C \geq -25\%$ for type III (Wang et al. 2012a). Carbon isotope values of black shale kerogen of the lower part of

Table 4.5 Kerogen meceral and type of black rocks in Lower Longmaxi Formation of Southern Sichuan Basin and its periphery

Layer	Original number	Lithology	Sapropelite %			Humic amorphous %	Vitrinite %	Inertinite %	Sapropel group color	Type index	Type
			Amorphous	phycoplast	Total						
S_1l	HCP-B11	Carbonaceous shale	61	0	61	0	0	39	Tan	22	II_2
S_1l	HJDP-B7	Carbonaceous shale	62	0	62	0	0	38	Brown	24	II_2
S_1l	QGP-S4	Carbonaceous shale	38	0	38	0	45	17	Brown	−12	III
S_1l	QLP-B21	Carbonaceous shale	66	0	66	0	0	34	Brown	32	II_2
S_1l	XMP-S8	Carbonaceous shale	65	0	65	0	4	31	Tan	31	II_2
S_1l	XMP-S12	Carbonaceous shale	72	0	72	0	2	26	Tan	44.5	II_1
S_1l	SBTP-B11	Carbonaceous shale	62	0	62	0	12	26	Brown	27	II2
S_1l	SQP-B8	Carbonaceous shale	71	0	71	0	0	29	Tan	42	II_1
S_1l	SQP-B9	Siliceous carbonaceous shale	68	0	68	0	0	32	Tan	36	II_2
S_1l	XDP-B4	Carbonaceous shale	68	0	68	0	0	32	Brown	36	II_2
S_1l	XJP-B22	Carbonaceous shale	73	0	73	0	0	27	Brown	46	II_1
S_1l	XJP-B13	Carbonaceous shale	67	0	67	0	0	33	Tan	34	II_2
S_1l	DHP-S4	Calciferous carbonaceous shale	89	0	89	0	0	11	Tan	78	II_1
S_1l	DJP-B29	Carbonaceous shale	69	0	69	0	0	31	Tan	38	II_2
S_1l	DJP-B38	Silty shale	73	0	73	0	0	27	Tan	46	II_1
S_1l	JDSP-B14	Silty shale	66	0	66	0	0	34	Tan	32	II_2
S_1l	JGP-B1	Calcium-bearing mud shale	78	0	78	0	0	22	Brown	56	II_1
S_1l	XHP-B5	Carbonaceous mud shale	80	0	80	0	0	20	Dark Brown	60	II_1
S_1l	LMRP-S4	Carbonaceous shale	72	0	72	0	0	28	Brown	44	II_1
S_1l	RZP-B7	Carbonaceous shale	55	0	55	0	12	33	Brownish Yellow	13	II_2
S_1l	STBP-B5	Carbonaceous mud shale	75	0	75	0	0	25	Brown	50	II_1
S_1l	XSP-S3	Black shale	62	0	62	0	0	28	Tan	34	II_2
S_1l	XQP-S4	Carbonaceous shale	82	0	82	0	0	18	Tan	64	II_1
S_1l	YYP-S2	Carbonaceous calcareous shales	73	0	73	0	0	27	Tan	46	II_1

Note See Fig. 4.5 for section code

Longmaxi Formation range from − 31.3 to − 24.3% in Southern Sichuan and its periphery, with an average of −29.4%. Samples with −29% account for 66% (Table 4.7), indicating that the organic matter in the black rock series of Longmaxi Formation is mainly type I kerogen, followed by type II_1 kerogen, which are all favorable types for shale gas generation, and very few samples are type III kerogen. According to the feature of the kerogen components of drilling samples (Table 4.6), the kerogen type is further verified as type I and is very constant in the study area. This feature shows that the organic matter type is mainly sapropel in the study area. This is accordant with the marine sedimentary environment (Bai 2012a).

In addition, the black shale of the lower part of Longmaxi Formation in Southern Sichuan and its periphery was deposited in a deep-water shelf-reducing environment. The organic matter is mainly algae and plankton (graptolite), and with bacteria metabolites in the process of microbial degradation. The organic matter is rich in lipids and proteins, and the sapropelic components with high potential hydrocarbon generation can be formed by further alkylation. In conclusion, the black shale of the lower part of Longmaxi Formation in Southern Sichuan and its periphery is dominated by Type I kerogen, which is also consistent with previous research results (Wang et al. 2000; Chen et al. 2013b; Guo et al. 2013). Vertically, the organic carbon content in the lower part of Longmaxi Formation gradually decreases from bottom-up, and usually kerogen type changes from type I to type II_1, which further confirms that the sedimentary water is gradually shallower.

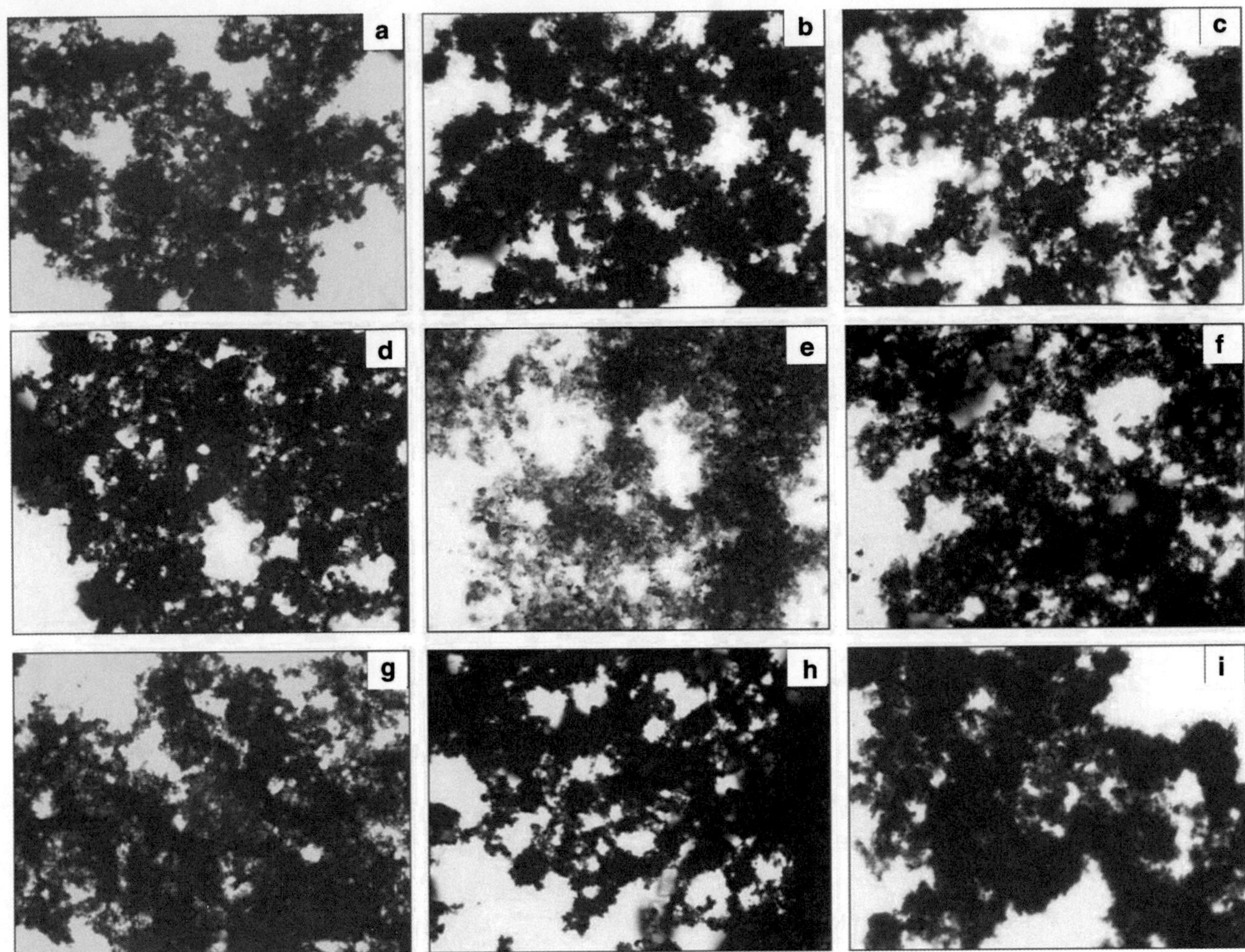

Fig. 4.34 Kerogen microscopic characteristics of black rocks in Longmaxi Formation of Southern Sichuan Basin and its periphery

4.3.2 Organic Matter Abundance of Black Rock Series

1. Ordovician Wufeng Formation

The organic content of black shale in the Wufeng Formation is comparatively high in general, with the content ranging from 1.25 to 6.73%. Compared with Longmaxi Formation, although the thickness of the Wufeng Formation is relatively small, its organic carbon content is relatively high, generally higher than 1%, except in the adjacent area of ancient uplifts. In the shale gas wells, the thickness of black shale of Wufeng Formation in Well Daoye 1 is about 12 m, and TOC is generally greater than 3%, with an average of 3.38%. There are seven samples of 1.8 m thick Wufeng Formation being analyzed from Guanyinqiao section in Qijiang, and the organic carbon content of seven samples is all greater than 5%, with an average of 6.65% (Fig. 4.35).

Regionally, the organic carbon content of the black shale in the Wufeng Formation is stable. According to the contour map of TOC > 0.5% and TOC > 1%, three high TOC regions occurred according to the organic carbon content of the Wufeng Formation (Figs. 4.36, 4.37, 4.38). In Western Sichuan, the organic carbon content mostly ranges from 2 to 3% with Luding-Hanyuan as the center. In Southeastern Sichuan, the organic carbon content generally ranges from 3 to 4% with Xingwen-Xuyong as the center, and the highest TOC is more than 5%. In addition, the Eastern Sichuan area centered on Qijiang has an organic carbon content of 3–5%, and the highest Guanyinqiao section has an organic carbon content of more than 6%. From the contour map of TOC > 0.5%, > 1%, > 2%, it can be seen that during the deposition of the Wufeng Formation, there were also three depositional centers with TOC greater than 2%. In the Western Sichuan area, the deposition thickness is mostly between 6 and 12 m. According to the drilling data of Southeast of Sichuan Province, the sedimentary thickness is

Table 4.6 Meceral and type of kerogen of black rocks in Lower Longmaxi Formation of DY1 well

Layer	Lithology	Original number	Sapropelite %				Humic amorphous %	Vitrinite %	Inertinite %	Type index	Type
			Debris body	Amorphous	phycoplast	Total					
S_1l	Carbonaceous shale with silt	DYI-S63	54	44	1	99	0	1	0	98	I
S_1l	Carbonaceous shale with silt	DYI-S66	71	27	0	98	0	2	0	97	I
S_1l	Carbonaceous shale with silt	DYI-S77	35	62	0	97	0	2	1	95	I
S_1l	Carbonaceous shale with silt	DYI-S82	62	36	0	98	0	1	1	96	I
S_1l	Carbonaceous shale with silt	DYI-S86	32	64	0	96	0	3	1	93	I
S_1l	Carbonaceous shale with silt	DYI-S91	32	65	0	97	0	1	2	94	I
S_1l	Carbonaceous shale with silt	DYI-S95	20	74	0	94	0	2	4	89	I
S_1l	Carbonaceous shale with silt	DYI-S99-1	36	59	0	95	0	3	2	91	I
S_1l	Carbonaceous shale with silt	DYI-S103	45	54	0	99	0	0	1	98	I
S_1l	Carbonaceous shale with silt	DYI-S106	9	87	0	96	0	0	4	92	I

Note See Fig. 4.5 for section code

mostly between 12 and 18 m, and this area also belongs to the thickest deposition area of the Wufeng Formation in the whole region. In addition, the TOC thickness of more than 2% in Southeastern Chongqing is mostly 6–9 m (Figs. 4.39 and 4.40).

2. Silurian Longmaxi Formation

The total organic content (TOC) of the black shale in the lower part of Longmaxi Formation of Southern Sichuan basin and its periphery is between 0.01 and 8.06%, mainly in the range of 0.5–4.0%. In the vertical column, it gradually decreases upward (Fig. 4.41). The variation of organic carbon content can be used as the most intuitive indicator of the boundary between organic-rich shale and non-favorable shale in Longmaxi Formation. The TOC content in South Sichuan is high and uneven; the highest is 8.1%, and the lowest is 0.26%. Horizontally, the average TOC value ranges from 0.02 to 5.37% and is above 1.0% in most of the study area. The average TOC value in South Sichuan and Southwest Sichuan is above 4.0%, and the average TOC value in the marginal area is 2.0–4.0%. TOC value of the surrounding area of the ancient uplift is less than 1%, which is non-favorable for shale gas. In the tidal flat sedimentary area near the uplift, black shale did not develop, TOC is low (Fig. 4.42). The area bounded by the Central Guizhou Uplift and Central Sichuan Uplift had the highest organic carbon content larger than 4% and gradually decreased outwards.

According to the actual situation of exploration and development, the content of organic carbon in Shizhu-Pengshui area in Northeast Chongqing is also high. Similar to TOC distribution features, the thickness of black shale with 0.5% TOC gradually increases from the uplift area to the center of the basin, in which the sedimentary thickness of the Shuifu-Jiangjin area is the largest, with an average thickness of > 100 m (Fig. 4.43).

And distribution pattern of black shale with TOC > 1% is similar. The TOC value in the basin center is higher and the thickness is larger, while the organic carbon content in the uplifting area is reduced and the thickness is smaller (Figs. 4.44 and 4.45). In the south of Wuchuan and east of Suiyang, organic-rich shales did not develop, which is mainly influenced by the Central Guizhou Uplift and the Xuefeng uplift, which provide a large amount of terrigenous clastic and are not conducive to the preservation of organic-rich shales. Close to the center of the basin, the average thickness of black shale with TOC > 1% is more than 100 m, and the hydrocarbon generation potential is very good. The thickness of organic-rich shale in Western Sichuan is relatively thin, and the thickness is more than 60 m, which is the secondary profitable hydrocarbon generation area.

Table 4.7 Characteristics of kerogen carbon isotope of black rocks in Longmaxi Formation of Southern Sichuan Basin and its periphery

Order Number	Original number	Sample Name	δ13C % (PDB)	Order Number	Original number	Sample Name	δ13C % (PDB)
1	DJP-B30	Calcium-bearing carbonaceous shale	−30.2	36	XMP-S9	Carbonaceous calcareous mudstone	−30.4
2	DJP-B32	Calcium-bearing carbonaceous shale	−30.7	37	XMP-S10	Carbonaceous calcareous mudstone	−30.9
3	DJP-B4	Carbonaceous siliceous shale	−30.2	38	XMP-S11	Carbonaceous calcareous mudstone	−31.3
4	DJP-B16	Carbonaceous siliceous shale	−30.8	39	XMP-S12	Carbonaceous calcareous mudstone	−31.1
5	DJP-B26	Carbonaceous shale	−30.2	40	XMP-S13	Carbonaceous calcareous mudstone	−30.9
6	DJP-B27	Carbonaceous shale	−30.1	41	XMP-S14	Carbonaceous calcareous mudstone	−30.9
7	HCP-B1	Carbonaceous shale	−30.8	42	XMP-S15	Carbonaceous calcareous mudstone	−30.7
8	HCP-B11	Carbonaceous shale	−30.8	43	XMP-S16	Carbonaceous calcareous mudstone	−30.8
9	HJDP-B2	Carbonaceous shale	−30.4	44	XMP-S17	Carbonaceous calcareous mudstone	−29.9
10	HJDP-B4	Carbonaceous shale	−30.6	45	XMP-S18	Carbonaceous calcareous mudstone	−29.5
11	HJDP-B7	Carbonaceous shale	−30.3	46	XMP-S19	Carbonaceous Calcareous mudstone	−29.6
12	HJDP-B8	Carbonaceous shale	−30.2	47	XMP-S20	Carbonaceous mudstone	−29.2
13	JGP-B1	Calcareous mud shale	−28.1	48	XMP-S21	Carbonaceous mudstone	−29.0
14	JGP-B5	Argillaceous limestone	−30.6	49	XMP-S22	Carbonaceous mudstone	−28.9
15	JGP-B6	Calcareous carbonaceous mudstone	−28.4	50	XMP-S23	Carbonaceous mudstone	−28.7
16	JKP-B2	Calcium-bearing carbonaceous shale	−29.8	51	XMP-S24	Carbonaceous mudstone	−28.9
17	JKP-B4	Silt-bearing carbonaceous shale	−27.9	52	SBTP-B11	Carbonaceous mudstone	−30.0
18	XHP-B1	Carbonaceous shale	−28.0	53	SBTP-B5	Carbonaceous shale	−30.1
19	XHP-B3	Carbonaceous shale	−27.5	54	SQP-B2	Carbonaceous shale	−30.0
20	XHP-B5	Carbonaceous shale	−27.0	55	SQP-B3	Carbonaceous shale	−29.7
21	XHP-B7	Carbonaceous mud shale	−27.1	56	SQP-B8	Carbonaceous mudstone	−28.8
22	LQP-B1	Carbonaceous shale	−28.4	57	SQP-B9	Silica-bearing carbonaceous shale	−28.4
23	QGP-S1	Carbonaceous mud shale	−31.0	58	STBP-B3	Calcium-bearing mud shale	−27.4
24	QGP-S6	Carbonaceous shale	−28.4	59	STBP-B4	Calcium-bearing mud shale	−28.3
25	QGP-S7	Siltstone mudstone	−28.5	60	STBP-B5	Calcium-bearing mud shale	−28.4
26	QLP-B4	Carbonaceous shale	−29.4	61	STP-B0-1	Carbonaceous mudstone	−28.4
27	QLP-B21	Carbonaceous shale	−28.3	62	STP-B6	Carbonaceous mudstone	−29.4
28	XMP-S1	Carbonaceous calceous shale	−29.7	63	XDP-B2	Carbonaceous shale	−28.1
29	XMP-S2	Carbonaceous calceous shale	−29.9	64	XDP-B4	Carbonaceous shale	−27.2

(continued)

Table 4.7 (continued)

Order Number	Original number	Sample Name	δ13C % (PDB)	Order Number	Original number	Sample Name	δ13C % (PDB)
30	XMP-S3	Carbonaceous shale	−30.5	65	XJP-B22	Carbonaceous shale	−29.4
31	XMP-S4	Carbonaceous shale	−30.4	66	XJP-B13	Carbonaceous shale	−29.3
32	XMP-S5	Carbonaceous shale	−30.5	67	XQP-S5	Carbonaceous mudstone	−29.6
33	XMP-S6	Carbonaceous shale	−30.6	68	XSQP-B3	Carbonaceous mud shale	−28.8
34	XMP-S7	Carbonaceous calcareous shale	−30.6	69	XSQP-B4	Carbonaceous mud shale	−29.2
35	XMP-S8	Carbonaceous calcareous shale	−30.5	70	XSQP-B11	Shale	−28.5

Note See Fig. 4.5 for section code

4.3.3 Maturity of Organic Matter in Black Rock Series

The maturity of kerogen can be described not only to predict the hydrocarbon generation potential of source rocks, but also to find possible fractured shale gas reservoirs in highly metamorphic areas, as an indicator of organic gas research in shale reservoir systems. The thermal maturity of kerogen also affects the natural gas content in the shale that can be adsorbed on the surface of organic matter. Generally, higher Ro indicates greater gas generation (greater gas generation), more fracture development, and larger shale gas production.

Current shale exploration in the USA indicates that shale maturity is generally over 1.3% in shale plays (Martineau 2007; Pollastro et al. 2007) and could up to 4.0% in Southern West Virginia of the Appalachian basin. Shale gas is produced only in high-maturity areas (Milici et al. 2006). Herein, the high maturity is favorable for shale gas production. The organic-rich shale in the study area in China has generally experienced complex multistage thermal history changes, and the maturity is generally high, basically in the high-overmature stage. Thus, the question arises whether the high maturity is still conducive to shale gas accumulation and enrichment. At present, in order to avoid the risk of extremely high maturity, the upper limit of maturity favorable value is set at 4% by scholars and industry.

Marine shales in the Sichuan basin basically experienced a long time of shallow burial in the early stage, a long-time uplift in the early-middle stage, deep burial again in the middle stage and rapid uplift in the late stage. It has undergone long-term structural and thermal evolution and is characterized by complex evolutionary history, high thermal maturity and early hydrocarbon generation time (Nie et al. 2012a). For shale, organic vitrinite reflectance (Ro) is an indicator of thermal maturity, reflecting the maximum paleogeodetic conditions of diagenesis and hydrocarbon generation conditions of shale and is the most suitable parameter to reflect diagenesis (Tang et al. 2012). Globally, vitrinite reflectance (Ro) is an important index to measure the thermal maturity of organic matter, but vitrinite is hard to find in Longmaxi Formation in the Southern Sichuan basin. It is derived from higher plants, but type I and type II_1 kerogen are mainly plankton or algal matter. Thus, using this parameter cannot directly evaluate the thermal maturity of Longmaxi Formation. Nowadays, bitumen reflectance (Rb) is adopted to evaluate the maturity of organic matter, which could be converted into equivalent vitrinite reflectance (Ro).

According to the research by previous scholars like Zhang et al., when Ro > 4.0%, kerogen is fully aromatized (graphitized) and no more natural gas could be generated. Chen et al. conducted a high-temperature simulation experiment on source rock, when Ro > 3.5%, the amount of gas generation decreases greatly, when the simulated temperature exceeds 600 °C, Ro reached 4.0%, dark materials appear, indicating the end of gas generation. Therefore, the termination line of hydrocarbon generation Ro = 4.0% is adopted, and the criteria for maturation stage classification of black shale is shown in Table 4.8.

1. The Ordovician Wufeng Formation

The organic-rich shale in the Wufeng Formation underwent multistage thermal history and reached the high-mature to the overmature stage. Regionally, maturity evolution gradually increases from south to north (Fig. 4.46; Table 4.9). In the south, close to the Central Guizhou Uplift belt, the thickness of the Wufeng Formation is relatively thin, and the maturity of mud shale is low. In the adjacent area of the Northern Central Sichuan Uplift, *R*o is generally low due to the upper layer thickness being thin by weathering and denudation. The Wufeng and its overlying strata in Nanchuan-Daozhen area have larger thickness and higher maturity of organic-rich mud shale, with *R*o values ranging between 1.48 and 2.39%, mostly more than 2.0%. Southward to Zhengan-Wuchuan-Yanhe area, *R*o values range

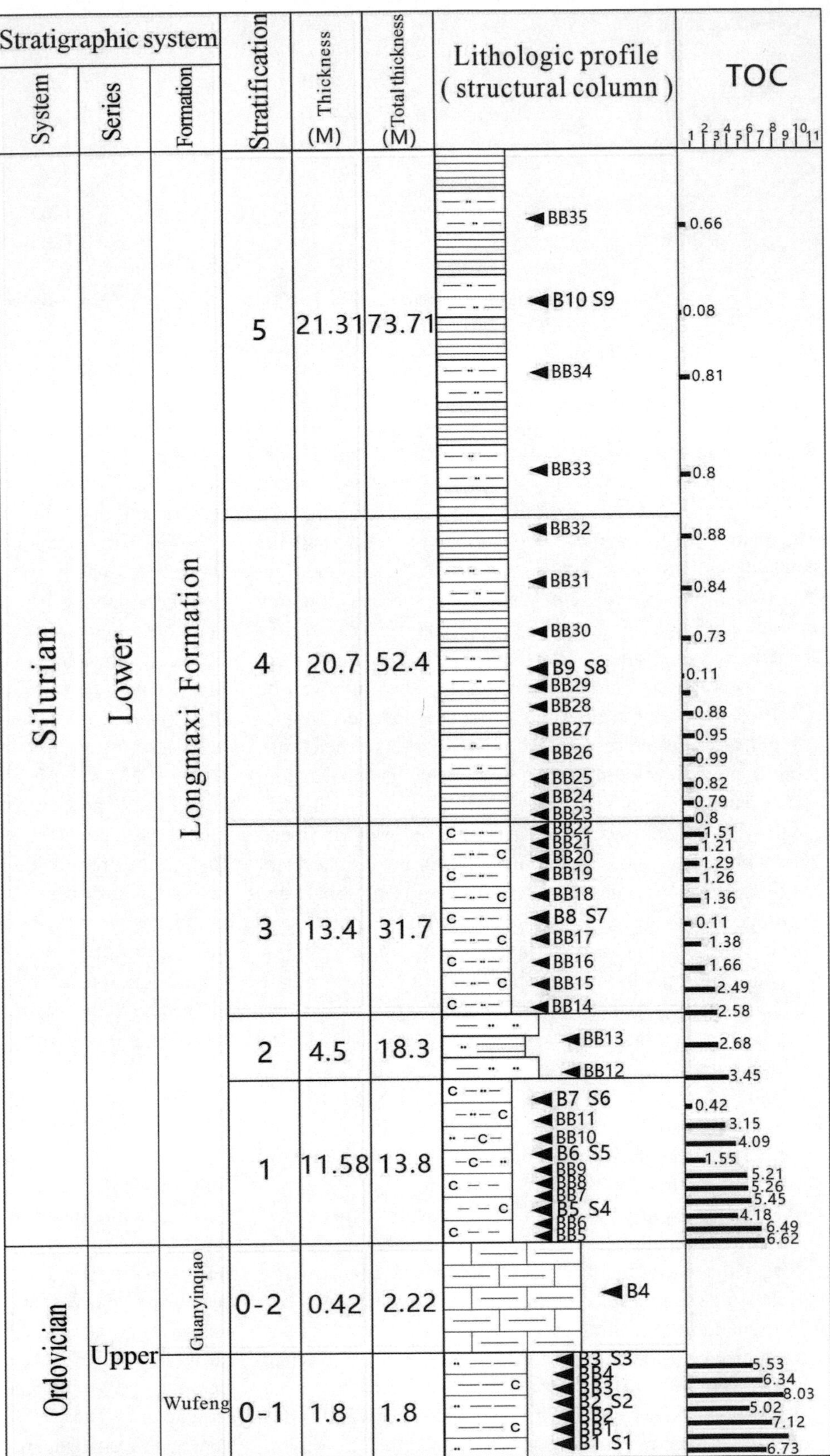

Fig. 4.35 Vertical evolution of organic carbon content in black shale of Wufeng Formation, Guanyinqiao, Qijiang

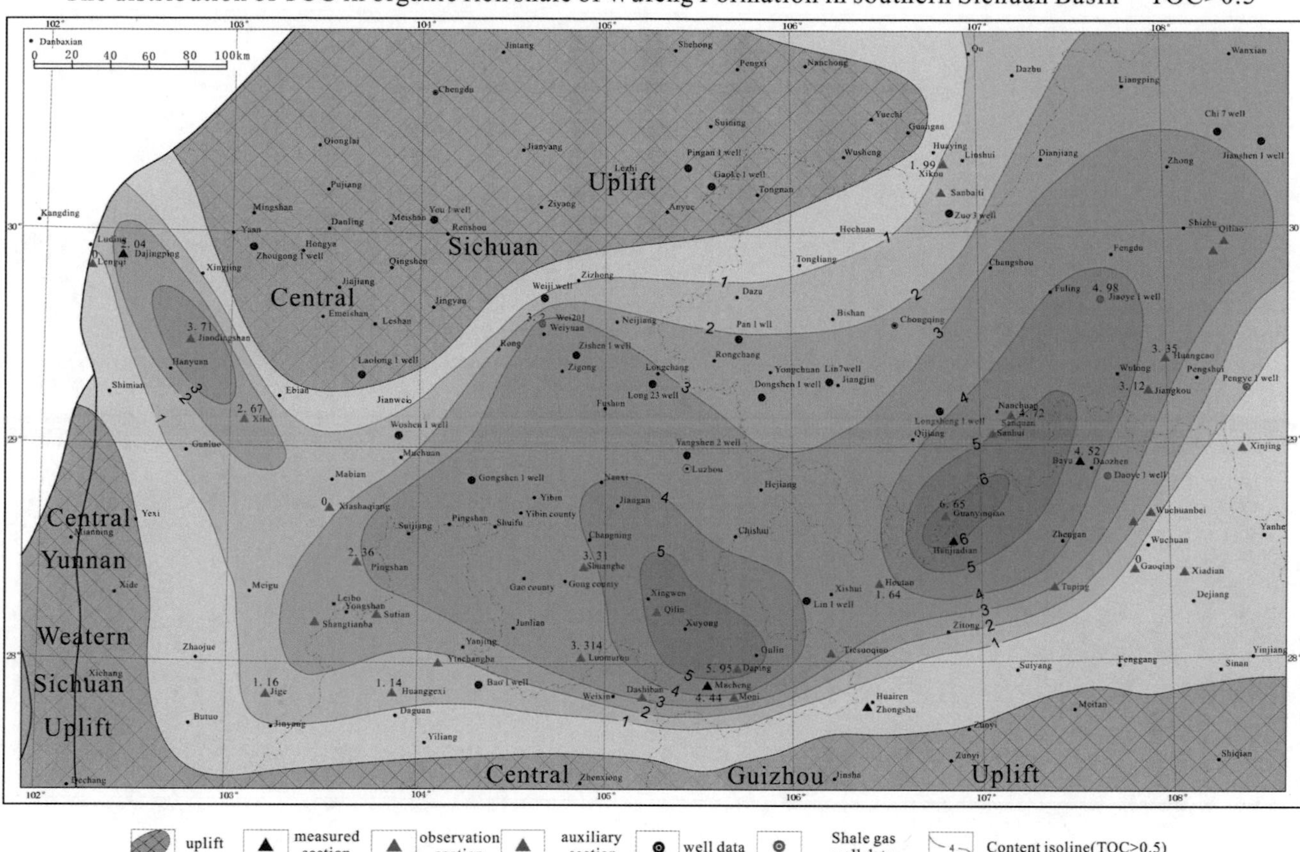

Fig. 4.36 Organic carbon content (TOC > 0.5%) of black rock series of Wufeng Formation in Southern Sichuan and its periphery

from 1 to 2.0%. In the south of Suiyang-Dejiang line, the maturity *R*o is generally less than 1.5%. In the Luding-Hanyuan area of Western Sichuan, the *R*o value is more than 2%, generally between 3 and 4%, belonging to the late stage of overmaturity, while in Yibin-Changning-Xingwen area, the *R*o value is roughly between 1 and 2%.

2. The Lower Silurian Longmaxi Formation

The vitrinite reflectance of organic matter of the Lower Silurian Longmaxi Formation in the Southern Sichuan and its periphery ranges from 1.63 to 4.54%, with an average of 2.62% (Table 4.10; Fig. 4.47). The Southern part of the study area is close to the Central Guizhou Uplift, and the Northern part is close to the Central Sichuan Uplift. And the thickness of Longmaxi Formation is thin, and the maturity of shale is low; meanwhile, *R*o is generally lower than 2% (Fig. 4.48). In the NE-SW direction of Shuifui-Jiangan-Qijiang-Fuling depositional center and its periphery, the thickness of Longmaxi Formation is larger, and the maturity of organic-rich shale is higher with *R*o value > 3%. The average *R*o value of well Yangshen 1 was 3.25%, and kerogen was in the high-overmaturity stage.

The thermal evolution degree of organic matter in the Western Sichuan is very high, up to 4.45% in Ebian of Western Sichuan, which is probably influenced by the Emei mantle plume in the later period. The maturity quickly reached the overmature stage and entered the shallow metamorphic stage (Table 4.10; Fig. 4.48). Zeng et al. (2011) also pointed out the influence of the Permian Emei mantle plume thermal regime. This accelerated the maturation process of the Longmaxi shale in the Southern Sichuan.

For different strata, a certain difference might exist between pyrolysis peak temperature (T_{max}) and different vitrinite reflectance (Ro). But the pyrolysis peak temperature (T_{max}) is generally not influenced a lot by buried depth or buried period. Thus, it can be used as an important proxy to describe the organic matter maturity (Yu 2008). The pyrolysis peak temperature depends on the kerogen structure; that is, the T_{max} is closely related to the activation energy of kerogen. When the maturity of sedimentary rocks is relatively high, the chemical bonds with lower activation energy in kerogen have been cleaved, and the remaining kerogen has higher activation energy and higher pyrolysis peak temperature (T_{max}). The evaluation criteria of maturity based

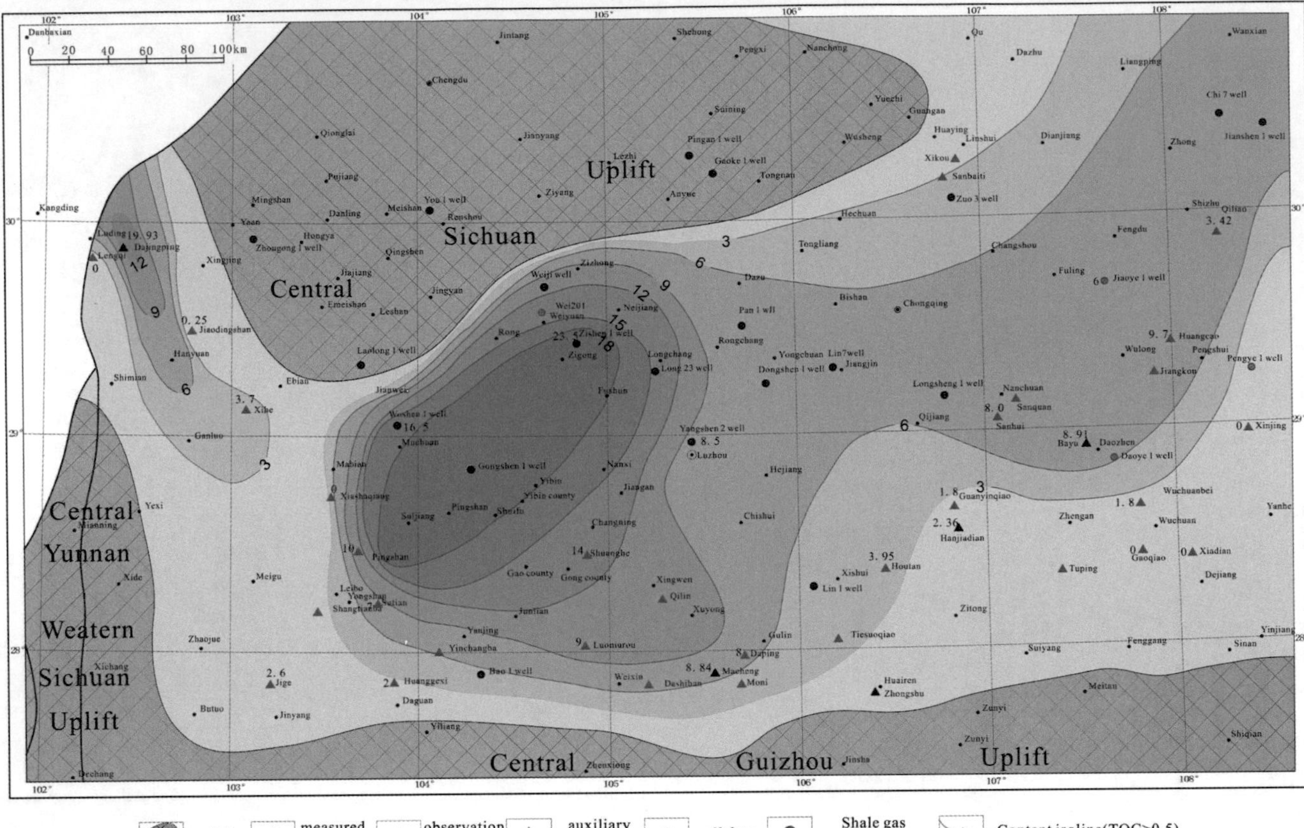

Fig. 4.37 TOC > 0.5% black rock series thickness contour map of Wufeng Formation in Southern Sichuan and its periphery

on the maximum pyrolysis peak temperature (T_{max}) of organic matter in shale are presented (Table 4.11), and the criteria are based on the classification scheme of the highest pyrolysis peak temperature on the maturity of organic matter (Wu et al. 1986), and according to the classification basis of the highest pyrolysis peak and the maturity of organic matter in the Criteria for The Classification of Diagenetic Stages of Clastic Rocks (SY/T 5477–2003). The maximum pyrolysis peak temperature (T_{max}) of organic matter in the black rock series of the lower Member of Longmaxi Formation in the Southern Sichuan and its periphery is 360–571 °C, mainly 360–450 °C, with an average of 456 °C, indicating that organic matter has evolved to the overmature stage (Table 4.12; Fig. 4.49). Regionally, the distribution of the maximum pyrolysis peak temperature (T_{max}) of organic matter of the lower Longmaxi Formation is characterized by local high values, and high values are distributed in Hanyuan, Lebo, Renhuai and Qijiang-Tongzi areas, all of which are over 510 °C. The T_{max} in Ebian, Yanhe, Yanjin, Xingwen and Wulong are all < 435 °C, 435 °C on average, and the organic matter is immature.

This may be related to the transformation of clay minerals during diagenesis. Luo et al. (2001a) pointed out that the transformation of clay minerals plays an important catalytic role in the generation of organic matter at the burial depth of 1000−2800 m, among which montmorillonite plays the strongest catalytic role, which can increase the organic hydrocarbon rate by 2−3 times and reduce the pyrolysis temperature by 50 °C. During this transformation process, a large amount of Al instead of Si occurs, resulting in the surface charge of montmorillonite to be unbalanced and acidic.

Considering the T_{max} has different characteristics, along with the change of maturity II type and III type kerogen change with maturity and obvious changes, I kerogen range is small, so the II type and III type kerogen, the highest pyrolysis peak temperature (T_{max}) is a good maturity parameter. The effect on type I kerogen was not obvious. The kerogen of black rock series of Longmaxi Formation in the study area is mainly type i, and the maximum pyrolysis peak temperature (T_{max}) of organic matter is not significantly correlated with vitrinite reflectance (Fig. 4.50). Therefore,

The distribution of TOC in organic rich shale of Wufeng Formation in southern Sichuan Basin(TOC>1%)

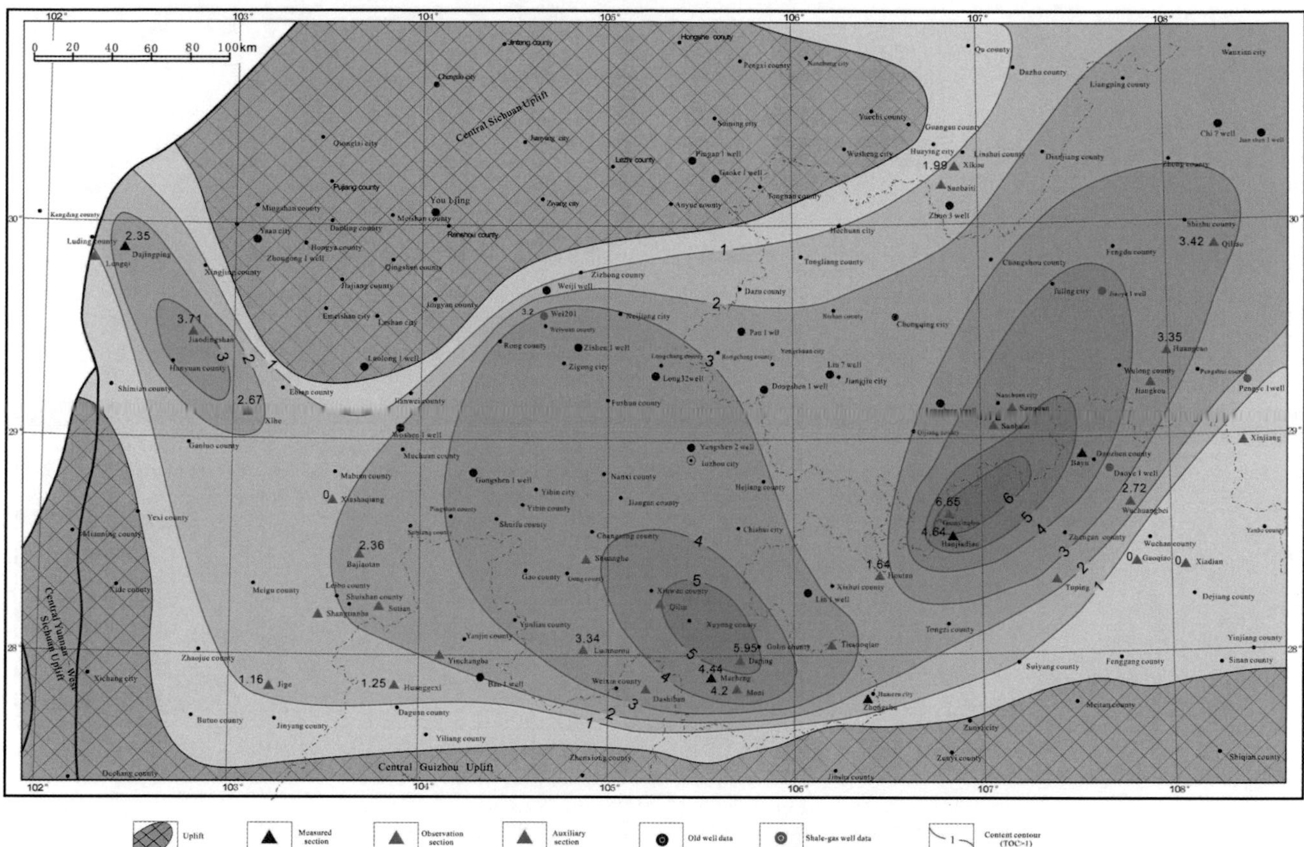

Fig. 4.38 Organic carbon content (TOC > 1%) of Wufeng black-rich rock series in Southern Sichuan and its periphery

the maximum pyrolysis peak temperature (T_{max}) of organic matter is only used for judging the evolution degree of organic matter, not for analyzing the evolution process.

The thermal metamorphism coefficient TAI of the comparison samples (DY1-24, 588.30 m) from the Longmaxi Formation in well Daoye 1 was as high as 3.3, which was determined by Weatherford laboratory, indicating that the organic matter is in the overmature evolution stage and has entered the dry gas window. Based on the analysis of vitrinite reflectance and maximum pyrolysis peak temperature (T_{max}), the organic-rich shale kerogen in the lower part of Longmaxi Formation in Southern Sichuan and its periphery is in the stage of high maturity-overmaturity evolution, and in some areas, it reaches the stage of low-level metamorphism.

In conclusion, the black shale of the Lower Silurian Longmaxi Formation in the Southern Sichuan and adjacent areas is mainly distributed in the Suijiang-Luding-Qijiang area in the center of the basin. The thickness of source rocks is 20–100 m, and the organic carbon content is mainly 2–4%. The kerogen macerals are mainly sapropelic amorphous, and the kerogen type is mainly type I and type II1. The thermal evolution degree of shale corresponds to the formation depositional center, and the Ro value is distributed in 1.63–4.45%, mainly in 2–2.5%, which is in the early stage of high maturation-overmaturation. In some areas, it reaches the stage of low-grade metamorphism, mainly producing cracking gas and dry gas.

4.3.4 Exception Investigate of T_{max} of Organic Matter

Both vitrinite reflectance and T_{max} are indicators of the evolution degree of organic matter. Generally, the accordances of these two indicators are poor. Besides, as the overlying strata of the Wufeng Formation, the Longmaxi Formation should be slightly less evolved than the Wufeng Formation. The vitrinite reflectance of the Wufeng Formation is 1.62−4.47%, with an average of 2.42%, slightly lower than that of the Longmaxi Formation (2.62% on average). For example, the vitrinite reflectance of Wufeng Formation in Wulong is 1.512% (B1) and 1.591% (B3), while that of upper samples in Longmaxi Formation is 1.830%. The vitrinite reflectance of Wufeng Formation in Guanyinqiao, Qijiang, is 2.247%, 2.262% and 2.235%, respectively, while

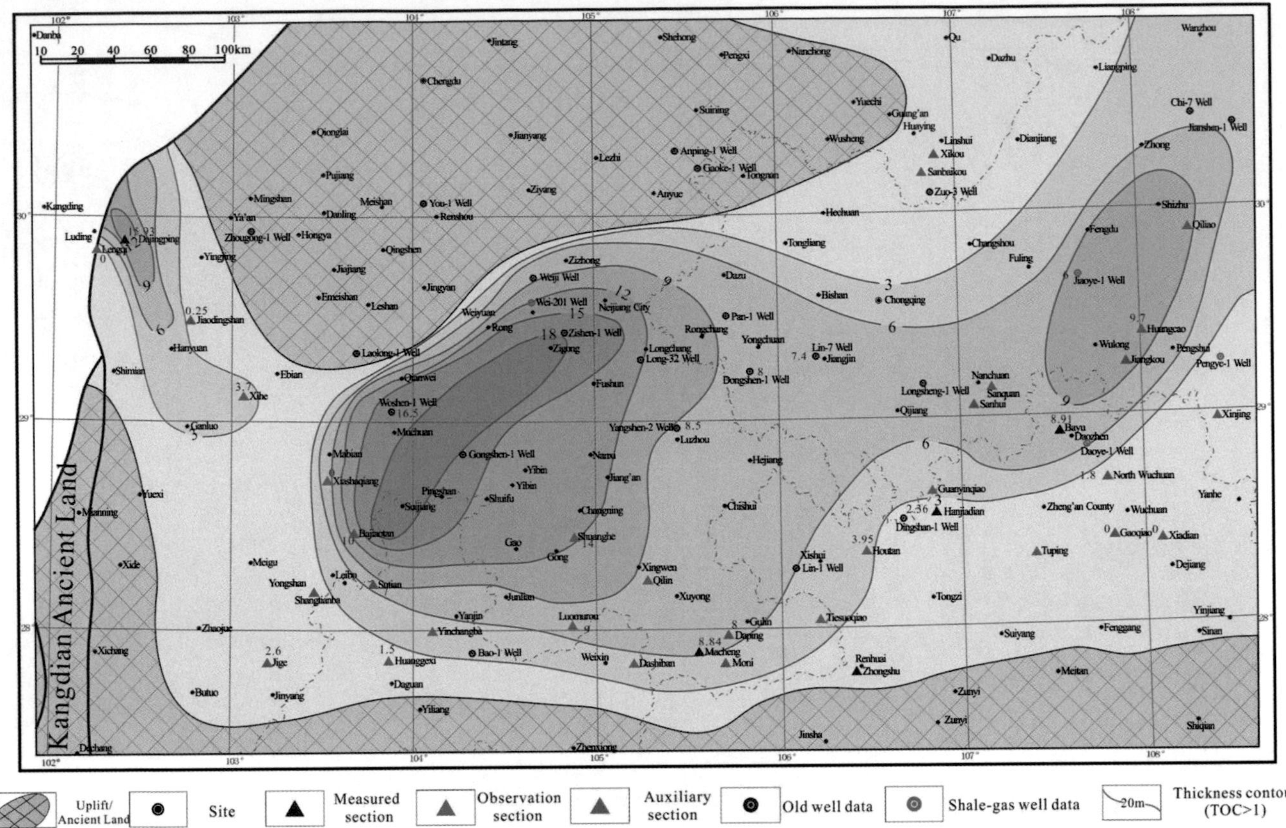

Fig. 4.39 TOC > 1% black rock series thickness contour map of Wufeng Formation in Southern Sichuan and its periphery

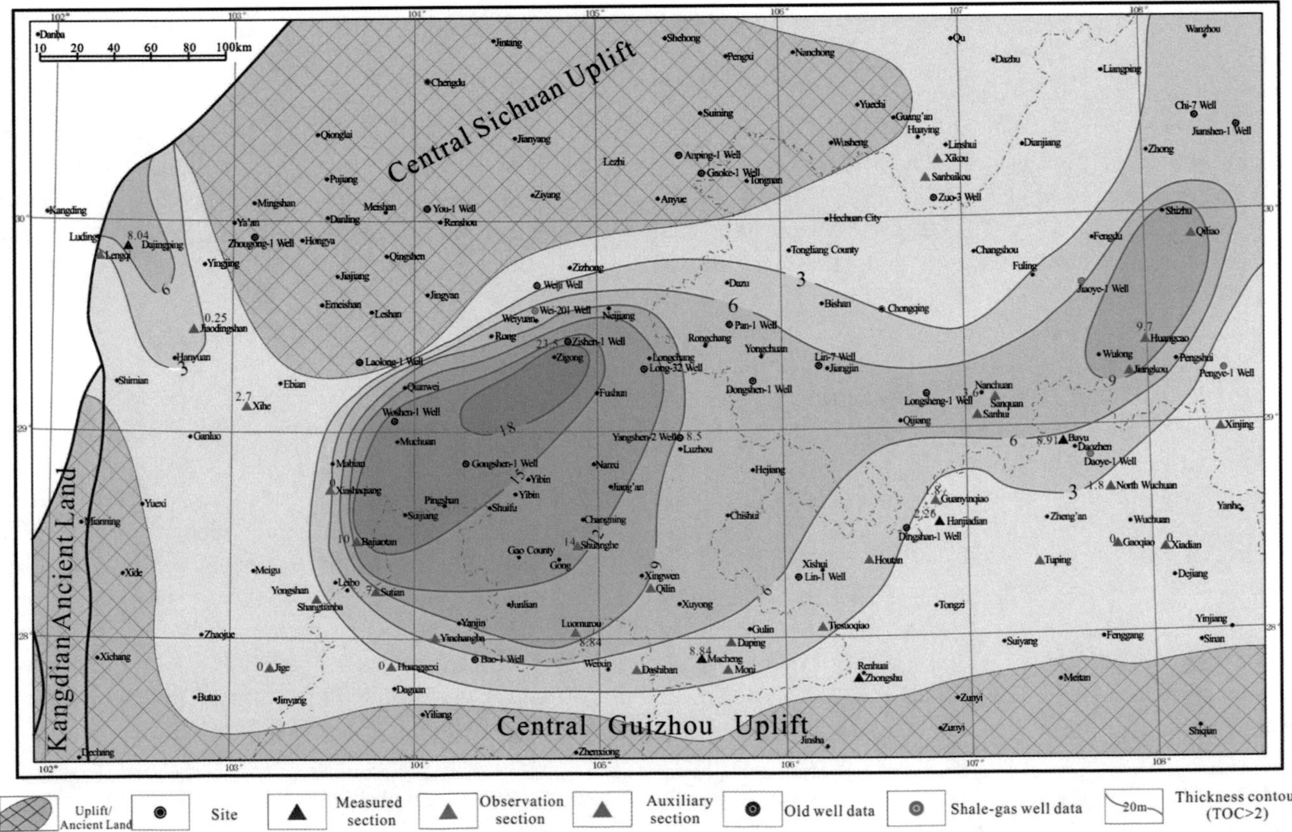

Fig. 4.40 TOC > 2% black rock series thickness contour map of Wufeng Formation in Southern Sichuan and its periphery

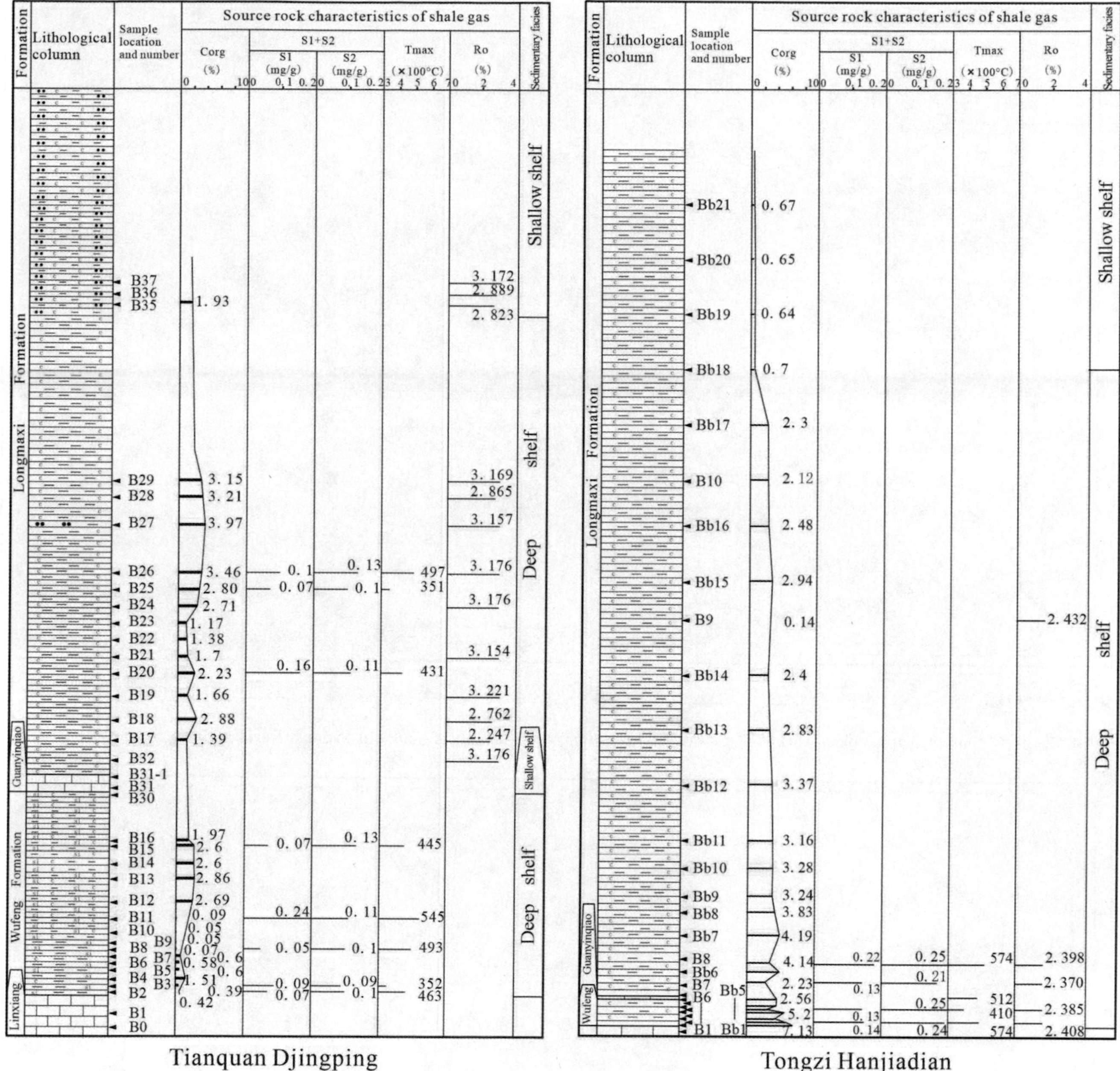

Fig. 4.41 TOC vertical distribution characterisctics of black rocks in Lower Longmaxi Formation of Southern Sichuan Basin and its periphery

that of Longmaxi Formation is 2.268%, 2.261%, 2.276% and 2.269%, indicating that the evolution degree of organic matter in of Longmaxi Formation is relatively high.

However, Wufeng Formation contains higher illite and lower mixed-layer minerals than Longmaxi Formation, and smectite was not found. The average content of illite in 31 samples of the Wufeng Formation is 41.9%, and the average content of Illite and mixed layer is 47%. The average content of illite in 29 samples from the Longmaxi Formation is 31.9%, and the average content of Iran-Mongolian-mixed-layer minerals is 55%. The clay minerals show that the Wufeng Formation has a high degree of evolution. The maximum pyrolysis peak temperature (T_{max}) is an indicator of the maturity of source rocks, but its value cannot effectively reflect the maturity of source rocks. As above analysis, the highest pyrolysis peak temperature (T_{max}) may be associated with the transformation of clay minerals content, low clay minerals content, and however, to stabilize the terrigenous clastic formation mainly consists of illite and for example TianQuan big well plateau region clay minerals content mainly illite, mineral rarely get layer and vitrinite reflectance of 2.99% (Table 4.6). The highest pyrolysis peak temperature (T_{max}) was only between 351 and 497 °C (Table 4.8). The maturity of source rocks in Guanyinqiao

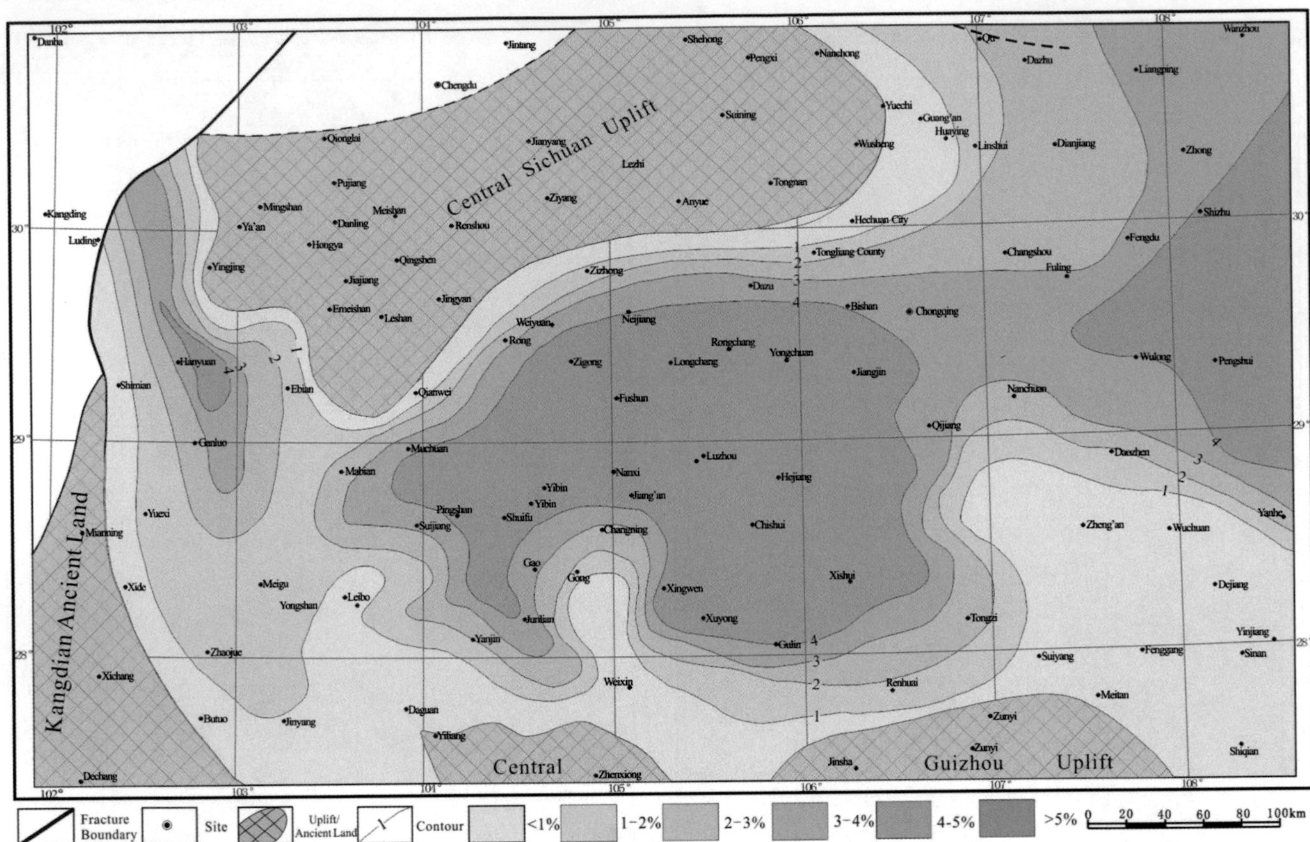

Fig. 4.42 TOC contour map of black rocks (TOC > 0.5%) in Longmaxi Formation of Southern Sichuan Basin and its periphery

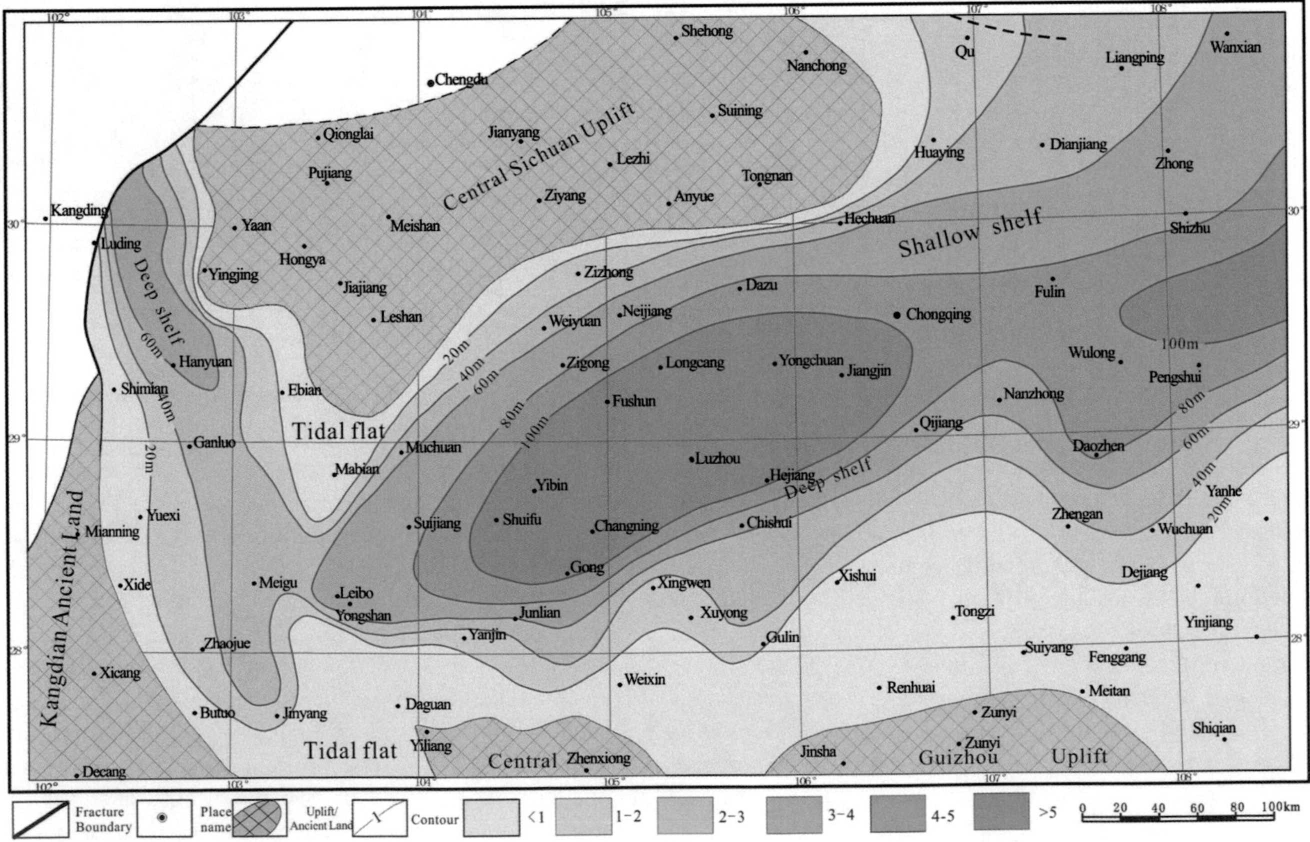

Fig. 4.43 Thickness contour map of black rocks (TOC > 0.5%) in Longmaxi Formation of Southern Sichuan Basin and its periphery

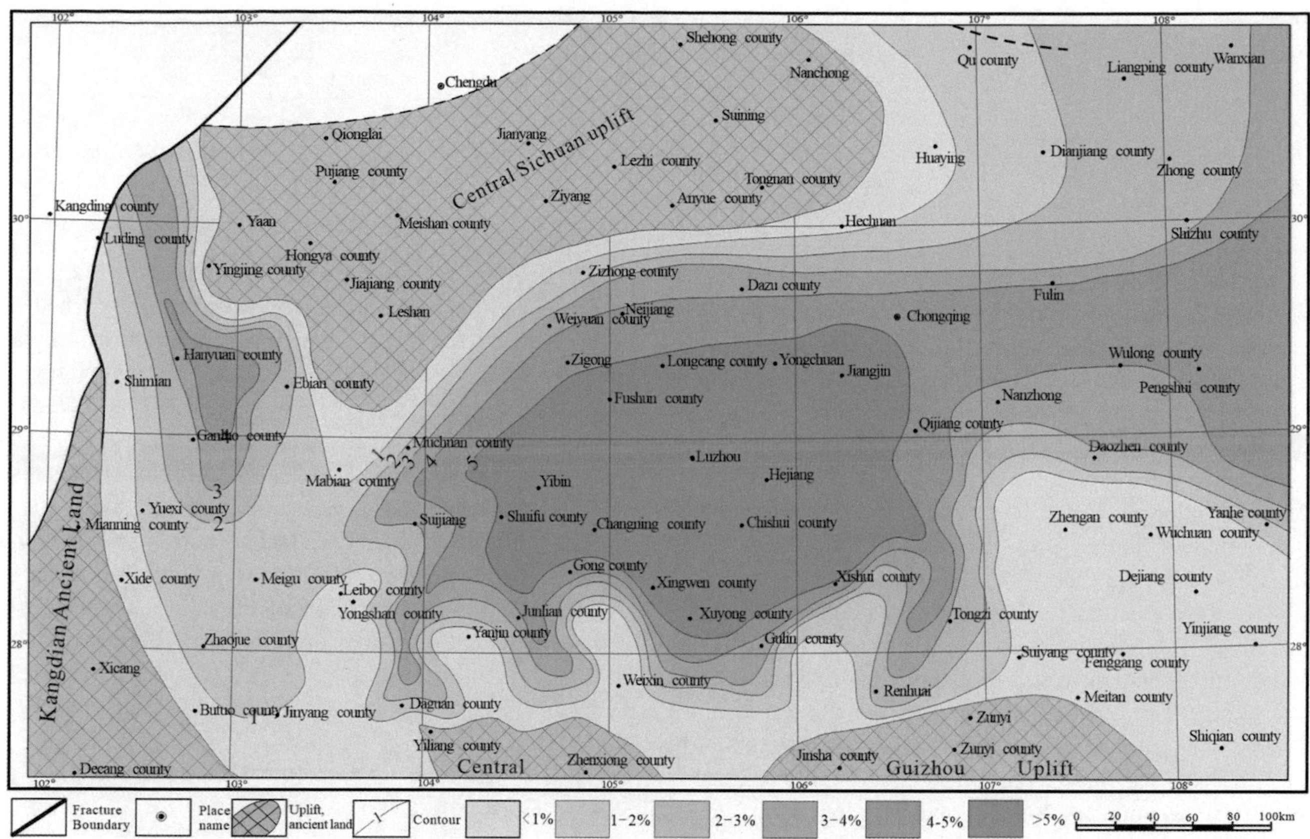

Fig. 4.44 TOC contour map of black rocks (TOC > 1%) in Longmaxi Formation of Southern Sichuan Basin and its periphery

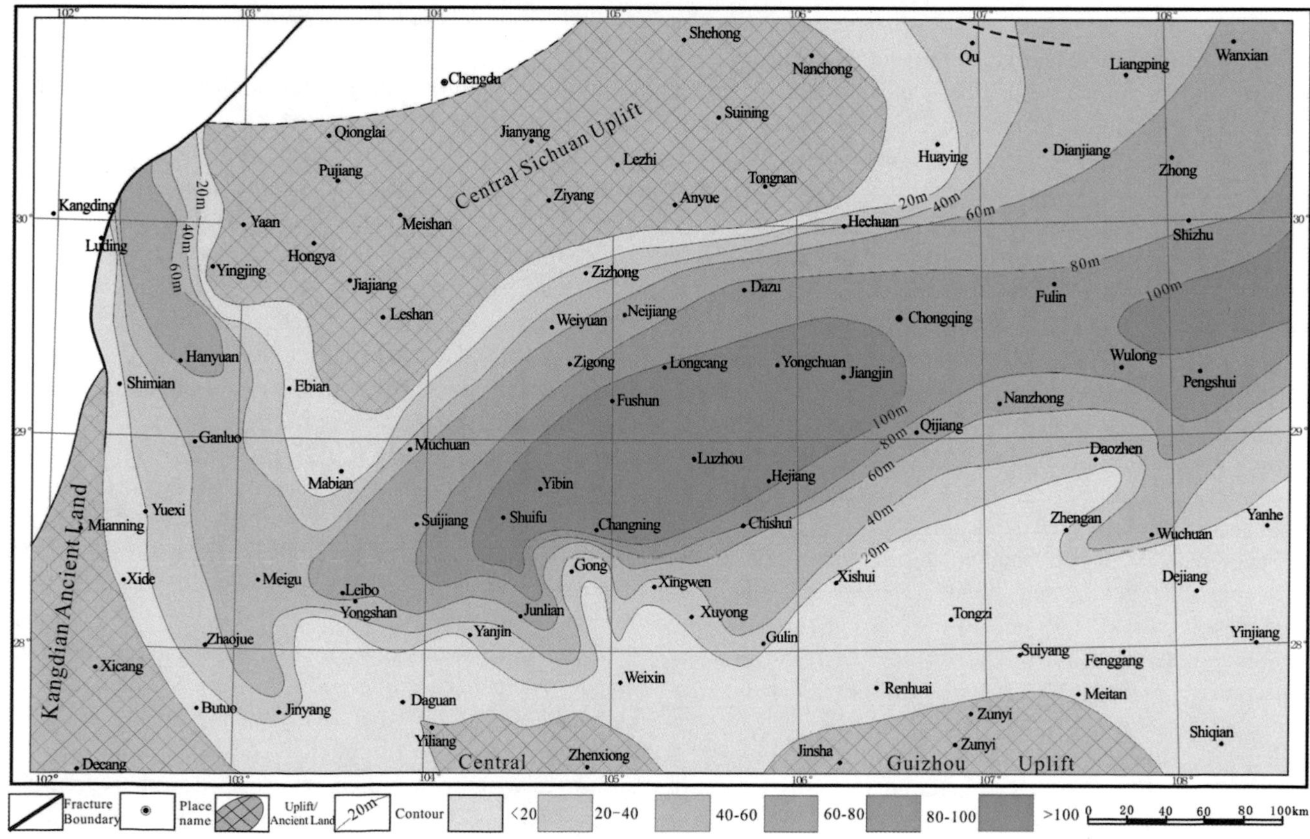

Fig. 4.45 Thickness contour map of black rocks (TOC > 1%) in Longmaxi Formation of Southern Sichuan Basin and its periphery

Table 4.8 Mature stage division standard of black shale in China (after Nie et al. 2012a)

Maturity stage	Immaturity	Maturity	High maturity stage	Early overmaturity	Late stage of immaturity	Expiry date
Ro (%)	<0.5	0.5–1.3	1.3–2	2–3	3–4	≥4
Hydrocarbon-forming stage	Biogas	Oil-forming period	Condensate-wet	Dry Gas		Biogenic hydrocarbon termination

Formation and Longmaxi Formation in Qijiang is higher than that in Wufeng Formation, and the highest pyrolysis peak temperature (T_{max}) is 576 °C (Table 4.8). The evolution degree of source rocks in Huangcao Longmaxi Formation is slightly higher than that in Wufeng Formation, and the maximum pyrolysis peak temperature (T_{max}) is about 100 °C higher than the capping temperature of Wufeng Formation. The clay minerals in the latter two areas are dominated by illite/montmorillonite, and the content of illite is only slightly lower than that of the mixed layer.

Therefore, the transformation of clay minerals in shale of the lower Longmaxi Formation in the study area is not enough to significantly reduce the maximum pyrolysis peak temperature (T_{max}) of organic matter, but the transforming effect of clay minerals is an important factor that cannot be ignored. At least nine layers of bentonite are developed in Wufeng Formation at Wulong, and the obvious uncorrelation of Wufeng-Longmaxi Formation may be due to the influence of clay minerals in the process of sedimentation-diagenesis of bentonite. Thus, we speculate that the uneven decrease of T_{max} of organic matter in Ordovician Wufeng Formations-Silurian Longmaxi Formation in the Southern Sichuan and its periphery may be related to the uneven and widespread distribution of bentonite in the strata.

4.4 Characteristics of Rock Mineral Composition and Its Impact on Shale Gas

Mineral composition not only affects the hydrocarbon generation mode and expulsion efficiency of source rocks (Fu et al. 2011), but also plays an important role in the development of shale gas reservoirs (Liang et al. 2012a; Yu et al. 2013; Curtis 2002; Jarvie et al. 2005, 2007; Liu et al. 2011; Zhang et al. 2013a, b). Bust et al. (2013) pointed out that the large and medium pores in shale gas reservoirs are related to the aggregate of organic matter and clay minerals or the aggregate of organic matter and carbonate minerals. The brittleness of shale is related to the content of quartz and carbonate minerals (Liu et al. 2012a), mineral composition and content determine the development degree of the fracturing of artificial fracture (Sui et al. 2007), which is affecting the drilling and hydraulic fracturing effect most this factor (Bai et al. 2013). It's to determine the key factors of shale gas wells with high quality (Bowker 2003). Bowker (2007) believes that Barnett shale can achieve such high shale gas production because its brittleness has reached the level of effective hydraulic fracturing. Without the characteristics of today's mineral composition, the exploitation of Barnett shale gas could not be successful under today's mining technology. Therefore, mineralogical study of shale gas resource evaluation, reservoir forming mechanism analysis and development measures of process design is of great significance (Chen, et al. 2011c). At the current level of research, rock mineral composition is an important index for the description and evaluation of organic-rich shale reservoirs in Wufeng Formations and Longmaxi Formations (Liu 2012).

4.4.1 Types and Characteristics of Mineral Components

1. Ordovician Wufeng Formation

According to the special geological characteristics of shale gas reservoir, the rock components of mud shale can be divided into two categories: brittle minerals and clay minerals. And the brittle minerals can be divided into siliceous minerals (quartz, feldspar, pyrite, etc.) and carbonate minerals (calcite, dolomite). There is no exact standard to classify shale in shale gas reservoir by rock mineral composition. Clay minerals in Barnett shale are generally less than 1/3, mainly siliceous minerals and generally contain carbonate minerals, which are considered as siliceous shale (Loucks and Ruppel 2007; Loucks et al. 2009a). In addition, Loucks et al. (2012) used 50% content as the boundary to illustrate the relationship between the three endmembers of quartz + pyrite, carbonate mineral + feldspar and clay minerals and shale stability. The organic-rich shale of the lower member of Longmaxi Formation in Southern Sichuan and adjacent areas is essentially a "fine-grained sedimentary rocks". Jiang, et al. (2013), with a major component of fine-grained sedimentary silt, clay and carbonate for three yuan, with their content of 50% is bounded into powder sandstone, clay rocks, carbonate rocks and hybrid fine-grained four categories of sedimentary rocks. Based on the above practical experience, considering

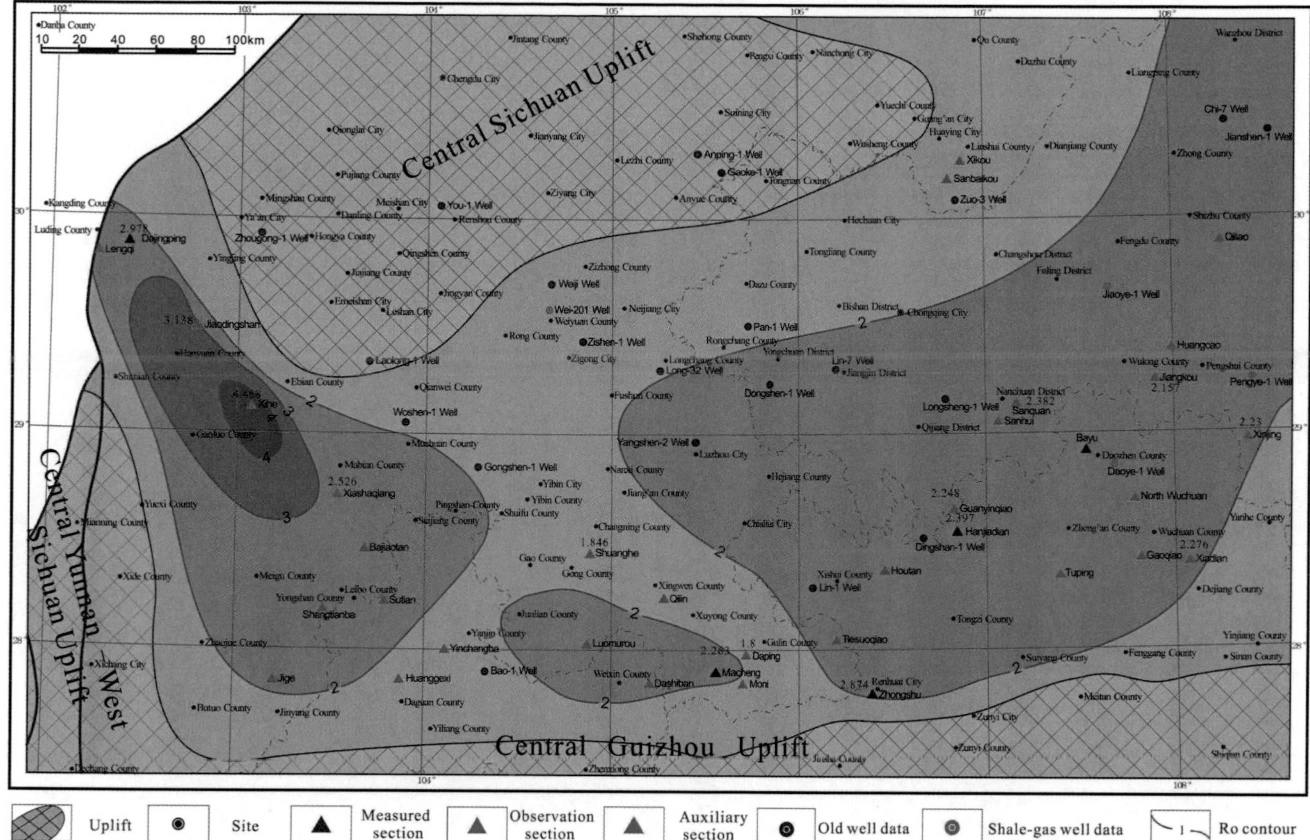

Fig. 4.46 Distribution map of thermal evolution degree of organic-rich shale of Wufeng Formation in Southern Sichuan and its periphery

Table 4.9 Vitrinite reflectance statistics of Wufeng Formation in Southern Sichuan and its periphery

Original number	Reflectance of vitrinite (Ro%)			Number of samples
	Highest	Minimum	Average	
LQP	2.068	2.068	2.068	1
XQP	2.433	2.587	2.587	1
DJP	3.142	3.115	3.135	6
DJSP	3.153	3.122	3.138	2
XHP	4.505	4.431	4.468	2
WDP	3.073	2.539	2.958	5
YYP	2.25	1.94	2.095	2
STBP	1.884	1.767	1.826	2
JGP	2.671	2.671	2.671	1
XSQP	2.671	2.362	2.526	2
STP	2.59	2.59	2.59	1
LMRP	2.327	2.312	2.320	2
RZP	2.903	2.834	2.874	3
YYP	3.251	3.251	3.251	1
DHP	2.708	2.386	3.37	3
SBTP	1.682	1.640	1.661	2
QLP	2.341	2.342	2.343	1
HCP	1.591	1.512	1.552	2

(continued)

Table 4.9 (continued)

Original number	Reflectance of vitrinite (Ro%)			Number of samples
	Highest	Minimum	Average	
QGP	2.262	2.235	2.248	3
XSP	2.237	2.237	2.237	1
JKP	2.157	2.157	2.157	1
HJDP	2.408	2.385	2.397	2
SQP	2.443	2.321	2.382	3
XMP	2.340	2.231	2.263	7
XJP	2.233	2.233	2.233	1
XDP	2.276	2.276	2.276	1
DZP	1.644	1.479	1.562	2

Note See Fig. 4.5 for section code

Table 4.10 Ro of black rocks in Longmaxi Formation of Southern Sichuan Basin and its periphery

Profile number	Reflectance of vitrinite (Ro%)			Number of samples
	Highest	Minimum	Average	
LQP	2.07	2.07	2.07	1
XQP	2.43	2.34	2.38	6
DJP	3.22	2.64	2.99	18
DJSP	3.18	3.12	3.15	3
XHP	4.46	4.44	4.45	3
WDP	3.54	3.07	2.97	6
YYP	2.25	1.94	2.10	2
STBP	4.54	1.84	2.78	3
JGP	2.61	2.32	2.47	3
XSQP	2.59	2.42	2.51	2
STP	3.6	2.35	2.73	6
LMRP	2.43	2.35	2.43	3
GTP	2.48	1.93	2.28	6
RZP	2.6	2.03	2.37	15
YYP	3.73	2.85	3.36	5
DHP	3.75	2.00	3.37	5
SBTP	1.63	1.63	1.63	1
XKP	1.75	1.75	1.75	1
QLP	2.34	2.34	2.34	1
HCP	1.83	1.82	1.83	2
QGP	2.28	2.18	2.25	6
XSP	2.19	2.08	2.15	4
JKP	2.46	2.46	2.46	1
HJDP	2.43	2.37	2.42	3
SQP	2.46	2.42	2.44	2
XMP	2.3	2.12	2.26	18
XJP	2.26	2.26	2.26	2
XDP	2.29	2.29	2.29	1
DY1	1.51	2.18	1.79	16

Note See Fig. 4.5 for section code

Table 4.11 Organic matter mature divided by T_{max}

Evolutionary Stages	Immaturity	Low maturity	Maturity	High maturity	Overmaturity
Mudstone/°C	<435	435–445	445–480	480–510	>510

the geological characteristics of shale gas and the development of shale mineral components in the study area, the shale gas reservoir can be divided into siliceous shale (siliceous > 50%, clay minerals < 40%, carbonate minerals < 30%), clayey shale (siliceous < 50%, clay minerals > 40%, carbonate minerals < 30%) and carbonate shale (carbonate mineral > 30%).

In general, the average content of brittle minerals in black shale of Ordovician Wufeng Formation in Southern Sichuan and its periphery is 79.24%, including 49.8% of siliceous minerals, 22.47% of carbonate minerals and 27.73% of clay minerals. Siliceous mudstones are mainly developed in the Ordovician Wufeng Formation (Fig. 4.51).

According to X-ray diffraction analysis results of black shale of Wufeng Formation, the content of clay minerals is 1–67%, mainly concentrated in 10–40%, accounting for 63.2% of all samples (Fig. 4.52). Among the clay minerals, illite content mostly ranges from 30 to 50%, accounting for 62.5% of all samples, followed by 50–70%, with an average relative content of 47.71%. The mineral content of the Iran-Mongolian-mixed layer is mostly 40–60%, accounting for 50% of all samples, followed by 30%, with an average relative content of 39.13%. Chlorite content is less than 10%, accounting for 53.6% of all samples, followed by 10–20%, with an average relative content of 10.18%. Fifty of the 56 samples tested did not contain kaolinite, and the content of montmorillonite in the mixed-layer minerals was mainly concentrated in 10%, which accounted for 62.5% of all the samples.

According to the scatter diagram between clay mineral content and TOC in the X-ray diffraction test results of 68 samples from Wufeng Formation in Southern Sichuan and adjacent areas in Fig. 4.53, the characteristic is: the TOC < 1%, the distribution of clay content is uneven, ranging from less than 10% to more than 60%, with the highest content reaching 67%; When 1% < TOC < 2%, the content of clay minerals ranges from 0–40%, and most samples are concentrated in 10–30%, accounting for 45% of all samples. When 2% < TOC < 4%, the content of clay minerals is mostly distributed between 0–50%, and the samples with 10–30% content account for about 85% of the total samples. When TOC > 4%, the clay mineral content is mostly concentrated between 10 and 30%, few samples exceed 30%, and most of them account for 76.9% of the total samples. By analyzing the relationship between the content of clay minerals and TOC in Wufeng Formation, it can be seen that the TOC content of corresponding rocks is higher when the content of clay minerals is between 10–30%.

1. Silurian Longmaxi Formation

(1) Types and characteristics of mineral components

The black rock series developed in the lower member of Silurian Longmaxi Formation in Southern Sichuan and its periphery have relatively uniform mineral composition types, but the content of black rock series shows certain heterogeneity in vertical direction and plane. According to the X-ray diffraction analysis test results, the mineral components include quartz, potassium feldspar, plagioclase, calcite, dolomite, pyrite, clay minerals and a small amount of siderite, gypsum, etc. Under polarizing microscope, they are fine silt-argillaceous structure, and the detrital particles are floating in argillaceous and cementing substrate. The content of detrital particles is 10–50%, including quartz, feldspar, mica, etc., and the particle size is usually < 0.05 mm (Fig. 4.54a, b). The cements and argillaceous clasts are associated with black organic matter (Fig. 4.54a–c). The cements include clay minerals, siliceous, carbonate minerals, and pyrite. Argillaceous clasts are black, gray-black, or brown clumps, mainly consisting of argillaceous quartz, feldspar, and clay minerals.

The content of quartz is high, ranging from 9 to 90%, with an average of 42.32%, including clastic quartz and siliceous cements. Quartz particles mostly account for more than 75% of the total amount of debris. They are clean and bright, generally well-sorted and poorly rounded. The particle size ranges from 0.02 to 0.05 mm. The distribution of feldspar is not uniform, and the content of plagioclase is high. The content of potassium feldspar and plagioclase ranges from 0–19% to 0–34%, respectively, and the average content is 2.37% and 6.47%, respectively. The grain size is larger than that of quartz, and the surface is clean. Weak clay or partially metasomatized by carbonate minerals shows irregular dissolution edges (Fig. 4.30d, e). Both biotite and muscovite are developed. The latter are thinner and have certain compaction deformation. Biotite is often corroded into chlorite and exhibits certain water absorption and expansibility along cleavage joints (Fig. 4.54f).

Pyrite, with an average content of 1.41%, is an authigenic mineral generated at the early stage of sediment formation (Zhang et al. 2013a, b). It is ubiquitous and often has euhedral

Table 4.12 T_{max} of black rocks in Lower Longmaxi Formation of Southern Sichuan Basin and its periphery

Analysis Number	Original number	Sample Name	S0 (mg/g)	S1 (mg/g)	S2 (mg/g)	T_{max} (°C)
1	DHP-S3	Argillaceous Limestone	0.02	0.06	0.10	491
2	JKP-B4	Silt-bearing Carbonaceous Shale	0.05	0.11	0.24	389
3	XDP-B4	Carbonaceous Shale	0.04	0.12	0.19	375
4	XJP-B13	Carbonaceous Shale	0.05	0.13	0.27	482
5	YYP-S3	Siltstone Mudstone	0.02	0.05	0.09	380
6	YYP-S4	Interbedding of mudstone and limestone	0.02	0.06	0.09	423
7	DJP-B30	Calcium-bearing Carbonaceous Shale	0.02	0.09	0.09	380
8	DJP-B32	Calcium-bearing Carbonaceous Shale	0.02	0.08	0.09	382
9	DJP-B38	Siltstone Shale	0.02	0.07	0.10	352
10	HCP-B11	Carbonaceous Shale	0.06	0.61	0.46	395
11	HCP-B13	Carbonaceous Shale	0.07	0.91	0.23	491
12	JGP-B1	Calcareous Mud Shale	0.02	0.03	0.07	513
13	JGP-B5	Argillaceous Limestone	0.03	0.11	0.08	427
14	QLP-B4	Carbonaceous Shale	0.04	0.07	0.20	510
15	QLP-B21	Carbonaceous Shale	0.04	0.06	0.15	455
16	QLP-B23	Carbonaceous Shale	0.04	0.07	0.16	451
17	RZP-B7	Carbonaceous Shale	0.03	0.08	0.22	539
18	RZP-B11	Carbonaceous Shale	0.02	0.05	0.13	543
19	RZP-B15	Siltitesiltstone	0.02	0.04	0.10	518
20	SBTP-B11	Carbonaceous Shale	0.05	0.09	0.23	571
21	XKP-S2	Carbonaceous Shale	0.05	0.09	0.20	458
22	XQP-S4	Carbonaceous Shale	0.02	0.05	0.09	383
23	XQP-S5	Carbonaceous Shale	0.02	0.07	0.09	422
24	HJDP-B7	Carbonaceous Shale	0.05	0.13	0.21	512
25	HJDP-B8	Carbonaceous Shale	0.05	0.22	0.25	574
26	JDSP-B14	Siltstone Shale	0.02	0.09	0.13	545
27	JDSP-B18	Carbon Siliceous shale	0.03	0.13	0.13	544
28	SQP-B8	Carbonaceous Shale	0.05	0.15	0.20	366
29	SQP-B9	Silica-bearing Carbonaceous Shale	0.05	0.18	0.18	454
30	STBP-B5	Carbonaceous Mud Shale	0.01	0.03	0.06	507
31	STBP-B6	Carbonaceous Mud Shale	0.02	0.03	0.06	508
32	STBP-B8	Carbonaceous Mud Shale	0.02	0.06	0.08	516
33	XHP-B5	Carbonaceous Mud Shale	0.02	0.10	0.11	360
34	XHP-B7	Carbonaceous Mud Shale	0.02	0.08	0.08	360
35	DJP-B20	Carbonaceous Shale	0.02	0.16	0.11	431
36	DJP-B25	Carbonaceous Shale	0.02	0.07	0.10	351
37	DJP-B26	Carbonaceous Shale	0.02	0.10	0.13	497
38	QGP-S4	Carbonaceous Shale	0.11	0.93	0.37	576
39	QGP-S5	Carbonaceous Shale	0.06	0.44	0.25	576
40	QGP-S6	Carbonaceous Shale	0.06	0.20	0.23	462
41	QGP-S7	Siltstone Shale	0.06	0.10	0.22	568
42	QGP-S8	Siltstone Shale	0.06	0.10	0.23	543

(continued)

Table 4.12 (continued)

Analysis Number	Original number	Sample Name	S0	S1	S2	T_{max}
			(mg/g)	(mg/g)	(mg/g)	(°C)
43	QGP-S9	Marlite	0.06	0.09	0.23	496
44	XMP-S8	Carbonaceous Calcareous Shale	0.05	0.17	0.27	574
45	XMP-S9	Carbonaceous Calcareous Shale	0.05	0.12	0.19	399
46	XMP-S10	Carbonaceous Calcareous Shale	0.05	0.09	0.20	465
47	XMP-S11	Carbonaceous Calcareous Shale	0.05	0.13	0.19	390
48	XMP-S12	Carbonaceous Calcareous Shale	0.04	0.14	0.19	412
49	XMP-S13	Carbonaceous Calcareous Shale	0.05	0.09	0.20	460
50	XMP-S14	Carbonaceous Calcareous Shale	0.05	0.10	0.21	417
51	XMP-S15	Carbonaceous Calcareous Shale	0.05	0.10	0.20	479
52	XMP-S16	Carbonaceous Calcareous Shale	0.05	0.12	0.20	402
53	XMP-S17	Carbonaceous Calcareous Shale	0.05	0.09	0.19	435
54	XMP-S18	Carbonaceous Calcareous Shale	0.05	0.13	0.19	384
55	XMP-S19	Carbonaceous Calcareous Shale	0.05	0.12	0.19	392
56	XMP-S20	Carbonaceous Shale	0.05	0.15	0.19	341
57	XMP-S21	Carbonaceous Shale	0.05	0.10	0.20	539
58	XMP-S22	Carbonaceous Shale	0.05	0.09	0.18	466
59	XMP-S23	Carbonaceous Shale	0.05	0.12	0.19	380
60	XMP-S24	Carbonaceous Shale	0.04	0.12	0.19	513

Note See Fig. 4.5 for section code

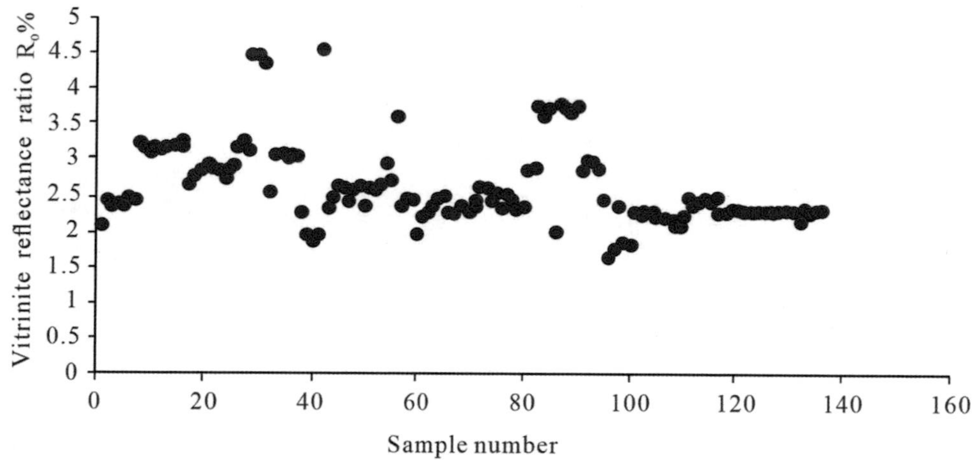

Fig. 4.47 Ro of black rocks in Lower Longmaxi Formation of Southern Sichuan Basin and its periphery

or semi-euhedral crystals or berry-shaped aggregates with intergranular pores (Fig. 4.54g). The average content of calcite and dolomite is 9.17% and 7.64%, respectively. The content of calcite is high, up to 70%, local enrichment, dolomite content ranges from 0 to 48%, and the content of individual profiles and samples is high. Under single polarizing light, sparry crystals are mainly found, with hedral and semi-hedral grain characteristics and irregular filling (Fig. 4.54h). Micrite calcite is founded in some samples in the form of lumps (Zhang et al. 2013a, b), often metasomatic quartz, feldspar and interstitial materials (Fig. 4.54c). The average contents of clay minerals and carbonate minerals are 30.29% and 16.81%, respectively. Clay minerals mainly consist of illite (48.54%) and Illite/montmorillonite (38.8%), followed by chlorite (11.63%). Local areas and sections contain a small amount of kaolinite, which is generally produced as a flake aggregate, in which authentically generated illite can be seen around the primary pores (Fig. 4.54i).

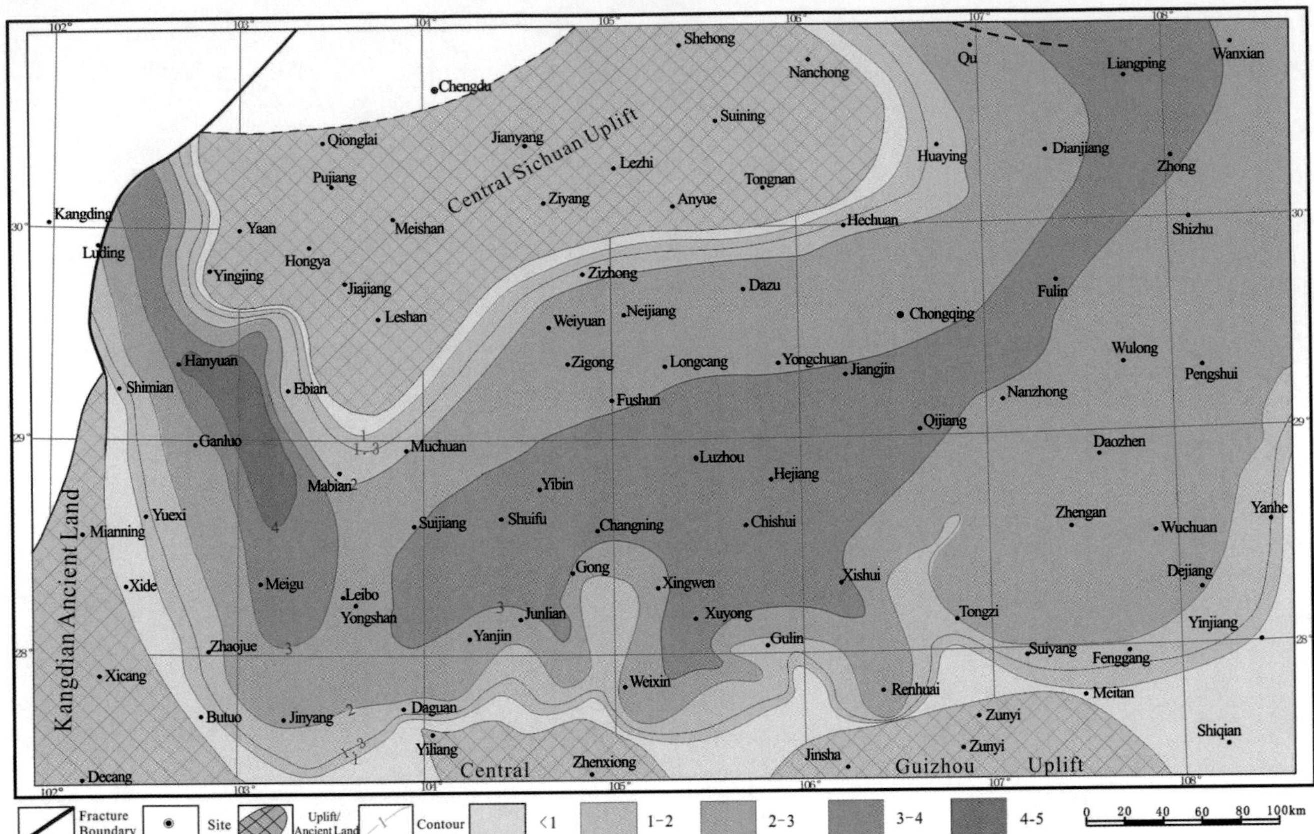

Fig. 4.48 Distribution of Ro of black rocks in Lower Longmaxi Formation of Southern Sichuan Basin and its periphery

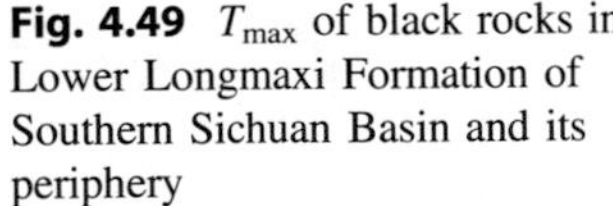

Fig. 4.49 T_{max} of black rocks in Lower Longmaxi Formation of Southern Sichuan Basin and its periphery

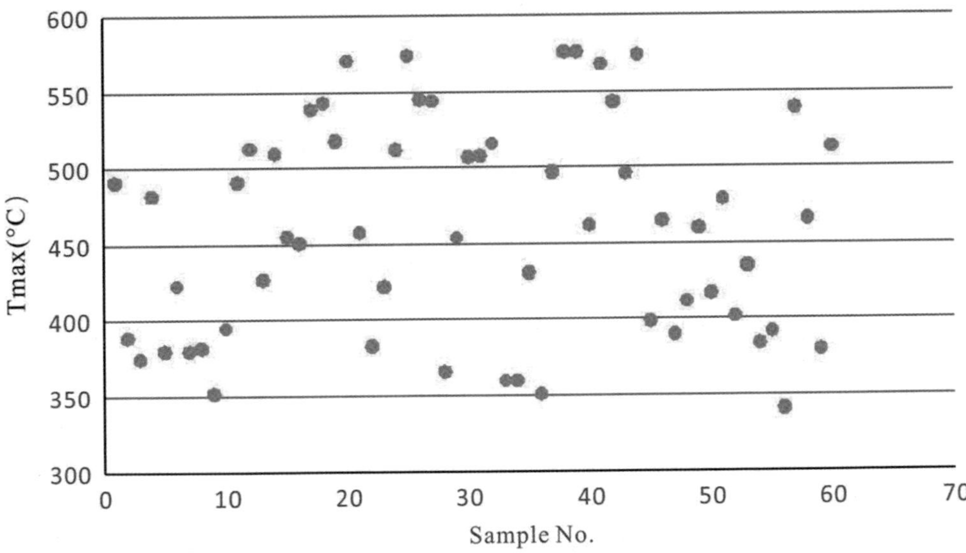

Elemental analysis shows (Table 4.13) that the chemical composition of the black rock series is mainly SiO_2, followed by Al_2O_3 and Fe_2O_3, respectively. K_2O is enriched and evenly distributed, and the content of CaO and MgO in some samples is relatively high, which verifies that the rock composition of the black rock series is mainly quartz and the clay minerals contain high illite and chlorite minerals. Carbonate minerals are not evenly distributed.

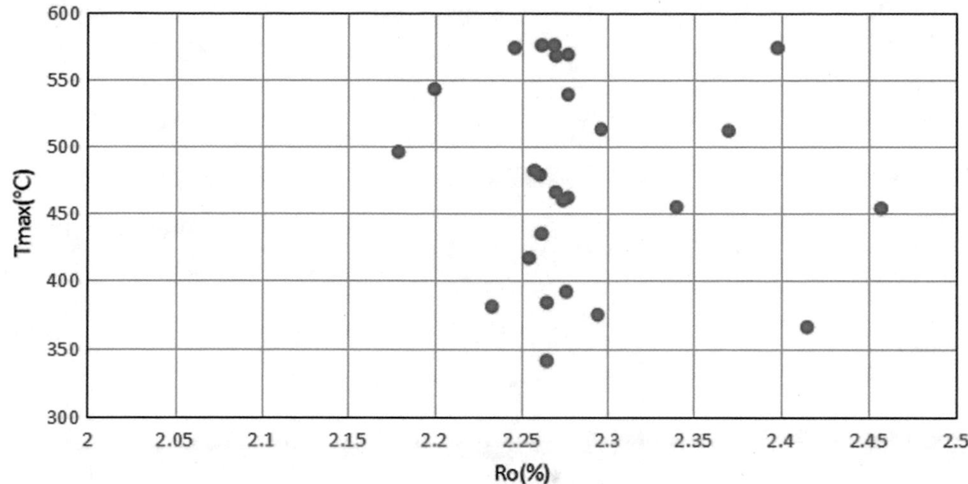

Fig. 4.50 Relationship of T_{max} and *Ro* of black rocks in Wufeng-Longmaxi

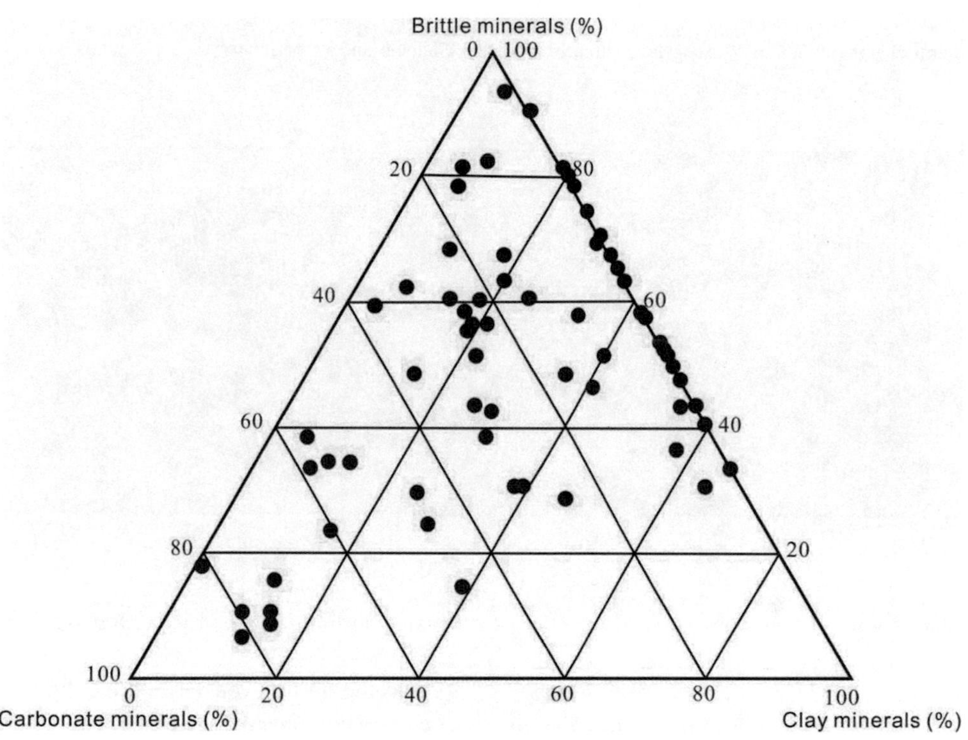

Fig. 4.51 Mineralogy diagram of black rock series of Wufeng Formation in Southern Sichuan and its periphery

(2) Classification of mineral rock types

The average content of siliceous brittle minerals and carbonate minerals in the black shale of Longmaxi Formation in Southern Sichuan and its periphery are 69.71% and 16.81%, respectively. Siliceous shale is mainly developed, and the content of clay minerals is relatively low (Fig. 4.55).

(3) Distribution characteristics of brittle minerals

According to the classification of rock types mentioned above, the plane distribution characteristics of carbonate minerals and siliceous (quartz + feldspar, etc.) brittle minerals were statistically analyzed. The black rock series of the lower member of Longmaxi Formation in Southern Sichuan

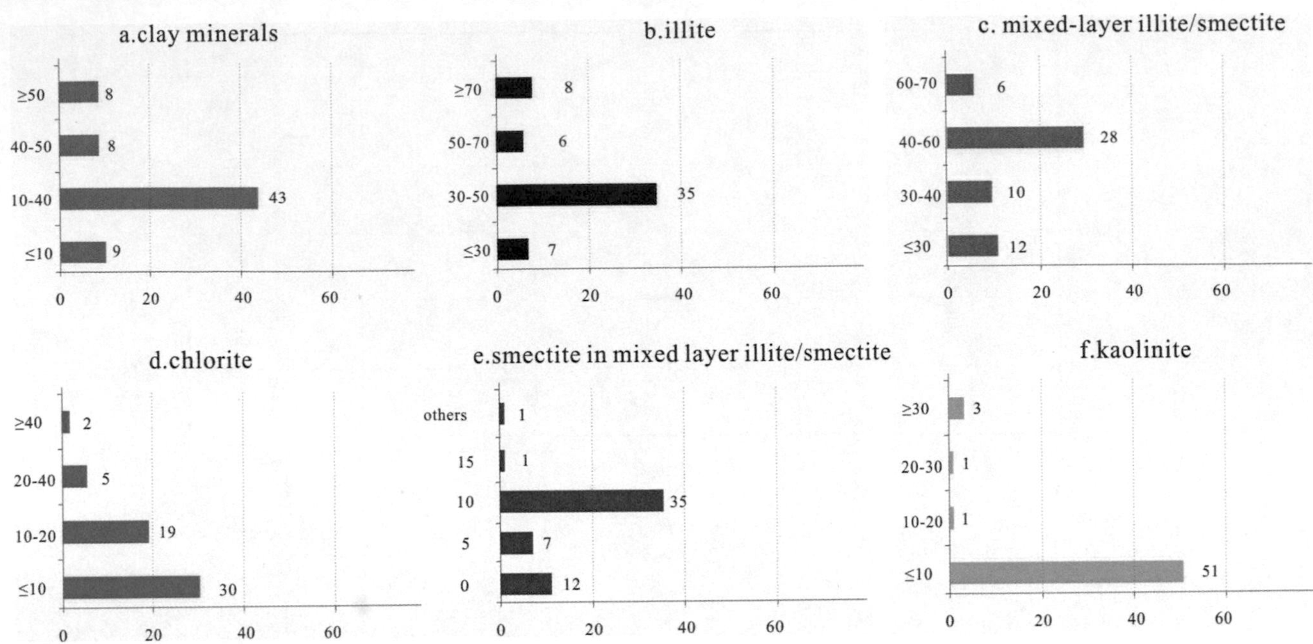

Fig. 4.52 Clay mineral characteristics of Wufeng Formation in Southern Sichuan and its periphery

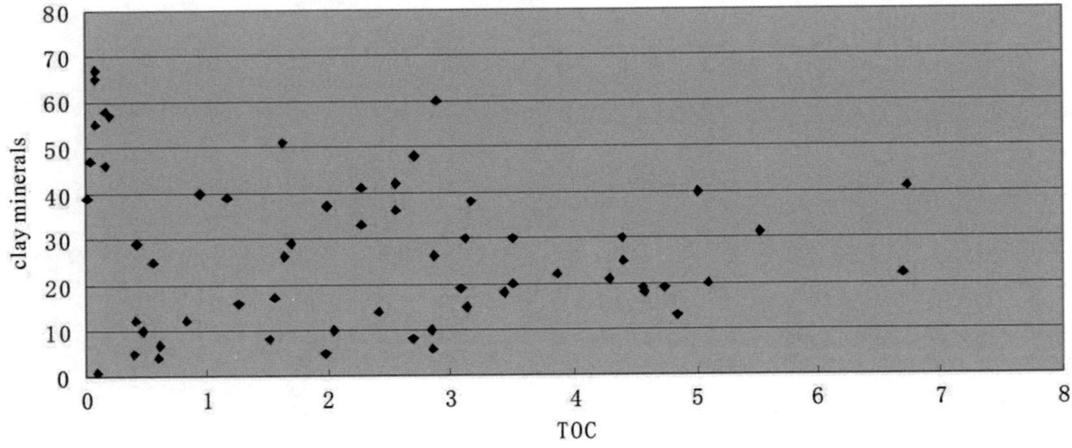

Fig. 4.53 Relationship between clay minerals and TOC in Wufeng Formation in Southern Sichuan and its periphery

and its periphery has high mineral content, such as quartz + feldspar 50%, locally. 30% (Fig. 4.56). The distribution of carbonate minerals is not uniform, and the average content is mainly less than 10%, local enrichment, its content is greater than 30% (Fig. 4.57). It can be seen from the above that siliceous minerals have a good correlation with the content of carbonate minerals, siliceous content less than 30% area and carbonate content over 30%. The distribution area is located at the edge of uplift and belongs to the tidal flat and marl shallow shelf sedimentary environment. In the Northern area of Central Guizhou Uplift, far away from the uplift area, the content of siliceous brittle minerals is bounded by 30% and increases rapidly to the Northward. The content of carbonate minerals decreases 25% to the north, even 10%. It can be seen that the content of carbonate minerals decreases with that of siliceous brittle minerals, which may be caused by the common depositional environment. It is further verified that siliceous shale is the most developed black rock series in the lower member of Silurian Longmaxi Formation in Southern Sichuan and its periphery.

(4) Characteristics of clay minerals

Clay minerals are important and main components of mud shale, and their types, occurrence, content, and variation characteristics are not only helpful to analyze the

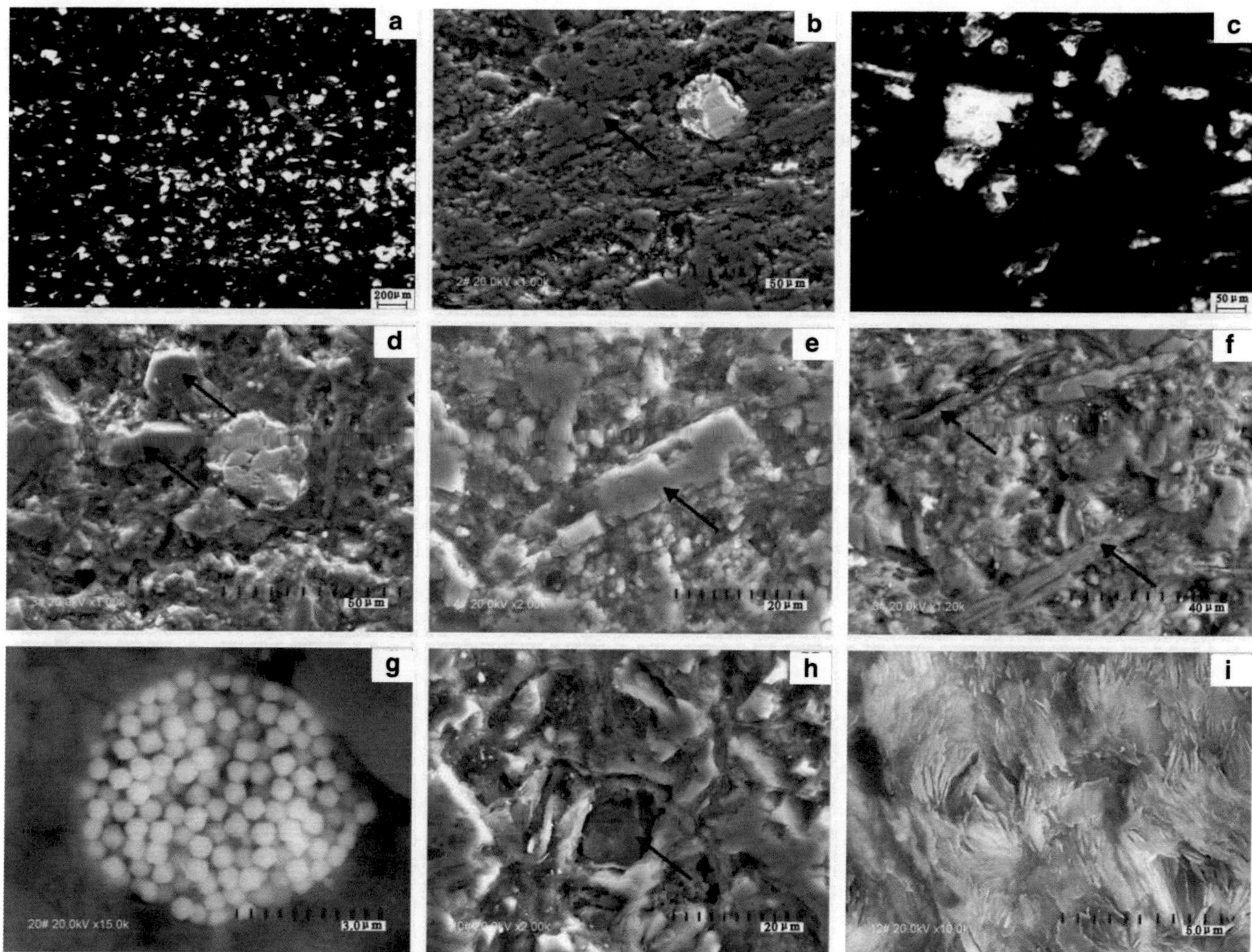

Fig. 4.54 Microscopic characteristics of black rocks in Longmaxi Formation of Southern Sichuan Basin and its periphery **a** rock composition characteristics, detrital particles (red arrow) and interstitial materials (green arrow), with heterobasal support, Xishu Liangcun (single polarized light: DPG); **b** microscopic characteristics of rock composition, clastic particles (red arrow) and interstitial materials (green arrow), silty and argillaceous structure (backscattering: BSE), Shizhu Qiliao; **c** quartz clasts (red arrows), edge replaced by calcite, Qijiang Guanyinqiao (DPG); **d** weak carbonation of albite (BSE), Shizhu Qiliao; **e** potassium feldspar with cracking and dissolution (BSE), Wulong Huangcao; **f** mica (red arrow) and potash feldspar (green arrow). Cleavage joints are filled with organic matter (BSE), Shizhu Qiliao; **g** pyrite complex (BSE), Xuyong Macheng; **h** calcite is encapsulated by clay minerals and is semi-idiomorphic crystal (BSE), Nanchuan Sanquan; **i** clay minerals with sheet structure (BSE), Tongzi Hanjiadian

paleoenvironment and diagenesis experienced by mud shale, but also considered as favorable tools to search for oil and gas. In the study of shale gas, clay minerals content is not only the impact of shale gas reservoir and hydrocarbon content of the main indexes, but also one of the main factors influencing the shale gas production fracturing effect. Therefore, the research for the analysis and understanding of clay minerals content in black rock series of diagenesis and reservoir evolution important tasks, for shale gas exploration and development has a certain guiding significance.

(i) Types and characteristics of clay minerals

According to X-ray diffraction analysis and test results, clay mineral content is 4–82%, mainly concentrated in 25–50%, accounting for 66.7% of all samples, 10–25% (Fig. 4.58a). In the clay mineral composition, almost every rock sample contains more illite, the average relative content is 48.21%. Most of the samples contained illite/montmorillonite minerals, accounting for 84.8% of the total samples, with an average relative content of 44.14%. Chlorite was found in most of the samples, accounting for 96.2% of the total samples, with an average relative content of 12.29%. Only some samples contain a small amount of kaolinite, and montmorillonite is rarely distributed.

Illite content is the highest and relatively evenly distributed, with the relative content ranging from 23 to 100%. There are two main distribution ranges of 50–70% and 30–50%, respectively, and the number of samples in these two ranges accounts for 81.3% of all clay mineral samples

Table 4.13 Composition of black rocks and the minerals in Longmaxi Formation of Southern Sichuan Basin and its periphery

Sample number	Sample name	SiO_2	Al_2O_3	Fe_2O_3	FeO	CaO	MgO	K_2O	Na_2O	TiO_2
SQP-B9	Carbonaceous shale	64.94	13.57	3.83	1.27	2.21	2.38	3.32	0.85	0.62
HCP-B10	Carbonaceous shale	72.81	10.34	2.05	0.09	0.37	0.65	2.79	1.0	0.56
XKP-S2	Carbonaceous shale	44.08	14.59	2.96	0.94	7.12	7.60	8.64	0.07	0.57
XJP-B12	Carbonaceous shale	73.23	10.62	1.68	0.10	0.40	0.66	3.07	0.85	0.66
XMP-S1	Carbonaceous calcareous mudstone	30.41	8.35	2.65	0.45	25.46	3.89	2.25	0.29	0.40
XMP-S5	Carbonaceous shale	62.10	5.80	1.82	0.10	10.54	1.84	1.58	0.16	0.31
XMP-B8	Carbonaceous shale	76.84	8.93	2.27	0.16	2.07	0.84	2.24	0.69	0.49
XMP-S10	Carbonaceous calcareous mudstone	58.39	11.69	3.34	0.19	6.33	2.38	3.09	0.40	0.52
XMP-S17	Carbonaceous calcareous mudstone	47.26	13.22	4.87	0.19	9.54	3.89	3.40	0.52	0.61
XMP-S21	Carbonaceous shale	65.20	13.35	2.91	0.83	3.80	1.83	3.40	1.46	0.62
QGP-B5	Carbonaceous shale	61.50	10.94	3.88	0.087	3.97	2.10	2.75	0.85	0.55
XLP-B9	Carbonaceous shale	77.05	7.40	0.90	0.11	0.31	0.43	1.90	0.60	0.42
XLP-S2	Carbonaceous shale	83.75	4.83	1.19	0.079	0.63	0.31	1.36	0.10	0.25

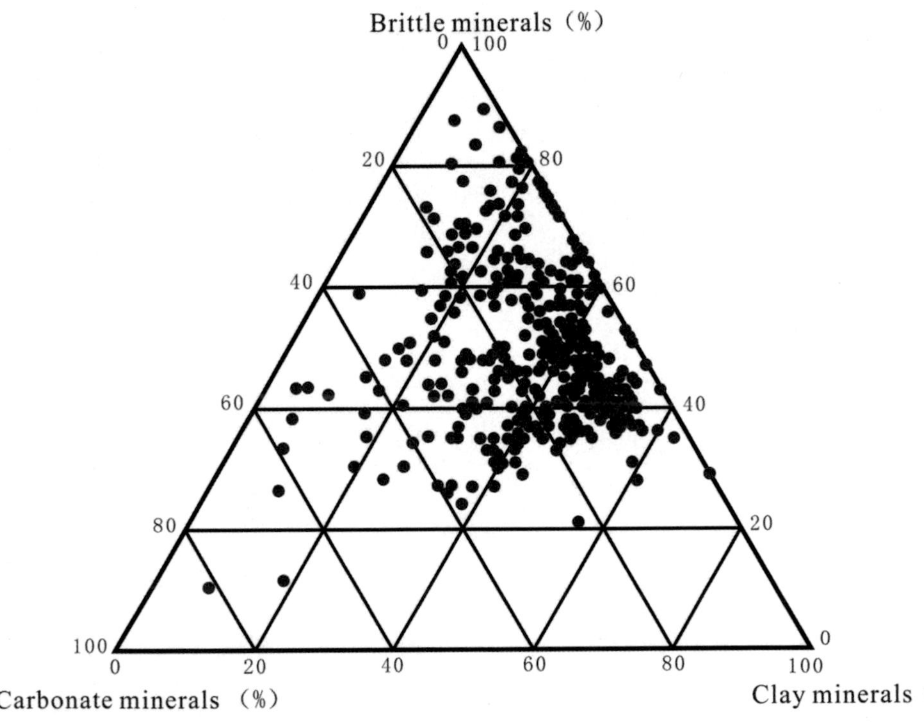

Fig. 4.55 Mineralogy ternary diagrams of black rocks in Longmaxi Formation of Southern Sichuan Basin and its periphery

analyzed (Fig. 4.58b). The illite/montmorillonite is a secondary rich clay mineral type with a relative content of 2–76%, mainly distributed between 30 and 50%, accounting for 79.1% of the total, followed by 50–70% (Fig. 4.58c). Chlorite is widely distributed in Longmaxi Formation, and its relative content ranges from 1 to 46%. 20%, accounting for 91.5% of all samples (Fig. 4.58d). The content of montmorillonite in Iran-Mongolian-mixed-layer minerals is mainly limited to 10%, accounting for 83.6% of all samples, followed by 15% and less than 5% (Fig. 4.58e). Among the 368 clay samples which analyzed by X-ray diffraction, 57 samples contained kaolinite, with uneven distribution and average content of 4.58%, usually less than 7%, and only one sample reached 10% (Fig. 4.58f). Montmorillonite is poorly developed and only found in a sample from the lower part of Longmaxi Formation in Xiashaqiang area, Mabian area, with a relative content of 6%, accounting for only 3.4% of the total sample content.

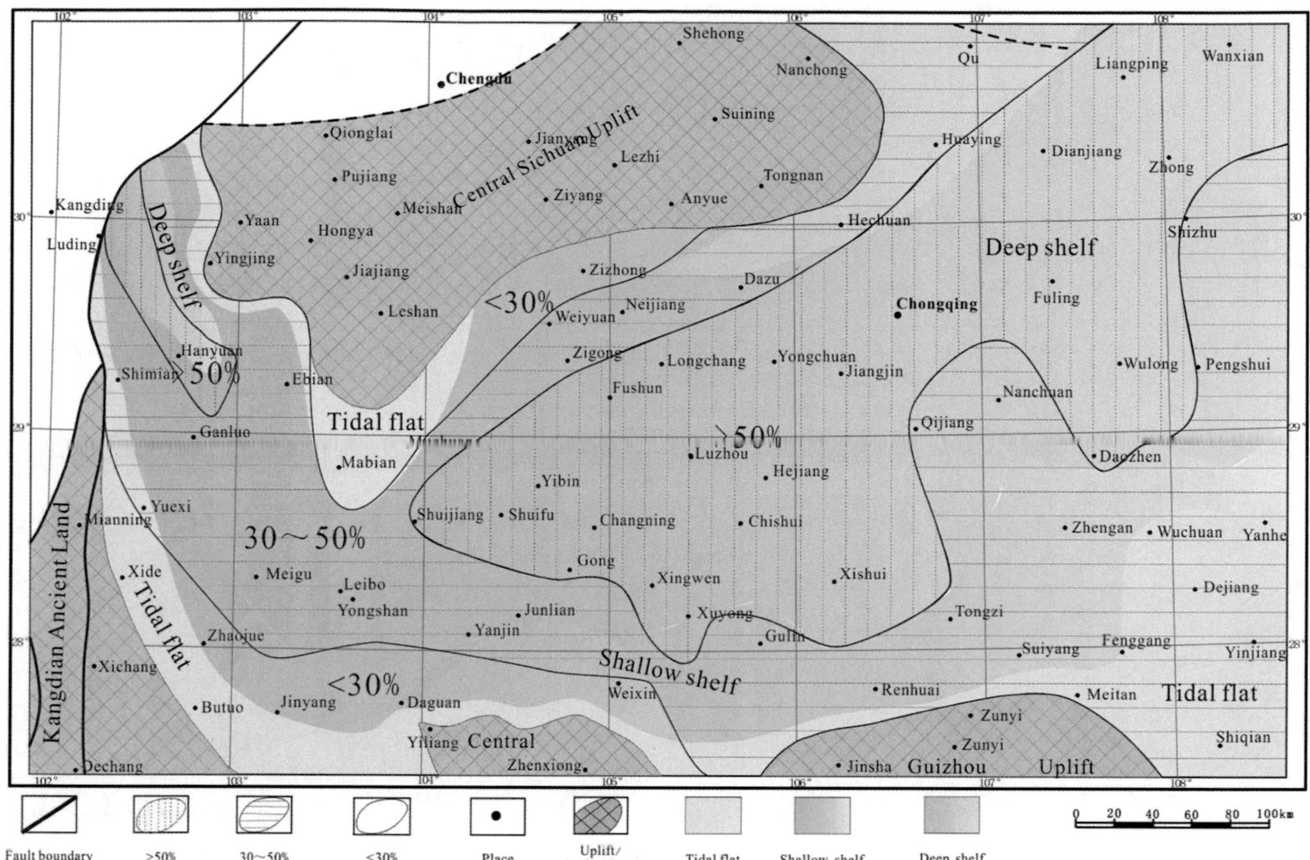

Fig. 4.56 Distribution characteristics of brittle mineral (quartz + feldspar) of black rocks in Longmaxi Formation of Southern Sichuan Basin and its periphery

From bottom to top, the clay mineral components of Longmaxi Formation are basically similar, and no green/montmorillonite minerals are found, and the content of each component is not very different. Only the profiles of Guanyinqiao in Qijiang, Liangcun in Xishui and Huanggexi in Daguan of Longmaxi Formation show a slight upward trend of chlorite, while the profiles of Lujiao in Pengshui (Li et al. 2012a) and Well Changxin 1 (Chen et al. 2013b) show a gradual increase of chlorite content with the increase of depth.

Statistical analysis also found that the clay mineral components of Longmaxi Formation in different regions were basically similar, but there were differences in content (Table 4.14). Kaolinite is found in the Southeastern part of the basin (Wuchuan, Daozhen, Nanchuan and other areas), in the SouthWestern margin (Jinyang, Yongshan and other areas) and in some layers of Well Changxin 1 (Chen et al. 2013b). In 57 samples from these areas, the total amount of clay minerals is 34.52% on average, and illite is found in each sample. There was no chlorite or illite/montmorillonite in some samples. The average relative content of illite and illite/montmorillonite was 54.68% and 28.82%, respectively, and the average relative content of chlorite was 11.92%. The average content of clay minerals in 65 samples from Southeast Chongqing-Northern Guizhou area is 35.2%, the highest content of Illite is 45.8%, the average relative content of illite is 43.3%, the average content of chlorite is 11.3%, and chlorite is not found in five samples. In well Jiaoye 1, the content of clay minerals in the lower high-quality shale of Longmaxi Formation (including Wufeng Formation) ranges from 16.6% to 49.1%, with an average content of 34.6%. The main clay minerals are illite, accounting for 63.5% and 31.4% of the total clay, respectively, followed by chlorite, accounting for 4.9% of the total clay. Montmorillonite and kaolinite were not found (Guo and Liu 2013). The content of clay minerals in well Yuye 1 ranges from 18.5 to 53.2%, with an average of 33%. Illite and Aemon mixed-layer minerals are also dominant, with a relative content of 75–91%, chlorite 7–20%, and a small amount of kaolinite with uneven distribution (Wu et al. 2013). The bottom 30 m of well Changxin 1 (including the lower part of Wufeng Formation and Longmaxi Formation) in Changning area is organic-rich shale, with an average clay mineral content of 24.7%, mainly illite with a relative average content of 60% (37–75%), and the relative average mineral content of illite/montmorillonite is 28.4% (12–49%). Chlorite is 11.2% (1–23%) (Chen et al. 2013b). In 223

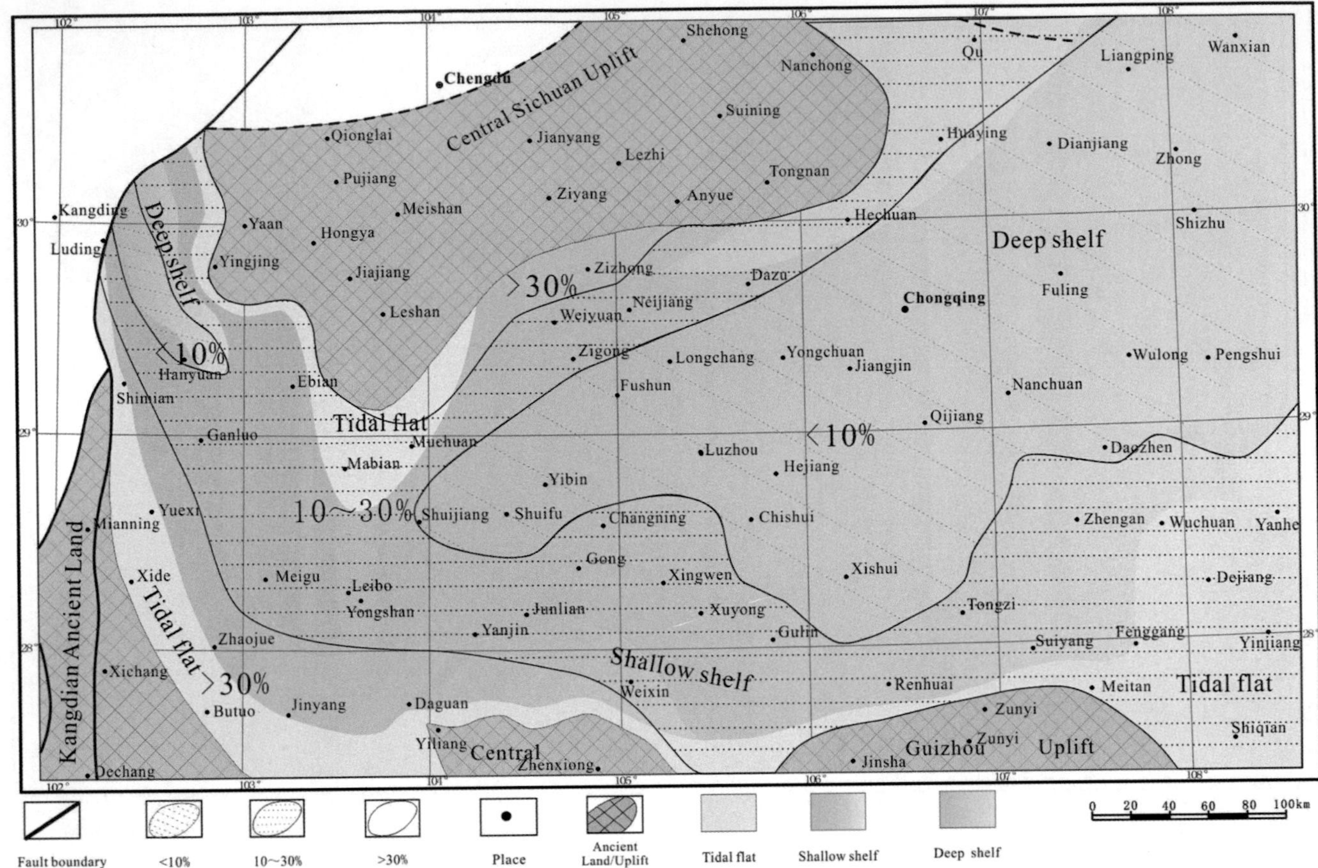

Fig. 4.57 Distribution characteristics of carbonate minerals of black rocks in Longmaxi Formation of Southern Sichuan Basin and its periphery

samples from Western Guizhou to Northern Yunnan, the average content of clay minerals is 44.3%, the highest content of illite is 54.6%, the average relative content of illite/montmorillonite is 33.1%, all samples contain chlorite, and the average relative content is 12.3%. Among the 18 samples from Western Sichuan, the composition is relatively complex, the average content of clay minerals is 23.1%, and kaolinite is not found in all of them. Among them, all the samples (3) from Jiaodingshan in Hanyuan and 6 of the 8 samples from Dajingping profile in Tianquan are illite, and the lithology is carbonaceous shale and silty shale, and the other 2 samples from Dajingping profile in Tianquan are free of chlorite. One sample in the profile of Mabian Xiashaqiang contains 6% montmorillonite, while only two of the other five samples contain illite, chlorite and illite/montmorillonite, and the other three samples have no illite/montmorillonite. The contents of illite and chlorite are relatively high, with the average relative contents of 60.6% and 24.4%, respectively.

(ii) Composition and distribution characteristics of clay minerals

It can be seen from above that in the black organic-rich shale in the Southern Sichuan Basin, the clay mineral assemblage is dominated by chlorite + Illite/illite (C + I/S + I), which is developed in the profiles of 20 different areas, mainly distributed in the center of the basin, Southeastern Chongqing-Northern Guizhou and Western Guizhou-Northern Yunnan. Kaolinite + chlorite + Illite/Illite (K + C + I/S + I) assemblage is mainly distributed in the Southeastern and South-Western margin of Sichuan Basin, near the Central Guizhou Uplift and the Kangdian Ancient Land. The clay mineral assemblages of a few samples are kaolinite + chlorite + illite (K + C + I), illite/montillite + illite (I/S + I), chlorite + illite (C + I) and all illite (I), mainly distributed in the Western Sichuan area, near the Kangdian Ancient Land. Among the six basic patterns of clay mineral distribution summarized by Zhao et al. (2008) based on the analysis data

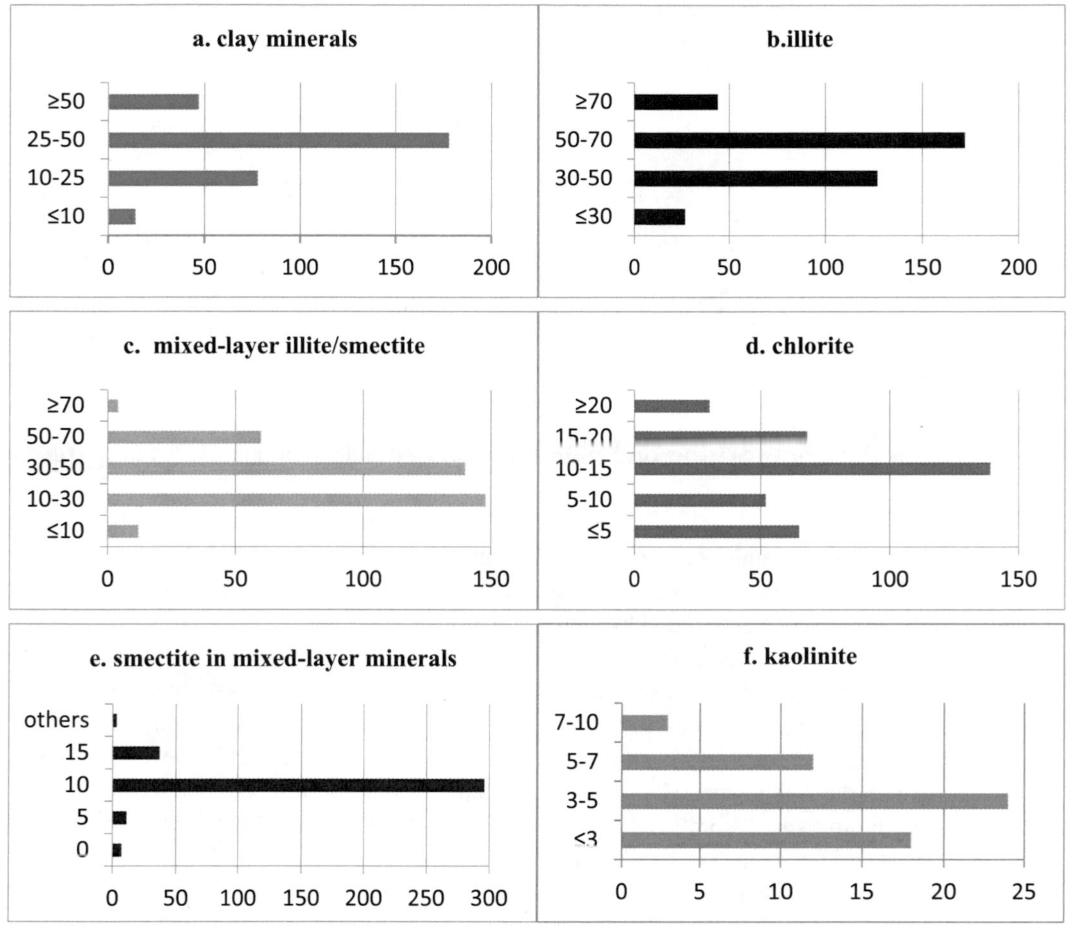

Fig. 4.58 Characteristics of clay minerals of black rocks in Longmaxi Formation of Southern Sichuan Basin and its periphery

of mud shale, they mainly belong to type I-normal transformation type. The vertical distribution of clay minerals in normal transformation mode is characterized by two sequences of smectite converting to illite and kaolinite converting to chlorite with increasing depth (Zhao et al. 2008). The conversion of smectite to illite is common, the part of smectite is completely transformed into illite, and a large number of Illite/montmorillonite minerals appear. However, chlorite content in Longmaxi Formation is relatively stable, with little change in depth, there may be a transformation from kaolinite to chlorite.

The content of clay minerals in the black organic-rich shale of Longmaxi Formation in the study area shows certain regularity (Fig. 4.59). The content of clay minerals is mainly < 40%, distributed near the center and edge of the basin, mainly shallow shelf and some deep-water shelf; However, near the uplift, the clay mineral content is usually lower 30% in transitional sedimentary area with shallow water. In the center of the basin, far away from the provenance, the sedimentary water is deep and quiet. Considering that the clay minerals in the black rock series of the lower Longmaxi Formation in the study area are mainly terrigenous clastic origin (discussed in detail later). According to Qin et al. (2010), there is a certain correlation between the total organic carbon content and mineral content of Marine shale source rocks in Southern China. The clay mineral content decreased with TOC content, while the quartz content increased with TOC content. This relationship may be caused by the stable environment of platform basin, platform depression and lagoon, which is favorable for the formation of marine high-quality shale, and is not conducive to the transport and deposition of terrigenous clastic clay with water. As the sedimentary water becomes shallower, the Longmaxi Formation in the study area is bottom-up and the content of clay minerals increases gradually (Fig. 4.60). Near the bottom of the Longmaxi Formation, the content of clay minerals ranges from 5 to 47%, with an average of 23.8% (Table 4.15). In summary, the closer to the deep-water area in the center of the basin, the clay mineral content is relatively reduced, and its average content should usually be < 50%, and the average < 30%, which is a deep-water sediment area in the open ocean. It can be seen that the distribution of clay minerals has an obvious correlation with the sedimentary environment, presenting a low–high-low characteristic as a whole.

Table 4.14 Characteristics of clay minerals in different areas of black rocks in Longmaxi Formation of Southern Sichuan Basin and its periphery

Area	Sample number	Clay mineral content (%)	Relative content of illite (%)	Relative content of illite/smectite mixed layer (%)	Relative content of chlorite (%)	Relative content of kaolinite (%)
Southeast and southwest edge of the basin	57	34.5	54.7	28.8	11.9	4.6
Southeast Chongqing-North Guizhou region	65	35.2	43.3	45.8	11.3	–
Western Guizhou—Northern Yunnan region	223	44.3	54.6	33.1	12.3	–
Western Sichuan	18	23.1	81.0	7.5	11.0	Local occurrence

4.4.2 Influence of Mineral Composition on Shale Gas

The black rock's material compositions of Ordovician Wufeng Formation and Silurian Longmaxi Formation in Southern Sichuan and its periphery can be divided into organic matter and inorganic minerals. Organic matter in source rocks is derived from biological residues sediments, which abundance of them is an important index for evaluating source rocks. Organic matter in shale gas is not only the parent material of hydrocarbon generation, but also an important adsorption medium for gas adsorption in shale gas. Shale gas reservoir is self-generated and self-stored unconventional gas reservoirs. High organic carbon content and high maturity are one of the main controlling factors for shale gas enrichment. Therefore, in the black rock of Wufeng and Longmaxi Formations with high maturity, organic carbon content is one of the basic geological conditions for shale gas, which is one of the important indexes for shale gas evaluation. And inorganic mineral components of black rock usually have different sources and causes, which not only show the different types of reservoir space and physical property, but also affect the occurrence state of organic matter and the content, and even affect the formation of air content. Therefore, on the basis of detailed analysis of the origin of clay minerals, siliceous minerals and carbonate minerals, it has the great guiding significance to judge their relationship with shale gas geological conditions.

1. Ordovician Wufeng Formation

Based on the X-ray diffraction analysis of 68 outlying samples from the Ordovician Wufeng Formation in Southern Sichuan and its periphery, the mineral composition includes quartz, potash feldspar, plagioclase, calcite, dolomite, pyrite, clay minerals and a small amount of siderite and gypsum. Quartz exists in the form of clastic particles and siliceous cements with an average content of 43.53%. Feldspar is not evenly distributed, and the content of plagioclase is high. The average content of potassium feldspar and plagioclase is 1.65% and 2.75%, respectively. The content of pyrite is 1.65%. The average content of clay minerals and carbonate minerals is 27.73% and 22.47%, respectively. The clay minerals are mainly illite (47.71%) and illite/smectite (39.13%), followed by chlorite (10.18%). A small amount of kaolinite (2.98%) is found in local areas and sections, and the average content of calcite and dolomite is 12.16% and 10.31%, respectively (Table 4.16).

According to Table 4.16, the black rock with different organic matter contents has differences in rock composition and content, while the clay minerals have little difference in composition and content. They are all dominated by illite and illite/montmorillonite with certain chlorite, and kaolinite is distributed in small quantity and unevenly. Among them, the black shale's (TOC > 4%) quartz content is high, mainly in the range of 25–68%, the largest up to 74%, average 51.23%. The clay mineral content mainly ranged from 13 to 41%, the lowest was 13%, average 25.31%. The average content of carbonate is 17.08%, and the average content of calcite (12.23%) is more than dolomite (4.85%). Siliceous rocks are generally dominant (Fig. 4.61a). Black shales with TOC value of 2–4% also have high quartz content, mainly in the range of 30–79%, and some samples can reach over 90%, with an average of 55%. The content of clay minerals ranges from 15 to 40%, and a few samples can be as low as 10%, with an average of 26.3%. The content of carbonate minerals is approximately 10–30%, the largest value can be up to 56%, with an average of 14%. The overall content of calcite and dolomite is basically the same, and the average content is 6.75% and 7.25%, respectively. Siliceous shales are generally dominate (Fig. 4.61b). Black shales with TOC value of 1–2% also have high quartz content, mainly in the range of 20–50%, and some samples can reach over 92%, with an average of 45.3%. The content of clay minerals ranges from 15 to 40%, with an average of 26.3%. The content of carbonate minerals is approximately 20–50%, with an average of 22.3%. Siliceous shales are generally dominate (Fig. 4.61c). The clay mineral content of black

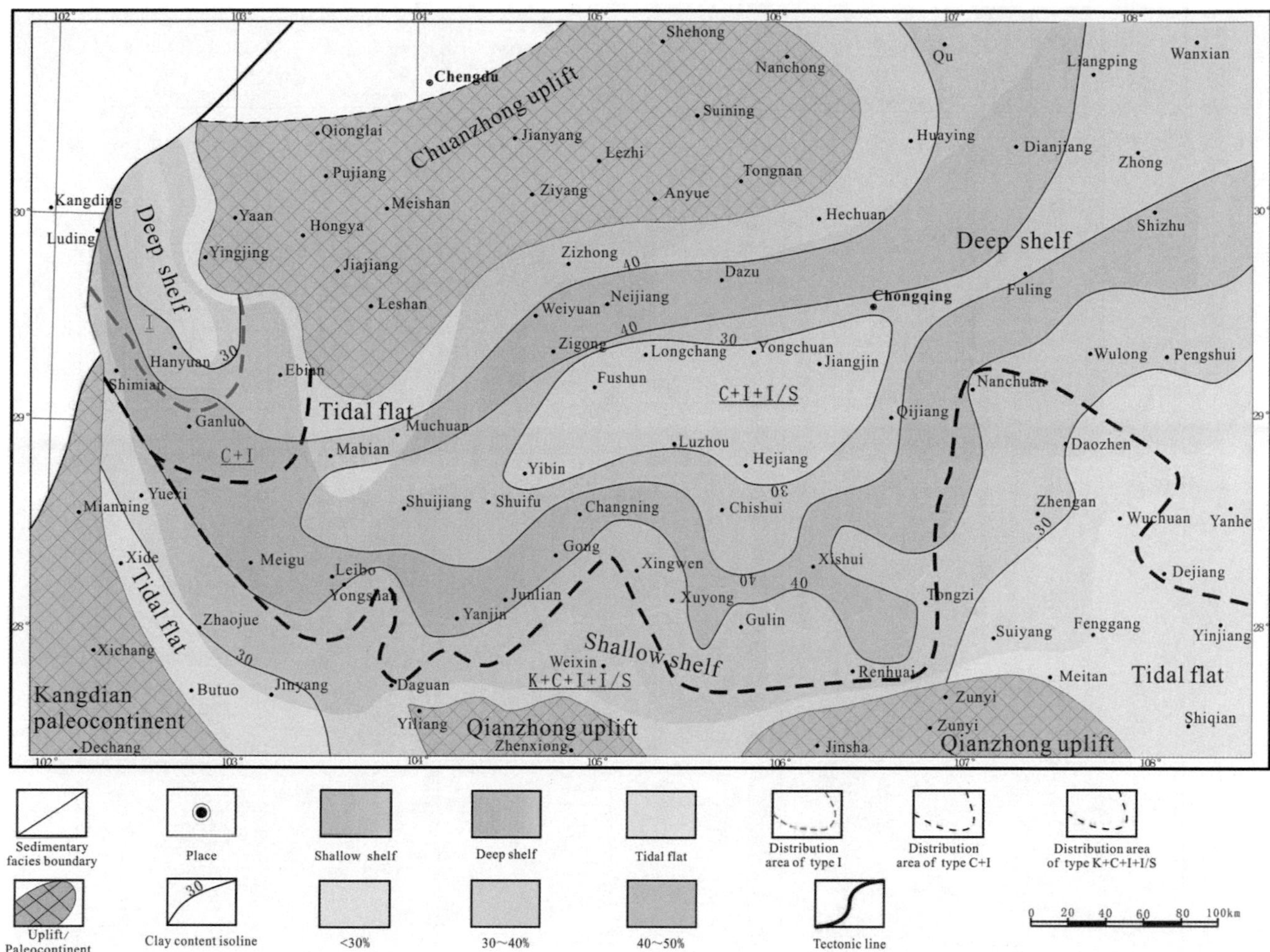

Fig. 4.59 Distribution characteristics of clay minerals in different areas of black rocks in Longmaxi Formation of Southern Sichuan Basin and its periphery

shales with TOC value less than 1% increases, mainly in the range of 10–50%, and some samples can reach over 58%, with an average of 32.17%. The content of quartz is low, mainly ranging from 10 to 40%, with an average of 30.89%. The average content of carbonate minerals is 28.5%, and some samples can reach over 82%. The rock types are mainly clayey and carbonate, with transitional types between them (Fig. 4.61d).

It can be seen from above that the rock composition of Wufeng black rock in Southern Sichuan and its periphery has the influence on the organic carbon content. While the carbonate mineral content is less than 30%, the quartz content is higher, the clay mineral content is lower, and the organic carbon content is higher. Siliceous shale has higher organic carbon content than clayey shale. While the carbonate mineral content is higher, the quartz and clay mineral content is lower, and the organic carbon content is lower. The organic carbon content of carbonate shale is low. In general, siliceous shale has the highest organic matter content, the second is clayey shale, and the carbonate shale has the lowest organic matter content (Figs. 4.62 and 4.63).

According to Figs. 4.62 and 5.11, the scatter diagram between the content of clastic minerals (quartz + feldspar + pyrite), carbonate minerals (calcite + dolomite) and TOC value in the X-ray diffraction test results of 68 samples of Wufeng Formation in Southern Sichuan and adjacent areas shows that: While the TOC value less than 1%, the content of detrital minerals is mostly about 20–40%, and few samples exceed 50%. The distribution of carbonate minerals is not uniform, but the content of most samples is less than 50%. While 1% < TOC < 2%, the content of detrital minerals in most samples is less than 40%, and only one or two samples have a corresponding content of more than 50%. The distribution of carbonate minerals is not uniform. While the TOC value is more than 2%, the content of detrital minerals mostly ranges from 40 to 70%, which accounted for 81.8% of all samples. In addition, a few other samples exceeded 80%, and the highest was 90%. The

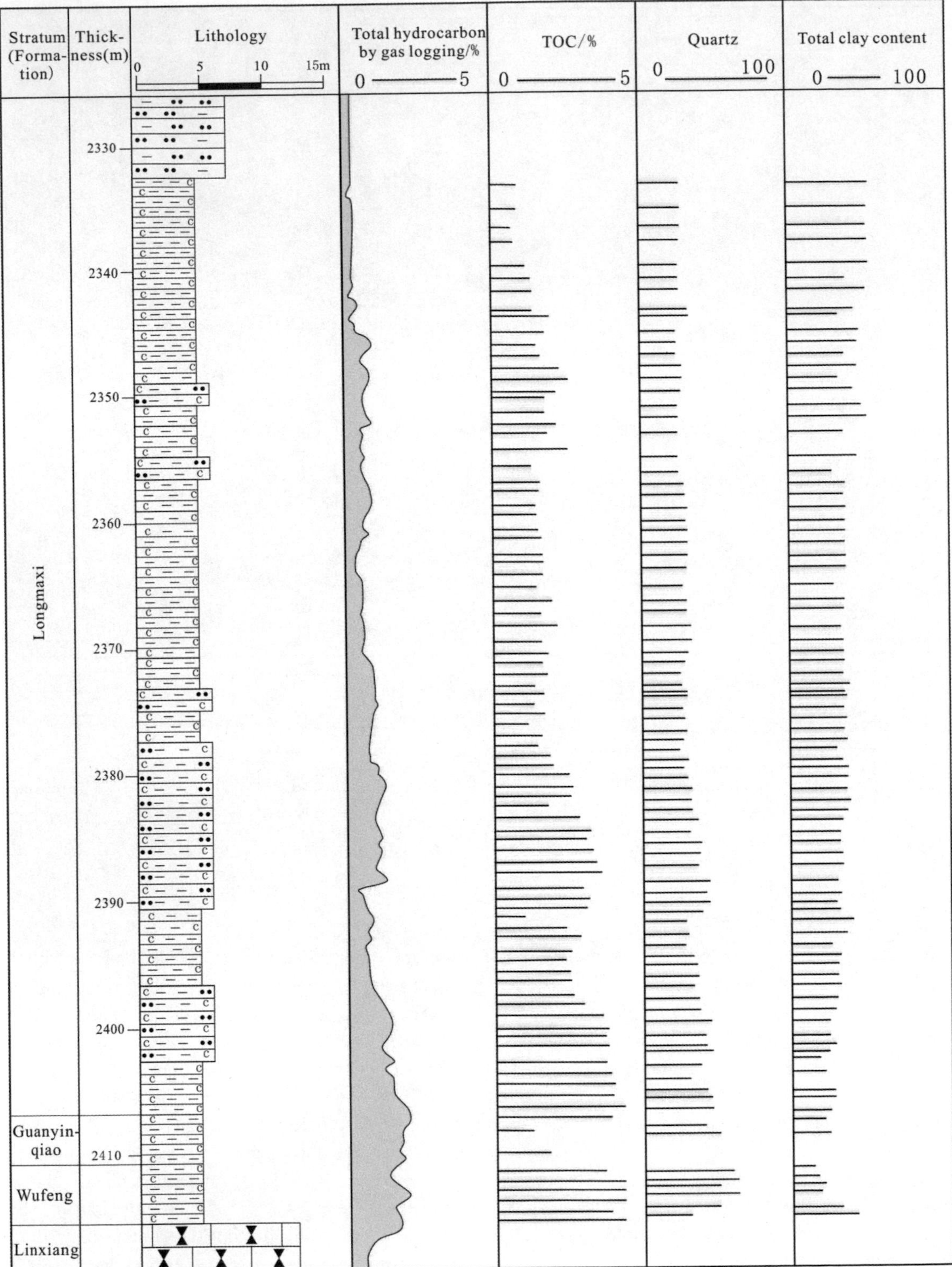

Fig. 4.60 Comprehensive geological histogram of black rocks in Lower Longmaxi Formation of well JY1 (modified from Guo et al. 2013)

Table 4.15 Mineral composition of black rocks and its minerals in bottom of Lower Longmaxi Formation in Southern Sichuan Basin and its periphery

Section	Sample name	Clay mineral relative content (%)						Whole-rock quantitative analysis (%)					
		K	C	I	I/S	Smectite	Clay minerals	Quartz	Potash feldspar	plagioclase	calcite	dolomite	Pyrite
Qijiangguanyinqiao	Carbonaceous shale	0	13	59	28	10	31	54	0	7	2	4	0
	Carbonaceous shale	0	15	45	40	10	35	45	1	7	5	3	3
Wulonghuangcao	Carbonaceous shale	0	0	69	31	10	23	68	1	8	0	0	0
Nanchuansanquan	Carbonaceous shale	0	12	29	59	10	40	38	2	8	3	8	1
	Siliceous carbonaceous shale	0	15	36	49	10	38	41	2	8	2	7	2
Xuyongmacheng	Carbonaceous calcareous shale	0	22	37	41	10	26	35.1	1	5	8	12	2
	Carbonaceous calcareous shale	0	13	32	55	10	25	37	1	2	16	17	2
Tongzihanjiadian	carbonaceous shale	0	23	39	38	10	37	50	0	6	6	0	1
Xishuiliangcun	Graptolite bearing shale	0	5	33	62	10	25	67	0	8	0	0	0
	Graptolite bearing shale	0	2	41	57	10	27	66	2	5	0	0	0
Daguanhuanggexi	Calciferous carbonaceous shale	0	6	64	30	5	8	9	0	1	45	36	1
	Silty shale	0	26	38	36	10	16	31	0	9	31	13	0
Dajingba	Carbonaceous shale	0	0	100	0	0	5	85	0	2	5	2	1
Gulintiesuoqiao	Silty carbonaceous shale	0	0	26	74	10	39	52	3	4	2	0	0
	Silty carbonaceous shale	0	1	31	68	10	34	61	1	3	1	0	0
Leiboshangtianbaba	Carbonaceous shale	0	29	35	36	10	60	32	0	4	4	0	0
	Carbonaceous shale	0	31	37	32	5	63	33	0	2	2	0	0
Yongshansutian	Carbonaceous shale	5	14	42	39	15	15	43	1	2	21	16	2
	Carbonaceous shale	0	22	41	37	10	31	31	2	4	12	18	2
	Carbonaceous shale	0	20	41	39	10	33	34	3	4	11	13	1
Renhuaizhongshu	Carbonaceous shale	0	13	29	58	10	42	33	3	12	10	0	0
	Carbonaceous shale	0	16	32	52	10	35	40	2	11	9	2	1
Xingwenqilin	Carbonaceous shale with silt	0	2	41	57	10	26	69	0	3	0	0	2
	Carbonaceous shale	0	8	51	41	5	17	81	0	2	0	0	0
Yanjinyinchangba	Carbonaceous calcareous shale	0	26	30	44	10	31	33	4	9	10	10	1
	Silty shale	0	17	33	50	10	47	26	1	5	10	10	1
Yunnanluomurou	Carbonaceous shale	0	6	43	51	5	27	38	1	2	14	17	1
	Carbonaceous shale	0	8	36	56	5	26	27	1	1	10	33	2
Yanhexinjing	Carbonaceous shale	0	6	29	65	10	28	63	2	7	0	0	0
Huayingsanbaiti	Carbonaceous shale	0	7*	41	52	10	8	90	0	0	0	2	0
Shizhuqiliao	Common shale	0	0	40	60	10	71	23	0	6	0	0	0
Daozhenbayu	Siliceous carbonaceous shale	0	0	0	0	0	19	79	1	1	0	0	0
	Siliceous carbonaceous shale	0	0	0	0	0	22	74	1	3	0	0	0
	Carbonaceous shale	0	0	0	0	0	24	70	2	4	0	0	0

(continued)

Table 4.15 (continued)

Section	Sample name	Clay mineral relative content (%)						Whole-rock quantitative analysis (%)					
		K	C	I	I/S	Smectite	Clay minerals	Quartz	Potash feldspar	plagioclase	calcite	dolomite	Pyrite
Daoye I well	Dark gray black shale	5	11	71	13	10	35	40	0	14	5	0	0
	Gray-black carbonaceous shale	0	0	0	0	0	31	42	4	13	6	4	0
	Gray-black carbonaceous shale	0	0	0	0	0	25	44	3	11	7	5	5
Wuchuanbei	Carbonaceous shale	/	7	76	17	0	10	39	9	24	14	4	0
	Carbonaceous shale	0	0	0	0	0	18	41	8	25	8	0	0
	Carbonaceous shale	3	10	52	35	0	17	44	9	20	8	2	0
Yanhexinjing	Carbonaceous shale	0	0	0	0	0	18	47	10	21	2	2	0
	Carbonaceous shale	/	1	39	60	0	18	64	6	10	2	0	0
	Carbonaceous shale	/	4	75	21	0	17	53	9	20	1	0	0
	Silty carbonaceous shale	0	0	0	0	0	16	41	9	19	15	0	0
Changningshuanghe	Siliceous carbonaceous shale	0	0	0	0	0	18	67	0	0	7	8	0
	pelitic siltstone	0	0	0	0	0	14	60	0	0	14	12	0
Pengshuilujiao	Black silty shale	0	17	50	33	10	29	45	4	15	6	1	0
	Black silty shale	0	16	49	35	10	31	43	4	14	7	1	0
	Black silty shale	0	15	49	36	10	25	44	5	17	8	1	0
Ebianxihe	Calcareous siltstone	0	26	74	0	0	34	40	2	3	0	21	0
Changxin I well	Shale	0	2	80	18	10	18	30	0	6	41	5	0
	Calcareous shale	0	1	82	17	10	7	27	0	2	25	34	5
	Calcareous shale	0	3	73	24	10	17	46	0	2	27	5	3
	Laminar shale	0	3	76	21	10	18	56	0	2	15	6	3
	Shale	0	3	69	28	10	27	30	0	2	17	20	4
	Shale	0	4	72	24	10	11	74	0	1	5	6	3
	Shale	0	3	68	29	10	14	68	0	1	8	9	0
	Shale	0	4	70	26	10	16	40	0	1	30	10	3
	Calcareous shale	0	4	68	28	10	20	46	0	2	16	12	4
	Shale	0	3	76	21	10	23	38	0	2	15	18	4
	Calcareous shale	0	2	74	24	10	13	42	0	0	13	28	4

K: Kaolinite; C: Chlorite; I: Illite; S: Montmorillonite

content of carbonate minerals mostly ranged from 0 to 40%, which is accounting for 87.9% of all samples. Therefore, when the clastic mineral content is 40–70% and the carbonate mineral content is 0–40%, the corresponding rocks have higher TOC content.

2. Silurian Longmaxi Formation

(1) Organic carbon content and its relationship with mineral composition

The clay mineral content of Paleozoic Marine source rocks in the middle and Upper Yangtze area tends to decrease with the increase of TOC value, while the quartz content tends to increase with the increase of TOC value (Fu et al. 2011). According to the classification basis (Li et al. 2013b) for shale organic matter abundance evaluation in Southern Sichuan Basin, it can be seen from Table 4.17 that the mineral components and contents of black rock with different organic matter contents to some extent, while the differences of clay minerals are not significant. While the black shale TOC value is more than 4%, there has high quartz content and the content of clay minerals is the least, the content of carbonate is generally small and unevenly distributed. The rock types are mainly siliceous shale, and a small amount of clayey shale (Fig. 4.64a). While

Table 4.16 Rock composition characteristics of Wufeng Formation black rock in Southern Sichuan and its periphery

Section	Sample	Clay mineral relative content (%)						Whole-rock quantitative analysis (%)					
		K	C	I	I/S	Smectite	Clay minerals	Quartz	Potash feldspar	plagioclase	calcite	dolomite	Pyrite
Qijiangguanyinqiao	Carbonaceous shale	0	13	59	28	10	31	54	0	7	2	4	0
	Carbonaceous shale	0	15	45	40	10	35	45	1	7	5	3	3
Wulonghuangcao	Carbonaceous shale	0	0	69	31	10	23	68	1	8	0	0	0
Nanchuansanquan	Carbonaceous shale	0	12	29	59	10	40	38	2	8	3	8	1
	Siliceous carbonaceous shale	0	15	36	49	10	38	41	2	8	2	7	2
Xuyongmacheng	Carbonaceous calcareous shale	0	22	37	41	10	26	35.1	1	5	8	12	2
	Carbonaceous calcareous shale	0	13	32	55	10	25	37	1	2	16	17	2
Tongzihanjiadian	Carbonaceous shale	0	23	39	38	10	37	50	0	6	6	0	1
Xishuiliangcun	Graptolite bearing shale	0	5	33	62	10	25	67	0	8	0	0	0
	Graptolite bearing shale	0	2	41	57	10	27	66	2	5	0	0	0
Daguanhuanggexi	Calciferous carbonaceous shale	0	6	64	30	5	8	9	0	1	45	36	1
	Silty shale	0	26	38	36	10	16	31	0	9	31	13	0
Dajingba	Carbonaceous shale	0	0	100	0	0	5	85	0	2	5	2	1
Gulintiesuoqiao	Silty carbonaceous shale	0	0	26	74	10	39	52	3	4	2	0	0
	Silty carbonaceous shale	0	1	31	68	10	34	61	1	3	1	0	0
Leiboshangtianbaba	Carbonaceous shale	0	29	35	36	10	60	32	0	4	4	0	0
	Carbonaceous shale	0	31	37	32	5	63	33	0	2	2	0	0
Yongshansutian	Carbonaceous shale	5	14	42	39	15	15	43	1	2	21	16	2
	Carbonaceous shale	0	22	41	37	10	31	31	2	4	12	18	2
	Carbonaceous shale	0	20	41	39	10	33	34	3	4	11	13	1
Renhuaizhongshu	Carbonaceous shale	0	13	29	58	10	42	33	3	12	10	0	0
	Carbonaceous shale	0	16	32	52	10	35	40	2	11	9	2	1
Xingwenqilin	Carbonaceous shale with silt	0	2	41	57	10	26	69	0	3	0	0	2
	Carbonaceous shale	0	8	51	41	5	17	81	0	2	0	0	0
Yanjinyinchangba	Carbonaceous calcareous shale	0	26	30	44	10	31	33	4	9	10	10	1
	Silty shale	0	17	33	50	10	47	26	1	5	10	10	1
Yunnanluomurou	Carbonaceous shale	0	6	43	51	5	27	38	1	2	14	17	1
	Carbonaceous shale	0	8	36	56	5	26	27	1	1	10	33	2
Yanhexinjing	Carbonaceous shale	0	6	29	65	10	28	63	2	7	0	0	0
Huayingsanbaiti	Carbonaceous shale	0	7*	41	52	10	8	90	0	0	0	2	0
Shizhuqiliao	Common shale	0	0	40	60	10	71	23	0	6	0	0	0
Daozhenbayu	Siliceous carbonaceous shale	0	0	0	0	0	19	79	1	1	0	0	0
	Siliceous carbonaceous shale	0	0	0	0	0	22	74	1	3	0	0	0
	Carbonaceous shale	0	0	0	0	0	24	70	2	4	0	0	0
Daoye I well	Dark gray-gray black shale	5	11	71	13	10	35	40	0	14	5	0	0
	Gray-black carbonaceous shale	0	0	0	0	0	31	42	4	13	6	4	0

(continued)

Table 4.16 (continued)

Section	Sample	Clay mineral relative content (%)						Whole-rock quantitative analysis (%)					
		K	C	I	I/S	Smectite	Clay minerals	Quartz	Potash feldspar	plagioclase	calcite	dolomite	Pyrite
	Gray-black carbonaceous shale	0	0	0	0	0	25	44	3	11	7	5	5
Wuchuanbei	Carbonaceous shale	/	7	76	17	0	10	39	9	24	14	4	0
	Carbonaceous shale	0	0	0	0	0	18	41	8	25	8	0	0
	Carbonaceous shale	3	10	52	35	0	17	44	9	20	8	2	0
Yanhexinjing	Carbonaceous shale	0	0	0	0	0	18	47	10	21	2	2	0
	Carbonaceous shale	/	1	39	60	0	18	64	6	10	2	0	0
	Carbonaceous shale	/	4	75	21	0	17	53	9	20	1	0	0
	Silty carbonaceous shale	0	0	0	0	0	16	41	9	19	15	0	0
Changningshuanghe	Siliceous carbonaceous shale	0	0	0	0	0	18	67	0	0	7	8	0
	pelitic siltstone	0	0	0	0	0	14	60	0	0	14	12	0
Pengshuilujiao	Black silty shale	0	17	50	33	10	29	45	4	15	6	1	0
	Black silty shale	0	16	49	35	10	31	43	4	14	7	1	0
	Black silty shale	0	15	49	36	10	25	44	5	17	8	1	0
Ebianxihe	Calcareous siltstone	0	26	74	0	0	34	40	2	3	0	21	0
Changxin I well	shale	0	2	80	18	10	18	30	0	6	41	5	0
	Calcareous shale	0	1	82	17	10	7	27	0	2	25	34	5
	Calcareous shale	0	3	73	24	10	17	46	0	2	27	5	3
	Laminar shale	0	3	76	21	10	18	56	0	2	15	6	3
	shale	0	3	69	28	10	27	30	0	2	17	20	4
	shale	0	4	72	24	10	11	74	0	1	5	6	3
	shale	0	3	68	29	10	14	68	0	1	8	9	0
	shale	0	4	70	26	10	16	40	0	1	30	10	3
	Calcareous shale	0	4	68	28	10	20	46	0	2	16	12	4
	shale	0	3	76	21	10	23	38	0	2	15	18	4
	Calcareous shale	0	2	74	24	10	13	42	0	0	13	28	4

K: Kaolinite; C: Chlorite; I: Illite; S: Montmorillonite

2% < TOC < 4%, the black shale also has high quartz content, low clay mineral content, the content of carbonate is unevenly distributed. The rock types are mainly clayey and carbonate, with transitional types between them (Fig. 4.64b). Black shale with TOC of 1–2% has significantly higher content of detrital particles and carbonate minerals, the distribution of carbonate minerals is uneven, while the black shale has the high content of quartz, and the low content of clay minerals. Siliceous shale, clayey shale and carbonate shale are all present, and carbonate shale is relatively developed (Fig. 4.64c). The black shale (TOC < 1%) is mainly distributed in the upper part of the black rock series, with high content of clay minerals and low content of quartz. The distribution of carbonate minerals is also uneven, mainly consisting of clay and carbonate types, and relatively few siliceous types (Fig. 4.64d).

It can be seen from above that the organic matter content of black rock series in the study area has obvious correlation with mineral composition. Considering the influence of sedimentary environment, the organic carbon content and quartz and clay mineral content of black rock in different sampling points also show the same rule. In the same profile, organic carbon content is positively proportional to quartz content (Fig. 4.65a, c–e) and mainly inversely proportional to clay mineral content (Fig. 4.65b–e). However, in the profile of Longmaxi Formation in Dajingping of Tianquan, Sichuan Province (Fig. 4.65a), and the clay mineral average content is 13%, which the distribution of carbonate minerals is very uneven. Influenced by high quartz content, the overall organic carbon content is approximately proportional to the clay mineral content. The black rock of Longmaxi Formation in Huanggexi, Yunnan (Fig. 4.65b), whose

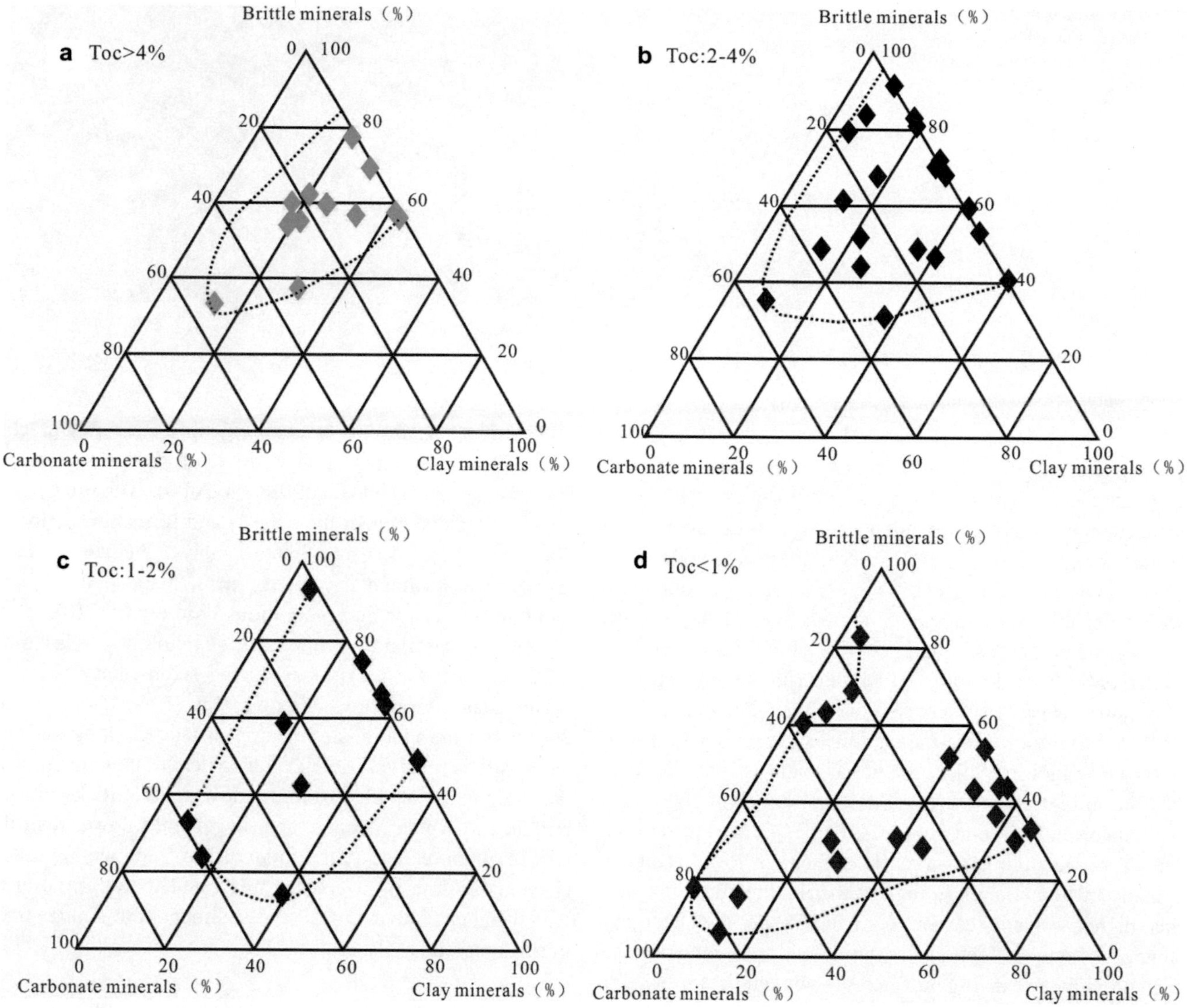

Fig. 4.61 Mineralogical diagram of black rock of Wufeng Formation in Southern Sichuan and its periphery

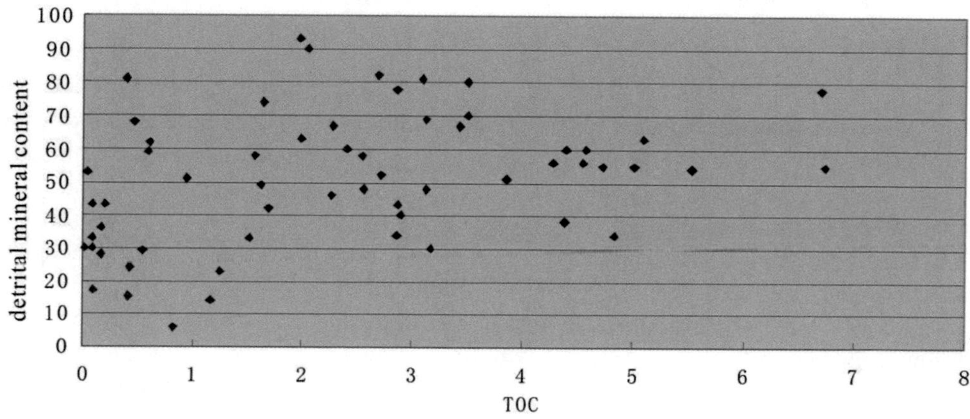

Fig. 4.62 Relationship between detrital mineral content and TOC value of Wufeng Formation in Southern Sichuan and its periphery

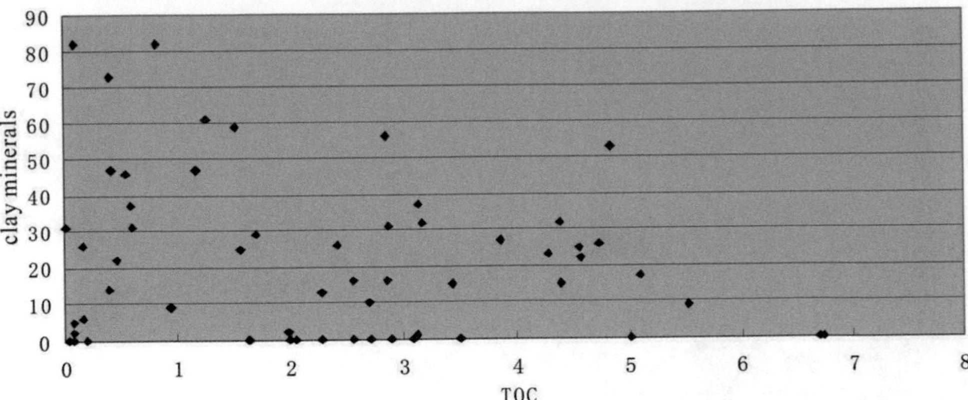

Fig. 4.63 Relationship between clay minerals and TOC value of Wufeng Formation in Southern Sichuan and its periphery

organic carbon content is inversely proportional to quartz content but positively proportional to carbonate mineral content; the content of clay minerals in black shale has a little change, which mainly due to its high carbonate mineral content ranging from 27 to 81%, with an average of 55.5%. The organic carbon content is very low, and the correlation with other mineral components is not obvious. As a whole, the organic carbon and quartz content of the black rock tend to decrease upward vertically, while carbonate minerals and clay minerals content increase upward.

Therefore, the rock composition of black rock of Longmaxi Formation in Southern Sichuan and its adjacent area has the influence on the organic carbon content. While the carbonate mineral content is less than 30%, the quartz content is higher, the clay mineral content is lower, and the organic carbon content is higher; that is, the siliceous shale has higher organic carbon content. While the carbonate mineral content is higher, the quartz and clay mineral content are lower; the organic carbon content is lower. The organic carbon content of carbonate shale is low. In general, siliceous shale has the highest organic matter content, the second is clayey shale, and the carbonate shale has the lowest organic matter content.

(2) Genetic analysis of mineral components

(i) The main controlling factors of the formation and distribution of clay minerals

Generally, clay minerals in sedimentary rocks can be divided into three types according to their genesis: primary, allogenic and secondary (Wu et al. 1997; Merriman 2005). In the research, allogenic clay minerals are used to analyze paleoclimate, provenance and sedimentary environment. Primary clay minerals can be used to invert sedimentary environment and water medium conditions, while secondary clay minerals can be used to analyze basin evolution characteristics and diagenetic environment (Fu 2000; Cai et al. 2008; Zhao et al. 2008; Liu et al. 2011). In addition, different types of clay minerals have different effects on hydrocarbon generation and storage (Kang et al. 1998; Fu 2000). Therefore, it is necessary to distinguish the genesis and formation period of clay minerals before using them to analyze the paleoclimate, provenance, sedimentary environment, diagenetic environment and water medium conditions (Xie et al. 2010).

Most of the major components of mud shale (clay minerals and silt) are transported to the sedimentary site in a clastic state through weathering (Liu 1980; Song et al. 2008), and the mud shale handing distance is long because its density is relatively small. The content of clay minerals in the sediments of Paleozoic muddy source rocks in the middle and Upper Yangtze area is generally lower than the terrestrial source rocks, the modern lake sediments, the terrigenous sediments in coastal areas, and the abysmal clay, which indicates that not all the sediments are source from terrigenous clays (Fu et al. 2011a).

(a) Characteristics of terrigenous clastic clay minerals

Terrigenous clastic clay minerals have been transformed by denudation, transport and deposition. And their original crystal forms have been damaged to varying degrees, such as wear and dissolution (Wu et al. 1997). The clastic clay minerals are often distributed in clumps or dispersions with poor crystalline morphology when observed by polarizing microscope or scanning electron microscope (Liu et al. 1998).

(1) Mass appearance of floccus

Under the influence of multiple provenances, clastic clay minerals show inhomogeneity and diversity in plane and vertical direction (Wu et al. 1997; Slatt and Rodriguez 2012). A wide range of millimeter and centimeter mineral compositions with an abrasive surface and a systematic distribution of coarse-to-tapering are interlace or parallel in

Table 4.17 Characteristics of composition of the black rocks in Longmaxi Formation of Southern Sichuan Basin and its periphery

Rock type (TOC%)	Sample size	Quartz (%)	Feldspar (%)		Carbonate Minerals (%)		Pyrite (%)	Clay minerals (%) Relative Amount (%)				Other minerals (%)
			k-feldspar	Plagioclase	Calcite	Dolo-mite		Illite	I/S	Chlorite	Kaolinite	Siderite Gypsum etc
All	224	42.32	8.84		16.81		1.41	30.29				0.26
			2.37	6.47	9.17	7.64		48.54	38.80	11.63	1.03	
> 4	29	54.87	9.00		9.86		0.9	24.14				0.87
			2.28	6.72	5.62	4.24		47.75	39.20	12.10	0.95	
2–4	65	49.09	6.51		15.43		1.89	26.86				0.22
			1.80	4.71	6.49	8.94		53.05	37.11	8.68	1.16	
1–2	37	39.19	12.70		19.82		1.14	27				0.16
			3.54	9.16	9.41	10.41		50.77	38.82	9.95	0.46	
< 1	93	34.91	8.89		18.75		1.33	35.90				0.14
			2.33	6.56	12.06	6.69		42.57	40.06	15.80	1.57	

I: Illite; S: Montmorillonite

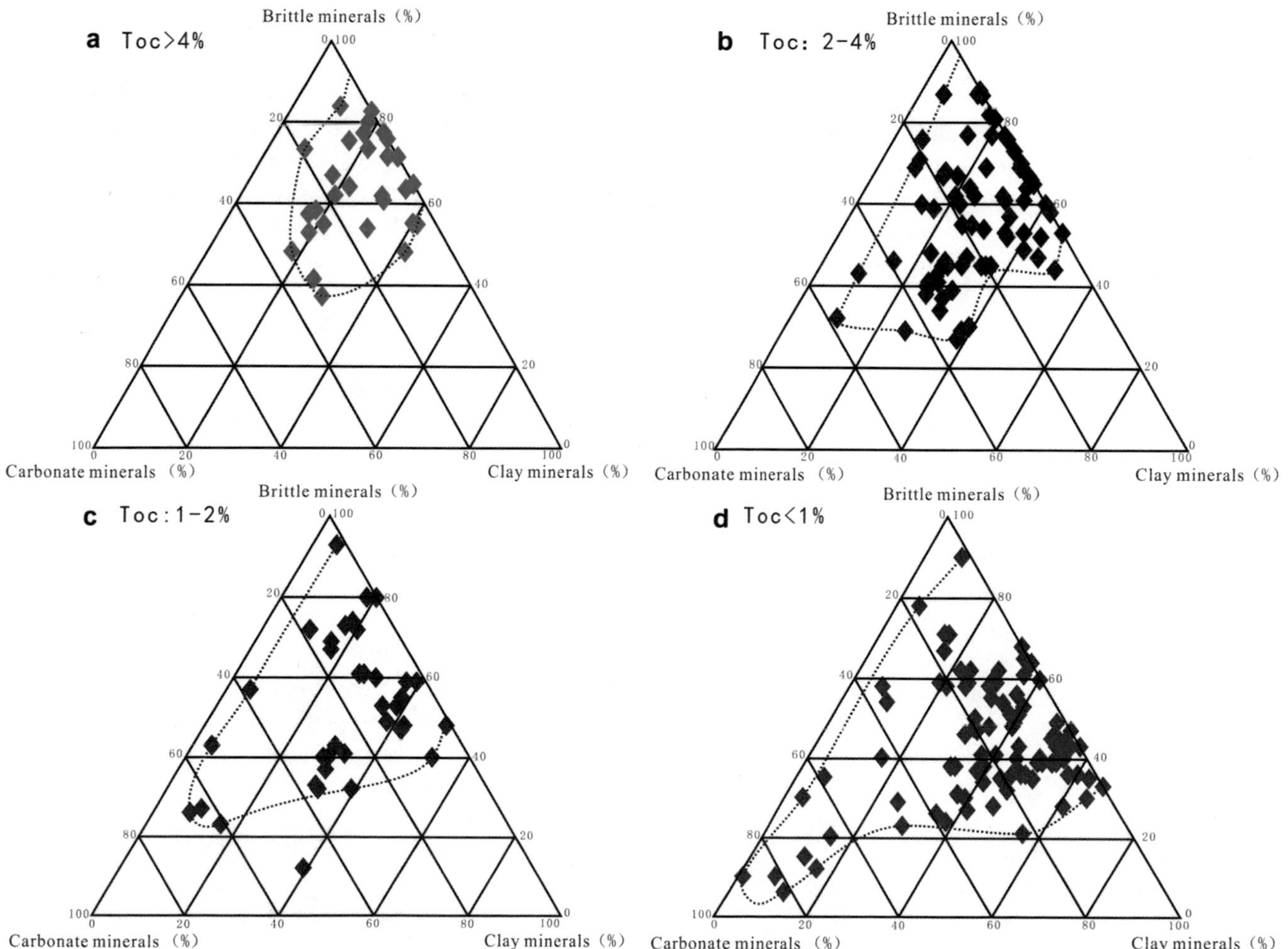

Fig. 4.64 Mineralogy ternary diagrams of different black rocks in Longmaxi Formation of Southern Sichuan Basin and its periphery

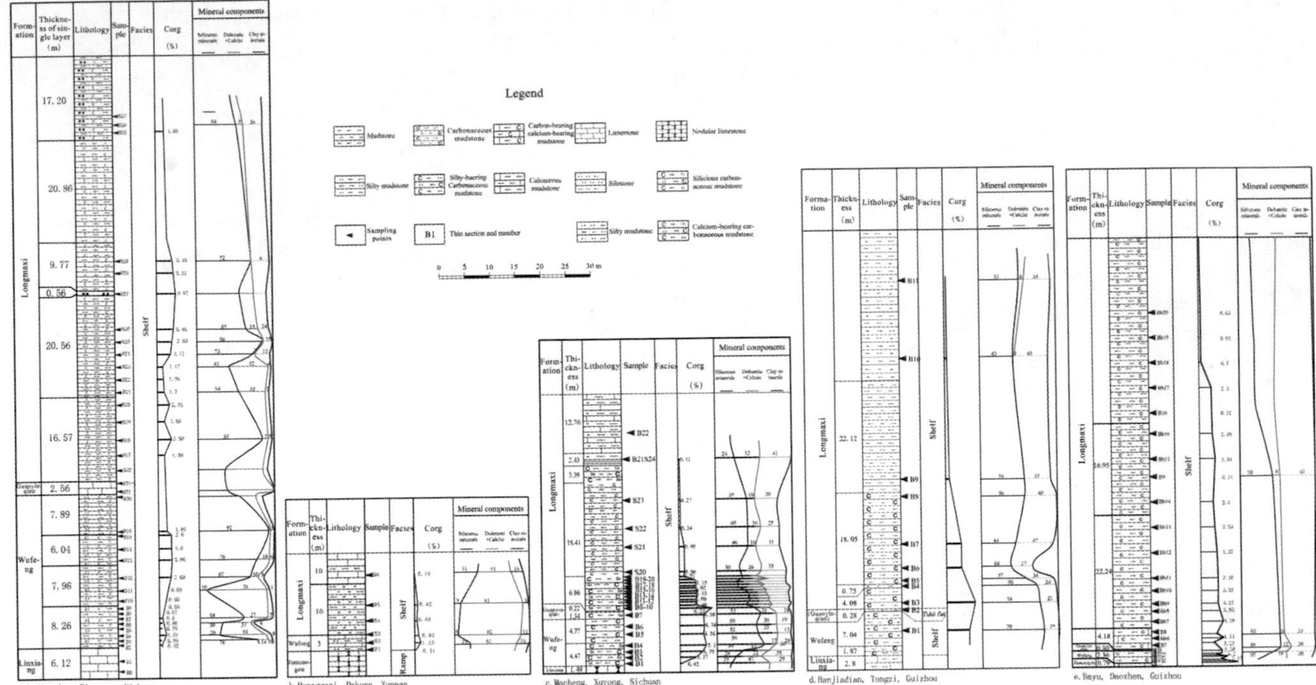

Fig. 4.65 Relationship between TOC and quartz or clay minerals of black rocks in Longmaxi Formation in Southern Sichuan Basin and its periphery

mudstone (Slatt and Abousleiman 2011). They are called flocculent structures. The floccule is thought to be transported and deposited by traction current under the same hydrodynamic conditions as coarse-grained debris. Minerals are in contact with each other in the form of "edge-face" or "edge-side" or "face-face" (O'Brien and Slatt 1990; Bennett et al. 1991). This formation process of flocs may facilitate the long-distance transport of a large number of clay minerals (Slatt and O'Brien 2011). It indicates the particularity and complexity of the long-distance suspended deposition of a large number of clay minerals in marine environment.

The characteristics of pore-rich clay minerals in organic-rich shale of Longmaxi Formation in the study area are very similar to those in Barnett and Woodford shales under the backscatter electron microscopy (Fig. 4.66a, b), which is also similar with the ancient strata flocs generated in laboratory (Slatt and O'Brien 2011). Therefore, there have a large number of terrigenous clastic genetic types in the Longmaxi Formation clay minerals in the study area.

(2) Stable clay minerals are present in abundance

According to the observation of a large number of rock thin sections, scanning electron microscopy and backscattering electron microscopy, the clay minerals in the mudstone of Longmaxi Formation in the study area have obvious characteristics of clastic clay minerals (Fig. 4.66c). Illite is mostly shaped as clastic sheet or curved sheet, with incomplete crystal shape, smooth and clear outline and slightly directional (Fig. 4.66a, b) (Xie et al. 2010). This illite is formed by the weathering of aluminosilicate minerals such as feldspar (Zhang 1992). The average content of chlorite is 4.82%, and the maximum value is 25.3%. He is mainly shaped as flake or thin plate shape, with irregular shape. The chlorite crystal edge is not straight, which has no lamellar aggregate structure. Generally, illite is the most stable phase of clay minerals. At present, it is generally believed that illite mainly comes from terrene and is formed in a climate with strong physical weathering (Jin et al. 2007). According to X-ray diffraction analysis, the illite of Longmaxi Formation in the study area is relatively stable in vertical and plane distribution, which average content is 19.03%. The maximum value of illite is 36%. Khormali and Abtahi (2003) also pointed out that the high content of illite and chlorite mainly came from parent rocks which is rich in these two kinds of minerals. Meanwhile, Li et al. (2012a) pointed out that chlorite and illite in the black shale were precipitated in the environment with saline and alkaline properties of ancient water medium, and the illite and chlorite from different places have been preserved up to now.

(3) Diagenetic inversion-smectite content is high

Smectite has good suspension power and is often deposited with illite. The yili fossilization process of smectite is a common feature of burial diagenesis (Daoudi et al. 2008).

According to X-ray diffraction analysis, the illite/smectite minerals in the study area lower Longmaxi Formation are distributed stably vertically and horizontally, with an average content of 13.3% and a maximum content of 52.7%. The relatively high content of illite–smectite-mixed-layer minerals (interlayer minerals for illite petrolization) indicates that there are more smectites in the original sediments.

(b) Characteristics of non-terrigenous clastic clay minerals

The diagenetic evolution of clays of clastic origin follows the general rule of smectite → Illite/smectite → Illite/kaolinite → illite (Kang et al. 1998). The mineral content of illite/smectite in the study area is high, so secondary clay minerals are developed in the study area.

Analysis shows that in acidic medium environment, sufficient colloidal SiO_2 and Al_2O_3 without interference of other elements (especially K+), and the relatively stable crystallization environment and sufficient crystallization space are the necessary conditions for the formation of false hexagonal kaolinite (Zhang 1992). However, in argillaceous rocks, it's difficult to form authigenic kaolinite even if there have abundant diagenetic components in acidic environment because of the lack of large enough crystallization space. Therefore, the kaolinite in the black organic-rich shale of Longmaxi Formation in the study area may come from clastic origin and secondary origin. Under acidic conditions, feldspar, smectite, muscovite and other minerals can be spontaneously altered into kaolinite (Wang et al. 2013c). Therefore, in coastal and shallow marine environments, feldspar, mica and other minerals occur alteration during the process of weathering and transportation. Kaolinite is eventually produced (Huang et al., 2009). Sampling in Well Changxin 1 is very dense, which sampling intervals usually less than 0.5 m. There is little difference in lithology and mineral component types and contents, while kaolinite is unevenly distributed vertically (Chen et al. 2013b). If kaolinite is formed during the diagenetic process, the properties of formation water changes frequent vertically. Under the condition of the same sedimentary background and little difference in mineral composition, the content of feldspar, quartz and other minerals should change correspondingly after kaolinization, which would not have the similar content characteristics at present. At the same time, 25% kaolinite (Chen et al. 2013b) in the 4% clay minerals which contained in the calcite veins should come from the later fluid detritus kaolinite. Therefore, it can be inferred that kaolinite is not a product of burial diagenesis.

However, there is a significant negative correlation between kaolinite and feldspar in Well Daoye 1 and Well Changxin 1 (Fig. 4.67), which indicates that kaolinite is the cause of feldspar erosion. In addition, the areas (Wuchuan, Daozhen, Nanchuan, Jinyang and Yongshan) where the clay minerals containing kaolinite were detected are in the Southeastern and Southwestern margin of the basin. They are located near the basin margin and uplift which is close to the provenance. To sum up, kaolinite is mainly formed by alteration of aluminosilicate minerals such as feldspar in the process of transportation, which presents inheritance precipitation in water bodies.

In alkaline burial diagenetic environment, smectite and kaolinite would occur the chloritization and illytization. Illyfossilization of smectite in the study area is very well-developed, but there are no green/montmorillonite minerals in all samples, which indicate that there has no chloritization. As long as there is abundant Supply of Fe^{2+} and Mg^{2+}, there would occur the transformation from the feldspar, kaolinite and smectite to chlorite. However, yili fossilization requires the participation of K^+ (Wang et al. 2013c). At the same time, since the Illite fossilization of kaolinite requires additional heat than its chloritization (Wang et al. 2013c). In the case of an adequate supply of diagenetic ions, there will occur the chloritization on kaolinite.

Kaolinite in the black shale of Longmaxi Formation in the study area is not evenly distributed, and the occurrence of a large number of minerals in the illite/montmorillonite mixed layer indicates the development of alkaline diagenesis. Li et al. (2013b) believed that the Longmaxi Formation was characterized by upper low salinity and lower high salinity. The paleowater medium with high salinity was rich in Fe^{2+} and Mg^{2+} plasma, which made kaolinite disappear in advance and more chlorite appeared. However, according to the Pengshui Lujiao profile's date (Li et al. 2012a), Well Changxin 1's date (Chen et al. 2013b) and the analysis data of this study, it can be seen that chlorite does not increase vertically with the increase of paleosalinity, and there is no obvious correlation between kaolinite and chlorite. Crystals are small and have uneven edges. Under backscatter electron microscopy, chlorite is in the form of flake or lamellar, with large thickness. Chlorite is a biotite alteration product (Fig. 4.66d), which is symbiotic with pyrite. And the chlorite has a fairly high Fe^{2+} and Mg^{2+} content (Table 4.18). Therefore, it can be inferred that there has no chloritization in clay minerals, which is mainly controlled by the content of Fe^{2+} and Mg^{2+}. Secondary chlorite is mainly derived from terrigenous clastic alteration.

The comparative analysis shows that the average relative content of illite and illite/smectite in the Wufeng Formation is 41.9% and 47%, respectively, in the same profile. The average relative content of Longmaxi Formation is 31.9% and 55%, respectively. It can be seen that with the increase of depth, the increase of illite is positively correlated with the decrease of illite/Mongolian mixed-layer mineral content. It can be inferred that illite is a secondary diagenetic type.

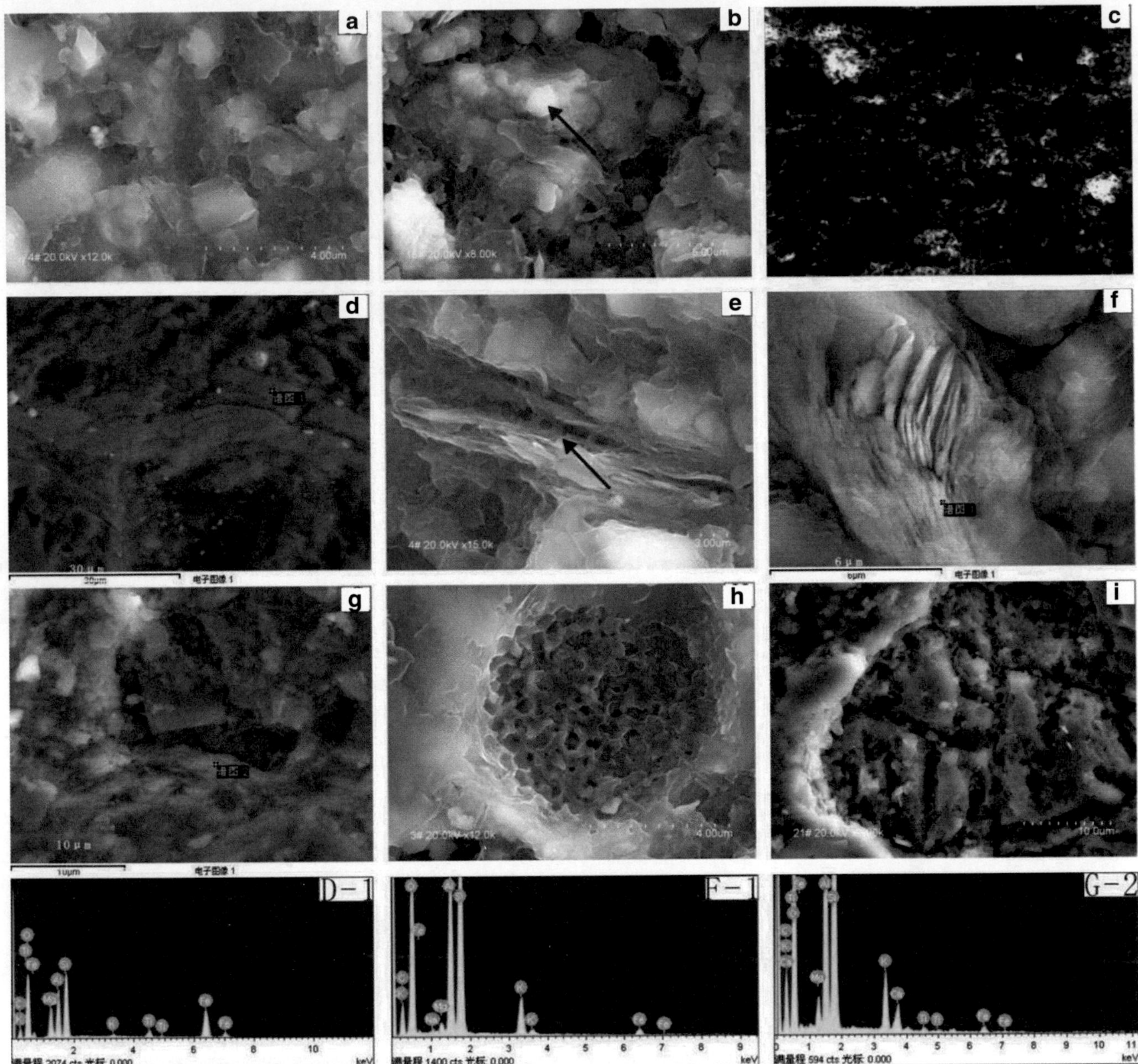

Fig. 4.66 Microscopic characteristics of clay minerals of black rocks **a** flocculent structure dominated by illite, Wulong Huangcao profile in Chongqing; **b** flocculent structure, Authigenic quartz (arrow) is symbiotic with clay minerals, Xishui Liangcun profile in Guizhou; C. Clumps of clay minerals, Xishui Liangcun profile in Guizhou; **d** chlorite is formed by alteration of biotite (Figure d-1 shows its X-ray diffraction pattern), Nanchuan Sanquan profile in Chongqing; **e** secondary illite fills the pores in a jointed manner, Wulong Huangcao profile in Chongqing; **f** altered illite (Figure f-1 shows its X-ray diffraction pattern), Wulong Huangcao profile in Chongqing; **g** illite/montmorillonite mixed mineral encapsulate pore margins (Figure g-2 shows its X-ray diffraction pattern), Authigenic calcite is found in the pores, Xuyong Macheng profile in Sichuan; **h** illite/montmorillonite mixed mineral, Shizhu Qiliao profile in Chongqing; **i** clay minerals filling clastic dissolution pores, Xuyong Macheng profile in Sichuan

According to the thermodynamic characteristics of feldspar, potassium feldspar is a relatively stable mineral. But the occurrence of illite fossilization would consumes K^+, which will promote the dissolution of potash feldspar (Wang et al. 2013c). The content of potassium feldspar in mud shale of Longmaxi Formation in the study area is obviously lower than that of plagioclase. It indicates that K^+ which is required by illite petrochemical is mainly provided by potassium-rich minerals such as potash feldspar dissolved with increasing depth and temperature, and a small proportion of potassium ions needs to be provided by water medium (Li et al. 2012a). Therefore, illite and smectite interbedded minerals and part

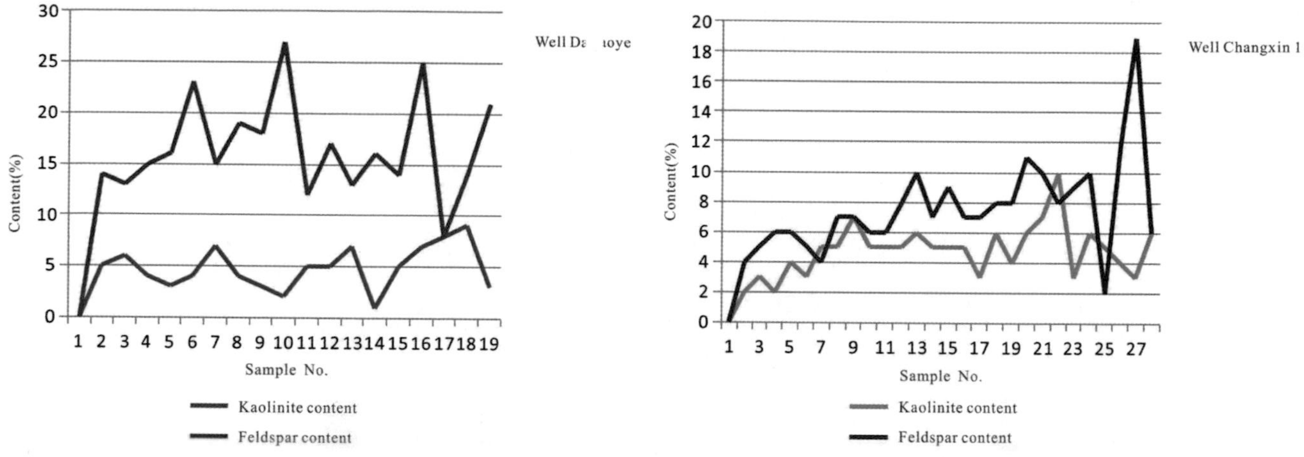

Fig. 4.67 Content comparison of kaolinite and feldspar in well DY1 (left) and CX1 (right)

of illite are the main diagenetic secondary minerals, which shape is irregularly lamellar and usually filled in micropores (Fig. 4.66e).

It can be found that illite contains higher K^+ and Al_2O_3, lower SiO_2 by using the X-ray diffraction spectroscopy. While the K^+ content is more than 6% (Table 4.18), the illite is the alteration product of mica, which appears as long strip under backscatter electron microscope (Fig. 4.66f). Illite/montmorillonite mixed-layer minerals are usually located at pore margins, which have the symbiosis with secondary quartz, pyrite grains and carbonate minerals filled in later stage, and their micromorphology falls in between illite and smectite (Fig. 4.66g). Illite/montmorillonite mixed-layer minerals are shaped like leaf-like or scaly, with serrated and curved edges, and there have the fine filamentous protrusions resembling "petals", which have the disordered and unoriented distribution (Zhang 1992). They have the intergranular pores (Fig. 4.66h). The secondary clay minerals are usually distributed along fluid migration channels, which fill the pores (Fig. 4.66i).

In conclusion, the clay minerals of Silurian Longmaxi Formation in Southern Sichuan and its periphery are mainly terrigenous clastic and some diagenetic secondary types. Because of lacking growth space, the content of authigenic clay minerals is scarce. Terrigenous clastic clay minerals are mainly smectite, illite, chlorite and kaolinite. Secondary clastic clay minerals are mainly illite/montmorillonite interbedded mineral, and there has a small amount of clastic-altered chlorite.

(ii) Genetic analysis of siliceous materials

The black rock of the Longmaxi Formation in the study area has a high content of quartz. Most scholars believe that the quartz of the Silurian Longmaxi Formation in Southern China is mainly imported from terrigenous sources, which is different from the biogenic silica of shale in North America (Zeng et al. 2011; Liu et al. 2011; Chen et al. 2011c; Zhang et al. 2012; Liang et al. 2012a). Zhang et al. (2013a) believed that the content of secondary quartz was high. Qin et al. (2010) and Fu et al. (2011) believed that siliceous materials were mainly biogenic by analyzing the petrology, mineralogy and biological characteristics of high-quality marine source rocks in the Middle and Upper Yangtze region. Quartz is an important component of organic-rich shale and one of the main minerals affecting the brittleness of shale. The different genetic types of quartz reflect the different genetic types of organic-rich shale.

Siliceous material has two distinct structures of granule structure and cement structure. From the vertical view, the content of organic carbon and siliceous material decreases upward, while the content of clastic silt increases upward. Therefore, the organic carbon content is inversely proportional to silty content, and the organic carbon content is proportional to the silica. Therefore, terrigenous clastic quartz is not the main type of silica. In this study, clastic quartz accounts for about 10–40% of the total silica.

Siliceous cementation is characterized by microcrystalline quartz and lamellar cements with high content. The lamellar cements are mainly formed during the process of the illite fossilization (Thyberg and Jahren 2011). Boles and Franks (1979) believed that 5% quartz will be produced when there the original rock contains 25% smectite. Kamp (2008) conducted a quantitative study on silica in shale. He believed that on the premise of the conservation of aluminum, 17–28 wt% and 17–23 wt% SiO_2 were released during the conversion of smeconite to illite and illite to sericite, respectively. In addition, Kamp also believed that quartz fragments in mud shale can be dissolved at 200 °C to produce SiO_2 of 6–9%. When the temperature is 200–500 °C, 10–15% SiO_2 can be further generated. Therefore, during the process of the burial diagenesis-hypometamorphic, a large amount of

Table 4.18 Component characteristics of clay minerals of black rocks in Longmaxi Formation in Southern Sichuan Basin and its periphery

Sample number	Clay minerals	MgO	Al_2O_3	SiO_2	K_2O	TiO_2	Fe_2O_3	CaO/Na_2O	Total amount
JKP-B1	Illite	1.12	26.79	42.40	7.48	0.30	1.44	–	79.53
QLP-B8	Illite	0.78	26.42	48.39	7.58	0.00	1.60	–	84.76
HCP-B1	Illite	0.86	28.29	45.50	7.82	0.45	1.07	0.51	83.99
SQP-B6	Illite	1.39	25.66	45.65	7.52	0.00	2.73	–	82.94
HJDP-B3	Illite	1.55	29.59	53.12	8.60	0.00	1.82	–	94.67
QGP-B6	Kaolinite	4.15	19.73	47.47	2.64	0.00	10.12	–	84.10
XMP-B3	Kaolinite	5.60	20.63	49.09	2.52	1.00	18.00	–	96.85
SQP-B4	Kaolinite	8.29	22.64	37.03	1.65	0.00	23.32	–	92.93
SQP-B4	I/S	2.17	21.79	48.28	5.26	0.47	2.89	0.67	80.86
XMP-B3	I/S	1.69	18.18	48.45	4.37	0.58	1.53	2.34	74.81
SQP-B2	I/S	1.76	19.35	60.84	6.08	1.00	4.22	–	93.25

Remarks: The profile code is shown in Fig. 4.5; I: Illite; S: Montmorillonite

silicon can be released from the transformation of clay minerals and dissolution of siliceous clastic mud shales. Considering the process of dissolution and conversion of feldspar into clay minerals and the process of the illyfossilization and chloritization of kaolinite all can produce an amount of silica. Therefore, although the content of secondary quartz cannot be measured, it is theoretically an important component of shale, which not only affects the diagenetic analysis of shale, but also has the restrictions on the porosity and permeability of shale gas reservoir. According to the characteristics of siliceous cements in the Upper Cretaceous mudstone given by Thyberg and Jahren (2011), there has the relatively high content of the silica. Meanwhile, Zhang et al. (2013a) believed that the secondary quartz content of Silurian Longmaxi Formation shale in Sichuan Basin is 15–80%. It is found that there have many Al^{3+} and K^{+} in the black rock according to the composition analysis of silica in the black rock, which also verified that it is a secondary genetic type.

There have three sources of SiO_2 in the ocean: biosiliceous shells and bones, weathering products of parent rocks from the terrestrial land, seafloor volcanoes erupt sediments and deep hydrothermal materials. Among them, the SiO_2 produced by marine organisms is 19.5 times of the amount injected by rivers, hydrothermal and other means (Zhao and Zhu 2001). Qin et al. (2010) believe that the main mineral sources of high-quality marine source rocks in Southern China (including Silurian Longmaxi Formation) are mainly formed benthic siliceous or calcalous frameworks, and the high-quality marine source rocks in Southern China should be mainly biogenic. In siliceous organisms developing marine strata, siliceous organisms form biological bodies according to directly decomposing and absorbing SiO_2 from seawater. When they die, SiO_2 reenters into aqueous solution or precipitates directly (Fu et al. 2011; Qin et al. 2010). The red algae, brown algae and acme creatures were identified in Longmaxi Formation of the study area (Liang et al. 2009). Organisms such as siliceous radiolarians and siliceous sponge spicules have also been identified (Zhang et al. 2012a). At the same time, there has no obvious correlation among the content of clay minerals, quartz content and the thermal maturity of organic matter in the black rock of Longmaxi Formation in the study area (Fig. 4.68), which is consistent with the characteristics of the Paleozoic marine source rocks in the Middle and Upper Yangtze (Fu et al. 2011) that indicates the content of secondary silica in the late diagenetic period is relatively few. Siliceous is mainly formed by the dissolution of siliceous microfossils in the early stage of biogenesis, which are related to biological processes. At the same time, energy spectrum analysis results show that some quartz briquettes have a high content of element C (Fu et al. 2011), which also reflects that they are affected by biological effects.

Under the marine sedimentary background, the under-compensated shallow-deep-water basin and the deep-water shelf are conducive to the formation of high-quality marine source rocks (Qin et al. 2010). Due to the small inputting of terrigenous materials in these sedimentary environments, the enough water depth, and the blossom of surface organism, the clay minerals mainly from terrigenous sources are diluted in the deposition process. All of these are conducive to the formation of siliceous shale (Fu et al. 2011). At the same time, the high organic matter content in the deep-water environment is also related to the low inputting of the terrigenous debris. If the siliceous material is dominated by terrigenous detritus, it would be contrary to the low terrigenous detritus inputting in the deep-water environment of the basin.

There have clastic quartz and the quartz which comes from the transformation and metasomatism in North American shales. But the content of microcrystalline quartz formed during diagenesis processes much more than the clastic quartz silt (Bowker 2003; Loucks and Ruppel 2007). Longmaxi Formation shale in Sichuan Basin is similar to Barnett shale in North America in geological background and environment, but the thermal evolution of organic matter is higher than that of Barnett shale (Zeng et al. 2011). The Longmaxi Formation shale has a relatively high degree of the diagenetic evolution. The content of clastic quartz is generally higher than that of Barnett shale, so there has more secondary silica formation during the strong thermal diagenetic evolution. However, there has the obvious positive correlation between organic carbon and silica in the study area, the silica is mainly biogenic, and the following are diagenetic secondary origin and clastic origin.

(iii) Origin of carbonate minerals

Carbonate minerals are very special rock components in shale gas research. They are brittle minerals which is the benefit of the hydraulic fracturing, and its dissolution can produce the dissolution pores. However, as the interstitial material which has the weak adsorption capacity for organic matter have the inhibiting hydrocarbon generation capacity (Jiang et al. 2009). The deposition, diagenetic effect and occurrence state of carbonate minerals have a significant influence on the formation of shale gas reservoir. Carbonate minerals in the study area are mainly distributed in tidal flat and shallow shelf environment at the edge of uplift. That is because mud shale contains more water when it is deposited, and they will be forming the thick cement-rich sediments (Cheng et al. 2013). At the same time, the Center Guizhou Uplift and the Xuefeng Uplift provide small amount of detritus to the sedimentary area, which is conducive to the formation of authigenic carbonate minerals (Liu et al. 2012a). Carbonate minerals have good shiny shape and high content in some areas, which can be metasomatic with quartz, feldspar and other characteristics. Therefore, the carbonate minerals in the black rock series of Longmaxi Formation in the study area should have two kinds of genesis: cementation and metasomatism.

According to the thin section observation description, carbonate minerals are more developed in the local microscopic area of clastic grain aggregation of mud shale and the sandstone, which is obviously controlled by the development of the primary intergranular pores, and there have no charged hydrocarbons be found in the carbonate cements. On the other hand, according to the development characteristics of clay mineral assemblages, it can be found that the water medium of Longmaxi Formation is characterized by high salinity and alkalinity during burial diagenesis. The sediments reflect the original sedimentary characteristics, which are weakly affected by water–rock reaction. That is consistent with the absence of dissolution pores in carbonate minerals formed in the early stage and the carbonate metasomatism to other minerals. Finally, the content of organic carbon is negatively correlated with the content of carbonate minerals, while there has little influence between the water–rock reaction of carbonate minerals and the hydrocarbon generation evolution of organic matter during late diagenesis. The negative correlation between them should be controlled by the same formation environment. In conclusion, the carbonate minerals were mainly formed before the large-scale hydrocarbon generation. The carbonate minerals are mainly the products of sedimentation-syngenetic cementation, and the second is the product of metasomatism.

In conclusion, the clay minerals in the black rocks of Longmaxi Formation in Southern Sichuan and its periphery are mainly terrigenous clastic, and there have some diagenetic secondary types and few autogenesis. The siliceous material is mainly biogenic type, and followed by the diagenetic secondary type and the clastic origin type. The carbonate minerals are mainly the products of sedimentation-syngenetic cementation, and the second is the product of

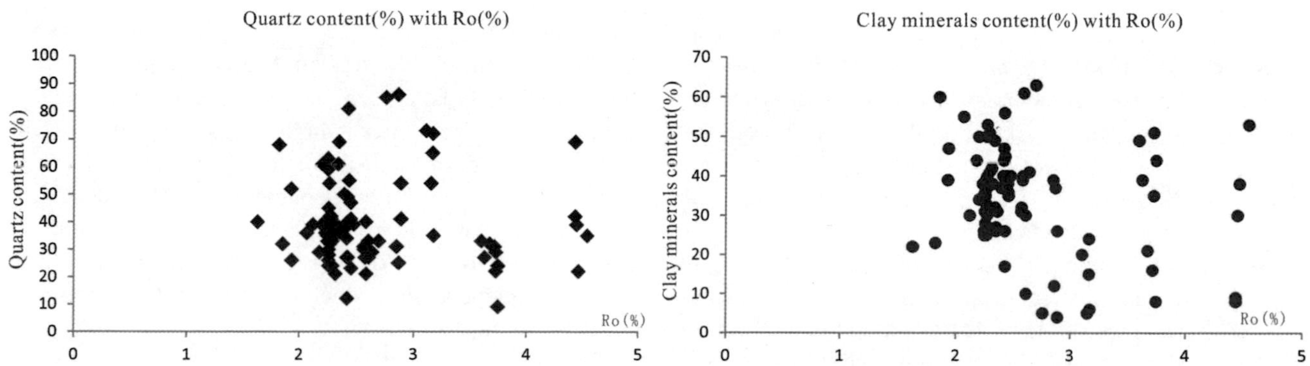

Fig. 4.68 Relationship between Ro and quartz or clay minerals of black rocks in Longmaxi Formation of Southern Sichuan Basin and its periphery

metasomatism. The genetic characteristics of three kinds of mineral components reflect that the sedimentation has the important influence for the black rock of Longmaxi Formation in Southern Sichuan and its periphery. However, the development of organic matter is more controlled by the sedimentary environment. The undercompensated and anoxic deep-water environment with low carbonate mineral content is more conducive to the preservation of organic matter (Zhang et al. 2012a; Cheng et al. 2013). Therefore, there has the correlation between mineral composition and organic matter abundance, which is because of the interaction of the mineral composition and organic matter development characteristics by the deposition.

(3) Sedimentary and diagenetic significance of mineral components and their relationship with shale gas

(i) The sedimentary and diagenetic significance of brittle minerals and their relationship with shale gas

According to the above analysis, siliceous is the main component of brittle minerals in black rock of Silurian Longmaxi Formation in Southern Sichuan and its periphery, which has three types: biogenesis, diagenesis and clastic origin, and the first type is dominant. Among them, the clastic quartz particles are well-sorted and poorly rounded. From the uplifted area to the basin, the content and grain size of clastic quartz decrease and the sorting becomes better. It can be seen that the content and structure characteristics of clastic quartz particles can reflect the distance between the sedimentary area and the source area. For example, the organic-rich shales in Tianquan Dajingping and Hanyuan Jiaodingshan in Western Sichuan contain more quartz silt than other areas in the study area, which is controlled by the long-term denudation of the Center Sichuan Uplift in the provenance area. The content of feldspar in clastic particles is relatively high. The content of relatively unstable plagioclase is higher than that of potassium feldspar due to weak acidic dissolution and strong clay mineral transformation during diagenesis. A small amount of short columnar Muscovite is widely distributed, and there have no unstable clasts such as carbonate and chlorite. That shows most of the unstable minerals were completely decomposed in the process of terrigenous detrital transport during the deposition of Silurian Longmaxi Formation in Southern Sichuan and its periphery (Zhang et al. 2013a). These factors lead to the decrease of terrigenous clastic particles in the sediments of black rock of Longmaxi Formation, which reduces the dilution of organic matter. Zhang et al. (2013a) pointed out that the decrease of sea level led to the increase of terrigenous detrital materials and the destruction of reducing environment, which was the main reason for the low organic carbon content of shale in the upper Longmaxi Formation. In conclusion, the closer to the center of the basin, the farther away from the provenance, the less terrigenous detrital particles, the more favorable it is to form shale with a higher abundance of organic matter.

As the main type of siliceous minerals, self-generated siliceous minerals are mainly controlled by the sedimentary environment. In the deep-water shelf environment, the high biological productivity of surface water leads to the gradual enrichment of biological siliceous in the bottom water with low activity. In addition, the enrichment of biological siliceous in local areas of the study area is also related to upwelling. For example, a large number of radiolaria silicides are developed in the black rock of Tianquan Dajingping and Hanyuan Jiaodingshan profiles in Western Sichuan, which are mainly related to upwelling. Paleozoic Yangtze plate was in low latitudes, which belongs to the tropical and subtropical climate. When cold water masses rich in nutrients (nitrate and phosphate) and the SiO_2 around Antarctica flowing from high latitudes to low latitudes along the ocean floor, upwelling can occur at the appropriate continental slope, which would make the area become a high biological production (Li et al. 2008a). Li et al. (2008a) pointed out that the late Ordovician was the peak of the development of the global ice age. And the rising ocean current of that time was the strongest which even affecting the interior of the Yangtze Craton. During the Silurian period, the ocean current caused by the temperature difference was weakened because of the global warming. The influence range of the ocean current was also greatly reduced. That indicated the richment of the siliceous radiolarians shales which formed by upwelling are only developed in the Western Sichuan area where the sedimentary slope is large. Both biological siliceous and organic matter of bacteria and algae are products of sedimentary period, which have the similar genetic mechanism. The silica is dominated by biological siliceous, which also verifies the positive correlation between silica and organic carbon. In addition, organic matter is often filled in the body cavity in the regions and intervals where radiolaria silicides are developed. They can be used as an important reservoir space for hydrocarbon, which is conducive to improving the abundance of organic matter. Although the black rock of Longmaxi Formation in Western Sichuan contains relatively more clastic silt, the content of clay minerals is low, and the content of organic matter is high. Conversely, the siliceous materials in the study area are mainly biogenic. It can be concluded that the Silurian Longmaxi Formation deep-water shelf sedimentary environment in Southern Sichuan and adjacent areas is easy to form shale rich in biological siliceous and organic matter, which is conducive to the formation of shale gas-rich areas.

Diagenetic secondary siliceous and carbonate cements are products of diagenesis. The siliceous cements are mainly lamellar, which indicate that they are mainly formed in the transformation process of clay minerals. Combined with the dissolution of potash feldspar, it indicates the development of illyfossilization of montmorillonite. The carbonate cements are mainly formed in the depositional syngenetic stage, which indicate that the sedimentary environment is weakly alkaline. The carbonate cements are distributed in one side of Central Guizhou Uplift, which indicate that the formation of carbonate minerals is also related to terrigenous area. Liu et al. (2012a) pointed out that Hannan Ancient Land and the Central Sichuan Uplift had been exposed for a long time and formed a marginal sedimentary facies belt dominated by terrigenous detrital supply as a stable denudation area. In contrast, the Center Guizhou-Xuefeng Uplift provides less detrital material to the sedimentary area which is favorable for the formation of authigenic carbonate minerals. In the process of hydrocarbon generation, a large number of organic acids are formed in source rocks, which resulting in a large amount of dissolution of acidic unstable minerals such as feldspar and carbonate. However, the high content of feldspar and the good preservation of early carbonate cements in the black rock of Longmaxi Formation indicate that the acidic medium is not developed in the diagenetic fluid. Combined with the extensive development of clay mineral transformation and the later generation of metasomatism carbonate minerals, it indicates that the diagenetic fluid is mainly weakly alkaline in the diagenetic process of the black rock of Longmaxi Formation. During the diagenetic process, the middle and late diagenesis is the peak of hydrocarbon generation after the influence of early strong compaction. The black rock is less affected by inorganic diagenetic transformation.

In addition, berrylike pyrite aggregates are widely developed in black rock of Silurian Longmaxi Formation in Southern Sichuan and its periphery that reflects the reductive depositional environment of the black rock. A large number of intergranular pores can be used as good storage space to enhance the reservoir property of shale.

The influence of brittle minerals on shale gas is mainly reflected in its relationship with organic carbon content and its effect on hydraulic fracturing of shale gas. Based on the analysis of the relationship between rock mineral components and organic matter, in the deep-water shelf sedimentary environment which is far away from the source area, the content of siliceous clastic is low, the content of biological siliceous is high, and the content of authigenic carbonate minerals and clastic clay is relatively low. The depositional environment is favorable for the formation of organic-rich shale. Siliceous and carbonate minerals are brittle minerals in black rock of Silurian Longmaxi Formation in Southern Sichuan and its periphery. Although they have different genetic mechanisms, both of them can improve the brittleness of shale gas reservoir and they are easy to form natural and induced fractures. It is conducive to the desorption and seepage of shale gas, and they can increase free storage space of the shale gas (Bowker 2007). In conclusion, black rock with high brittle mineral content in Silurian Longmaxi Formation in Southern Sichuan and its periphery are favorable for forming shale gas-rich areas.

(ii) Sedimentary and diagenetic significance of clay minerals and their relationship with shale gas

(a) Sedimentary significance of clay minerals

Terrigenous clastic clay minerals were deposited to the seafloor which is carried by wind and water over long distances. The composition and characteristics of the clastic clay minerals can indicate the paleoclimate and tectonic setting of the source area. Paleoclimate is the most important factor which affects the formation of smectite and kaolinite (Daoudi et al. 2008). In terrestrial weathering environments, smellite and kaolinite are generally considered to have been formed by chemical weathering in humid thermal-tropical environments (Wu et al. 1993; Long et al. 2007). Illite and chlorite are formed by physical weathering of sedimentary and metamorphic rocks in a cold and dry environment (Jeong and Yoon 2001). Yuan et al. (2007) believed that kaolinite + illite (K + I) clay mineral assemblage represents the wettability climate. Kaolinite + illite + smectite (K + I + S) clay mineral assemblage represents the semiarid climate. And illite + chlorite + kaolinite (rarely) (I + C + K) clay mineral assemblages often represent the arid climate.

The terrigenous clay minerals in black organic-rich shale of Silurian Longmaxi Formation in Southern Sichuan and its periphery are smellite + illite + chlorite + kaolinite (S + I + C + K), which indicate that the sedimentary period was semiarid to arid climate. Clastic chlorite is unstable and often decomposed after long-term transportation, and its content changes can reflect the distance of the terrigenous areas (Liu 1985). The content of chlorite in the section of Longmaxi Formation in Renhuai Zhongshu near the Central Guizhou Uplift is 6.23%. The chlorite content of Longmaxi Formation in Well Changxin 1 in Changning area which is relatively close to the center of the basin is 4.91%. Similarly, the content of chlorite in the section of Longmaxi Formation in Pengshui near uplift area is 4.62% (Li et al. 2012a). However, the content of chlorite in the Daozhen Bayu profile relatively close to the center of the basin is 3.72%. The difference of chlorite content indicates that the provenance of the Southern Sichuan Basin and its margin came from the marginal uplift area. In addition, kaolinite exists near the edge of Central Guizhou Uplift and disappears near the center basin, which shows the difference of sediment transport and mechanical differentiation.

The clay minerals of Longmaxi Formation in the study area are mainly terrigenous clastic origin, which is greatly affected by the depth of sedimentary water and the distance from the source area. Therefore, the type and content of clay minerals have the facies significance. The higher the content of stable clay minerals is, the deeper the sedimentary water is. Therefore, the clay mineral content of shallow sea shelf is higher than that of the tidal flat sedimentary area. Clay minerals, siliceous and carbonate minerals are the main components of black rock of Silurian Longmaxi Formation in Southern Sichuan and its periphery, and the content of clay minerals is also affected by the latter two contents. According to the above analysis, the center of the basin in the study area is affected by sedimentary water and climate, which have many biological siliceous sediments, and relatively few clays mineral content. Therefore, the clay mineral content is relatively reduced as water depth increase in shallow shelf facies. Its average content should normally be less than 50%, the average is less than 30%. The areas with clay mineral content between 30 and 50% are usually shallow shelf deposits. The sedimentary water is relatively deep, and the content of terrigenous clastic clay minerals is high, while the siliceous content of primary sediments is low. The clay mineral assemblages are mainly chlorite + Illite/Smectite + Illite (C + I/S + I) in the shallow shelf sedimentary environment. The terrigenous clastic content is high where is near the paleo-uplift. The clay mineral content is less than 30%, and the sedimentary facies is marine-continental transitional facies. The clay mineral assemblage is mainly kaolinite + chlorite + Illite/Smectite + Illite (K + C + I/S + I). It can be seen that the distribution of clay minerals has an obvious correlation with the sedimentary environment, which presents a low–high-low characteristic.

(b) Diagenetic significance of clay minerals

(1) Clay minerals reflect the characteristics of diagenetic water bodies

Terrigenous clastic clay minerals are affected by water medium in diagenetic environment during diagenesis, which is occurred the diagenetic changes. The properties of diagenetic fluid and the evolution characteristics of the basin can be inverted according to the transformation characteristics of clay minerals. Chlorite + Illite + Illite/smectite type (C + I + I/S) is the main clay mineral assemblage in the black organic-rich shale of Longmaxi Formation in Southern Sichuan and its periphery. The content of the illite and the illite/smectite is the highest, the content of the kaolinite is few and the content of smectite is rare, which indicates that the water medium in the burial diagenetic process of Longmaxi Formation is characterized by high salinity, alkalinity and rich K^+. The contents of Illite are the highest while the contents of potassium feldspar are low in Longmaxi Formation mud shale in Western Sichuan, which indicates that the salinity of water medium in Longmaxi Formation in Western Sichuan is higher than that in other areas. The water medium tends to be more alkaline.

(2) Secondary clay minerals reflect diagenetic components

Illite and illite/smectite mixed minerals are the main components of clay minerals, which content is significantly negatively correlated with quartz (Fig. 4.69a) that indicates the quartz mainly came from the detrital transport. However, the relative content of illite and illite/smectite is approximately positively correlated with quartz (Fig. 4.69b). The

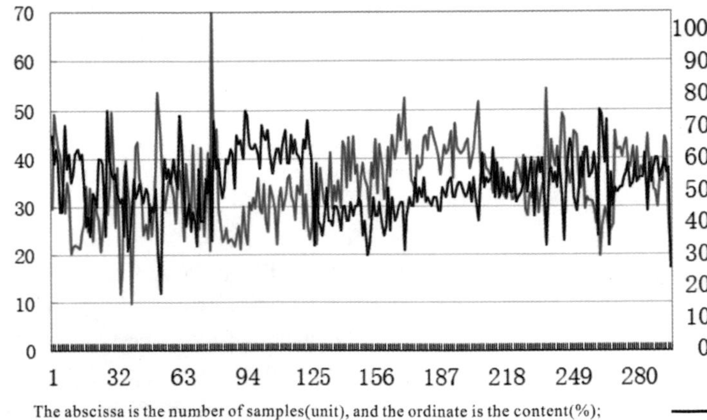

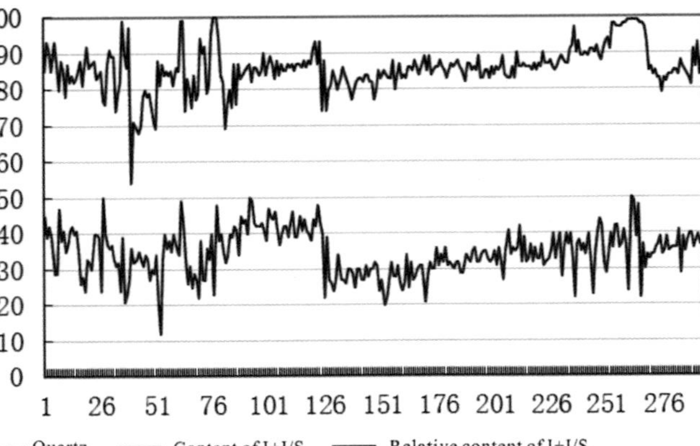

Fig. 4.69 Content relationship between I/S and quartz in black rocks of Longmaxi Formation in Southern Sichuan Basin and its periphery. **a** Illite + Illite/smectite (I + I/S) is negatively correlated with quartz; **b** The relative content of illite + Illite/smectite (I + I/S) is approximately positively correlated with quartz

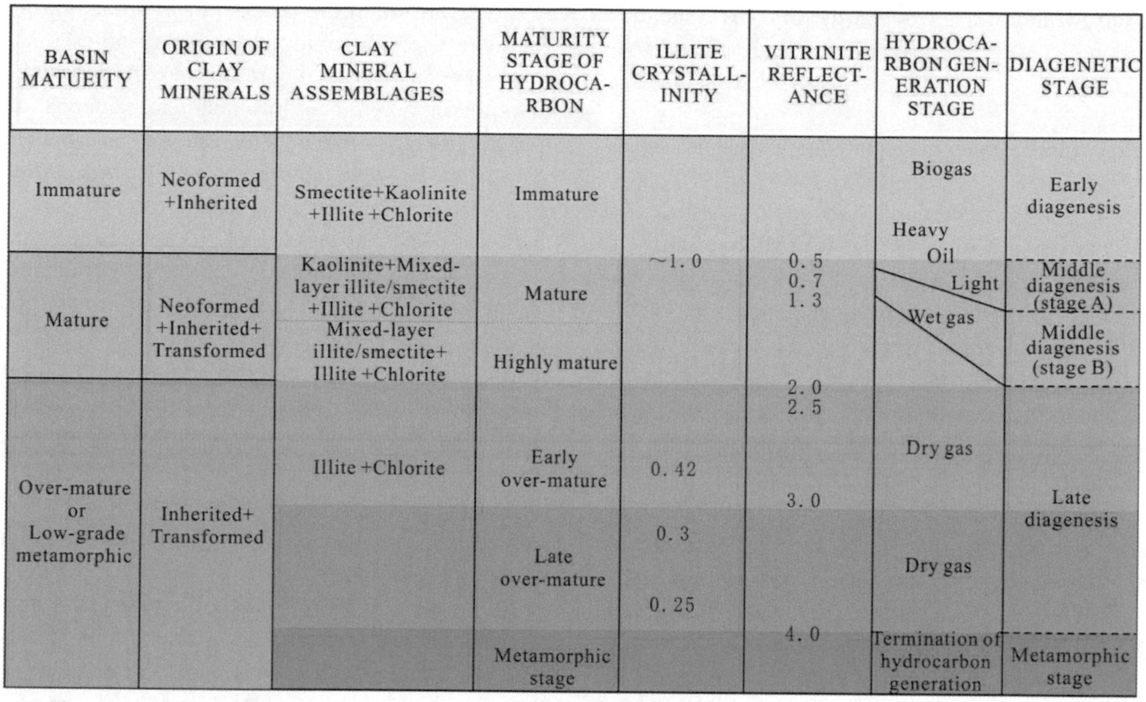

BASIN MATUEITY	ORIGIN OF CLAY MINERALS	CLAY MINERAL ASSEMBLAGES	MATURITY STAGE OF HYDROCA-RBON	ILLITE CRYSTALL-INITY	VITRINITE REFLECT-ANCE	HYDROCA-RBON GEN-ERATION STAGE	DIAGENETIC STAGE
Immature	Neoformed +Inherited	Smectite+Kaolinite +Illite +Chlorite	Immature			Biogas Heavy Oil	Early diagenesis
Mature	Neoformed +Inherited+ Transformed	Kaolinite+Mixed-layer illite/smectite +Illite +Chlorite	Mature	~1.0	0.5 0.7 1.3	Light	Middle diagenesis (stage A)
		Mixed-layer illite/smectite+ Illite +Chlorite	Highly mature			Wet gas	Middle diagenesis (stage B)
Over-mature or Low-grade metamorphic	Inherited+ Transformed	Illite +Chlorite	Early over-mature	0.42	2.0 2.5	Dry gas	Late diagenesis
			Late over-mature	0.3 0.25	3.0	Dry gas	
			Metamorphic stage		4.0	Termination of hydrocarbon generation	Metamorphic stage

Fig. 4.70 Relationship among clay mineral and other indicators of basin maturity (modified from Merriman 2005)

higher the content of these two minerals is, the higher the quartz content is. Illite fossilization of feldspar and smectite can release a certain amount of SiO_2 during diagenesis (Formulas 4.1 and 4.2) (Wang et al. 2013c), which indicates that quartz is not only come from the terrigenous, but also part of them come from feldspar alteration and clay mineral transformation during diagenesis. Zhang et al. (2013a) believed that quartz has two types: clastic quartz and secondary quartz according to the intensive research on the rock characteristics and mineral composition of Longmaxi Formation in Shuanghe area, and he believed that the content of quartz decreases significantly from bottom to top, and the content of secondary quartz is 15–80%. It is not easy to diffuse because of the large radius of the silicate ion (Zhang et al. 1986). Therefore, the silica produced by the transformation of clay minerals usually recrystallizes in situ, and they form microcrystalline fine-grained quartz which is included in the clay minerals (Fig. 4.66b). Although the existence of secondary quartz can reduce the micropores in clay minerals, which is not conducive to the reservoir of shale gas (Li et al. 2012a), secondary quartz can effectively enhance the fracturing effect of shale gas as a brittle mineral, which is conducive to the exploration of shale gas.

$$Al_2Si_2O_5(OH)_4 + KAlSi_3O_8(\text{potash feldspar}) + 2H^+ = K\,Al_3Si_3O_{10}(OH)_2(\text{illite}) + 2SiO_2 + H_2O \quad (4.1)$$

$$3NaAl_7Si_{11}O_{30}(OH)_6(\text{smectite}) + 7K^+ = 7\,KAl_3Si_3O_{10}(OH)_2(\text{illite}) + 4H^+ + 12SiO_2 + 3Na^+ \quad (4.2)$$

(3) Clay minerals as an indication of diagenetic evolution

Shale gas accumulation is closely related to maturity. The evolution and assemblage of some iconic clay minerals can be used as indicators to characterize diagenetic stage and maturity (Chen et al. 2011c), which can indicate the evolution characteristics of their basins (O'Brien and Slatt 1990; Merriman 2005). The evolution of sedimentary basins has been comprehensively classified according to the types and assemblages of clay minerals and the relative maturity of organic matter (Fig. 4.70). The characteristics of clay mineral assemblage and the large number of ordered illite/smectite minerals indicate that the rock of the Longmaxi Formation in the study area has at least entered the middle-diagenetic stage. The vitrinite reflectance of organic matter mainly ranges from 1.63 to 2.46%, with an average of 2.23%, which indicates that the maturity of source rocks has reached high maturate-overmaturity stage (Wang et al. 2012a; Huang et al. 2012b). Therefore, the diagenetic evolution of black organic-rich shale in Longmaxi Formation in the study area has reached the late diagenetic stage, and the hydrocarbon evolution has reached the dry gas stage, all of these are conducive to the formation of shale gas reservoir.

(c) Relationship between clay minerals and shale gas

As one of the most important components of shale, clay minerals have the influence on the formation of shale gas reservoirs, which can guide the exploration and exploitation of shale gas to some degree. In general, femic montmorillonite clays and volcanic-origin clays have problems with clay swelling and poor gas recovery during drilling and hydraulic fracturing. Kaolinite and granite illite have the least negative impact on drilling and water fracturing fluids (Zhang et al. 2011). The clay minerals of black organic-rich shale in Longmaxi Formation in the study area are mainly illite and illite/smectite, while the content of smectite with high water absorption and expansion is very few. All of these reflect that the black organic-rich shale in the study area has little influence on shale gas recovery. The hydrocarbons in the black organic-rich shale of Longmaxi Formation have evolved to the dry gas stage. The black organic-rich shale has the characteristics of large formation thickness and high organic carbon content, which have the gas source conditions for shale gas formation (Mou et al. 2011; Zuo et al. 2012; Wang et al. 2012a). Therefore, reservoir space is the key factor which determines the formation of favorable shale gas reservoir. Longmaxi Formation shale gas reservoir is characterized by low porosity and ultralow permeability, and the interlayer pores of clay minerals are one of the most important reservoir spaces (Wang et al. 2012a).

In the same diagenetic environment, illite petrolization is well-developed, while chloritization is not, which indicates that there lacking the f Fe^{2+} and Mg^{2+} in diagenetic fluid. The reason for the above phenomenon is that the porosity and permeability of reservoir become worse with the restriction of fluid migration in the early stage of diagenesis. The water–rock reaction is not strong under the influence of strong diagenesis. At the same time, the early deposition rate of Longmaxi is low, which is conducive to the formation of favorable source rocks. In the late Longmaxi period, the sedimentation rate increased rapidly (Zhang et al. 2012a), which made the organic-rich shale sediments of Longmaxi Formation undergo strong and rapid compaction at the early stage of diagenesis under the influence of the overbearing strata's gravity. All of these resulted in the nearly complete loss of primary pores in the shale gas reservoirs. Therefore, the activity of diagenetic fluids is limited because of the rapid densification of organic-rich shale in Longmaxi Formation at the early stage of diagenesis. As a result, kaolinite, feldspar and other acid-soluble minerals are weakly affected by organic acids in the process of hydrocarbon generation. There have few dissolutions secondary pores.

Terrigenous clastic clay minerals with floccule are rich in nano-micron pores. The flocculent clay minerals in the study area are strongly compacted by diagenesis, which have good orientation. However, there still retain some primary pores (Fig. 4.66 a, b), which provide some reservoir space for shale gas. The appearance of flocs indicates that a large number of clay minerals in marine environment are suspended over a long distance. In the regions which contain kaolinite, there has the characteristic of the nearby provenances, shallow water, and the less development of porous flocs. All of these restrict the migration of the diagenetic fluid which makes the kaolinite preserved. In addition, the regions which contain kaolinite always belong to the transitional facies in which clay content is less than 30%. This environment is not conducive to the formation of black shale rich in organic matter.

As the stratum burial depth increases, the ground temperature increases and the formation water gradually becomes alkaline, a large amount of interlayer water precipitates from clay minerals because of the dehydration transformation. And the microcracks are formed between layers (Wang et al. 2012a). At the same time, the illite/smectite minerals which are formed by dehydration transformation of clay minerals usually have certain intergranular pores. These can increase the secondary pores in mud shale. Therefore, under the background of high terrigenous clastic clay mineral content in the study area, the area with high content of illite/smectite is conducive to the growth of shale gas reservoir. Considering that the deep-water sedimentary environment is conducive to the development of black organic-rich shale, and the increase of clay mineral content is conducive to the increase of shale gas adsorption, while it is not conducive to the gas diffusion in the process of the fracturing (Zuo et al. 2012). At the same time, the smaller the porridge is while the closer it is to the center of the deposit, which is influenced by the burial depth (Nie et al. 2012). Therefore, shallow shelf sedimentary areas with clay content ranging from 30 to 50%, which have high illite/smectite content and have no kaolinite should be favorable areas for shale gas. The shallow shelf deposition area with high clay content is the secondary favorable area for shale gas because of the development of black organic-rich shale, which also contains high organic carbon and organic micropores.

According to the influence of clay minerals on shale gas and the distribution characteristics of organic carbon, the Yibin–Jiangjin–Fuling deep-water shelf sediments which exhibits a NE to SW distribution area is the favorable areas for shale gas. That area has the features of the high silica content, rich in organic matter, the reservoir space is dominated by organic micropores, and the clay content is less than 40%. There are the favorable exploration areas for shale gas in Qijiang–Wulong–Pengshui area in the Southeast of Chongqing, Leibo–Junlian–Xingwen–Renhua–Tongzi area in the west of Guizhou to the north of Yunnan, and the west of E'bian Xihe area in the Northwest of the study area. The clay mineral assemblage is mainly chlorite + illite + illite/smectite (C + I/S + I) in these areas; in these areas the organic carbon content is more than 2%, and the clay content ranges from 30 to 50%. Among them, the Qijiang–Wulong–Pengshui area in the Southeast of Chongqing has the highest relative mineral content of illite/smectite, which should be the relative most favorable area.

(4) Evaluation of mineral composition on favorable area of shale gas

(i) Evaluation method

Li et al. (2013a) believed that the lower limit of silica content in Longmaxi Formation shale should reach 35% that makes the shale have better brittleness, which is conducive to the formation of fractures and later shale reservoir reconstruction. The black rock series of Silurian Longmaxi Formation in Southern Sichuan and its periphery are siliceous shale with good hydraulic fracturing conditions.

Organic matter content is often proportional to the gas generation rate and adsorbed gas content of shale gas (Wang et al. 2009, 2012a), which always influence the development of shale gas reservoir space (Jarvie et al. 2005). The mineral composition of black rock has the correlation with the development characteristics of organic matter. Therefore, mineral composition analysis of black rock can be used as one of the indirect means to judge its hydrocarbon potential and favorable shale gas exploration areas. The prediction of favorable shale gas areas by mineral composition should be based on the distribution characteristics of sedimentary facies and organic matter. Combined with the distribution characteristics of clay minerals, it should be selected the brittle minerals such as quartz and carbonate minerals to predict the favorable shale gas areas. These are three important factors for judging the development characteristics of organic matter by using the mineral composition. On the plane, the average value of quartz, clay mineral and carbonate mineral content at each sampling point was calculated, and the reasonable evaluation standard was determined by comparing them with the organic carbon content. Furthermore, the plane distribution map of the quartz content, carbonate minerals content, and the clay minerals content are drawn. On the basis of the isogram map of organic carbon content, the distribution areas of the favorable source rocks are finally drawn under superimposing the mineral components shale types. The above is one aspect of shale gas favorable area evaluation.

(ii) Distribution characteristics of mineral components

There have many differences in thickness, organic carbon content and distribution characteristics of quartz, clay minerals and carbonate minerals of black rock at different sampling points in the study area (Fig. 4.65; Table 4.19). For example, the siliceous shales are obviously developed in the Jiaodingshan profile in Hanyuan, Sichuan, while the carbonate shales are mainly developed in Jige profile in Jinyang, Yunnan. The organic carbon content of the latter is obviously lower than that of the former. From the plane distribution characteristics of quartz, clay minerals and carbonate minerals, it can be known that the content of quartz and carbonate minerals has a good negative correlation on the plane. There has a good match between the area in which the quartz content is less than 30% and the area which the carbonate mineral content is more than 30%. This area is located at the

Table 4.19 Characteristics of mineral components and TOC of black rocks in Longmaxi Formation in different sampling sites of the study area

The sampling position	Sample size	Quartz content (%)	Clay minerals content (%)	Carbonate minerals content (%)	TOC content (%)	Favorable zone type
Sichuanhanyuan Jiaodingshan	9	41–86	4–26	0–52	1.18–3.97	IV
		68	13	16	2.73	
Yunnanjinyang Jige	3	21–27	10–39	38–63	0.3–1.89	NO
		23.7	30.5	46.7	0.84	
Changxin I well	44	22–74	6–50	0–59	2.41–8.1	III
		43.6	31.2	22.3	4.46	
Daoye I well	12	43–79	19–47	0–7	2.21–7.02	I_1
		56.4	33.3	1.8	3.42	
Chongqingshizhu Qiliao	3	38–48	38–49	0	1.73–6.63	I_5
		43	43.5	0	3.63	

edge of the uplift, which the organic carbon content is usually less than 1%, and that area is belonged to the tidal flat and marl shallow shelf sedimentary environment. There has a good match between the area which the carbonate mineral content is less than 10% and the area in which the quartz content is more than 50%. The distribution area is mainly located in the deep-water shelf sedimentary environment, which the organic carbon content is usually more than 3%. In the Northern area of Central Guizhou Uplift, the content of clay minerals increases gradually with the increase of the distance from the uplift area. The quartz content is bounded by 30% isoline which increases rapidly to the 50% in northward. The carbonate minerals are approximately bounded by this line which decreases rapidly to 25% in northward, and some content even decreases to 10%. It can be seen that the black rock of Longmaxi Formation in Southern Sichuan and its periphery are organic-rich strata which is dominated by the siliceous shale. In plane, the content of carbonate minerals and clay minerals is the important factor determining the development of organic matter, and the former is more prominent than the later.

In the vertical direction, the content of quartz in Longmaxi Formation decreases, while the content of clay minerals and carbonate minerals increases as the sedimentary water becomes shallower (Fig. 4.65; Chen et al. 2011c; Zhang et al. 2013a). In the center of the code position basin, the clay minerals from the terrigenous sources are diluted during the deposition because of the addition of a large number of marine-derived siliceous sediments as the sedimentary water becomes darker. Finally, siliceous shale or carbonate shale is formed (Fu et al. 2011). Zhang et al. (2012) and Liu et al. (2012a) believed that the depositional centers of the early Longmaxi Formation were located in Luzhou-Yongchuan and Shizhu-Pengshui areas based on the study of the Longmaxi Formation in the Upper Yangtze region. The depositional centers have the highest organic carbon content and the largest thickness of organic-rich shale. According to the development characteristics of mineral components, there has more quartz content (average = 30%) in the depositional center, while the content of carbonate minerals is the least, and the siliceous shales are well-developed.

(iii) Preliminary prediction of favorable area for shale gas by mineral composition

In this study, the basic evaluation criteria for mineral composition of favorable source rocks are as follows: (1) the average content of quartz is more than 30%, (2) the average content of clay minerals is less than 50%, (3) the average content of carbonate minerals is less than 30%. In this range, the black rock is mainly composed of siliceous shale, while the second component is clayey shale, and the carbonate shale is not developed. The composition and content of brittle minerals in organic-rich shale of Longmaxi Formation and the thickness and the enrichment of organic-rich shale are controlled by palaeogeographic pattern, and it is related to the location of the sedimentary and the ability of the provenance providing (Liu et al. 2012a).

Based on the distribution map of sedimentary facies and organic carbon content, the wide area from Suijiang to Shizhu in the Northeast is a favorable area for shale gas according to the content of carbonate minerals. There has no carbonate shale developing, and the average content of carbonate minerals is less than 10%, the average content of clay minerals is less than 50%, the average content of carbonate minerals is more than 50%. They have the high organic carbon content (TOC > 2%), which are located in the depositional center and its margin (Fig. 4.71). The average contents of clay minerals in Yibin-Jiangjin area (I1) at the center of the distribution area are less than 30%, which is the most favorable regions. The average contents of clay minerals in Suijiang-Shuifu area, Chishui-Qijiang area, Fuling-Zhongxian area and Suijiang-Shuifu area (I2) (Fig. 4.65e; Table 4.19 Chongqing Daozhen Bayu profile) are range from 30 to 40%, which is the second favorable regions. And all of them organic carbon content is more than 3%. The Yibin-Jiangjin area (I1) is in the center of the basin which lacks surface data. Huang et al. (2012b) found that organic-rich shale is developed in Longchang-Yongchuan area and their organic carbon content is more than 3%. It can be verified that the stratum is favorable for the formation and preservation of the shale gas which has the characteristic of the high brittle mineral content and the abnormal high pressure. The area (I3) is also a favorable distribution area for shale gas, which content of the clay minerals is higher (40–50%). Although there have many clayey shales, the average content of organic carbon is more than 3%, and it also contains many brittle minerals.

Changning-Xuyong area (II) is the secondary favorable area. The organic carbon content of them is mostly more than 2%, and there began to develop the carbonate shale. The average content of quartz is more than 50%, which the average content of clay minerals is range from 30 to 50%. And the content of the carbonate rock mineral average is 10–30% (Fig. 4.65c, Table 4.19 Well Changxin 1). An experimental shale gas well of Longmaxi Formation in that area which produced 10,000 m^3 of natural gas per day confirms that area has a good exploration prospect (Huang et al. 2012b). According to the content of clay minerals, it can be divided into two sub-distribution areas: Xingwen-Xuyong (II1) and Gongxian-Changning (II2). The former has lower clay mineral content (average content < 40%), and the clay mineral content of the latter is higher (average content > 40%).

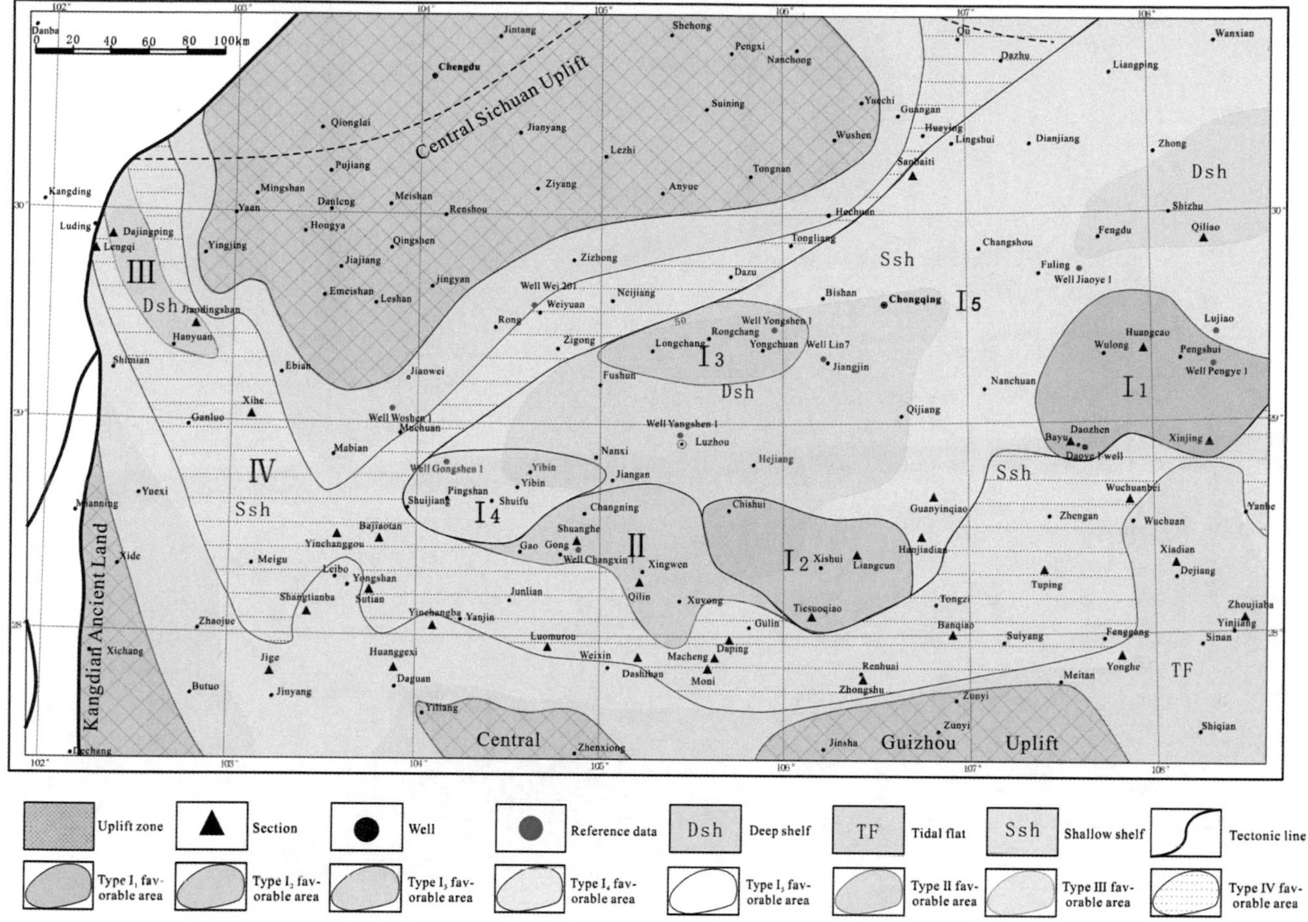

Fig. 4.71 Evaluation map of development profitable areas for shale gas of Longmaxi Formation in Southern Sichuan Basin and its periphery

Another favorable area is Luding-Hanyuan area (III) (-Fig. 4.65a; Table 4.19 Hanyuan Jiaodingshan, Sichuan). Its organic carbon content is high (TOC > 3%), and the quartz content is more than 50%. Carbonate minerals are unevenly distributed, and clay minerals are low in content (<30%, average = 15%). Low clay mineral content is a negative factor for shale's ability to adsorb natural gas (Fu et al. 2011; Ross and Bustin 2008). Therefore, the shale gas reservoir is limited in this area. The Jinyang-Daguan areas and the tidal flat facies area at the edge of the uplift (Fig. 4.65b, Table 4.19 Jinyang Jige, Yunnan) are carbonate shale development areas, which are not favorable areas for shale gas. The average carbonate content of these areas is more than 40%, while the average content of quartz and clay minerals is less than 30%. And the organic carbon content is usually less than 1%. These conditions are not conducive to the occurrence of the shale gas.

The shallow shelf sedimentary environment which is between the favorable area and the unfavorable area is also a favorable area. The quartz and clay minerals content of these areas are all range from 30 to 50%, while the carbonate mineral content is ranging from 10 to 40%. The organic-rich shale is relatively thin, in which the content of the organic carbon is more than 2%. According to the clay mineral content of 40%, it can be divided into two sub-distribution areas: Ganluo-Yanjin-Junlian, Weiyuan-Dazu (IV_1), Weixin-Renhuai-Zheng 'an-Huaying-Quxian (IV_2). As one aspect of the evaluation of shale gas favorable area, the mineral composition should comprehensively evaluate based on the influence of sedimentary microfacies, reservoir characteristics and diagenesis on shale gas reservoir. Finally, the favorable area of the shale gas is comprehensively evaluated.

4.5 Reservoir Space and Property

4.5.1 Reservoir Property

Previous studies have found that the porosity, permeability and pore types of black shale in the Wufeng Formation-Longmaxi Formation are similar to those in North America. The porosity of black shale reservoirs in the Silurian Longmaxi Formation in the Sichuan Basin and its

Table 4.20 Reservoir properties of black rock samples from the Longmaxi Formation in Southern Sichuan Basin and its periphery determined by mercury intrusion analysis

Sample	Lithology	Porosity (%)	Permeability$10^{-3}\mu m^2$)	Sample	Lithology	Porosity (%)	Permeability ($10^{-3}\mu m^2$)
DY1-2	Calcareous shale	1.17	2.4600	SHP7-CH5	Calcareous carbonaceous shale	4.64	0.0016
DY1-6	Calcareous shale	1.46	0.6000	SHP8-CH6	Carbonaceous siliceous shale	9.53	0.0155
DY1-9	Carbonaceous shale	1.18	3.3200	SHP9-CH7	Carbonaceous shale	6.25	0.0044
DY1-16	Carbonaceous shale	0.84	–	SHP9-CH8	Carbonaceous shale	6.45	0.0395
DY1-23	Carbonaceous shale	1.27	0.6890	SHP12-CH9	Calcium-containing silty sandy shale	8.11	0.0095
DY1-25	Carbonaceous shale	1.52	–	SHP14-CH10	Silt shale	13.09	0.0059
63-CH3	Silt shale	0.60	0.0200	SHP17-CH11	Silty carbonaceous shale	5.50	0.0025
145-CH5	Carbonaceous shale	0.30	0.0100	SHP2-BC1	Carbonaceous siliceous shale	5.69	0.0014
214-CH15	Carbonaceous shale	0.40	0.0200	SHP4-BC2	Argillary siltstone	2.69	0.0015
251-CH22	Carbonaceous shale	0.50	0.0400	SHP5-BC3	Argillary siltstone	8.17	0.0025
350-CH37	Carbonaceous shale	1.00	0.0100	SHP6-BC4	Carbonaceous shale	12.48	2.0496
355-CH40	Calcareous carbonaceous shale	0.20	0.0100	SHP7-BC5	Calcareous carbonaceous shale	6.54	0.0012
JDP17-CH10	Silty carbonaceous shale	7.60	0.0400	SHP8-BC6	Carbonaceous siliceous shale	9.60	0.0117
JDP16-CH9	Silty carbonaceous shale	7.40	0.0200	SHP9-BC7	Carbonaceous shale	5.44	0.0032
JDP15-CH8	Silty carbonaceous shale	6.60	0.0200	SHP9-BC8	Carbonaceous shale	6.57	0.0028
JDP12-CH7	Silty carbonaceous shale	8.90	0.0300	SHP12-BC9	Calcium-containing silty sandy shale	9.41	0.0022
JDP11-CH6	Carbonaceous shale	4.80	0.0100	SHP14-BC10	Silt shale	7.50	0.0420
JDP8-CH5	Silty carbonaceous shale	9.00	0.0300	MCP8-CH2	Carbonaceous shale	8.40	0.0300
JDP7-CH4	Silty carbonaceous shale	11.00	0.0500	MCP5-CH1	Carbonaceous shale	4.80	0.0200
JDP6-CH3	Silty carbonaceous shale	9.90	0.1000	MNP3-CH2	Calcareous carbonaceous shale	18.23	0.0155
SHP2-CH1	Carbonaceous siliceous shale	6.87	0.0018	MNP5-CH3	Calcareous carbonaceous shale	6.63	0.0012
SHP4-CH2	Argillaceous siltstone	3.86	0.0015	MNP7-CH4	Calcareous carbonaceous shale	6.99	0.0021
SHP5-CH3	Argillaceous siltstone	5.51	0.0021	DPP4-CH1	Carbonaceous shale	8.96	0.0066
SHP6-CH4	Carbonaceous shale	11.71	0.0061	DPP4-CH2	Carbonaceous shale	19.05	0.0513
HYP3-CH2	Carbonaceous shale	4.00	0.0013	LMRP1-BC1	Carbonaceous shale	6.82	0.3778

Remarks: The profile code is shown in Fig. 4.5.

surrounding areas mainly ranges from 0.22 to 12.75%, and the reservoirs with porosity > 2% account for more than 80% of the total reservoirs (Huang et al. 2012b; Nie et al. 2012a, b; Chen et al. 2013a; Guo et al. 2013). In addition, the permeability is typically small 0.00025–1.737 × $10^{-3}\mu m^2$, with an average value of 0.422 × $10^{-3}\mu m^2$ (Huang et al. 2012b). Mercury injection techniques were used to analyze the physical properties of the black rock series of the Ordovician Wufeng Formation and Silurian Longmaxi Formation in Southern Sichuan and its periphery. The porosity of the black rock series (2–10%) is minor difference from the reported values, and the permeability is generally low, mainly ranging from 0.001 to 0.05 × $10^{-3}\mu m^2$. The maximum permeability of the region with localized fractures is 24.6 × $10^{-3}\mu m^2$ (Table 4.20). The outcrop samples and core samples exhibit different

Table 4.21 Reservoir properties of black rock samples from the Longmaxi Formation in Southern Sichuan Basin and its periphery, determined by other analysis techniques

Number	Sample number	Lithology	Porosity (%)	Permeability ($10^{-3}\mu m^2$)
1	DZP-CH1	Carbonaceous siliceous shale	6.9	< 0.04
2	DZP-CH2	Calciferous siltstone	2.2	< 0.04
3	DZP-CH3	Carbonaceous shale	4.4	< 0.04
4	DZP-CH4	Silty carbonaceous shale	5.2	0.91
5	DZP-CH5	Silty carbonaceous shale	9.9	0.67
6	DZP-CH6	Silty carbonaceous shale	5.2	0.1
7	DZP-CH7	Silty carbonaceous shale	5.0	< 0.04
8	DZP-CH8	Calciferous carbonaceous shale	2.7	< 0.04

Remarks: The profile code is shown in Fig. 4.5

physical properties. The porosity of the core samples is small, ranging from 0.2 to 1.71%, while the porosity of the outcrop samples is more than 2%, with minor differences in the permeability, likely because of the influence of modern hypergenesis on the surface samples. This finding supports those of Clarkson et al. (2013), who used low-angle and extremely low-angle neutron scanning (SANS and USANS), low-pressure adsorption (N_2 and CO_2) and high-pressure mercury injection technologies to study the pore structure of shale gas reservoirs in North America, and indicated that the porosity is determined by the pore size. Due to the lower limit of the physical properties detected by mercury injection analysis is 3 nm, this technique cannot be used to detect micropores. Therefore, in this study, ultrapore-200A helium porosity meter and Ultra-PerMTM200 permeability meter were, respectively, used to analyze the physical properties of black rock series of samples of certain profiles in this study (Table 4.21). The results show that the permeability of the black rock series is relatively increased, but the porosity is similar. Therefore, micropores contribute little to the porosity of shale reservoir. In general, the Silurian Longmaxi Black Rock Reservoirs in Southern Sichuan and its periphery are characterized by a low porosity and ultralow permeability.

4.5.2 Types of Reservoir Spaces

According to the classification standard of the International Union of Theoretical and Applied Chemistry, pores with diameters < 2 nm, 2–50 nm and > 50 nm are termed micropores, mesopores and macropores, respectively (IUPAC 1994). Loucks et al. (2009, 2012) and Slatt et al. (2011) analyzed the pore genesis and noted that the main pores are intra- and intergranular pores formed by the hydrocarbon generation evolution of organic matter, intergranular pores of berrylike pyrite, intergranular pores of floccular minerals, and dissolved pores and microfractures of minerals and clastic particles. Based on the shale gas reservoir pore classification scheme proposed by Yu (2013), the pore types of black rock series in the Wufeng Formation-Lower Longmaxi Formation in southern Sichuan and its adjacent areas were divided into two categories and ten types (Table 4.22).

1. Intergranular pores

(1) Intergranular skeleton pores

Intergranular skeleton pores, with sizes typically on the micron scale, are formed by rigid mineral particles such as quartz and feldspar supporting one another. Similar to conventional clastic rocks, in the depositional diagenetic process, quartz, feldspar and other clastic grains support and overlap with one another, forming pores with irregular shapes when they contact one another. Moreover, intergranular skeleton holes exist between soft and plastic clay minerals, mica and hard and brittle particles owing to differences in the physical properties. Intergranular skeleton pores are abundant in shallowly buried sediments and typically exhibit a high connectivity to form efficient (permeable) pore networks. With increasing burial depth and overburden pressure and diagenesis, the intergranular pore water is gradually discharged, and the particles tend to be densely arranged. Plastic particles can deform and close the intergranular pore space and squeeze into the pore path, resulting in a large number of original intergranular pores. Thus, in older and deeply buried fine-grained sedimentary rocks, the number of intergranular skeleton pores is significantly reduced by compaction.

Intergranular skeleton pores are triangular and occupy the residual pore spaces among rigid particles through compaction and cementation (Fig. 4.72b). When the soft flake clay minerals contact with the abovementioned particles and compaction intensifies during the diagenetic process, the clay minerals undergo plastic deformation owing to the

Table 4.22 Classification of reservoir pore in black rock samples of the Longmaxi Formation in the Southern Sichuan Basin and its periphery (modified by Yu 2013)

Classification basis and category	Occurrence and type of pores				Genesis	Porosity character
	Main class		Class	Subclass		
Types of porosity	Rock matrix pore	Mineral matrix pore	Interparticle pores	Intergranular skeleton pores	Intergranular pores formed by particle accumulation	Macropores (>50 nm); complex shape, micron scale, high connectivity
				Intercrystal pore	Pores between crystals	Macropores (>50 nm) and medium pores (2–50 nm); less developed and low connectivity
				Flocculation pore	Pores formed by the accumulation of clay minerals	Micropores (<2 nm) and medium pores (2–50 nm); well-developed with high connectivity
				Rigid granular edge dissolution pores	Edge pores formed by dissolution	Micropores (<2 nm) and medium pores (2–50 nm); relatively developed with low connectivity
			Intragranular pore	Intergranular pores in aggregates	Intergranular pores in aggregates such as pyrite	Micropores (<2 nm) and medium pores (2–50 nm); relatively developed with low connectivity
				Interlayer pores of clay minerals	Micropores between lamellar clay mineral layers	Micropores (<2 nm) and medium pores (2–50 nm); well-developed with low connectivity
				Intragranular corrosion hole	Intragranular pores produced by dissolution	Micropores (<2 nm), middle pores (2–50 nm) and macropores (>50 nm); less developed and low connectivity
			Organic pore	Fossil coelomere	Formed after decay and dissolution of soft tissue in organisms	Micropores (<2 nm) and medium pores (2–50 nm); less developed and high connectivity
				Hydrocarbon-generating pores in organic matter	Caused by organic matter hydrocarbon generation	Medium pores (2–50 nm) and macropores (>50 nm); well-developed with low connectivity
			Fissure	Tectonic fracture	Fractures formed by or associated with local tectonics	size of the order of millimetres; locally developed with high connectivity
				Microfracture	Fractures formed by diagenesis and abnormal pressure fractures formed by organic matter evolution	Micropores (<2 nm), middle pores (2–50 nm) and macropores (>50 nm); well-developed with low connectivity

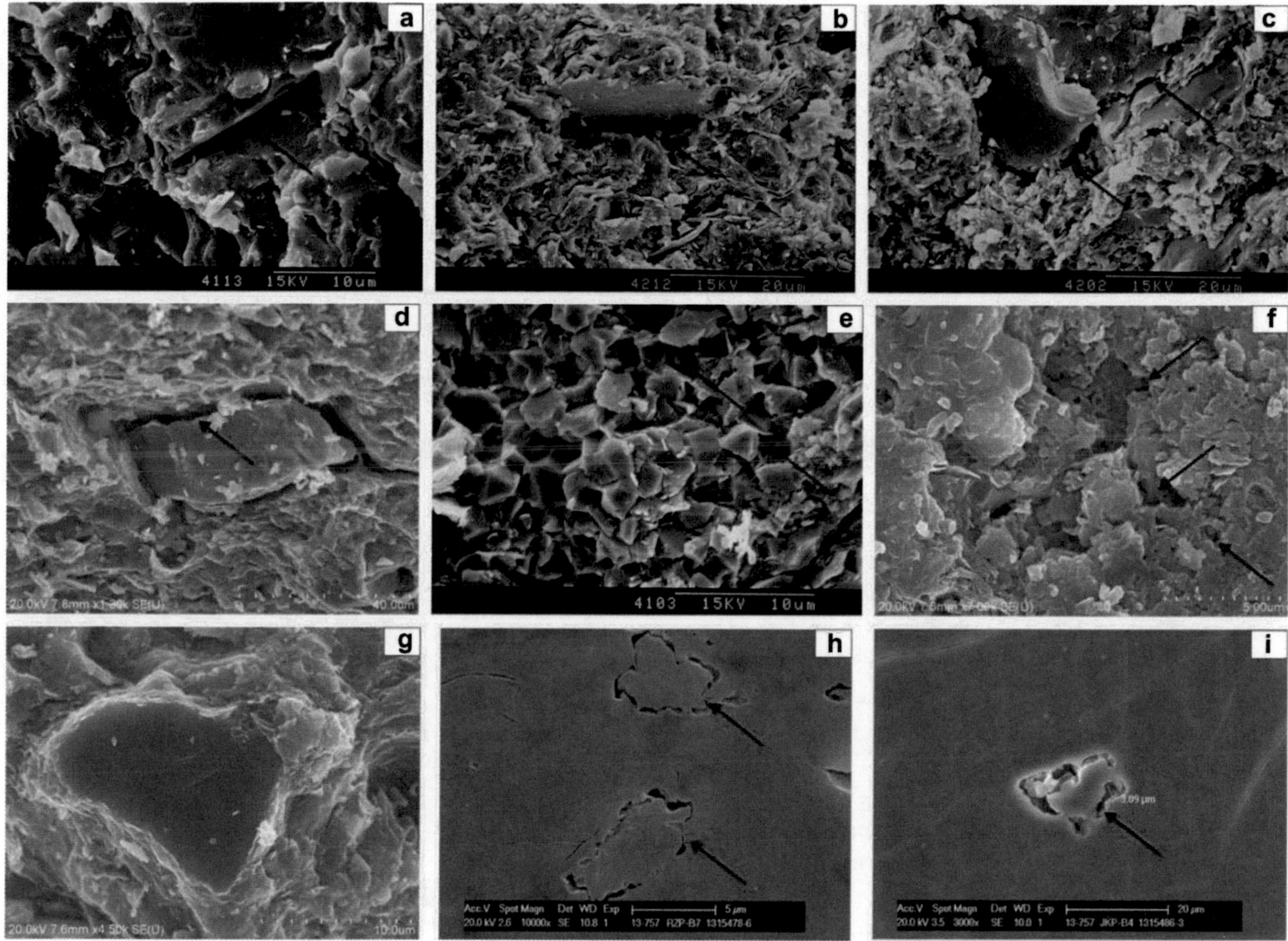

Fig. 4.72 Characteristics of intergranular pores in black rocks of Longmaxi Formation in Southern Sichuan Basin and its periphery **a** rigid intergranular pores, irregular strip, DY1-S53; **b** pores between rigid particles and mud, formed by particle support, DY1-S99; **c** pores between rigid particles and clay minerals, similar in shape to particles, DY1-S97; **d** porosity between rigid particles and clay minerals, similar in shape to particles, Junlian; **e** frontal pores between crystalline calcite, DY1-S47; F. clay mineral flocculation pores, Junlian; **g** clays are densely wrapped by the particles, Junlian; **h** edge dissolution pores of feldspar particles, Renhuai Zhongshu; **i** edge dissolution pores of feldspar particles, Wulong Jiangkou

difference in the hardness of the minerals and form certain support holes at the edges of the contact with brittle minerals, mostly in the linear form (Fig. 4.7c, d). The pores formed by the accumulation of various detritus particles in mud shale vary in diameter from several hundred nanometers to dozens of microns, and their length is less than 1 μm, although certain pores are sized 50 nm to several millimetres with a complex morphology and high connectivity. These pores are the main storage sites and seepage channels of free gas. The intergranular pores decrease by not only compaction but also cementation around particles such as quartz, calcite and feldspar. However, in fine-grained sedimentary rocks, brittle minerals such as quartz and feldspar are mostly dispersed in clay minerals and organic matter, and most of these minerals cannot support particles. Therefore, only a few residual skeleton pores exist among grains, mainly within the few brittle mineral particles or between brittle particles and clay minerals. Intergranular skeleton pores are rarely distributed in the Wufeng Formation and Longmaxi Formation black rock series in southern Sichuan and its adjacent areas, and the detrital particles exhibit dominant orientations except in the local distributions.

(2) Intergranular pores

The pores are developed by authigenic calcite, dolomite and authigenic siliceous cement and are formed by brittle mineral grains such as quartz and feldspar. The pore structure is related to the shapes of mineral crystals and brittle mineral grains, such as triangles, polygons and elongated minerals, with sizes > 50 nm (macropore types), typically more than 1–5 microns (Fig. 4.72e). Such pores are not developed and are mainly influenced by early compaction, causing the number of primary pores to rapidly decrease. Early formation of carbonate cement occurs in micritic aggregate. The cement formed by the late formation owing to increased space and contents of shale and clay minerals limits the development of better intergranular pores.

(3) Flocculation pore

Fine-grained sedimentary rocks have a high clay mineral content. Results of analyses indicate that the clay minerals in the black rock series of the Lower Longmaxi Formation in southern Sichuan and its adjacent areas are mainly terrigenous clastics, and thus, the flocculation pores of clay minerals are relatively developed. In the interior of clay mineral flocs, grid-like or banded pores are often formed, which are known as floc pores (Fig. 4.72f). O 'Brien and Slatt reported the presence of flocculent clay minerals in ancient microstratified structural shales in 1990 but could not explain how these open pores were preserved after hundreds of millions of years of burial and diagenesis (O'Brien and Slatt 1990). Although the interpretation of flocculation-derived pores remains questionable, such pores in floccule clay minerals (diameter of more than 3.8 nm) provide sites for methane molecules and are interconnected to form permeable channels. Thus, open or partially collapsed flocs in fine-grained sedimentary rocks can be considered sites of the intergranular pores between clay sheets. The black rock series of the Wufeng Formation and Longmaxi Formation in Southern Sichuan and its adjacent areas exhibit widely developed clay mineral flocculation pores, especially in the parts in which the clay minerals are locally enriched and connected, which represent non-negligible reservoir spaces for shale gas.

(4) Rigid grain edge corrosion holes

The black rock series of the Wufeng Formation and Lower Longmaxi Formation in Southern Sichuan and its adjacent areas are rich in aluminosilicate minerals such as feldspar. During the generation of organic matter, a large number of organic acids are formed that fill or flow through the pores, and the aluminosilicate minerals such as feldspar dissolve in acidic media. However, due to the influence of strong compaction in the early diagenetic process of fine-grained sediments, clastic particles often densely contact the clay minerals (Fig. 4.72g), and the original sedimentary water is weakly alkaline. The continuous process of hydrocarbon generation results in a short contact time between the acidic medium and feldspar, and the acid dissolution is limited. Irregular dissolution pores are often formed at the edges of particles and are mostly filled with organic matter or clay minerals (Fig. 4.72h, i). In the early and late diagenetic stages, the diagenetic fluid exhibits weak alkalinity, leading to the dissolution of unstable alkaline minerals such as quartz. Because the occurrence of quartz dissolution requires certain thermodynamic conditions, the dissolution pores at the edge of quartz grains are often developed in the late diagenetic stage. Such pores have an irregular strip shape and low connectivity and are less developed.

The primary types of intergranular pores are intergranular skeleton pores and flocculation pores, and the secondary types are intergranular pores and rigid grain edge dissolution pores, formed during diagenesis. The primary pores are relatively developed.

2. Intragranular pores

(1) Intergranular pores in the aggregate

Such pores are mainly concentrated in the aggregate of pyrite grains. The pyrite in black rock series of the Silurian Longmaxi Formation in Southern Sichuan and its adjacent areas are generally developed as framboids in addition to the partially disseminated framboids with diameters ranging from a few microns to dozens of micrometers and are composed of many homogenous pyrite grains. These pyrite intergranular grains exhibit several pores with sizes of the order of microns to nanometers (Fig. 4.73a, b) and low connectivity. After the oxidation of pyrite aggregates, regular round and oval pores are formed (Fig. 4.73c).

(2) Interlayer pores of clay minerals

With increasing stratum burial depth and paleotemperature and the presence of alkaline diagenetic fluid, a large amount of interlayer water precipitates due to the dehydration transformation of clay minerals, and microcracks form in the cambium. The clay minerals in the black rock series of the Longmaxi Formation in southern Sichuan and its adjacent areas exhibit the following transformations: montmorillonite → illite/montmorillonite lSll → illite, and the chloritization of kaolinite is not developed. The secondary lamellar clay minerals form irregular aggregates, often with elongated strips of pores with sizes ranging from a few nanometers to tens of nanometers between the crystal lamellar layers (Fig. 4.73d, f). Therefore, the pores between the clay mineral layers are relatively developed, with low specific volumes, large quantities and large surface areas. The connectivity is not high, but the pores exhibit a high adsorption ability and can thus serve as adsorption sites for natural gas.

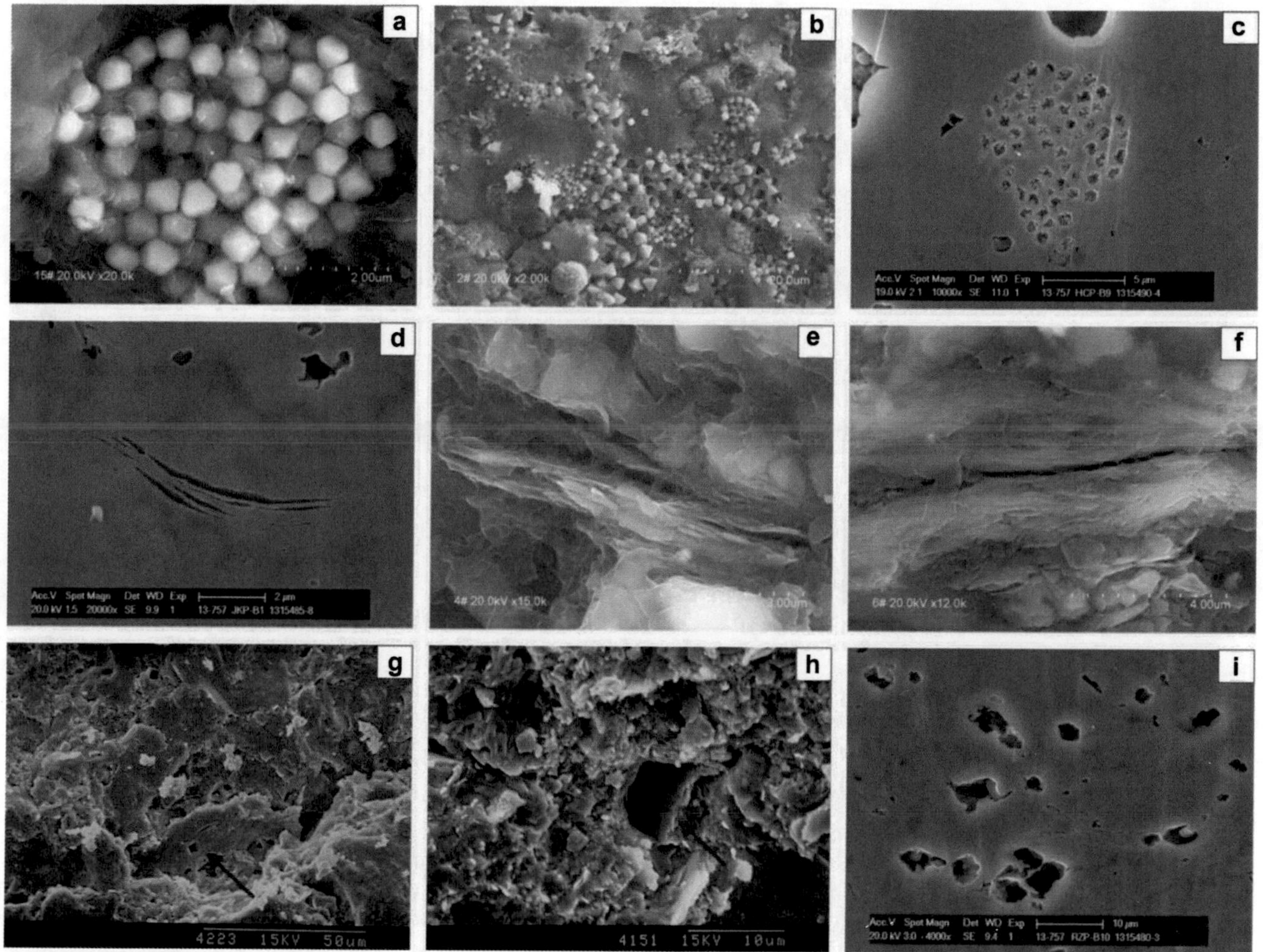

Fig. 4.73 Characteristics of intragranular pores in black rocks of Longmaxi Formation in Southern Sichuan Basin and its periphery **a** intergranular pore of pyrite aggregate, Qijiang Guanyinqiao; **b** pyrite grains, intergranular pore development and die hole formation by dissolution, Xuyong Macheng; **c** pyrite oxidation residual pores, Wulong Huangcao; **d** Micropores between clay mineral layers, Wulong Jiangkou; **e** Micropores between clay mineral layers, Wulong Huangcao; **f** Micropores between clay mineral layers, Wulong Huangcao; **g** Dissolution pores in feldspar grains, DY1; **h** dissolution-casting holes, DY1; **i** Feldspar dissolution to form the mold holes, Renhuai Zhongshu

(3) Dissolution pores in grains

The formation of dissolution pores in the granular matrix is related to the characteristics of the diagenetic fluid, similar to the formation of dissolution pores in rigid grain margins in intergranular pores. The dissolution pores of feldspar, calcite, dolomite and quartz grains can be observed in the black rock series of the Longmaxi Formation in southern Sichuan and its adjacent areas (Fig. 4.73g, h). The quartz grain dissolution pores are relatively developed, attributable to the development of alkaline diagenetic fluids. The outcrop samples exhibit a large number of dissolution pores in the feldspar grains and casting holes (Fig. 4.73i), with pore sizes of several microns. Chen et al. (2013a) and Ran et al. (2013) observed abundant dissolution pores and dissolution-casting holes in the outcrop samples. Such pores and holes are less developed in core samples. Wang et al. (2012c) pointed out that the abovementioned unstable mineral dissolution micropores are more common in several outlying samples (Shizhu, Xiushan, etc.), and 14 samples have been identified. According to the electron microscopy analysis, the diameter of the dissolution pores is 2–10 μm, and these pores are rarely observed in underground samples. Isolated dissolved pores in grains are occasionally observed (only 4 samples have been found). The burial characteristics, shale porosity evolution and forecast model exploration indicate that the production of organic acids and corresponding dissolution do not considerably influence the pores, and the contribution to the increase in the mud shale porosity is smaller than expected. This phenomenon occurs because the shale pores inside are smaller than those in the conventional reservoir, the permeability is low, and the fluid exchange is

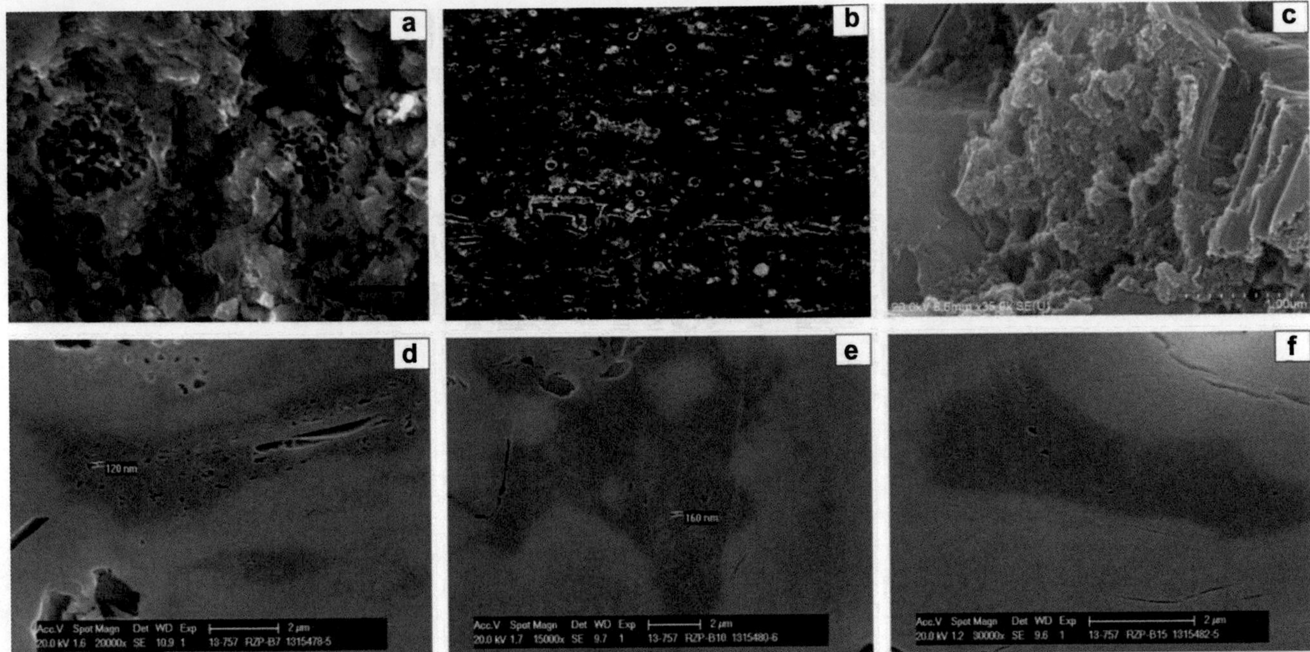

Fig. 4.74 Characteristics of organic matter intragranular pores in black rocks of Longmaxi Formation in Southern Sichuan Basin and its periphery **a** diatom fossil hole, Shizhu Qiliao; **b** Siliceous radiolarians developed coelomere, Tianquan Dajingping; **c** hydrocarbon-generating pores of organic matter, Changning; **d** hydrocarbon-generating pores of organic matter, oval and long strip shape, Renhuai Zhongshu; **e** the organic matter aggregate between grains is irregular, with elliptic hydrocarbon-generating pores, Renhuai Zhongshu; **f** massive organic matter with a few hydrocarbon-generating pores, Renhuai Zhongshu

insignificant. Diagenesis, such as dissolution and metasomatism, is not as strong as that in conventional reservoirs (Guo et al. 2013). Moreover, the analysis of drilling core data reported by Wang et al. (2013d), Long et al. (2012) for Yuye 1 well, and Guo and Liu (2013) for Jiaoye 1 well (159 samples, apertures between 3.8 nm and 36 nm, containing a small amount of dissolution and intergranular pores) and the analysis of the observations of core data in the Daoye 1 well in the black rock series of the Wufeng and Longmaxi Formations in southern Sichuan and its adjacent areas show that the dissolution pores in grains generated by dissolution are not developed, and the connectivity is low; thus, the contribution of these pores to the reservoir is negligible.

3. Mechanical pores

Organic pores are generated in the process of hydrocarbon generation of organic matter. Such pores are generally small, consisting mainly of medium and macropores of the nanometer scale, with a large surface area, partial enrichment and distribution, and a certain connectivity, thereby providing space for storing a large amount of adsorbed gas.

(1) Fossil coelomeres

The formation of fossil coelomere is the organic fossil material preserved in rock, which is formed after the decay and dissolution of soft tissue under the influence of rising underground temperature and pressure during late burial diagenesis. A few diatom fossils holes have been observed in the research of black rock series in the Longmaxi Formation. The diatom shell consists of rigid amorphous silicon, and the internal organic matter decomposition forms many regular arrangement cribriform small round hole, the nano-sized pores are well connected with each other, but the connectivity with the outside pores is low (Fig. 4.74a). In addition, siliceous radiolarian fossils such as the body cavity aperture are visible, aperture diameter can reach millimetre, and are filled by black organic matter or argillaceous (Fig. 4.74b). Because the siliceous radiolarian is distributed inhomogeneously, such pores are not highly developed. The siliceous radiolarian content of the Wufeng Formation is higher than that of the Longmaxi Formation, which is beneficial for the Wufeng Formation.

(2) Hydrocarbon-generating pores of organic matter

The organic matter in the study area is dispersed and banded in certain areas. Results of scanning electron microscopy analyses indicate that a large number of honeycombed nanoscale pores are generated in the organic matter in the black rock series of the Longmaxi Formation in Southern Sichuan and its periphery (Fig. 4.74c–f). These pores are formed by hydrocarbon generation residues generated by

Fig. 4.75 Tectonic seam features of Longmaxi Formation in Southern Sichuan Basin and its periphery **a** oblique crack, scratch developed, DY1; **b** high-angle fracture, unfilled, DY1; **c** high-angle fracture, calcite filled, DY1

cracking of organic matter with thermal evolution level. Generally, the formation, distribution and size of these pores are related to the abundance, type and evolution degree of organic matter (Huang et al. 2012b). Studies have shown that organic pores develop only when the thermal maturity level of organic matter exceeds approximately 0.6%, marking the beginning of peak oil generation (Dow 1977). With thermal evolution, more micropores are generated after the hydrocarbon generation of organic matter, and the pore size increases. An organic mass can contain hundreds to thousands of nanopores, most of which are irregularly round or oval and increase the adsorption energy of shale. Organic hydrocarbon generating pores not only provide important storage space for adsorbed and free gas and serve as storage sites (Wang et al. 2012c).

4. Fracture

(1) Structural fractures

Structural fractures refer to fractures formed or accompanied by local tectonics, mainly related to faults and folds, whose direction, distribution and formation are all related to the formation and development of local tectonics (Wu et al. 2005; Li et al. 2005a). Structural fractures are the main fracture types in the Wufeng Formation and Longmaxi Formation in the Southern of Sichuan and its periphery. Horizontal bedding fractures and high-angle structural fractures commonly occur in underground cores, with large concentrations in certain areas. The largest length of a single fracture is 1.5–2 m, and the largest width is several centimeters, mostly filled with calcite. According to the differences in mechanical properties, the filled fractures can be divided into tensile, shear and extrusion joints (Wu et al. 2005). Structural fractures are generated by the tectonic influence of the Wufeng Formation and Longmaxi Formation in the long-term burial evolution process. These fractures limit the preservation of shale gas and cannot function as storage space. The structural microfractures formed in the early stage may become seepage channels. Two tectonic fractures can be seen in the black rock series of the Wufeng Formation and the Lower Member of Longmaxi Formation in Southern Sichuan and its periphery. The early fractures are high-angle and vertical fractures that are generally flat and calcite filled, with a width of 0.5–2 mm. Both high-angle and low-angle fractures can be observed in the later stage. Argilly filled or unfilled fractures can be seen with sliding scratches and coal vitreous luster (Fig. 4.75).

In summary, the calcite filling fracture study area is relatively well-developed, which can be observed in outcrops, underground cores and under the microscope. Most of these fractures are completely filled, and the storage and permeability provided by intergranular pores of calcite or pyrite are of little significance. Conventional tests show that the physical properties are inferior. The surface of the carbon clay filled crack is mirrorlike shiny and sliding scratches exist. Most of these fractures are controlled by late structures and provide a channel for oil and gas migration, thereby contributing to shale gas storage and permeability.

(2) Microcracks

Microfractures in fine-grained sedimentary rocks of the Wufeng Formation and Longmaxi Formation in Southern Sichuan and its periphery not only provide sufficient storage space and migration channels for shale gas, but more importantly, facilitate the later development of shale gas (Long et al. 2012). Generally, these microfractures are formed during diagenesis, which are different from macro-tectonic fractures and smaller in scale. These microcracks are typically no more than 0.5 μm wide, providing ample space for methane molecules to permeate. In the diagenetic process, the fractures formed due to the shrinkage rock volume are nearly parallel to the plane. The main causes of these fractures are drying shrinkage, dehydration, mineral transformation or thermal contraction, unrelated to tectonics (Wu et al. 2005; Li et al. 2005a). Diagenetic contraction fractures commonly occur in the muddy intercalation of shale and

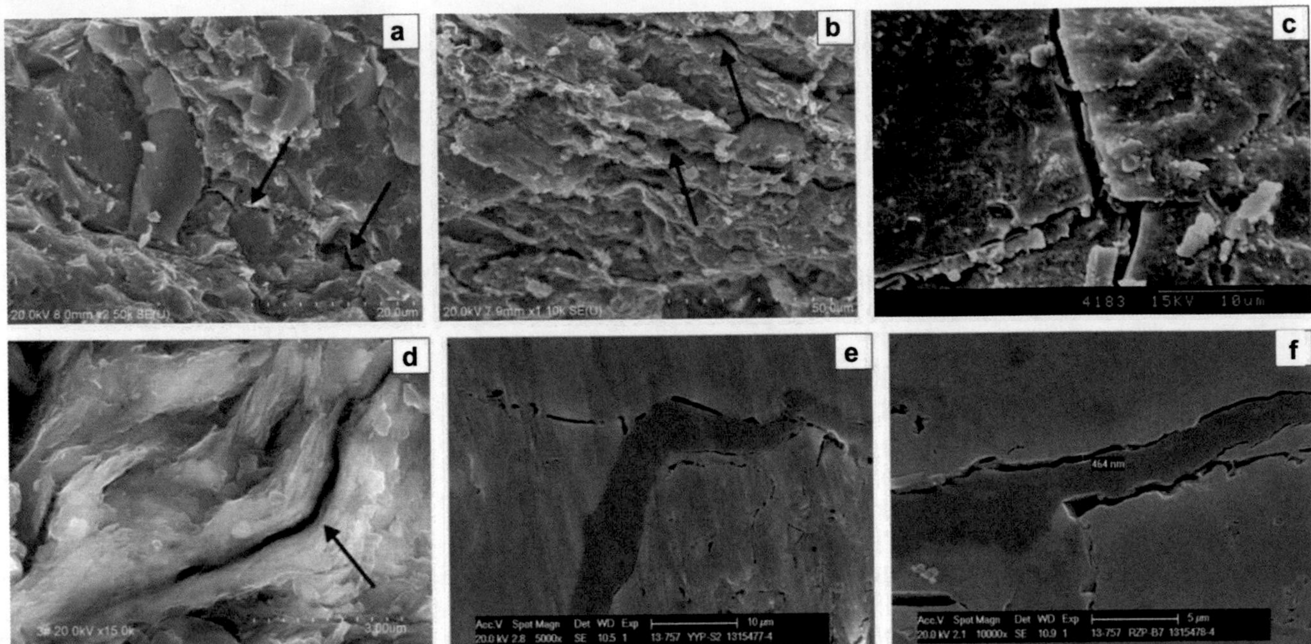

Fig. 4.76 Microfracture characteristics of black rocks in Longmaxi Formation in Southern Sichuan Basin and its periphery **a** microcracks distributed at the edge of Junlian particles, Junlian; **b** shrinkage joints between clay mineral aggregates, Junlian; **c** intergranular microcracks, DY1; **d** shrinkage joints between clay mineral aggregates, Shizhu Qiliao; **e** microcracks between organic matter and particles, Yanjin Yinchangba; **f** Microcracks between organic matter and particles, Renhuai Zhongshu

horizontal bedding marl, with small extension length but high connectivity, considerable variation in the opening degree and partial filling (Fig. 4.76a–d). Generally, fine-grained sedimentary rocks with high siliceous content at the time of deposition shrink due to chemical changes during diagenesis, resulting in the formation of widely distributed diagenetic contraction fractures (Zhang and Yuan 2002). Organic evolutionarily abnormal pressure fractures refer to fractures formed by rock fracture due to local abnormal pressure generated during organic matter evolution (Wu et al. 2005). These fractures are generally developed in organic-rich shale, with irregular fracture surfaces, and are mostly filled with organic matter (Fig. 4.76e, f).

In general, intergranular pores and intracrystalline dissolved pores are commonly developed in black shales of the Wufeng Formation, followed by fractures. Interlayer fractures and oblique fractures are commonly developed in the shales of the Longmaxi Formation, mostly in the lower part of the shale. Organic matter pores and interlayer pores of clay minerals are dominant matrix pores in black rock series of the Wufeng Formation-Lower Longmaxi Formation in Southern Sichuan and its periphery, while microfractures provide the main seepage channels.

4.5.3 Analysis of Factors Influencing Reservoir Space Development Characteristics

According to preliminary studies, shale gas occurring in medium-sized micropores mainly exists in the adsorbed state, while shale gas occurring in macropores and microfractures mainly exists in the free gas form (Daniel et al. 2009). A significant feature of fine-grained shale is its tiny pore structure, which is dominated by micro- and medium-sized pores and determines the occurrence state of shale gas (mainly in the adsorbed form) (Wu et al. 2013). Wu et al. (2013) pointed out that the pore size of Well Yuye 1 samples ranged from 3.51–6.76 nm, with a median radius of 1.75–3.38 nm and an average of 2.26 nm. The volume of the mesopores accounts for approximately about 70% of the total volume, and the volumes of the micropores and macropores are approximately 10% and 20%, respectively. Chen et al. (2012) found that in the pore volume of the Longmaxi Formation mud shale in Southern Sichuan is dominated by mesopores, followed by micropores and macropores. Mesopores provide the main pore volume space, while the micropores and mesopores provide the main pore-specific surface area. The black rock series of the Wufeng Formation and Longmaxi Formation in Southern

Sichuan and its periphery are mainly composed of organic matter pores and interlayer micropores of clay minerals, which are characterized by medium pores and micropores of < 50 nm, further proving that the shale gas in the Wufeng Formation is mainly in the adsorbed form.

According to the development characteristics of pore types in black rock series of the Ordovician Wufeng Formation and Silurian Longmaxi Formation in Southern Sichuan and its periphery, the development of reservoir space is mainly affected by the mineral composition, lithofacies type, type and content of organic carbon, maturity of organic matter and diagenesis (Liang et al. 2012a).

Although many factors influence the pore structure, the key factor is the material composition determined by the sedimentary environment and diagenetic evolution (Chen et al. 2013a). The diagenesis and original composition of mud shale must be considered in reservoir evaluations (Ross and Bustin 2007). Brittle minerals are mainly associated with intergranular skeleton pores, dissolution pores and intergranular pores, characterized by macropores. Clay minerals mainly form flocculation pores and interlayer micropores, with sizes < 50 nm. Therefore, the quartz content is negatively correlated with the micropore and mesopore volumes and positively correlated with the macropore volume, indicating that the increase in the terrigenous debris content can help increase the macropore volume. The relationship between the clay mineral content and pore volume is opposite to that of quartz. The clay mineral content is positively correlated with the micropore volume and negatively correlated with the macroholes, indicating that the clay mineral content controls the pore volume and size, with the control effect on the pore volume being more notable (Yu 2013; Wu et al. 2013). Chen et al. (2013a) noted that the organic carbon content and brittle mineral content considerably influence the pore formation, while clay minerals have the opposite effect, and the impact of clay minerals is one order of magnitude smaller than that of the organic carbon content and brittle mineral content. A slight increase in the organic carbon content considerably increases the porosity. As the content of brittle minerals increases, the microcracks formed by tectonic processes are easily extended, the shale is more easily broken, and the semiclosed and closed pores become open pores. Although a large number of inorganic pores can be formed from clay minerals during diagenesis and mineral transformation, the present results indicate that such pores contribute little to shale porosity, and thus, high clay areas cannot be considered promising regions for future exploration.

As the main reservoir space of shale gas, the formation and existence of organic pores are not only related to not only their organic carbon content, but also their thermal maturity. The influence degree of organic carbon content is more than 4 times that of brittle minerals, indicating that organic carbon content is the most critical and significant factor influencing the shale porosity. The organic carbon content is the main internal factor controlling the nano-pore volume and specific surface area of shale gas reservoirs in the Longmaxi Formation, and the corresponding sites provide the main storage space for shale gas (Chen et al. 2012). The TOC of the Lower Silurian Longmaxi Formation in Southeastern Chongqing is positively correlated with the total pore volume. The pore volume of the samples is dominated by mesopores and thus exhibits the strongest positive correlation and highest fitting degree with mesopores. Moreover, the pore volume exhibits a certain positive correlation with the micropore and macropore volume. For a certain TOC (less than 1.0%), the micropore and mesopore volumes increase with increasing clay mineral content; however, the relationship between the macropore volume and TOC is not clear. When the TOC is greater than 1.0%, the pore volume is controlled by the TOC (according to internal data, 2012). Mastalerz et al. (2013) pointed out that the porosity is directly related to organic matter and mineral composition, and the maturity is the main factor affecting organic matter and mineral composition. Organic carbon content at both high maturity and high maturity is positively correlated with the macropore volume, attributable to the increase in the macropore volume by the development of nanoscale microfractures in mature organic matter (Wu et al. 2013). In addition, the existence of brittle minerals such as quartz and calcite facilitates the formation of microcracks, and their contents determine the number of microcracks (Wang et al. 2013d).

In general, the development characteristics of the black rock series reservoir space of the Ordovician Wufeng Formation and Silurian Longmaxi Formation in Southern Sichuan and its periphery are mainly affected by the mineral composition and organic matter content and their evolution degree. Mud shale is rich in clay minerals and organic matter during evolution to high maturity and postmaturity, leading to a substantial decline in the hole porosity. Therefore, the number of macropores gradually decreases and that of micropores increases (Chalmers et al. 2011), leading to the formation of a reservoir space dominated by organic matter micropores and clay mineral interstitial holes, both of which are produced during diagenesis. Furthermore, the hydrocarbon generation and migration of organic matter lead to the transformation of organic matter, which further leads to porosity variations (Mastalerz et al. 2013). Therefore, the evolution of porosity is mainly related to diagenesis.

4.5.4 Rock Mechanical Characteristics of Rock

Rock mechanical properties and parameters provide the basis for the drilling design of oil and gas wells, establishment of reservoir reconstruction measures and scheme design. At present, three main research methods exist: laboratory measurement of rock samples; use of geophysical logging data; and hydraulic fracturing calculations.

The rock mechanical parameter profile of a continuous formation can be obtained based on laboratory tests, fracture calculations and correction logging methods. At present, only array sample tests have been conducted, and the mechanical properties of two sets of organic-rich shale in the research area have been determined, which can promote future work.

1. Compressive and tensile strength

Under normal temperature and pressure, the compressive strength of organic-rich shale is 28.97–153.73 MPa, and the tensile strength is 3.11–9.70 MPa.

In the case of little difference in rock mineral composition, the compressive strength and tensile strength differ greatly, and there is no obvious correlation with rock density. It is speculated that it is related to the degree of microfracture development of samples. When the rock is damaged, the crack will become the weak surface. Water medium has little effect on rock strength and "softening effect" is not obvious.

2. Elastic modulus and Poisson's ratio

Under normal temperature and pressure, the elastic modulus E of organic-rich shale ranges from 1.45 to 6.02 $\times$ 10^3 MPa, with an average value of 4.15 $\times$ 10^3 MPa. Poisson's ratios ranged from 0.23 to 0.35, with an average of 0.29 (Table 4.23).

Under saturated water conditions, the water content of rock is 0.53–1.57%. Under these conditions, the elastic modulus is similar to that under in natural conditions, which also indicates that the water medium does not considerably influence the mechanical properties.

4.6 Diagenetic Study of the Silurian Longmaxi Formation

The Ordovician Wufeng Formation and Silurian Longmaxi Formation are a set of continuous sedimentary strata, expected to belong to a similar diagenetic system, and the deposition thickness of the Ordovician Wufeng Formation is relatively small. Therefore, this book focuses on the diagenesis of the Silurian Longmaxi Formation.

4.6.1 Types of Diagenesis and Diagenetic Minerals

Compared with conventional clastic rock reservoirs, shale gas formations function as both source rocks and reservoirs and involve a more complex diagenesis. During burial diagenesis, the shale gas formation underwent the joint transformation of inorganic and organic diagenesis. Notably, inorganic diagenesis mainly includes compaction, cementation, dissolution and metasomatism, while organic diagenesis refers to the hydrocarbon generation evolution of organic matter.

1. Compaction

Compaction results in sediment volume reduction, drainage and dehydration. The fine-grained sediments were in the oozy state after deposition, and their original porosity was gt. 50%, with the pores being filled with free water. With the increase in the burial depth, the sedimentary material points

Table 4.23 Test results of rock mechanical properties of Wufeng Formation and Longmaxi Formation black rock series in Southern Sichuan and its periphery

Sample number	Lithology	Formation	Density	Test results		
				Young modulus/MPa	Poisson's ratio	compression strength/MPa
JDSP-B13	Dolomitic limestone	Wufeng formation	2.61	13,510	0.10	38.00
STP-B10	Carbonaceous mudstone	Longmaxi formation	2.58	19,320	0.10	62.10
RZP-B7	Limestone	Longmaxi formation	2.63	13,630	0.10	35.80
XMP-YL6	Carbonaceous mudstone	Wufeng formation	2.48	5890	0.09	26.4
XMP-YL1	Carbonaceous mudstone	Wufeng formation	2.42	6990	0.08	31.4
XMP-YL4	Carbonaceous mudstone	Wufeng formation	2.47	4850	0.07	21.4
HCP-B3	Carbonaceous mudstone	Wufeng formation	2.39	20,940	0.11	101.0
SQP-B4	Carbonaceous mudstone	Wufeng formation	2.61	5260	0.07	18.2
QLP-B10	Carbonaceous mudstone	Longmaxi formation	2.31	16,590	0.10	76.4

Note See Fig. 4.5 for profile code

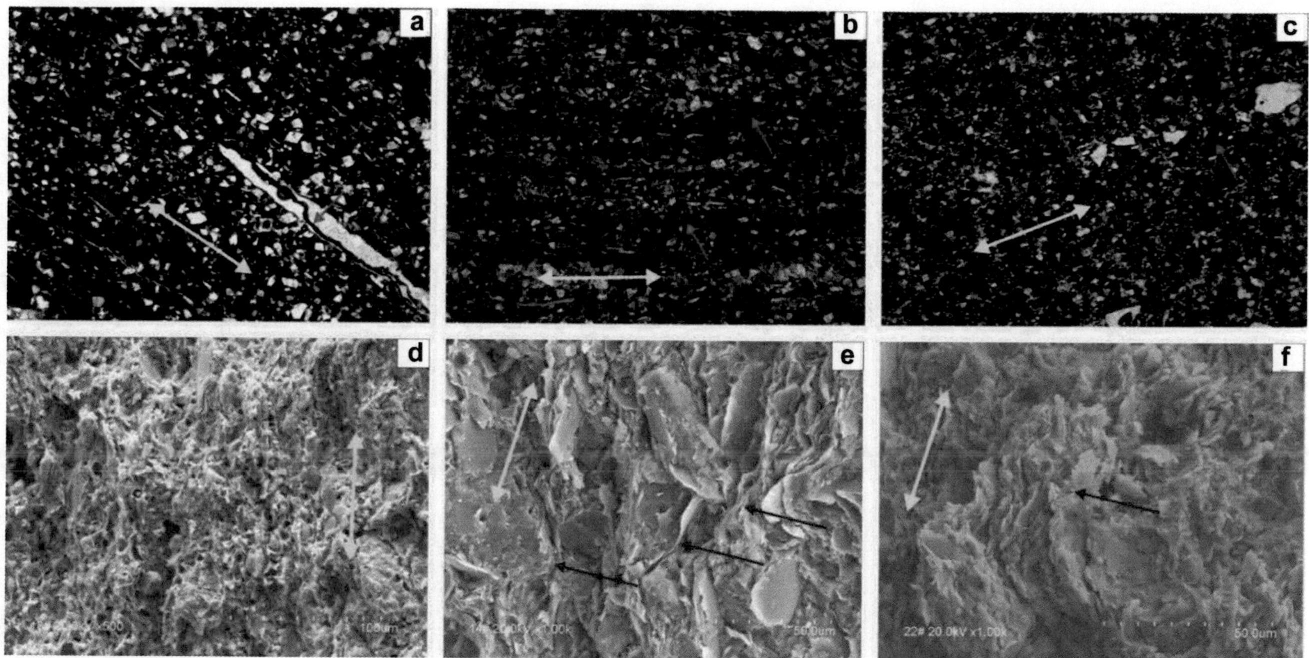

Fig. 4.77 Compaction of black rocks in the Longmaxi Formation in the Southern Sichuan Basin and its periphery **a** silty carbonaceous shale, short columnar mica (orange arrow), directional arrangement, organic matter filling fractures (green arrow); E'bian Xihe, Sichuan Province; single polarized light. **b** silty calcareous carbonaceous shale, short columnar mica (orange arrow), symbiotic organic matter and mud (green arrow); Gulin Tiesuoqiao, Sichuan Province; single polarized light. **c** silty calcareous carbonaceous shale, oriented mica (orange arrow), organic matter (green arrow) and argillaceous; Hanyuan Jiaodingshan, Sichuan Province; single polarized light. **d** oriented clay minerals; Renhuai Zhongshu, Guizhou Province. **e** oriented clay minerals, concave and convex contacts with clastic particles (orange arrows); Renhuai Zhongshu, Guizhou Province. **f** curved clay mineral assemblage (orange arrow); Yanjin Yinchangba, Yunnan Province

were rearranged, deformed or broken under the weight of the overlying water and sediment load, and the pore water was continuously discharged. The porosity and volume of the original fine-grained sediments considerably decreased, and compaction and consolidation occurred. The fine-grained sediments were rich in clay minerals and small clastic particles and exhibited a low compaction resistance. In the syngenetic-early diagenetic stage, owing to the effects of the overlying water and sediment, the mica and organic matter were oriented and stratified well (Fig. 4.77). The parts with high contents of clastic particles could resist compaction, and the detrital particles exhibited concave and convex contacts with mud (Fig. 4.77e). Mica, with a large particle size and soft texture, exhibited notable obvious extrusion deformation (Fig. 4.77f). Under the action of compaction, fine-grained sediments in shale underwent rapid dehydration in the first stage of early diagenesis, resulting in a substantial decrease in the pore water and excess interlayer water (Jiang 2003).

2. Cementation

Cementation mainly includes siliceous cementation, carbonate cementation and pyrite formation. Silicic cementation is characterized by microcrystalline quartz and lamellar cements, and microcrystalline quartz is symbiotic with clay minerals and organic matter (Fig. 4.78a). Cementation is related to the removal of interlaminar water by smectite under thermal action (70–90 °C) and can also be considered the dissolution process of smectite (Thyberg and Jahren 2011). This process likely occurred in early diagenetic stage B. Lamellar siliceous cements are microcrystalline quartz aggregates that are unevenly distributed in interstitial fillers (Fig. 4.78b). These materials were likely formed during the continuous removal of interlayer water by smectite and kaolinite under thermal action (above 90–100 °C) (Thyberg and Jahren 2011), that is, in the process of the transformation of illite from a chaotic mixed layer to an orderly mixed layer to a non-mixed layer. The transformation of clay minerals produced a large amount of silica (Thyberg and Jahren 2011). Boles and Franks (1979) noted that 5% quartz can be produced if the original rock contains 25% smectite. Notably, the transformation of illite, chloritization of kaolinite and kaolinization of feldspar may produce a certain amount of silica. Although the content of secondary quartz cannot be measured, it represents an important component of shale that affects the diagenetic analysis of shale and physical properties of the shale reservoirs. The compositional analysis of silica in the black rock series indicated high contents of Al^{3+}

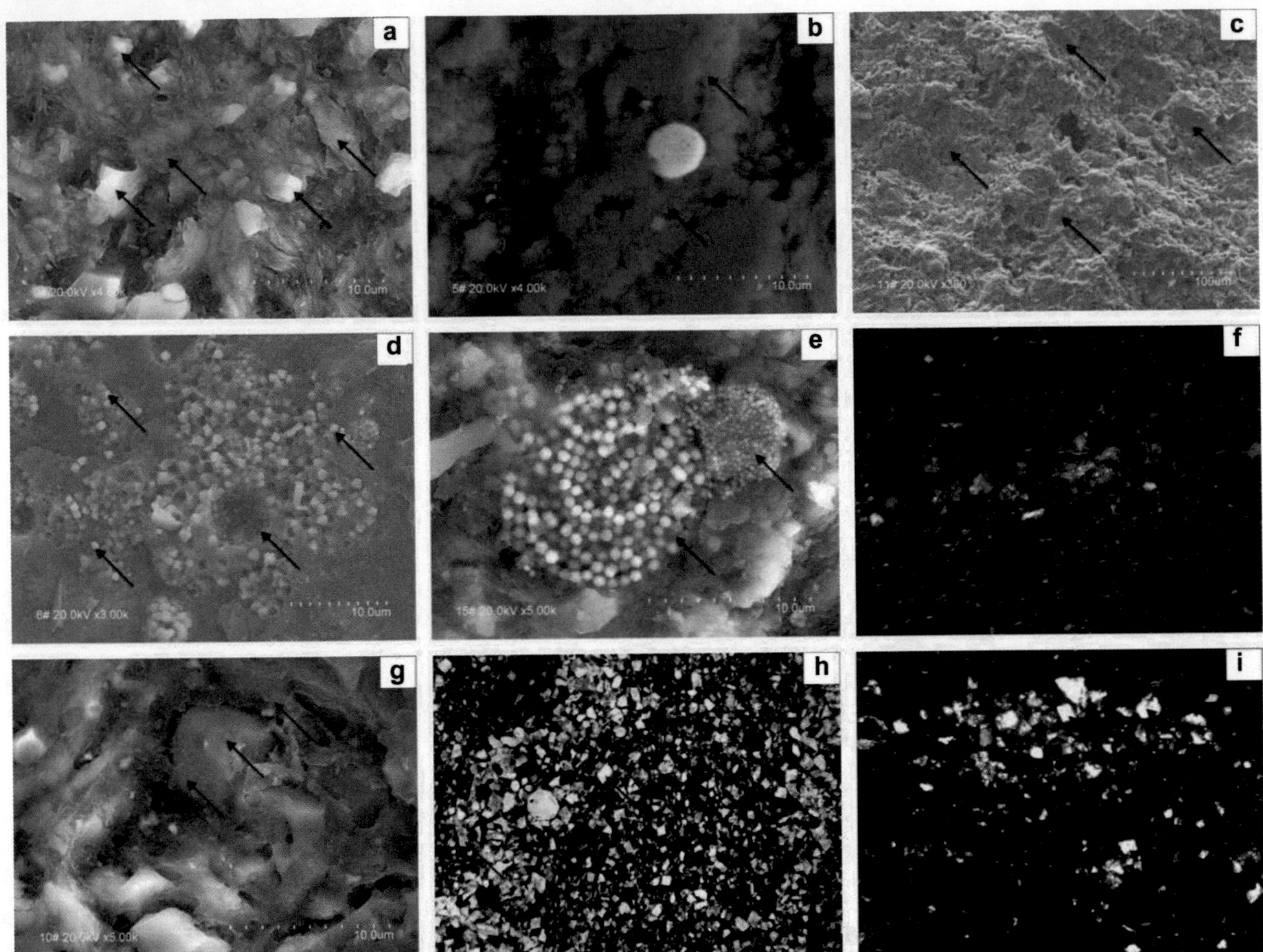

Fig. 4.78 Cementation of black rocks in the Longmaxi Formation in the Southern Sichuan Basin and its periphery **a** microcrystalline quartz (red arrow) symbiotic with Iran–Mongolian mixed clay minerals (blue arrow); Nanchuan Sanquan, Chongqing. **b** lamellar siliceous cement (red arrow)-coated pyrite (blue arrow); Wulong Huangcao. Chongqing. **c** detrital quartz particles; Renhuai Zhongshu, Guizhou Province. **d** Scattered pyrite; Tianquan Dajingping, Sichuan Province. **e** pyrite aggregate; Qijiang Guanyinqiao, Chongqing. **f** early carbonate cements; Tongzi Hanjiadian, Guizhou Province; orthogonal polarization. **g** early dolomite cementation (red arrow), lateral flake clay minerals (blue arrow); Nanchuan Sanquan, Chongqing. **h** rhomboid dolomite crystal; Ebian Xihe, Sichuan Province; single polarized light. **i** cementation and metasomatism of quartz (blue arrow) from carbonate minerals (red arrow); Xuyong Macheng, Sichuan Province; orthogonal polarization

and K^+ (Table 4.24), consistent with secondary genesis. Siliceous cements are indistinguishable under polarized light microscopy but can be distinguished from quartz particles by their distinct grain boundaries and uniform surface lacking clastic quartz particles (Fig. 4.78c) when observed through backscatter electron microscopy.

The secondary clay minerals are mainly ordered and interbedded in the Iran–Mongolian mixed layer, with illite as the secondary material and a small amount of clastic alteration chlorite. Kaolinite, smectite, a large amount of chlorite and part of illite are terrigenous clastics. The formation of clay cements was closely related to siliceous cements, which are products of the transformation process of clay minerals, and intergranular pores and interlayer micropores of clay minerals were simultaneously produced. Under the influence of compaction in the syngenetic and early diagenetic stages,

Table 4.24 Mineral composition of quartz in black rocks of Lower Longmaxi Formation in the Southern Sichuan Basin and its periphery

Sample number	Sample name	SiO_2	Al_2O_3	Fe_2O_3	FeO	CaO	MgO	K_2O	Na_2O	TiO_2
HCP-B12	Silicoide	97.87	1.51	–	–	–	–	0.53	–	–
HJDP-B3	Silicoide	91.69	3.04	–	–	–	–	0.81	–	–
XMP-B7	Silicoide	92.86	1.93	–	–	–	–	0.48	–	–

the amount of pore water considerably decreased, and siliceous and clay minerals were cemented in situ and often symbiotic (Fig. 4.78a).

In the black rock series of the Longmaxi Formation in the study area, pyrite exhibits a dispersed distribution (Fig. 4.78d) or a globular aggregation distribution (Fig. 4.78e) with a defined crystal shape. Moreover, a small amount of gypsum is observed, indicating that the sedimentary water was saltier and more alkaline than normal seawater, and the sedimentary environment was a strong reducing environment under the lower part of the stratified water and sedimentary interface (Li et al., 2012a; Qin et al. 2010). The pore water of the original sediments was rich in Ca^{2+} and CO_3^{2-}, and early carbonate cements were easily formed in the syngenic-early diagenetic stage, with high content and suspended particles (Fig. 4.78f, g), consistent with the characteristics of early carbonate cements described by Lee and Lim (2008).

During the late diagenetic process, as organic acids were reduced and consumed, the formation fluid became alkaline (Wang et al. 2013c), and cementation and metasomatism of the carbonate minerals occurred. Due to the influence of continuous strong compaction, the porosity and permeability of shale in the late diagenetic stage were extremely low, leading to the limited formation of carbonate cements in this stage; mainly dolomite with fine crystal shapes was produced in the form of fracture filling (Fig. 4.78h).

When observed through polarized light microscopy, the early carbonate cements (Fig. 4.78f) were evenly distributed around the sediments or in symbiosis with the argillite, distinguishing them from the carbonate minerals formed by metasomatism in the later stages (Fig. 4.78g). Therefore, the carbonate cements in the black rock series of the Longmaxi Formation in the study area corresponded to at least two periods and were mainly syngenetic and early diagenetic.

3. Dissolution and metasomatism

The dissolution of feldspar, typically in the form of partial dissolution, was notable. An irregular dissolution edge could be observed, or dissolution occurred along cleavage fractures. The dissolution of carbonate minerals was not obvious. Wang et al. (2012b, c) speculated that the presence of unstable mineral dissolution pores may be the result of freshwater leaching after uplifting of the formation, and the acidic dissolution in the organic-rich shale members of the Longmaxi Formation in the Southern Sichuan Basin did not develop during diagenesis. Notably, unstable mineral dissolution pores could be commonly observed in several outcrop samples, although only a limited number could be observed in the drilling samples. The dissolution of potash feldspar was related to the transformation of clay minerals. The formation of secondary illite consumed K^+ and promoted the dissolution of potash feldspar (Wang et al. 2013c). Therefore, the dissolution of potash feldspar was more developed than that of plagioclase. The dissolution of feldspar mainly produced certain intragranular dissolved pores (Fig. 4.79a, b). In addition, the alkaline dissolution of quartz particles could be observed, resulting in irregular and bayed erosion edges (Fig. 4.79c) and a few internally dissolved pores (Fig. 4.79d), which influence the characteristics of shale gas reservoirs. The diagenetic fluid was mainly alkaline due to the dissolution of feldspar and quartz.

Quartz and clay minerals in the black rock series of Longmaxi Formation in the study area were mostly metasomatic by carbonate to varying degrees (Figs. 4.78i and 4.79c), indicating a correlation between metasomatism and the alkaline diagenetic fluid. The widespread development and high content of carbonate cements, clay mineral assemblage characteristics of illite + illite/smectite + chlorite (I + I/S + C) and dissolution of feldspar and quartz particles indicate the influence of the saline-alkali water medium during diagenesis.

4. Relationship and characteristics of organic and inorganic diagenesis

The different types of diagenesis in source rocks were influenced by each other. Compaction and siliceous and clay mineral cementation were related to the clay mineral transformation. The dissolution and transformation of clay minerals were closely related to the hydrocarbon evolution. Therefore, the diagenesis of shale gas reservoirs must be examined by comprehensively analyzing the inorganic and organic diagenesis processes.

The dehydration and transformation of clay minerals were associated with the maturation and hydrocarbon generation of organic matter (Jiang 2003). The research results of many petroliferous basins show that the period of rapid transformation of smectite into illite was the most favorable period for oil and gas migration and accumulation, likely because the transformation of clay minerals played a catalytic role in organic gas generation at depths of 1000–22,800 m. In this context, smectite had the strongest catalytic effect, which could increase the organic hydrocarbon yield by 2–3 times and decrease the pyrolysis temperature by 50 °C (Luo et al. 2001). In addition, dehydration, which refers to the process of shrinking of mineral particles and increase in the number of pores, facilitated the discharge of free water and hydrocarbons from shales, thereby providing a channel for the initial migration of oil and gas (Jiang 2003).

The organic matter of the black rock series of the lower Member of the Longmaxi Formation in Southern Sichuan and its periphery appeared in four forms: (1) around clastic

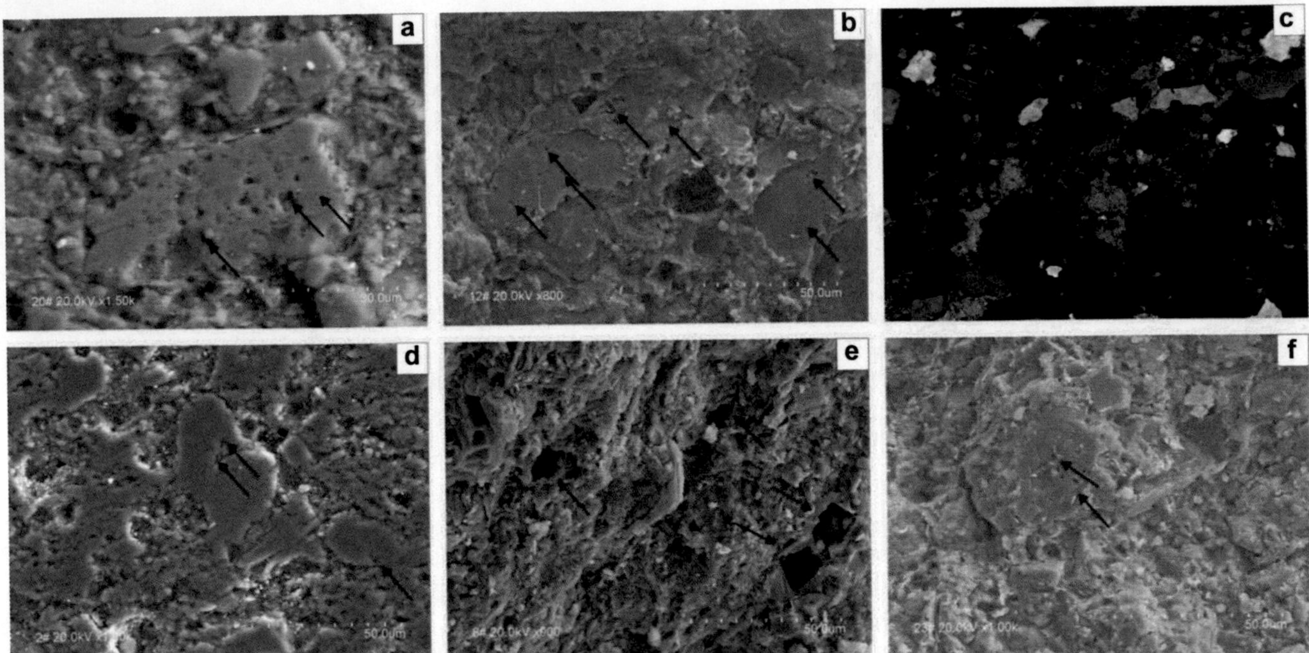

Fig. 4.79 Dissolution of black rocks in the Longmaxi Formation in the Southern Sichuan Basin and its periphery **a** dissolved potassium feldspar (red arrow), intragranular dissolved pores (blue arrow); Xuyong Macheng, Sichuan Province. **b** dissolved potassium feldspar and plagioclase (red arrow), intragranular dissolved pores (blue arrow); Renhuai Zhongshu, Guizhou Province. **c** quartz dissolution, margin estuary (red arrow), and incomplete metasomatism by carbonate (blue arrow); Han Yuan Jiaodingshan, Sichuan Province; orthogonal polarization. **d** dissolution of quartz particles, dissolution pores (red arrow), harbor dissolution margin; Shizhu Qiliao Chongqing. **e** rhomboid dissolution pores (red arrow), formed by dissolution of carbonate crystalline grains; Tianquan Dajingping, Sichuan Province. **f** partial dissolution of calcite, intragranular dissolved pores (red arrow); Yanjin Yinchangba, Yunnan Province

particles and symbiotic with clay minerals (Figs. 4.77a–c and 4.80a, d); (2) in the filling veinlets (Fig. 4.80c, b) and symbiotic with the discontinuous filling cryptocrystalline quartz and illite (Fig. 4.80e); (3) in quartz and calcite cements or in clastic grains (Zhang et al. 2013b) (Fig. 4.80f); and (4) in the cleavage cracks of primary calcite and muscovite. The first type of occurrence was the most developed. The hydrocarbon generation evolution of organic matter had two main effects on the water–rock reaction. Mu and Zhang (1994) believed that organic matter continuously generated organic acids during the whole evolution process, and CO_2 generated by the decarboxylation of organic acids controlled the pH of the aqueous solutions such that it was conducive to dissolution. The accumulation of hydrocarbons in the pores effectively inhibited diagenesis because hydrocarbons are a mixture of hydrocarbon compounds and cannot dissolve and precipitate inorganic salts that constitute authigenic minerals (Luo et al. 2001). The continuous expulsion of hydrocarbons during diagenesis promoted the water–rock reaction in shale gas reservoirs. However, after the cessation of hydrocarbon expulsion, the pores were filled with a large number of hydrocarbons, resulting in the termination of the water–rock reaction.

The decarboxylation of organic matter in the diagenetic process was expected to produce a large number of organic acids, resulting in extensive dissolution of aluminosilicate minerals and carbonate minerals. However, extensive dissolution was not observed for the black rock series of the Longmaxi Formation, likely because of the following aspects. First, the acidity was neutralized by the saline and alkaline primary pore water. Second, under the strong influence of hydrocarbon generation and compaction as well as the first-stage dehydration of clay minerals, the acidic fluid was rapidly discharged and not allowed to act on the soluble minerals in the source rocks. Third, the production of organic acids decreased in the middle-diagenetic stage, and the decomposition of binary organic acids and unary organic acids occurred, resulting in a higher partial pressure of CO_2. Fourth, the organic acids continued to be discharged outward due to the second-stage dehydration of clay minerals. Therefore, due to the lack of sufficient action time, the organic acids and soluble substances were not prone to widespread and strong corrosion. The higher partial pressure of CO_2 and discharge of organic acids produced cracks in the shale.

The organic diagenetic process, that is, the hydrocarbon evolution process, is an irreversible maturation process with a specific sequence. The evolution of organic matter can be inverted, and the influence on the diagenetic environment can be analyzed considering the characteristics of the present organic matter. The residual organic carbon content

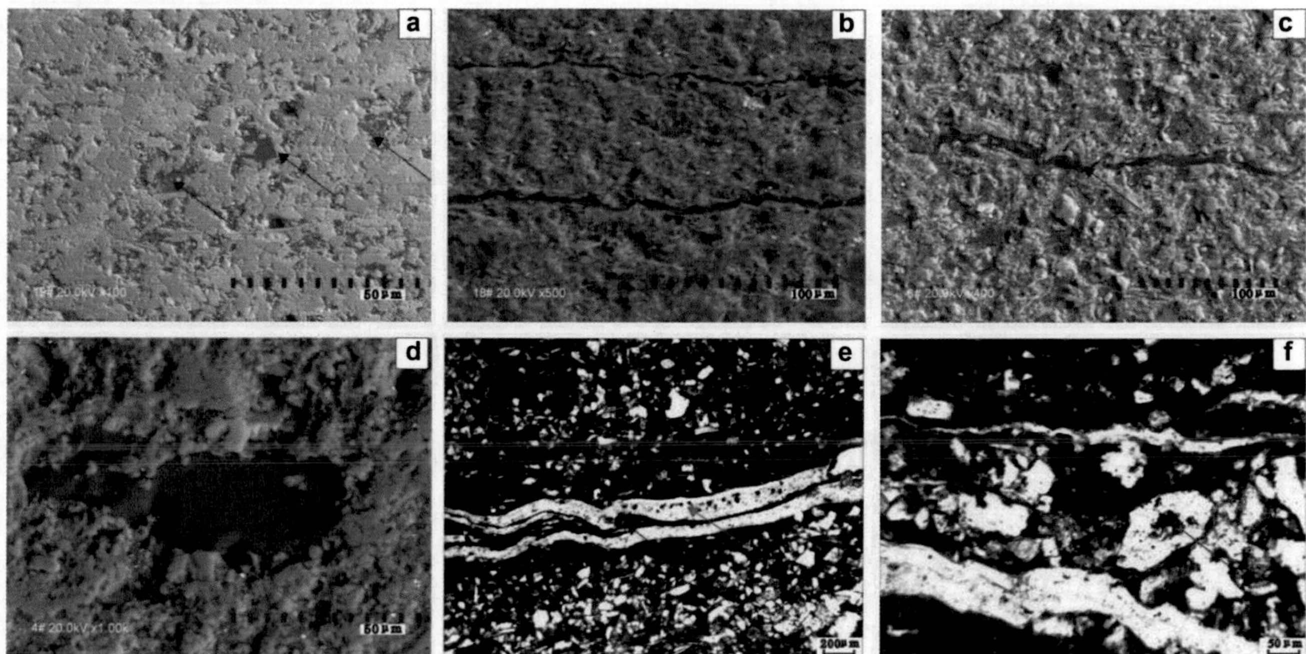

Fig. 4.80 Characteristics of organic matter in black rocks of Longmaxi Formation in the Southern Sichuan Basin and its periphery **a** organic matter symbiotic with clay minerals; Shizhu Qiliao, Chongqing. **b** organic aggregate distributed along the fissure; Wulong Huangcao, Chongqing. **c** organic matter aggregate distributed in thin strips; Wulong Huangcao, Chongqing. **d** organic matter (red arrow) symbiotic with clay minerals, organic matter with micropores (green arrow) (BSE); Wulong Huangcao, Chongqing. **e** organic matter (red arrow) filled in silica (green arrow) cracks. Xuyong Mcheng, Sichuan Province (DPG). **f** organic matter filled in the dissolution cavity and dissolution edge. Xuyong Mcheng, Sichuan Province (DPG)

(TOC) of the black rock series of the Longmaxi Formation in the Southern Sichuan Basin and its periphery ranged between 0.01 and 8.06%, with most values lying between 0.5 and 4.0%. Most of the organic matter was type I kerogen, followed by type II. The organic carbon content decreased from the bottom to the top, and the kerogen changed from type I to type II (Fig. 4.81). The interior of the Sichuan Basin was characterized by long-term shallow burial in the early stage, long-term uplift in the early and middle stages, secondary burial depth in the middle stage and rapid uplift in the late stage. Marine shale in the Sichuan Basin underwent long-term structural and thermal evolution and is characterized by a complex evolutionary history, high thermal maturity and early hydrocarbon generation time (Nie et al. 2012b) (Fig. 4.82). The Longmaxi Formation in the study area was in a low maturity stage (Ro: 0.5–0.7%) at the end of the early Permian, reached the peak of hydrocarbon generation (Ro: 0.9–1.1%) at the end of the Triassic, and entered the wet-condensate stage (Ro: 1.1–1.3%) in the early Jurassic. At present, this formation is in the late overmature stage, and all liquid hydrocarbons have decomposed into dry gas (Huang et al. 2012b).

4.6.2 Diagenetic Sequence

1. Classification of the diagenetic stage

The vitrinite reflectance (Ro) and maximum pyrolysis peak temperature (T_{max}), as important reference indices of the organic maturity, can reflect the degree of diagenetic evolution. The crystallinity of illite, a clay-like mineral formed spontaneously in the process of sedimentary diagenesis, increased irreversibly with increasing diagenesis degree and temperature (Liu et al. 2009b). Based on the criteria for the classification of black shale maturation stages in Southern China (Nie et al. 2012b), characteristics of organic matter evolution and hydrocarbon generation models (Zhang et al. 1999), and sedimentary and diagenetic environment characteristics of Longmaxi Formation in the study area, the diagenetic stages were divided according to the standards of the oil and gas industry (SY/T5477-2003). To this end, the indictors included the maturity of organic matter, highest pyrolysis peak temperature of organic matter (T_{max}), and crystallinity of the diagenetic minerals and illite in the black rock series of the target layer.

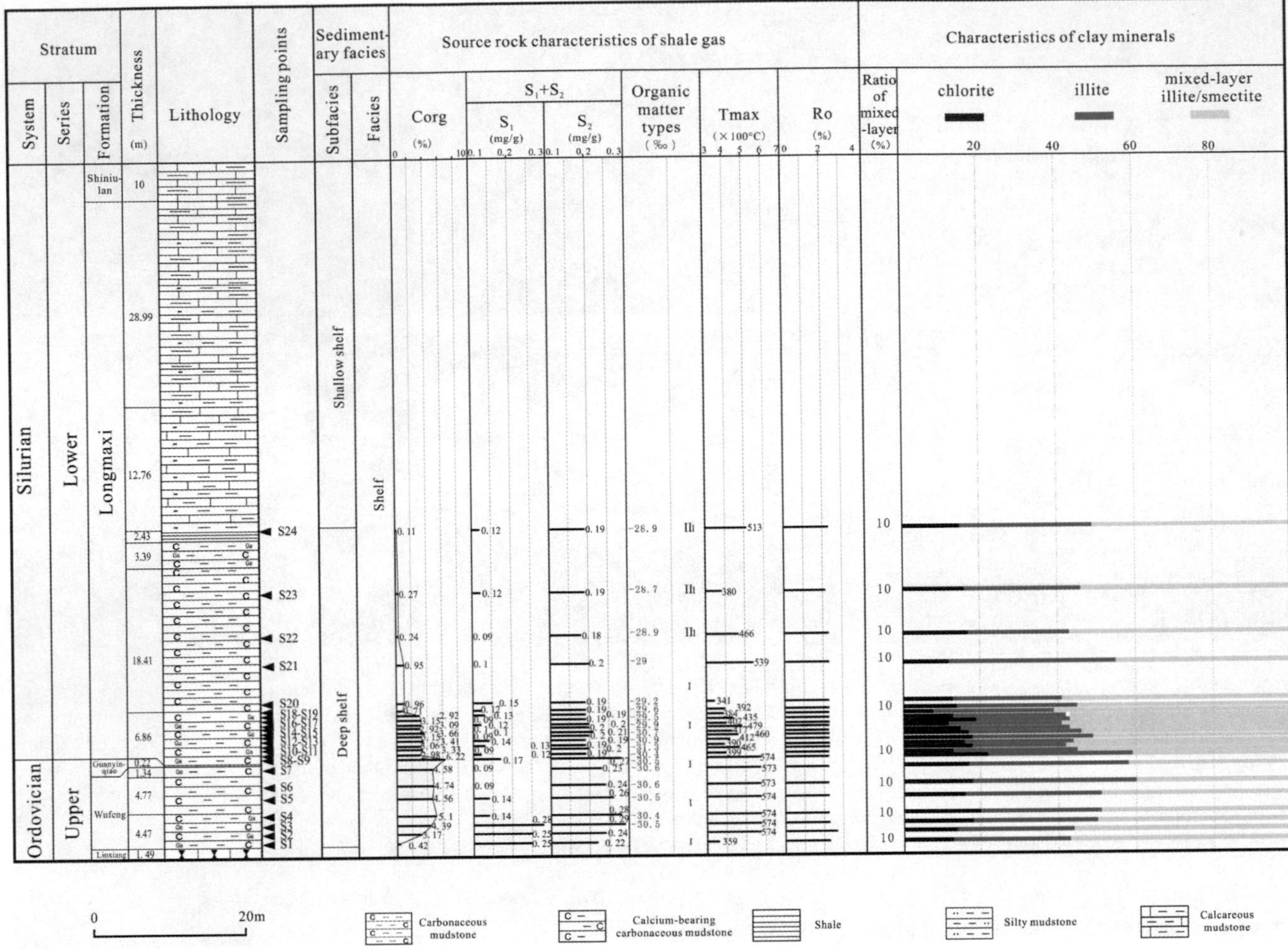

Fig. 4.81 Comprehensive geological histogram of the Longmaxi Formation in Macheng village of Xuyong County, Sichuan Province

The classification basis can be summarized as follows: (1) According to the evolution history of the Sichuan Basin, the burial depth of the black shale of the Longmaxi Formation in the study area was the largest after the Cretaceous, exceeding 5000 m (Zeng et al. 2011). The paleotemperature gradient for the Silurian to Jurassic periods mainly ranged from 2.5 °C/hm to 3.5 °C/hm (Wang and Liu 2011), indicating that the maximum paleotemperature was over 170 °C, and the diagenetic evolution had reached the late diagenetic stage. (2) The equivalent vitrinite reflectance of organic matter in the black shale of the Longmaxi Formation ranged from 1.63 to 4.54%, with most values lying between 2 and 2.5% (Fig. 4.81; Table 4.25), indicating that the organic matter in the black shale of the Longmaxi Formation was in the high-overmature stage, equivalent to the late diagenetic stage, and several layers had undergone low-level metamorphism. (3) The maximum pyrolysis peak temperature (T_{max}) of organic matter had a wide range of 360–571 °C, with most values lying in the range of 360–450 °C, with an average value of 456 °C. These values could be attributed to the catalysis of clay minerals and bentonite strata, indicating that the organic matter evolved to an overmature stage, corresponding to the late diagenetic stage (Table 4.25). Two vertical thermal peaks (Fig. 4.81) were observed, indicating that the thermal fluid movement was episodic. (4) Chlorite + illite/illite (C + I/S + I) was the dominant clay mineral assemblage in the black shale of the Longmaxi Formation, followed by kaolinite + chlorite + illite/illite (K + C + I/S + I). Several areas were characterized by illite/smectite + illite (I/S + I), chlorite + illite (C + I) and all illite (I). The characteristics of the clay mineral assemblage indicated that the diagenetic evolution had reached the late diagenetic stage. The content of smectite (S) in I/S was 0–15%, mainly 5% or 10% (Fig. 4.81; Table 4.25), indicating that the diagenetic evolution in this area had at least entered middle-diagenetic stage B, and the late diagenetic stage had been attained in several areas. The features of the clay mineral content combinations indicated that the black rock series of the Longmaxi Formation in Southern Sichuan and its periphery had evolved to the late diagenetic stage. (5) The illite crystallinity of clay minerals in the black rock series of the Longmaxi Formation ranged from 0.39 to 0.69° $\Delta 2\theta$ (Table 4.25), and the crystallinity for approximately 81% of

Table 4.25 Division of diagenesis stages of black rocks in Longmaxi Formation of the Southern Sichuan Basin and its periphery

Division marks	Number of samples (pieces)	Distribution range/Primary type	Average value/Minor type	The highest diagenetic stage
Vitrinite reflectance (Ro)	132	1.63–4.54%	2.61%	Low-grade deterioration stage
Highest peak pyrolysis (T_{max})	95	360–571 °C	456 °C	Late diagenetic stage
Clay mineral combination	368	(C + I/S + I)	(K + C + I/S + I);(I/S + I); (C + I);(I)	Late diagenetic stage
Smectite (s) Content ratio	368	0–15%	Dominated by 10%	Late diagenetic stage
Crystallinity (Illite)	26	0.39–0.69° Δ2θ	Mostly > 0.42° Δ2θ (Sample number of 21)	Dominated by the late diagenetic stage A, and the local strata reached early phase in the Late diagenetic stage B

C: Chlorite; I: illite; K: Kaolinite; S: Smectite

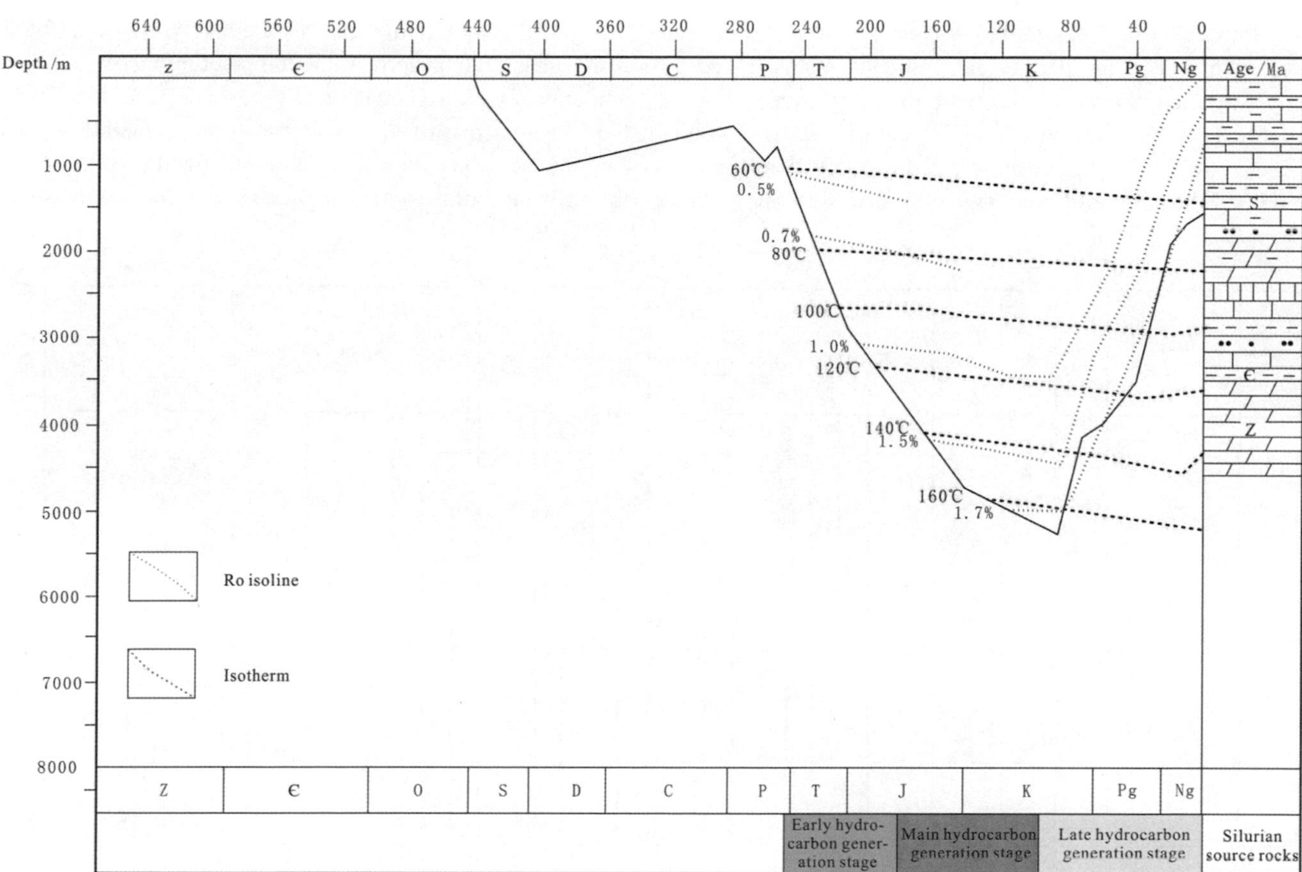

Fig. 4.82 Sedimentary evolution of the sediments in the Sichuan Basin (rectified by Zeng et al. 2011)

illite was more than 0.42° $\Delta 2\theta$. According to Merriman (2005)'s criterion for the classification of organic maturity and diagenetic evolution stage based on illite crystallinity, the diagenetic evolution had reached late diagenetic stage A. In several areas, the strata had reached the early stage of late diagenetic stage B but had not entered the lower metamorphic stage. In conclusion, the diagenetic evolution of the black rock series of the Longmaxi Formation in the study area had reached the late diagenetic stage.

2. Diagenetic evolution and diagenetic sequence

The inorganic and organic diagenesis processes in shale gas reservoirs influence one another. Therefore, the diagenetic sequence and evolution characteristics of the black rock series of the Longmaxi Formation in Southern Sichuan and its periphery were determined (Fig. 4.83) according to the diagenetic mineral assemblage and diagenetic phenomena. The organic matter evolution characteristics, mineral response relations and characteristics of the diagenetic environment were comprehensively considered. The black rock series of the Longmaxi Formation in the study area are favorable source rocks (Su et al. 2007), rich in organic matter and widely developed pyrite and formed in deep shelf facies of anaerobic-anoxic-reducing environments (Li et al. 2012a; Qin et al. 2010). As mentioned, the sedimentary water was saltier and more alkaline than ordinary seawater (Qin et al. 2010). According to the boron content in clay minerals, the sedimentary water of the Longmaxi Formation in Southeastern Chongqing was α-medium saline and polysaline (Li et al. 2012a) and vertically characterized by upper light and lower salinity, with the paleosalinity ranging from 6 to 49%, with an average value of 22.5%. The diagenetic fluid was influenced by sedimentary water, which rendered the original pore water salty and alkaline.

(1) Cogenesis stage

The sediments were loose and included abundant primary pores. The chemical properties of pore water were derived from those of the bottom water that was rich in Ca^{2+}, Na^{+}, Mg^{2+} and other metal ions as well as reduced Fe^{2+} and Mn^{2+}. The diagenetic environment remained an anoxic reductive environment, forming a small amount of siderite, micritic microgranular calcite and powdery and strawberry pyrite. Silicate minerals such as quartz were weakly dissolved in the alkaline medium. Earthy yellow and purple bentonite were commonly observed in the black rock series of the Longmaxi Formation in the study area. A small amount of smectite was likely formed from bentonite under marine hydrolysis in the syngenetic stage.

(2) Early diagenesis stage A

Organic matter was immature under a paleotemperature lower than 65 °C. Organic matter was selectively dissolved by the active anaerobic bacteria as the temperature increased. Small amounts of CH_4, CO_2, H_2S and H_2O were generated. Part of the CO_2 was dissolved in water to form a small amount of carbonic acid. The diagenetic fluid remained alkaline. Under the influence of strong compaction, the rock was weakly

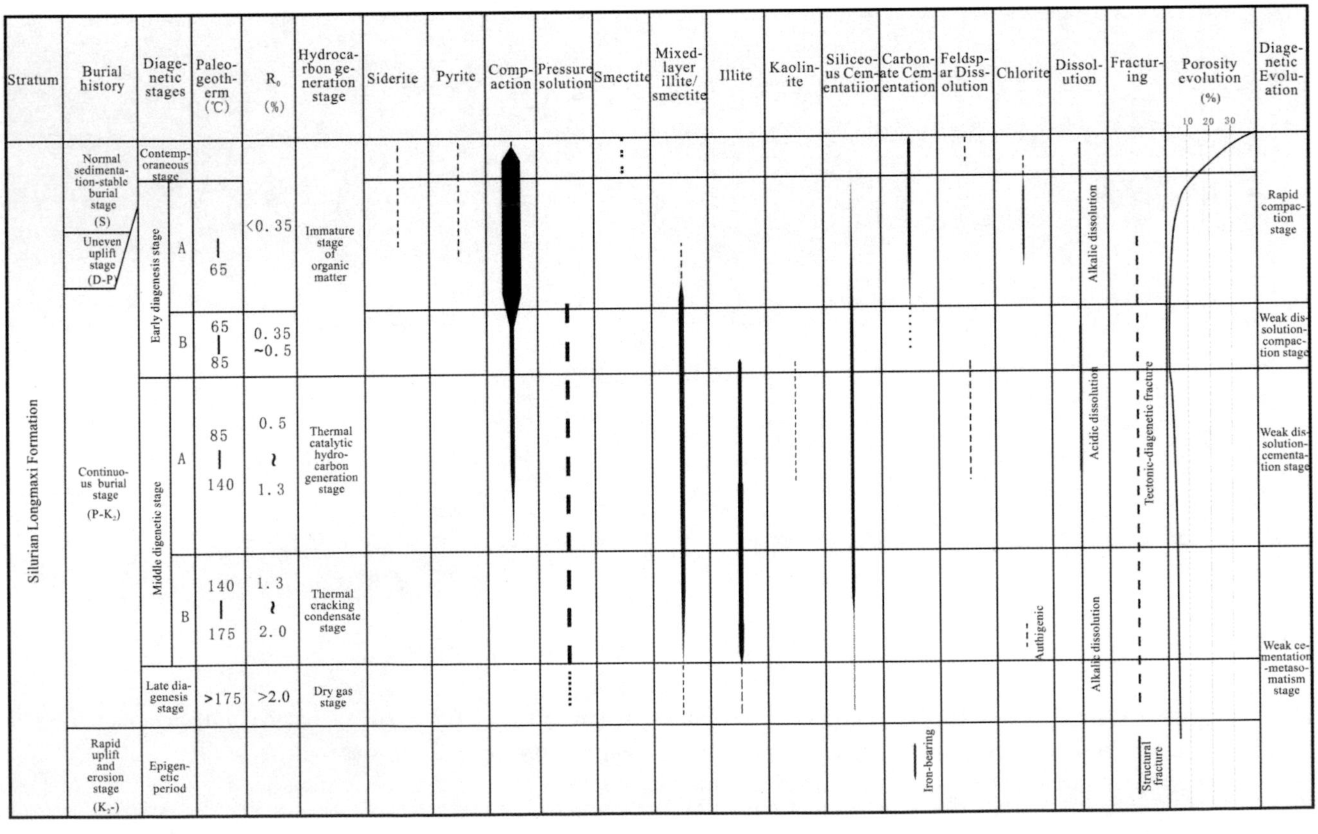

Fig. 4.83 Diagenetic sequence and evolution of black rocks in Longmaxi Formation of Southern Sichuan Basin and its periphery

consolidated and semiconsolidated, resulting in the rapid decrease in the number of primary intergranular pores. A small amount of siderite and pyrite formed continuously, and microcrystalline calcite-cemented under the action of the alkaline medium-rich in Fe^{2+}. Clay minerals were mainly smectite, containing a certain amount of kaolinite (unevenly distributed). Aluminosilicate minerals such as kaolinite, smectite and feldspar underwent spontaneous chloritization in the water rich in Fe^{2+} and Mg^{2+} (Wang et al. 2013c). The secondary chlorite was mostly leaf-shaped because of the sufficient growth space in this stage.

(3) Early diagenesis stage B

The organic matter was semi-mature and still in the biochemical generation stage. A few carboxylic anions were generated because of the carboxylic acids released from the kerogen as the temperature increased, which partly reduced the alkalinity of the diagenetic fluid. Under the influence of continuous strong compaction, the rock gradually changed from a semiconsolidated to consolidated state, resulting in a continuous decrease in the primary intergranular pores. Smectite removed the interlayer water and transformed into the Iran–Mongolian mixed layer. The Iran–Mongolian mixed-layer minerals belonged to the disordered mixed-layer zone because of the 50–70% smectite content. The silica formed several small quartz crystals that migrated and cemented in situ or in close proximity. Secondary enlarged edges of silica cements began to form. However, due to the production of a large amount of biochemical gas, the resulting displacement pressure caused the fluid in the primary pores to be expelled. Consequently, the water–rock reaction was weaker than that in early diagenetic stage A. The alkalinity of the diagenetic fluid gradually decreased with the carbonic acid accumulation, carboxylic acid release and OH^- consumption, and the fluid likely transformed from weakly alkaline to neutral to weakly acidic.

(4) Middle-diagenetic stage A

The rock was completely consolidated, and the primary intergranular pores were almost completely microporous. The burial depth of the sediments exceeded 150–2500 m, and organic matter entered the stage of thermocatalytic hydrocarbon generation, that is, the oil generation window. With the temperature and clay minerals acting as catalysts, a large amount of carboxylic acids was released from kerogen, forming a large amount of water-soluble monobinary short-chain organic acids. The organic acids were continuously generated and accumulated and neutralized with OH^- in the formation fluid, transforming the diagenetic fluid from alkaline to weakly acidic. A considerable amount of hydrocarbons remained in the source rocks in this stage, which decreased the available open pores and limited the fluid flow. Because types I and II kerogens were dominated by liquid hydrocarbons in the oil generation window (Ro = 0.5–1.0%), only a small amount of gaseous hydrocarbons was present, and the molecular diameter of the liquid hydrocarbons was larger than the pore throat diameter, which limited the hydrocarbon expulsion (Liu et al. 2011). Due to the inhibition of authigenic mineral generation, metasomatism and transformation of minerals in inorganic diagenesis (Luo et al. 2001); the expulsion of pore water by strong hydrocarbon generation; and compaction-induced critical damage to shale gas reservoirs, the water–rock reactions were weak, and authigenic minerals were not generated. In contrast, the acid-soluble components such as feldspar and carbonate were not susceptible to the action of acidic media, resulting in weak dissolution and a lack of dissolution of secondary pores.

Affected by the temperature, smectite continuously removed the interlayer water and transformed it into the Iran–Mongolian mixed layer. The smectite content in the Iran–Mongolian mixed-layer minerals was 15–50%, indicating that disordered mixed-layer minerals gradually changed into ordered minerals. Autochthonous silica cementation formed during the transformation of smectite into illite and was characterized by symbiosis with secondary clay minerals.

(5) Middle-diagenetic stage B

The shale gas reservoir exhibited ultralow porosity and low permeability under the influence of compaction and cementation. The burial depth exceeded 3500–4000 m, and the organic matter entered the pyrolysis stage to produce condensate gas, forming condensate oil and moisture gas. The carboxylic acid deacidified and decomposed to form CH_4 and CO_2 under the action of high temperature and high pressure. The formation of the acidic medium was terminated, and the medium was partly consumed by the dissolution in middle-diagenetic stage A, causing the diagenetic fluid to gradually become weakly alkaline. Metal ions such as Na^+, Ca^{2+}, Fe^{2+} and Mg^{2+} entered the pore solution under the action of dissolution and produced anhydrite and other minerals. Authrogenic chlorite was rarely formed because of the limited amount of dissolved aluminosilicate minerals. Andolomite and ferric calcite were not formed in large quantities because carbonate minerals were prone to dissolution at higher temperatures and greater depths (Merriman 2005; Yan et al. 2010b). No additional smectite was observed, and the contents of secondary minerals of illite and the Iran–Mongolian mixed layer increased. The smectite content in the Iran–Mongolian mixed-layer minerals was lower than 15%, indicating that the mixed-layer minerals were completely ordered, and part of the mixed-layer minerals was converted to illite.

(6) Late diagenetic stage

The rocks were extremely dense, and the organic matter evolved to an overmature stage. The black rock series of the Longmaxi Formation in the study area exhibited the early stage of overmaturity. Specifically, a large amount of dry gas was generated, and the water-rock reaction was terminated. The Iran–Mongolian mixed-layer minerals almost disappeared under a sufficient K+ supply. The clay mineral assemblage was illite + chlorite (I+C).

The K^+ and Al^{3+} contents in the formation fluids considerably influenced the transformation of smectite to illite (Sui et al. 2007). The organic matter of the Longmaxi Formation in the study area evolved into the late diagenetic stage; however, the mineral characteristics indicated that the Longmaxi Formation had not entered the late diagenetic stage due to the high content of Iran–Mongolian mixed-layer minerals in most areas with Smectite content ratios of 10% and 5%. The reason could be the insufficient K^+ supply within and outside the shale gas reservoir. A small amount of K^+ formed because of the weak water–rock reaction and incomplete dissolution of potassium feldspar in the process of hydrocarbon generation. At present, the potassium feldspar content is within 20%. Moreover, the pore water was expelled due to the displacement pressure during hydrocarbon generation, and the K^+ in the pore water could not participate in the water–rock reaction, thereby contributing insufficiently to the transformation of smectite to illite. In general, the K^+ required for the transformation of smectite to illite in shale may be derived from the albitization and metasomatism of sandstone in open systems (Barbera et al. 2011). However, the black rock series of the Longmaxi Formation did not involve any directly contacting sandstone layers, and the diagenetic and accumulation system was relatively closed, rendering it challenging to obtain abundant K^+ replenishment. Therefore, the transformation of smectite to illite was not complete in most of the black rock series of the Longmaxi Formation in the study area. According to the results of whole-rock chemical analyses (Table 4.24), the K_2O content varied irregularly with the increase in the burial depth, suggesting the lack of an external source of K^+ (Liu et al. 2012a). However, the Hanyuan and Luding areas in Sichuan Province near the Central Sichuan Uplift have a large supply of clastic materials (Liu et al. 2012a). In the diagenetic process, part of the K^+ in the sandstone likely entered the clay mineral transformation process of the source rock. A large amount of smectite transformed to illite, and the clay mineral assemblage was mainly illite + chlorite (I + C) and illite (I). Owing to the low concentration of K^+, the transformation of kaolinite to illite was inhibited in the alkaline diagenetic environment with a palaeogeothermal of more than 130 °C, and kaolinite was preserved.

In general, the black rock series of the Longmaxi Formation in Southern Sichuan and its periphery mainly experienced four stages of diagenetic evolution (Fig. 4.83), according to the characteristics of the diagenetic environment and burial history of the strata (Huang et al. 2011). (1) The long-term shallow burial in the early stage and rapid compaction in the initial phase of the secondary burial stage corresponded to the stage of syngenic to early diagenetic A. (2) The weak alkalinity and weak dissolution and compaction in the early phase of the secondary burial stage occurred in early diagenetic stage B. (3) The weakly alkaline and weakly acidic dissolution and cementation in the middle phase of the secondary burial stage corresponded to the stage of thermocatalytic hydrocarbon generation in middle-diagenetic stage A. After the termination of the water–rock reaction, the formation pressure was released in the process of uplifting the strata. A large number of microcracks formed due to tectonic action. A large number of free hydrocarbons and partially adsorbed hydrocarbons were dispersed along the fractures. The residual Ca^{2+}, CO^{3-}, Fe^{2+} and Mg^{2+} in the fluid formed ferric calcite and andolomite when the paleotemperature decreased to less than 105–1130 °C. The diagenetic evolution of mesodiagenetic stage B and later stages corresponded to the fourth evolution stage, that is, (4) the stage with weak alkalinity, weak cementation and metasomatism between the late burial stage and the rapid uplift stage. In addition, the measured samples were mainly extracted from outcrop sections in the field, and a small amount of smectite was found in several sections, attributable to the K^+ removal of illite during the process of superformation after stratigraphic uplift (Khormali and Abtahi 2003).

4.6.3 Impact of Diagenesis on Shale Gas

The organic diagenetic evolution in diagenesis influenced the hydrocarbon characteristics and abundance of shale gas.

The diagenetic evolution of the black rock series of the Silurian Longmaxi Formation in Southern Sichuan and its periphery reached late diagenetic stage A, with highly mature organic matter and a large amount of dry gas. A large number of pores were generated during the hydrocarbon generation of organic matter, which provided valuable reservoir space for shale gas. The diagenetic evolution of inorganic minerals also influenced the reservoir space of shale gas. Therefore, the influence of diagenesis on shale gas was mainly reflected in the development of the shale gas reservoir space.

1. Influence of inorganic diagenesis on the shale gas reservoirs

Shale reservoir, as a special type of oil and gas reservoir, is characterized by extremely low porosity and permeability and diverse types of reservoir spaces (Wang et al. 2012b). Diagenesis influences the shale gas reservoir properties. Strong compaction was the main reason for the loss of primary pores and induction of ultralow porosity and low permeability. Moreover, with the development of early cementation, the primary pores almost disappeared. The transformation of clay minerals in shale gas reservoirs was representative of highly developed diagenesis, and the development of siliceous and clay mineral cements was closely related to these processes. The cementation of carbonate, siliceous and clay minerals and pyrite led to the development of intergranular pores and interlayer micropores, which enhanced the storage properties of shale gas reservoirs. Due to the mechanical compaction and cementation in the early stage, inhibition of the hydrocarbon generation and influence of saline-alkali water medium in the diagenetic process, the development of late dissolution was highly limited. Several micron-scale dissolution pores with low connectivity could be observed in the feldspar and carbonate minerals. These secondary pores could partly enhance the storage properties of the reservoirs. The metasomatism of carbonate minerals in late diagenesis decreased the number of micropores in clay minerals and clogged the throats, resulting in a further decrease in the shale gas reservoir porosity.

The brittleness of shale formation considerably influences the microfractures and hydraulic fracturing, and the content of brittle minerals affects the pore formation (Wang et al. 2012b). The contents of quartz and carbonate are the material basis for the brittleness and fracture development of shale. Under the same stress, shale with a high content of carbonate minerals and silica is strongly brittle and prone to fracture (Long et al. 2012). The content of brittle minerals, such as siliceous minerals, increased in the Longmaxi Formation from the top to the bottom. The amount of siliceous matter was positively correlated with the content of residual organic carbon. The abundance of organic matter and brittle minerals represented an important geological factor for the enrichment and high yield of shale gas in Longmaxi Formation. The reticular fractures formed by multistage bedding slip and tectonism in the lower part of the Longmaxi Formation provided the reservoir conditions for shale gas enrichment and high yield (Guo and Liu 2013a). Therefore, compaction and cementation in the early stage of diagenesis caused shale to become brittle, which enhanced the shale gas reservoir properties.

In general, compaction and cementation resulted in the loss of primary pores. The dissolution and transformation of clay minerals produced partial dissolution pores and interlayer micropores, which enhanced the shale gas reservoir. The diagenesis mainly occurred in early diagenetic stage B to middle-diagenetic stage A. Inorganic diagenesis resulted in brittle transition of shale, which promoted the formation of microfractures and hydraulic fracturing effects in the process of exploitation and had a favorable influence on the shale gas reservoirs.

2. Influence of organic diagenesis on shale gas reservoirs

Shale gas reservoirs represent source rocks. The evolution of organic matter is a key aspect of shale gas reservoirs and influences the physical properties. The transformation of organic matter in the process of hydrocarbon generation was the main cause of porosity changes (Jarvie 1991). Shale with an organic matter content of 7%, accounting for 14% by volume, could increase the porosity of shale by 4.9% in the process of hydrocarbon generation, assuming that 35% organic carbon was consumed (Liu et al. 2011). Mastalerz et al. (2013) examined New Albany shale and indicated that with the maturation of organic matter during diagenesis, the porosity was considerably decreased owing to the inorganic diagenesis. In the late maturation stage of organic matter, the early pores were filled with oil or solid bitumen, thereby decreasing the available open pores and limiting the fluid flow. As the thermal evolution progressed, the porosity increased as oil and bitumen converted to dry gas, resulting in microcracks that opened the original pores. The formation of highly overmature shale gas was consistent with the evolution of the reservoir space. As a key component of shale reservoir space (Wang et al. 2012b), the organic pores mainly included those related to biological fossils and those generated during the thermal evolution of organic matter hydrocarbon generation. The black rock series of the Longmaxi Formation in the study area was dominated by organic matter hydrocarbon generation pores. Considering the influence of organic matter evolution on reservoir space, the organic matter hydrocarbon-generating pores were mainly formed in the process of maturation (high maturity and overmaturity) of organic matter from the middle-diagenetic stage to the late diagenetic stage, specifically, the stage with weak dissolution and cementation and stage with weak alkalinity, weak cementation and metasomatism.

3. Pore evolution process of shale gas reservoirs

Considering the diagenetic evolution characteristics and development characteristics of the reservoir space, clay minerals began to transform in early diagenetic stage B, with the removal of the interlayer water and formation of secondary interlayer micropores of clay minerals. Moreover, the organic matter gradually matured to produce carboxylic anions and introduce acidic ions into the pore water. A large number of clay minerals underwent transformation, and the organic matter reached maturity with the increasing temperature in middle-diagenetic stage A. Under the catalytic action of clay minerals, especially smectite, the organic matter underwent thermal degradation and was transformed into oil and moisture. A large number of organic acids were generated and entered the pore fluid, forming a weakly acidic diagenetic environment. These two diagenetic stages corresponded to the weak dissolution and compaction stage of diagenetic evolution and the weak dissolution and cementation stage. The latter stage corresponded to the formation of clay mineral interlayer microcracks and organic matter hydrocarbon-generating pores. Simultaneously, certain dissolution micropores were produced, which increased the reservoir space of the shale gas reservoir. In addition, microfractures, as one of the important reservoir spaces of shale gas reservoirs, are associated with a high content of brittle minerals such as quartz, feldspar and carbonate (Nie et al. 2012b). The presence of siliceous cement, produced during the clay mineral transformation, increased the brittleness of shale. In the process of late diagenesis and formation uplift, microcracks were easily generated owing to diagenesis and fracturing. The development of microfractures in shale provided sufficient storage space and migration channels for shale gas and facilitated the later development of shale gas (Long et al. 2012). The generation of carbonate minerals increased the brittleness of mud shale and provided soluble materials for acidic dissolution. However, the intergranular pores and later dissolution micropores were blocked by carbonate cements in general, which decreased the number of matrix pores. Overall, the development of shale gas reservoirs was attributable to the synergistic effect of weakly alkaline and weakly acidic dissolution in middle-diagenetic stage A, clay mineral transformation and hydrocarbon generation from organic matter in the cementation stage.

The evolution of pores in the black rock series of the Longmaxi Formation in the study area can be divided into two stages. The early stage involved abundant and rapid loss of primary pores, and the middle and late stages corresponded to the formation of secondary micropores and microfractures, associated with the syngenic-early diagenetic stage of (1) and (2) diagenetic evolution and middle-diagenetic stages of (3) and (4) diagenetic evolution. According to the characteristics of diagenetic evolution, the reservoir porosity was the lowest in early diagenetic stage B because the primary pores almost disappeared owing to the action of compaction and cementation, and the transformation of dissolution and clay minerals was weak. With the maturation of organic matter and transformation of smectite into illite, the number of clay mineral interlayer microcracks and organic matter hydrocarbon-generating pores gradually increased after the evolution entered middle-diagenetic stage A. Some micropores were produced by weak dissolution, and some diagenetic fractures were produced under the simultaneous action of high temperatures and pressures, resulting in a gradual increase in the porosity. In the later uplifting process of the formation, certain tectonic microfractures were produced under the influence of tectonic action, which enhanced the physical properties of shale gas reservoirs to a certain extent. The developed shale reservoir exhibited an ultralow porosity and permeability.

4.7 Influence of Sedimentary Phase on Shale Gas

The developed layers of shale gas are all source rocks, and the sedimentary environment is the main factor controlling the development of source rocks, which not only controls the abundance of organic matter, but also affects the quality of organic matter, and can comprehensively reflect many factors such as sedimentary rate and original productivity (Shuang et al., 2008; Liu et al. 2013; Zhang et al. 2012a; Zhang et al. 2013a); Qin (2010a) especially pointed out that the sedimentary environment in Southern China is the main factor to control the organic matter of high-quality source rocks, so it is particularly important to study the paleoenvironment of the development of black and dark mud shale of Silurian Longmaxi Formation in the study area. The shelf sedimentary environment of Ordovician Wufeng Formation-Silurian Longmaxi Formation in the Upper Yangtze area is favorable for the formation of organic-rich shale, and the main evaluation criteria of shale gas include organic carbon content, brittle mineral content, organic matter maturity, reservoir physical properties and preservation conditions, etc. (Wang et al. 2012c; Hu and Xu 2012), the former two are controlled by sedimentary facies, while the latter two are related to sedimentation. Lithofacies, as one of the basic elements of shale gas research, affects the evaluation and exploration and development of shale gas. At the same time, as an important material expression of sedimentary environment, lithofacies reflects the characteristics of sedimentary environment. According to previous studies (Zeng et al. 2011; Liu et al. 2011; Liang et al. 2012a; Chen et al. 2013b; Wang et al. 2014), according to different classification standards, Wufeng Formation and Longmaxi Formation develop different lithofacies types in different areas and influenced by sedimentary environment, lithofacies has certain correlation with organic carbon content.

To sum up, the sedimentary environment of Ordovician Wufeng Formation-Silurian Longmaxi Formation in

Southern Sichuan and its periphery controls the distribution characteristics of black shale and organic carbon content. Different sedimentary environments develop different lithofacies combinations, and lithofacies division is an important means to identify sedimentary environments. There are also different lithofacies assemblages in the same sedimentary environment, which further reflects different shale gas geological conditions. Therefore, through the detailed study of sedimentary environment and lithofacies, the influence of both on the geological conditions of shale gas is given. On the basis of favorable sedimentary environment, favorable lithofacies combination is superimposed, and the evaluation of shale gas, that is, the division of favorable areas, has important guiding significance.

4.7.1 Relationship Between Lithofacies and Geological Conditions of Shale Gas

Different lithofacies have different petrophysical and mechanical properties and different organic matter content (Hickey and Henk 2007). Wang et al. (2014) pointed out that the lithofacies of Wufeng-Longmaxi Formations in Sichuan Basin has a good relationship with organic carbon content. Organic matter can occur in cleavage fractures of primary calcite and muscovite and is absorbed by clay minerals in large quantities (Zhang et al., 2013a), which also indicates that there is a certain correlation between lithofacies and organic matter. According to Li et al. (2013b), the lower limit of total organic carbon content was set as 1.0% in the evaluation of shale organic matter abundance in Southern Sichuan Basin, and the organic matter abundance was divided into four levels, TOC < 1% was poor, 1.0–2.0% was average, 2.0–4.0% was good, and > 4.0% was very good. In order to understand the relationship between rock microfacies and shale gas content, the rock microfacies types of Ordovician Wufeng Formation and Silurian Longmaxi Formation were classified and counted according to the four-level classification standard of organic matter abundance.

1. Ordovician Wufeng Formation

According to the developmental characteristics of rock microfacies, tidal flat facies are characterized by the development of mixed deposition of tidal flat. The main rock lithology is dolomitic mudstone and muddy limestone in Chuanzhong Uplift. The main rock lithology of tidal flat facies in Western Sichuan, Yunnan-Guizhou uplift peripheral area is carbonaceous silty micritic limestone and calcareous siltstone. Due to the water body is shallow around the edge of the uplift, the organic matter content is significantly lower, mostly less than 0.5%. Only a few samples contain more than 0.5% organic carbon content, but none exceed 1%. There are two sets of rock assemblages in shallow shelf facies. In the area of Ebian and Mabian of Western Sichuan, the main rock assemblages are carbonaceous siltstone and carbonaceous mudstone containing dolomite and siltstone, and silty mudstone developed in some areas. The organic carbon content of these microfacies rock assemblages is generally between 1 and 2%, normally less than 3%.

In the area of Southern Sichuan and Yanjin-Weixin-Dejiang-Yinjiang of Northern Guizhou, the main rock assemblages are silty calcareous carbonaceous mudstone and carbonaceous silty mudstone, with a small amount of carbonaceous calcareous mudstone and silty carbonaceous mudstone. The organic carbon content of these microfacies rock assemblages is generally between 0.5 and 2% and no more than 2%. There are also three types of rock assemblages in the deep-water shelf facies. In the Luding-Hanyuan area of Western Sichuan, the rock assemblages contain dolomitic siliceous mudstone and carbonaceous siliceous mudstone. The organic carbon content of carbonaceous siliceous mudstone is 2.60–2.86%, which is generally more than 2%, and the organic carbon content of dolomitic siliceous mudstone is less than 0.5%. The rock assemblages developed in Pengshui-Huangcao-Shizhu area in Southeastern Chongqing are carbonaceous mudstone and siliceous mudstone. The organic carbon content of carbonaceous mudstone is generally more than 2%, and most of the organic carbon content is 3–6%. In Changning and Xuyong area, the main rock assemblages are calcareous carbonaceous silty mudstone and silty carbonaceous calcareous mudstone. The organic carbon content of the two rock types is high, ranging from 2.04 to 4.74%.

The rock assemblage with TOC > 4.0% is mainly carbonaceous siliceous mudstone, carbonaceous silty mudstone and carbonaceous calcareous argillaceous siltstone, followed by carbonaceous calcareous silty mudstone and carbonaceous calcareous argillaceous siltstone, in which carbonaceous siliceous mudstone has the highest organic carbon content (Table 4.26). The rock assemblage with TOC of 2–4% is mainly carbonaceous siliceous mudstone, silty carbonaceous mudstone, carbonaceous mudstone, carbonaceous silty mudstone, silty carbonaceous calcareous mudstone and calcareous carbonaceous argillaceous siltstone (Table 4.27). The rock assemblage with TOC of 1–2% is mainly carbonaceous dolomitic siliceous mudstone, carbonaceous siliceous mudstone, carbonaceous silty mudstone, followed by carbonaceous dolomitic siliceous mudstone, carbonaceous dolomitic micritic limestone and carbonaceous silty calcarenite due to increase of dolomite or calcite content (Table 4.28). The main types of rock with TOC < 1% are carbonaceous and dolomitic siliceous mudstone, silty calcareous mudstone, silty mudstone and

Table 4.26 Mineral Shale TOC of the Wufeng Formation in Southern Sichuan and its periphery > 4% of the lithofacies

No	Sample	Thin section naming	Terrigenous clast (%)	Siliceous + feldspar (%)	Carbonate minerals	TOC (%)
1	QGP-S1	Carbonaceous silty mudstone	26	8	–	6.73
2	QGP-S2	Carbonaceous silty mudstone	25	8	–	5.02
3	QGP-S3	Carbonaceous silty mudstone	26	8	–	5.53
4	HCP-B3	Carbonaceous silty mudstone	7	30	–	3.51
5	XMP-S3	Calciferous carbonaceous silty sandy mudstone	26	8	13	4.39
6	XMP-S4	Calciferous carbonaceous silty sandy mudstone	52	7	13	5.1
7	XMP-S6	Calciferous carbonaceous silty sandy mudstone	53	7	13	4.74
8	XMP-S7	Calciferous carbonaceous silty sandy mudstone	53	7	13	4.58
9	WDP-S5	Carbonaceous silty mudstone	26	7	–	4.56
10	WDP-S6	Carbonaceous silty mudstone	26	7	–	4.4
11	DZP3-S3	Carbonaceous silty mudstone	26	7	–	6.7
12	MNP3-CH2	Carbonic calcareous argillaceous siltstone	52	6	26	4.84
13	SHP5-CH3	Calciferous carbonaceous silty sandy mudstone	52	6	14	4.29

Remarks: The profile code is shown in Fig. 4.5

Table 4.27 Mechanistic mud shale TOC of the Wufeng Formation in Southern Sichuan and its periphery are 2–4% lithographic type

NO	Sample	Thin Section Naming	Terrigenous Clast (%)	Siliceous + Feldspar (%)	Carbonate Minerals	TOC (%)
1	DJP-B12	Carbonaceous siliceous mudstone	7	35	–	2.69
2	DJP-B13	Carbonaceous siliceous mudstone	7	35	–	2.86
3	HJDP-B5	Dolomitic silty sandy carbonaceous mudstone	7	35	12	2.56
4	JDSP-B10	Carbonaceous siliceous mudstone	7	35	–	2.85
5	JKP-B1	Silty carbonaceous mudstone	7	8	–	3.12
6	HCP-B4	Carbonaceous siliceous mudstone	7	40	–	3.51
7	HCP-B3	Carbonaceous siliceous mudstone	7	33	–	3.51
8	XMP-S2	Silty carbonaceous cloud mudstone	12	6	26	3.17
9	SQP-B2	Silty carbonaceous mudstone	13	6	–	2.55
10	STP-B0-1	Carbonaceous mudstone	8	–	–	3.13
11	WDP-S1	Carbonaceous silty mudstone	26	7	–	2.27
12	LMRP-S1	Silty carbonaceous calcareous mudstone	14	7	25	2.87
13	DZP2-S2	Carbonaceous silty mudstone	26	7	–	3.09
14	HYP2-CH1	Silty carbonaceous mudstone	13	7	–	2.28
15	HYP3-CH2	Silty carbonaceous mudstone	13	7	–	2.04
16	MNP1-CH1	Silty carbonaceous calcareous mudstone	13	7	26	2.89
17	SHP2-CH1	Silty carbonaceous calcareous mudstone	13	7	26	3.44
18	SHP4-CH2	Silty carbonaceous calcareous mudstone	13	7	26	2.41
19	SHP6-CH4	Silty carbonaceous calcareous mudstone	53	6	13	3.87
20	LMRP1-CH1	Silty carbonaceous calcareous mudstone	13	6	25	2.71

Remarks: The profile code is shown in Fig. 4.5

Table 4.28 Mechanic mud shale-rich TOC of the Wufeng Formation in Southern Sichuan and its periphery are rock facies of 1–2%

No	Sample	Thin section naming	Terrigenous clast (%)	Siliceous + feldspar (%)	Carbonate minerals	TOC (%)
1	DJP-B4	Carbonaceous dolomitic siliceous mudstone	7	30	12	1.51
2	DJP-B16	Carbonaceous siliceous mudstone	7	30	–	1.97
3	JGP-B6	9/5000 Carbonaceous silty arenaceous limestone	12	7	57	1.16
4	XLP-B1	Carbonaceous mudstone	26	7	–	1.64
5	SBTP-B6	Carbonaceous mudstone	13	7	–	1.99
6	XMP-S5	Carbonaceous mudstone	13	7	–	1.56
7	SQP-B3	Carbonaceous and dolomitic silty mudstone	26	8	12	1.69
8	SQP-B5	9/5000 Carbonaceous and dolomitic silty mudstone	26	7	12	1.62
9	DHP-S2	Carbonaceous dolomitic mud micritic limestone	7	7	57	1.25

Remarks: The profile code is shown in Fig. 4.5

Table 4.29 Mechanistic mud shale TOC of the Wufeng Formation in Southern Sichuan and its periphery are < 1% of the lithofacies

No	Sample	Thin section naming	Terrigenous clast (%)	Siliceous + feldspar (%)	Carbonate minerals	TOC (%)
1	DHP-S1	Silty mud crystal limestone	13		57	0.41
2	DJP-B3	Dolomitic siliceous mudstone	7	30	12	0.39
3	DJP-B6	Carbonaceous and siliconeous dolomitic mudstone	7	15	27	0.59
4	DJP-B7	Carbonaceous and siliconeous cloud mudstone	7	30	12	0.6
5	DJP-B11	Siliceous micritic limestone	7	14	57	0.09
6	DJP-B30	Siliceous cloud mudstone	7	14	26	0.46
7	LQP-B1	Mudstone	9	9	–	0.01
8	XMP-S1	Silty calcareous mudstone	13	5	26	0.42
9	STBP-B3	Silty calcareous mudstone	13	5	26	0.16
10	STBP-B4	Silty calcareous mudstone	13	6	26	0.16
11	XDP-B2	Silty calcareous mudstone	26	6	–	0.94
12	XQP-S1	Silty mudstone	26	6	–	0.08
13	SQP-B3	Silty mudstone	26	6	–	0.08
14	SQP-B4	Silty mudstone	26	6	–	0.08
15	XQP-S2	Silty mudstone	26	6	–	0.2
16	DHP-S3	Carbonaceous dolomitic limestone	9	5	57	0.82
17	YYP-S1	Carbonaceous silty calcareous mudstone	12	5	26	0.54
18	DZP1-S1	Silty mudstone	26	5	–	0.04

Remarks: The profile code is shown in Fig. 4.5

mudstone and so on. (Table 4.29). With the increase of carbonate minerals and terrigenous clasts input, the organic carbon content of organic-rich shale in Wufeng Formation decreases. In general, the organic matter content of carbonaceous siliceous mudstone is the highest, followed by silty carbonaceous mudstone. The organic matter content of silty calcareous carbonaceous mudstone is also locally high, and the organic matter content of silty mudstone, siltstone, calcareous mudstone or carbonate is very low.

2. Silurian Longmaxi Formation

The organic carbon content of the lower member of Silurian Longmaxi Formation in Southern Sichuan and its periphery is high and unevenly distributed, with the average organic carbon content ranging from 0.02 to 5.37% and above 1.0% in most areas. The shale with TOC > 4.0% is mainly carbonaceous siliceous shale and carbonaceous shale with silty (calcium), followed by carbonaceous calcareous shale with silty shale. All of the rocks are shelf facies sediments, and the carbonaceous siliceous shale has the highest organic carbon content (Table 4.28). The silty carbonaceous shale and silty calcareous carbonaceous shale with TOC of 2–4% mainly developed, followed by carbonaceous siliceous shale, which is also all continental shelf sediments and obviously dominated by shallow-water continental shelf (Table 4.28). The lithofacies with TOC of 1–2% are diverse, mainly consisting of carbonaceous calcareous silty shale, carbonaceous calcareous shale, and carbonaceous argillaceous limestone or dolomite. The carbonate minerals are relatively increased, showing a small amount of tidal flat facies sediments (Table 4.28). Lithology with lower TOC is mainly characterized by calcareous silty shale, calcareous shale, limestone, mudstone, argillaceous siltstone and siltstone. The content of carbonate minerals and terrigenous clasts is high, and the sediment of tidal flat facies increases significantly (Table 4.28).

In conclusion, the organic carbon content of black organic-rich shale of the Ordovician Wufeng Formation and Silurian Longmaxi Formation decreased with the increase of carbonate minerals and terrigenous debris in the study area. In general, the organic matter content of carbonaceous siliceous shale and carbonaceous shale is the highest, followed by silty carbonaceous shale. The organic matter content of calcareous carbonaceous silty shale is also high, while the organic matter content of siltstone, calcareous shale or carbonate is very low. According to the development characteristics of lithofacies assemblages in different

Table 4.30 Lithofacies types of black rocks (TOC > 4%) of Lower Longmaxi Formation in Southern Sichuan Basin and its periphery

No	Sample	Thin section naming	Terrigenous clast (%)	Siliceous + feldspar (%)	Carbonate minerals	TOC (%)
1	XJP-B13	Carbonaceous silty sandy shale	30	72	0	4.79
2	XJP-B16	Banded clay, calcareous quartz siltstone	70	–	–	3.74
3	XJP-B17	Silty calcareous shale	15	–	–	3.68
4	XQP-B3	Carbonaceous Siliceous shale	10	83	0	5.39
5	XQP-B4	Carbonaceous Siliceous shale	10	72	0	4.87
6	JDSP-B16	Calcium-carbonaceous siliceous shale	5	–	–	11.12
7	XLP-B9	Carbonaceous silty shale	10	75	0	5.69
8	XLP-B11	Carbonaceous silty shale	10	68	0	4.86
9	QGP-B5	Carbonaceous shale	6	61	6	4.18
10	RZP-B7	Silty calcareous carbonaceous shale	13	47	10	6.73

Remarks: The profile code is shown in Fig. 4.5

sedimentary environments, it can be concluded that the shallow marine deep-water shelf facies is favorable to the enrichment of organic matter, while the tidal flat environment is not favorable to the formation of shale gas. Therefore, the study on the relationship between sedimentary facies and shale gas geological conditions of the Upper Ordovician Wufeng Formation and Lower Silurian Longmaxi Formation in Southern Sichuan and its periphery is mainly focused on the analysis of shallow marine shelf facies (Tables 4.30, 4.31, 4.32, and 4.33).

4.7.2 Relationship Between the Sedimentary Environment and Geological Conditions of Shale Gas

One of the main controlling and basic factors conducive to shale gas accumulation is the enrichment of organic matter, which is mainly related to the superposition of high biological yield and anoxic environment (Jin 2001). A higher productivity of living organisms providing the organic matter and the anoxic environment providing preservation conditions are necessary conditions for the formation of high-quality hydrocarbons source rocks, considerably influenced by the sedimentary environment (Li et al. 2008b). The sedimentary environment controls the development of organic matter and affects the characteristics of shale lithofacies and mineral composition, thereby controlling the explgoration and development of shale gas. The geochemistry of sediments indicates the nature and evolutionary information of the paleoenvironment (Chen et al. 1996), and the sedimentary environment is an important factor affecting the geochemical characteristics of sediments. Therefore, the geochemical analysis of sediments can be performed to invert the characteristics of the sedimentary environment.

1. Well Huadi No.1

(1) Total organic carbon content

The total organic carbon contents of 34 samples from the Wufeng-Longmaxi Formation are listed in Table 4.34. The contents can be divided into three categories. Across the Ordovician–Silurian boundary, the TOC contents of the black shales from the bottom of Wufeng Formation range from 2.66 to 2.73%. In the upper regions, the TOC contents gradually increase to 3%, and the TOC values in the middle and lower part of Wufeng Formation stabilize between 3 and 4.31% (avg.3.42). At the top of the formation, the values decrease to 3%, and upward, in the Guanyinqiao Formation, the TOC values vary from 2.96 to 3.2% (avg. 3.05%). Further upward, at the base of the Longmaxi Formation, the TOC rapidly returns to a high value (3.06%). At the bottom of the Longmaxi Formation TOC values range from 3.16 to 3.34% (avg. 3.23%). In the upward direction, the TOC values of Longmaxi Formation first an increase and then a decrease. At the beginning the TOC increases gradually to 4.43%, ranging from 3.62 to 4.43% (avg. 4.05%), and then decreases to approximately 3% (avg. 3.25%). During the depositional period of Wufeng Formation, black shales gradually replaced the nodular limestones of Linxiang formation in carbonate ramp environment. Fluctuations in TOC contents of Wufeng Formation show a growing trend, and at the top of this formation, they begin to fall, the fluctuations decrease. Upward, in the Guanyinqiao formation, the TOC values decrease to a stable value (2.96%). In the base of the Longmaxi Formation, the TOC increases gradually from 3 to 4.43%, and in the upward direction, the value decreases to 3% or lower. Overall, the TOC variations from the Wufeng to Longmaxi Formation exhibit a high-low–high-low tendency. The middle and lower parts of the Wufeng and

Table 4.31 Lithofacies types of black rocks (TOC = 2–4%) of Lower Longmaxi Formation in Southern Sichuan Basin and its periphery

No	Sample	Thin section naming	Terrigenous clast (%)	Siliceous + feldspar (%)	Carbonate minerals	TOC (%)
1	JDSP-B18	Dolomitic carbonaceous siliceous shale	5	71	18	2.63
2	XLP-B13	Silty carbonaceous shale	15	43	0	2.34
3	DJP-B18	Calcareous and bioclastic carbonaceous siliceous shale	10	87	7	2.88
4	DJP-B25	Siliceous shale	5	77	3	2.77
5	HCP-B12	Silty carbonaceous shale	42	77	0	3.53
6	HCP-B14	Silty carbonaceous shale	15	–	–	3.44
7	QLP-B28	Argillaceous quartz siltstone	75	–	–	2.28
8	XHP-B4	Dolomitic silty carbonaceous shale	15	45	27	2.98
9	XHP-B5	Argillaceous silty carbonaceous dolomitic limestone	30	27	35	2.96
10	XMP-B9	Silty calcareous carbonaceous shale	15	41	20	2.98
11	XMP-B10	Silty calcareous carbonaceous shale	15	40	33	3.33
12	XMP-B11	Silty calcareous carbonaceous shale	15	45	25	3.06
13	XMP-B12	Silty calcareous carbonaceous shale	15	46	25	3.41
14	XMP-B13	Silty calcareous carbonaceous shale	15	45	25	3.15
15	XMP-B14	Silty calcareous carbonaceous shale	15	45	21	3.66
16	XMP-B15	Silty calcareous carbonaceous shale	15	39	28	2.92
17	XMP-B16	Silty carbonate carbonaceous shale	13	30	29	3.09
18	XMP-B17	Silty calcareous carbonaceous shale	16	34	32	3.15
19	XMP-B18	Silty calcareous carbonaceous shale	20	30	26	2.92
20	LMRP-B4	Dolomitic calcareous carbonaceous shale	6	41	31	2.92
21	LMRP-B5	Calcareous carbonaceous shale	5	29	43	2.27

Remarks: see Fig. 4.5 for profile code

Longmaxi Formations exhibit high and stable TOC concentrations and are the most promising strata for shale gas exploration.

(2) Major, trace and rare earth elements

The major, trace and rare earth element concentrations of 34 samples from the Well Huadi No. 1 are listed in Chapter Five (Hirnantian glaciation). Certain (V, Cr, Co, Ni, U, and Mo) trace elements are considered to assess the redox conditions of depositional environments.

(3) Redox conditions

Several redox-sensitive elements, such as U, Mo, V, Ni, Co, and T, are soluble in oxic environments. In anoxic conditions, these elements are insoluble and preserved in the sediments. The content or ratio of these trace elements can be used to indicate the redox state of the ancient sea. In oxic environments, V exists as V(V) in vanadate ionic species, whereas under mildly reducing conditions, V(V) is converted to V(IV) that may be removed from the sediments when combined with the organic matter. Cr, in the form of terrigenous clastics (chromite and clay minerals), is exported to the sediments by replacing Al and Mg. Because of their stable geochemical properties, V/Cr is typically considered an index of redox conditions, along with Ni/Co and V/V + Ni (Algeo and Maynard 2004; Algeo and Lyons 2006; Tribovillard et al. 2006; Algeo and Tribovillard 2009). In general, ratios of Ni/Co > 7, 5–7, and < 5 indicate anoxic, dysoxic, and oxic environments, respectively. Ratios of V/Cr > 4.25, 2–4.25 and < 2 indicate anoxic, dysoxic and oxic environments, respectively. Ratios of V/(V + Ni) > 0.60, 0.46–0.60 and < 0.46 indicate anoxic, dysoxic and oxic environments, respectively. In the Well Huadi No. 1, the Ni/Co ratios of samples B1–B11 from the Wufeng Formation range from 1.74 to 4.31, indicating oxic conditions, and upward, the depositional environment evolves to have dysoxic-anoxic conditions, with values increasing gradually. The Ni/Co ratios of samples B12–B16 vary from 7.76 to 10.38, which implies anoxic conditions, but on the top of this formation, the ratios decrease to 5.74. The Ni/Co ratios of three samples from the Guanyinqiao Formation are low, ranging from 3.38 to 4.52, consistent with oxic

Table 4.32 Lithofacies types of black rocks (TOC = 1–2%) of Lower Longmaxi Formation in Southern Sichuan Basin and its periphery

No	Sample	Thin section naming	Terrigenous clast (%)	Siliceous + feldspar (%)	Carbonate minerals	TOC (%)
1	DJP-B17	Silty calcareous carbonaceous shale	18	–	–	1.39
2	DJP-B19	Carbonaceous calcareous siliceous shale	10	–	–	1.66
3	DJP-B21	Carbonaceous silty carbonate shale, interbedded with calcareous bands	10	57	35	1.70
4	DJP-B22	Micritic lime dolostone	5	–	–	1.38
5	DJP-B23	Micritic lime dolostone	5	43	52	1.173
6	DJP-B36	Carbonaceous silty shale	32	61	9	1.93
7	DJP-B40	Carbonaceous silty calcareous shale	30	46	42	1.75
8	GTP-S1	Carbonaceous silty shale	33	59	2	1.14
9	JGP-B5	Argillaceous micritic limestone	10	27	63	1.89
10	QGP-B6	Carbonaceous silty shale	35	53	8	1.55
11	QLP-B18	Silty carbonaceous shale	13	–	–	1.71
12	QLP-B22	Carbonaceous shale	10	49	0	1.30
13	QLP-B26	Argillaceous quartz siltstone	80	–	–	1.76
14	QLP-B27	Argillaceous quartz siltstone	80	–	–	1.95
15	RZP-B9	Silty carbonaceous carbonate shale	17	48	12	1.68
16	RZP-B10	Silty carbonaceous carbonate shale	17	49	10	2.00

Remarks: see Fig. 4.5 for profile code

Table 4.33 Lithofacies types of black rocks (TOC < 1%)of Lower Longmaxi Formation in Southern Sichuan Basin and its periphery

No	Sample	Thin section naming	Terrigenous clast (%)	Siliceous + feldspar (%)	Carbonate minerals	TOC (%)
1	XQP-B5	Carbonaceous silty shale	15	43	11	0.71
2	YYP-B2	Carbonaceous calcareous silty shale	25	46	20	0.80
3	YYP-B3	Calcareous argillaceous siltstone	60	32	20	0.06
4	YYP-B4	Calcareous shale	5	38	21	0.39
5	YYP-B6	Calcareous shale	5	35	14	0.14
6	DHP-B5	Micritic limestone	0	10	81	0.62
7	DHP-B6	Calcareous shale	5	40	44	0.10
8	DJP-B32	Lime siliceous dolostone	5	35	50	0.43
9	DJP-B35	Carbonaceous argillaceous calcareous quartz siltstone	49	–	–	0.70
10	DJP-B37	Carbonaceous argillaceous silty micritic dolomitic limestone	55	–	–	0.67
11	DJP-B39	Carbonaceous argillaceous silty micritic dolomitic limestone	20	20	63	0.46
12	GTP-S2	Carbonaceous silty shale	45	65	1	0.71
13	GTP-S3	Carbonaceous silty shale	50	64	0	0.82
14	GTP-S4	Carbonaceous silty shale	25	68	0	0.67
15	GTP-S5	Carbonaceous silty shale	50	60	0	0.56
16	HJDP-B9	Shale	5	46	8	0.14
17	JGP-B1	Calcareous shale	7	24	38	0.30
18	JGP-B3	Calcareous shale	5	26	39	0.34
19	QGP-B7	Silty shale	32	47	0	0.42

(continued)

Table 4.33 (continued)

No	Sample	Thin section naming	Terrigenous clast (%)	Siliceous + feldspar (%)	Carbonate minerals	TOC (%)
20	QGP-B8	Silty shale	15	46	4	0.11
21	QGP-B9	Silty shale	15	45	5	0.11
22	QGP-B10	Silty shale	5	35	21	0.08
23	QGP-B11	Shale	7	33	26	0.70
24	QGP-B12	Finely crystalline limestone	95	–	–	0.63
25	QLP-B16	Silty carbonaceous shale	15	–	–	0.71
26	QLP-B20	Silty argillaceous limestone	15	–	–	0.97
27	QLP-B24	Argillaceous quartz siltstone	75	–	–	0.92
28	QLP-B25	Argillaceous quartz siltstone	80	–	–	0.92
29	RZP-B13	Carbonaceous shale	10	–	–	0.46
30	RZP-B14	Silty carbonaceous shale	15	–	–	0.34
31	RZP-B15	Silty carbonaceous shale	15	49	9	0.29
32	RZP-B17	Silty shale	14	–	–	0.16
33	RZP-B20	Calcareous argillaceous siltstone	60	–	–	0.11
34	RZP-B21	Silty shale interbedded with argillaceous siltstone	35	–	–	0.34
35	SBTP-B14	Argillaceous siltstone	40	–	–	0.79
36	SBTP-B18	Argillaceous siltstone	8	–	–	0.94
37	SBTP-B20	Argillaceous siltstone	45	–	–	0.95
38	XDP-B4	Carbonaceous silty shale	15	39	8	0.11
39	XDP-B5	Calcareous siltstone	50	–	–	0.10
40	XDP-B6	Calcareous siltstone	70	–	–	0.10
41	XMP-B19	Calcareous carbonaceous silty shale	30	41	31	0.71
42	XMP-B20	Calcareous carbonaceous silty shale	30	41	19	0.96
43	XMP-B21	Argillaceous micritic limestone	0	27	32	0.95
44	XMP-B22	Argillaceous micritic limestone	0	–	–	0.24
45	XMP-B23	Argillaceous micritic limestone	0	–	–	0.27
46	XMP-B24	Micritic limestone	0	–	–	0.11
47	LMRP-B6	Silty shale	18	37	16	0.27
48	SQP-B8	Silty calcareous carbonaceous shale	20	40	11	0.91
49	SQP-B9	Calcareous carbonaceous silty shale	25	43	10	0.90

Remarks: see Fig. 4.5 for profile code

conditions. The Ni/Co ratios start to increase at the base of the Longmaxi formations (between 2.67 and 5.29), consistent with oxic-dysoxic conditions. Further upward, the Ni/Co ratio of sample B34 from the Longmaxi Formation increases to 6.32, suggesting anoxic conditions. The V/Cr ratios have trends similar to those of Ni/Co across the Well Huadi No. 1. The V/Cr values in the lower part of the Wufeng Formation range from 1.22 to 2.64 and indicate an oxic-dysoxic environment. However, on the top of the Wufeng Formation, the V/Cr values reach 4.13–11.01, the redox conditions turn dysoxic-anoxic. The V/Cr values of the Guanyinqiao Formation vary between 2.8 and 4.14 and reflect a dysoxic sedimentary environment, and the V/Cr values from the lower part of the Longmaxi Formation range from 1.48 to 3.16, indicating oxic-dysoxic conditions. The increasing V/Co ratios show that bottom water conditions shift from oxic-dysoxic to anoxic. All the V/(V + Ni) values are generally high, reflecting a dysoxic-anoxic environment. The V/(V + Ni) ratios of the Wufeng Formation (0.66–0.90) indicate an anoxic water environment. The increasing trend of V/V + Ni values indicates that the anoxic environment of the Wufeng Formation is dominant. The V/(V + Ni) ratios of the Guanyinqiao Formation (0.51–0.61, avg. 0.58) suggest dysoxic conditions. The V/(V + Ni) ratios of the black shales from the Longmaxi Formation (0.61–0.72) indicate that the Yangtze Sea generally exhibited anoxic conditions.

Table 4.34 Total organic matter, paleoproductivity, redox condition and terrigenous inputs indices of Well Huadi No.1

Sample (HDP)	B1	B2	B3	B4	B5	B6	B7	B8	B9	B10
Formation Name	Wufeng Formation									
TOC	2.66	2.73	3	3.06	3.22	3.28	3.12	3.01	3.36	3.78
Ba (ppm)	472	452	497	450	433	477	383	460	610	566
SiO_2 (wt%)	57.02	54.12	56.19	56.27	54.15	53.71	38.26	73.24	69.45	68.59
Cu (ppm)	80	67.3	96	118	122	102	76.2	97.5	77	82.1
TiO_2 (wt%)	0.769	0.726	0.761	0.725	0.693	0.684	0.482	0.407	0.619	0.609
Al_2O_3 (wt%)	15.36	15.09	15.48	15.39	14.06	14.23	10.21	8.69	13.35	9.81
Zr (ppm)	165	134	147	146	136	136	108	83.1	177	208
Th (ppm)	22.4	20.4	22.7	21.1	18.8	27.4	12.5	11.7	12.8	16.1
V (ppm)	99.1	89.9	93.8	136	120	96.8	102	94.5	116	80.9
Ni (ppm)	46.3	43.8	47.3	53.9	50.2	49.1	38.8	37.2	38.8	40.3
Co (ppm)	17.6	20.5	27.2	21.5	17.1	19.9	11.2	9.87	10.8	11.3
Ba_{bio}	471.94	451.94	496.94	449.94	432.94	476.94	382.96	459.96	609.95	565.96
Si_{bio}	1.32	0.41	0.73	0.92	2.12	1.64	1.04	19.87	10.43	15.86
Ti/Al	0.06	0.05	0.06	0.05	0.06	0.05	0.05	0.05	0.05	0.07
Zr/Al	20.29	16.77	17.94	17.92	18.27	18.05	19.98	18.06	25.04	40.05
Th/Al	2.75	2.55	2.77	2.59	2.53	3.64	2.31	2.54	1.81	3.10
V/V+Ni	0.68	0.67	0.66	0.72	0.71	0.66	0.72	0.72	0.75	0.67
Ni/Co	2.63	2.14	1.74	2.51	2.94	2.47	3.46	3.77	3.59	3.57
V/Cr	1.31	1.22	1.24	1.59	1.40	1.23	1.70	1.72	2.05	2.64

Sample (HDP)	B12	B13	B14	B15	B16	B17	B18	B19	B20	B21
	Wufeng Formation						Guanyinqiao formation			S_{11}
TOC	3.98	4.02	4.08	4.31	3	4.21	3.2	2.96	2.99	3.16
Ba (ppm)	567	545	405	274	596	568	339	531	547	569
SiO_2 (wt%)	68.55	91.18	82.35	91.57	76.47	69.01	30.09	56.61	55.44	63.69
Cu (ppm)	107	28.5	62.3	27.5	93.4	58.5	35.3	47.5	38.3	53.2
TiO_2 (wt%)	0.299	0.06	0.16	0.06	0.35	0.31	0.14	0.25	0.45	0.548
Al_2O_3 (wt%)	8.36	1.57	3.48	1.49	7.69	6.36	3.22	4.82	9.49	13.43
Zr (ppm)	88	18.1	42.8	16.2	117	90.5	50.9	71.3	222	225
Th (ppm)	9.8	1.55	3.9	1.46	9.64	9.02	5.91	7.05	17.1	17.3
V (ppm)	558	131	293	132	619	271	79.1	135	106	107
Ni (ppm)	88.5	26.8	50.7	33.3	65.9	88.9	74.6	87.2	67	69.3
Co (ppm)	9.66	3.11	6.53	3.49	6.35	15.5	22.1	19.3	17	13.1
Ba_{bio}	566.97	544.99	404.99	273.99	595.97	567.97	338.99	530.98	546.96	568.95
Si_{bio}	18.23	39.97	32.70	40.28	23.02	21.73	8.74	18.48	10.25	7.61
Ti/Al	0.04	0.05	0.05	0.04	0.05	0.05	0.05	0.06	0.05	0.05
Zr/Al	19.88	21.78	23.23	20.54	28.74	26.88	29.86	27.94	44.19	31.65
Th/Al	2.21	1.86	2.12	1.85	2.37	2.68	3.47	2.76	3.40	2.43
V/V+Ni	0.86	0.83	0.85	0.80	0.90	0.75	0.51	0.61	0.61	0.61
Ni/Co	9.16	8.62	7.76	9.54	10.38	5.74	3.38	4.52	3.94	5.29
V/Cr	9.82	11.01	10.93	4.13	8.29	5.47	4.14	4.10	2.80	2.14

(continued)

Table 4.34 (continued)

Sample 样品 (HDP)	B22	B23	B24	B25	B26	B27	B28	B29	B30	B31	B32	B33	B34
Formation name	Longmaxi formation												
TOC	3.20	3.34	4.30	4.23	4.43	3.69	3.89	4.23	4.06	3.62	3.39	3.34	3.02
Ba (ppm)	516	537	547	496	517	564	657	1043	540	541	897	646	567
SiO_2 (wt%)	64.54	49.45	53.36	56.3	65.72	59.6	50.91	42.09	62.85	64.25	51.52	58.53	48.02
Cu (ppm)	47.4	33.7	67.8	56.3	39.6	53.2	69.1	26.2	40.6	44.5	28.4	53.7	51.9
TiO_2 (wt%)	0.579	0.495	0.601	0.621	0.581	0.575	0.546	0.448	0.587	0.619	0.442	0.556	0.557
Al_2O_3 (wt%)	10.78	13.64	12.93	11.55	11.15	11.12	10.47	8.88	11.32	12.14	8.14	11.43	11.84
Zr (ppm)	243	226	201	225	208	189	171	179	244	212	211	203	151
Th (ppm)	15.8	21.4	18.7	16	15.7	16.6	14	13.9	16.8	17.1	12.6	17.2	15.5
V (ppm)	126	94.4	121	101	111	110	102	63.5	112	121	85.9	129	225
Ni (ppm)	61.4	48.7	73.5	56.4	46	54.7	50.4	34.6	55.7	59.9	51.3	50.4	94.8
Co (ppm)	12.3	11.1	14.9	14.8	14.3	14.3	14.3	10.3	14.3	15.3	11.5	18.9	15
Ba_{bio}	515.96	536.94	546.95	495.95	516.95	563.95	656.96	1042.96	539.95	540.95	896.97	645.95	566.95
Si_{bio}	12.37	0.62	3.61	7.26	12.31	9.50	6.52	5.02	10.69	10.00	10.64	8.49	2.92
Ti/Al	0.06	0.04	0.05	0.06	0.06	0.06	0.06	0.06	0.06	0.06	0.06	0.06	0.05
Zr/Al	42.58	31.30	29.36	36.80	35.24	32.10	30.85	38.08	40.71	32.99	48.96	33.55	24.09
Th/Al	2.77	2.96	2.73	2.62	2.66	2.82	2.53	2.96	2.80	2.66	2.92	2.84	2.47
V/V+Ni	0.67	0.66	0.62	0.64	0.71	0.67	0.67	0.65	0.67	0.67	0.63	0.72	0.70
Ni/Co	4.99	4.39	4.93	3.81	3.22	3.83	3.52	3.36	3.90	3.92	4.46	2.67	6.32
V/Cr	2.31	2.07	1.93	1.61	2.11	1.75	1.68	1.48	1.84	1.81	2.13	1.99	3.16

The increasing tendency of the V/(V + Ni) ratios from the lower part of the Longmaxi Formation is similar to that of the Wufeng Formation, and the anoxic environment of the black shale in the lower part of the Longmaxi Formation is expected to be dominant. The redox indices (V/V + Ni, V/Cr, and Ni/Co) indicate that the water experienced a dysoxic-anoxic process during the sedimentary period of the Wufeng Formation in the Well Huadi No. 1. The subsequent low values of the redox indices are suggestive of an oxic-dysoxic environment in the Guanyinqiao Formation. For the black shales of the lower part of the Longmaxi Formation, the bottom water conditions shift from oxic-dysoxic to anoxic.

(4) Paleoproductivity

(i) Copper

Coper can be buried and precipitated when combined with organic matter, or it may form organic complexes. High levels of Cu reflect high organic matter inputs and high paleoproductivity. The Cu content (27.5–126.00 μg/g; avg. 83.65 μg/g) of the Wufeng Formation in the Well Huadi No. 1 is high; the content first increases and then decreases from the bottom. The Cu content of the Guanyinqiao Formation is 35.3–47.5 μg/g, significantly lower than that of the Wufeng Formation. The Cu content of the black shale from the Lower Longmaxi Formation is 26.2–69.10 μg/g, higher than that of the Guanyinqiao Formation but lower than that of the Wufeng Formation.

(ii) Biogenic Silica content (Si_{bio})

Biogenic silica refers to chemically determined amorphous silicon content, also known as biological opal or opal, which is formed in surface seawater by siliceous plankton such as diatoms and radiolarians via photosynthesis. The enrichment of biogenic silica on the seafloor is related to the primary productivity of the overlying water and typically used as an indicator of marine productivity. Si_{bio} is determined as the difference of the total Si concentration in the sediment and the Si associated with terrigenous material:

$Si_{bio} = Si_{sample}\text{-}Al_{sample}\ (Si/Al)_{PAAS}$, where Si_{sample} and Al_{sample} are the total contents of Si and Al in the measured rock samples, respectively. PAAS is the post-Archean average shale (Taylor and McLennan 1985), and the value of $(Si/Al)_{PAAS}$ is 3.11.

According to the vertical distribution of biogenic silica in the Well Huadi No. 1, Si_{bio} of samples B1–B7 at the bottom of the Wufeng Formation ranges from 0.41 to 2.12%, indicating high productivity. From sample B8, Si_{bio} suddenly increases to 19.84%. The Si_{bio} content exhibits a wide range,

5.57–40.28%, with a median value of 22.77% to the top of the Wufeng Formation. The Si_{bio} content of the Guanyinqiao Formation ranges from 8.74% to 18.48% (avg. 12.49%). The Si_{bio} contents of the lower part of the Longmaxi Formation show significant variability (2.92–12.37%), except for sample B13 with a value of 0.62%, and the average value is 8.23%. Fluctuations in the Si_{bio} content reflect that the Wufeng Formation has the highest paleoproductivity, and the productivity of the Guanyinqiao Formation significantly decreases. Although the average Si_{bio} from the Longmaxi Formation is lower than that of the Guanyinqiao Formation, the Si_{bio} contents of many samples of the Longmaxi Formation are higher than those of the Guanyinqiao Formation, and thus, the productivity of the Longmaxi Formation is generally higher than that of the Guanyinqiao Formation and lower than that of the Wufeng Formation.

(iii) Biogenic barium content (Ba_{bio})

Barium is a key index to evaluate the paleoproductivity. The organic matter produces SO_4^{2-} ions during the decomposition process, which react with Ba^{2+} in water to form $BaSO_4$ and are preserved in the sediments. The two main sources of Ba are biogenesis and terrigenous material. Biogenic barium is closely related to the marine productivity (Yan et al. 2009c). Ba_{bio} is determined as the difference in the total Ba concentration in the sediment and the Ba associated with terrigenous material.

The Ba_{bio} content (273.99–609.95 ppm; avg. 487.49 ppm) of the Wufeng Formation in the Well Huadi No. 1 corresponds to a high productivity. The value increases from the bottom to the top and reflects increasing productivity. The Ba_{bio} content of the Guanyinqiao Formation varies from 338.99 ppm to 546.956 ppm with an average value of 472.31 ppm, which suggests a lower productivity than that of the Wufeng Formation. The Ba_{bio} content of the Longmaxi Formation ranges from 495.95 ppm to 1042.96 ppm (avg. 616.88 ppm), reflecting the highest productivity in the three formations. However, two aspects must be considered. First, the redox environment of water affects the content of biogenic barium (Schoepfer et al. 2015; Shen et al. 2015; Li 2017): When the water body is in the oxidation state, barium sulfate is in a saturated state, and thus, Ba can be well-preserved. However, when the water is in the reduction state, the sulfate is easily reduced by sulfide bacteria, leading to the decomposition of barium sulfate and influencing the content of biogenic barium (Schoepfer et al. 2015; Shen et al. 2015).

The redox conditions of the black shales from the Wufeng-Longmaxi Formations examined in this stage are dysoxic-anoxic, and thus, the sediment productivity of the Wufeng-Longmaxi Formations cannot be reflected by the content of biogenic barium.

According to the Cu, biogenic silica and biogenic barium contents, the productivity of the Wufeng Formation in the Well Huadi No. 1 is the highest and increases gradually from the bottom to the top, while the paleoproductivity of the Guanyinqiao Formation is significantly lower than that of the Wufeng Formation. The productivity of black shale in the lower Longmaxi Formation is higher than that of the Guanyinqiao Formation but lower than that of the Wufeng Formation.

(5) Terrigenous inputs

Most trace elements in the sediments originate from terrigenous detrital influx and autogenic sources. Terrigenous detritus is the main source of sediments and considerably influences the sedimentary environment. Before they transform into sedimentary rocks, terrigenous clasts are influenced by the transport of atmospheric and running water and weathering. Among the four elements Al, Ti, Th and Zr, Al exhibits stable geochemical properties during transportation and weathering, while the other three elements are mostly concentrated in coarse terrigenous debris (Lézin et al. 2013). Therefore, we use the ratios of Al with other elements as an indicator of terrigenous inputs (Tribovillard et al. 2006; Lézin et al. 2013; Young and Nesbitt 1998).

The Ti/Al ratios of 34 samples from the Wufeng-Longmaxi Formations vary in a narrow range from 0.04 to 0.07. The Ti/Al ratios of the Wufeng Formation change slightly, with an average of 0.05, and the average values of the three samples from the Guanyinqiao Formation are 0.55, 0.06 higher than that of the Longmaxi Formation and slightly higher than that of the Wufeng Formation. The Zr/Al and Th/Al ratios vary considerably from the bottom to the top. The Zr/Al (avg. 22.24) and Th/Al (avg. 2.47) ratios of the Wufeng Formation are 16.77–40.05 and 1.81–3.64, respectively. The Zr/Al (avg. 34.00) and Th/Al (avg. 3.21) ratios of the Guanyinqiao Formation are 27.94–44.19 and 2.76–3.47, respectively. The Zr/Al (avg. 34.87) and Th/Al (avg. 2.73) ratios of the lower part of the Longmaxi Formation are 24.09–48.96 and 2.43–2.96, respectively. According to the ratios of Ti/Al, Zr/Al and Th/Al, the input of the terrigenous detritus of the black shales from the Wufeng-Longmaxi formations in the Well Huadi No. 1 increased owing to the continuous intensification of the Caledonian tectonic movement. Specifically, as the compression and collision between the Yangtze and Cathaysia blocks intensified, the Central Sichuan, Central Guizhou, and Xuefeng uplifts constantly expanded and provided more terrigenous clastic material for the ocean. Therefore, the terrigenous input in sediments increased from the end of the Late Ordovician to the early Silurian period.

2. Huangyingxiang section, Wulong District, Chongqing City

(1) Total organic carbon content

For the 15 samples from the Huangyingxiang section, the TOC contents of the middle and lower parts of the Wufeng Formation range from 3.76 to 4.85% (avg. 4.46%), suggesting that the organic matter content of this part is high. However, the TOC content begins to decrease at the top of the Wufeng Formation, and the TOC contents of the B12 and B13 samples at the top are 2.96 and 2.75%, respectively. The TOC of the Guanyinqiao Formation is 2.82%, which increases to 2.98% near the bottom of the Longmaxi Formation. The TOC of the black shales in the Longmaxi Formation increases constantly from the bottom to the top, ranging from 2.98 to 4.48%, with an average value of 3.76%. These variation characteristics are similar to those of the Huadi No. 1 well, and the TOC varies in a high-low–high manner from the Wufeng to Guanyinqiao to Longmaxi formations.

(2) Major, trace and rare earth elements

The major, trace and rare earth element concentrations of 15 samples from the Wufeng-Longmaxi Formations in the Huangyingxiang section are listed in Tables 4.35 and 7.3. For the black shales in the Wufeng Formation, SiO_2 (74.19–85.17%) and Al_2O_3 (3.9–8.73%) are the most abundant oxides. The second-most abundant oxides are Fe_2O_{3T} (0.404–1.69%), K_2O (1.08–2.52%) and CaO (0.064–0.629%). The MgO concentration in the shales is 0.278–0.717%. The Na_2O content is similar to that of TiO_2, with values of 0.126–0.543% and 0.205–0.452%, respectively. The contents of P_2O_5 and MnO are lower than 1.0%.

Only one sample, B9, in the Guanyinqiao Formation, is a biological muddy limestone. The highest content is SiO_2, 26.07%, followed by CaO (18.46%) and MgO (10.75%). The Al_2O_3 and Fe_2O_{3T} contents are similar, 4.07% and 4.84%, respectively. The contents of the remaining major elements Na_2O, P_2O_5, TiO_2, K_2O and MnO are lower than or close to 1%.

The shale samples in the Longmaxi Formation are characterized by a dominance of SiO_2 (17.26–80.58%) and Al_2O_3 (3.69–8.67%). The second-most abundant oxide is Fe_2O_3 (1.2–5.76%). The CaO and MgO contents are less than 1%, except for sample B10 (CaO, 22.13%; MgO, 12.9%). The K_2O (0.73–2.49%, avg. 1.81%) content is slightly higher than that for the Wufeng Formation. The concentrations of Na_2O, MnO, TiO_2 and P_2O_5 are lower than 1%.

The total rare earth element contents (ΣREE) of the Wufeng Formation exhibit significant variability ((68.78–211.60) × 10^{-6}), with an average value of 110.48 × 10^{-6} (Table 4.35), which is lower than that of the PAAS (184.77 × 10^{-6}; Taylor and McLennan 1985) in Table 4.36. The light rare earth element (LREE) contents ((62.93–196.33) × 10^{-6}, avg. 98.67 × 10^{-6}) are higher than those of the heavy rare earth elements (HREEs) ((5.84–25.96) × 10^{-6}, avg. 11.80 × 10^{-6}). The LREE/HREE, (LaN/YbN) and δEu ratios are 3.43–12.85, 3.26–12.60, and 0.57–0.68, respectively. According to the chondrite-normalized REE pattern diagram, the LREE and HREE contents are slightly lower than those of PAAS. All the samples exhibit similar REE distribution curves, with slight LREE enrichment and flat HREE patterns with negative Eu anomalies.

Table 4.35 Major elements of the rock samples from the Wufeng-Longmaxi Formations in the Huangyingxiang profile, Wulong District, Chongqing City

Sample	B1	B2	B3	B4	B5	B6	B7	B8	B9	B10	B11	B12	B13	B14	B15
Formation name				Wufeng formation				Guanyinqiao formation				Longmaxi formation			
SiO_2	85.07	76.26	80.91	84.95	84.08	87.3	85.17	74.19	26.07	17.26	76.40	76.60	74.48	80.58	78.26
Al_2O_3	5.11	8.73	6.92	4.13	4.59	3.93	3.9	7.83	4.07	3.69	7.51	7.00	8.67	5.64	6.70
CaO	0.064	0.078	0.085	0.085	0.073	0.144	0.119	0.629	18.46	22.13	0.20	0.62	0.34	0.20	0.16
Fe_2O_{3t}	0.425	1.11	0.873	0.805	0.498	0.404	0.607	1.69	4.84	5.76	1.68	1.70	1.71	1.20	1.79
FeO	–	–	–	–	–	–	–	–	–	–	–	–	–	–	–
K_2O	1.49	2.52	1.95	1.15	1.28	1.08	1.08	2.23	0.92	0.73	2.19	2.00	2.49	1.60	1.90
MgO	0.415	0.699	0.543	0.314	0.341	0.29	0.278	0.717	10.75	12.92	0.57	0.55	0.71	0.44	0.52
MnO	0.003	0.003	0.004	0.004	0.004	0.004	0.004	0.004	0.43	0.52	0.00	0.00	0.00	0.00	0.00
Na_2O	0.126	0.421	0.458	0.324	0.348	0.317	0.329	0.543	0.44	0.41	0.63	0.61	0.72	0.49	0.56
P_2O_5	0.025	0.031	0.029	0.018	0.019	0.014	0.018	0.036	0.10	0.11	0.05	0.08	0.09	0.05	0.09
TiO_2	0.271	0.399	0.347	0.213	0.279	0.205	0.22	0.452	0.19	0.17	0.40	0.36	0.46	0.30	0.36
LOI	6.91	9.71	7.84	7.48	8.49	6.29	7.91	11.3	33.73	36.14	10.34	10.01	10.32	9.03	9.10

Table 4.36 Rare earth elements of the rock samples from the Wufeng-Longmaxi Formations in the Huangyingxiang profile, Wulong District, Chongqing City

Sample ($\times 10^{-6}$)	B1	B2	B3	B4	B5	B6	B7	B8	B9	B10	B11	B12	B13	B14	B15
Formation name				Wufeng formation				Guanyinqiao formation				Longmaxi formation			
La	21.90	48.30	28.20	16.00	21.90	16.20	17.70	30.00	19.10	19.80	29.60	26.60	32.70	20.70	25.10
Ce	43.60	88.00	46.50	28.30	36.60	28.30	30.90	52.70	33.20	35.30	51.50	45.50	55.80	34.80	41.90
Pr	5.99	10.90	5.34	3.43	4.22	3.45	3.79	6.22	4.43	4.86	6.21	5.56	6.74	4.19	4.98
Nd	25.00	41.30	18.80	12.80	15.40	12.60	14.60	23.40	20.50	22.40	22.70	20.80	24.80	15.80	18.20
Sm	4.82	6.73	2.87	2.01	2.54	2.01	2.31	3.73	4.86	5.44	3.50	3.28	3.93	2.61	2.97
Eu	0.90	1.10	0.54	0.39	0.47	0.38	0.43	0.63	1.19	1.31	0.67	0.59	0.74	0.52	0.56
Gd	4.07	4.76	2.37	1.65	2.17	1.60	1.75	3.08	5.86	6.55	2.95	2.63	3.19	2.07	2.39
Tb	0.68	0.69	0.40	0.28	0.38	0.25	0.28	0.52	1.09	1.22	0.48	0.42	0.51	0.35	0.40
Dy	3.48	3.37	2.39	1.49	2.12	1.35	1.52	2.82	6.24	6.93	2.62	2.21	2.82	1.84	2.29
Ho	0.68	0.69	0.53	0.32	0.45	0.29	0.32	0.59	1.33	1.46	0.55	0.46	0.57	0.38	0.47
Er	2.11	2.18	1.73	0.99	1.39	0.91	0.95	1.78	3.78	4.13	1.73	1.43	1.85	1.18	1.50
Tm	0.38	0.41	0.34	0.19	0.26	0.17	0.17	0.33	0.64	0.69	0.31	0.26	0.33	0.21	0.28
Yb	2.53	2.75	2.26	1.16	1.65	1.11	1.12	2.12	3.91	4.35	2.01	1.62	2.17	1.40	1.88
Lu	0.40	0.42	0.35	0.18	0.24	0.17	0.15	0.31	0.59	0.63	0.30	0.24	0.32	0.23	0.28
$\sum$REE	116.53	211.6	112.62	69.17	89.80	68.78	75.98	128.22	106.72	115.07	125.12	111.59	136.47	86.27	103.20
LREE	102.21	196.33	102.25	62.93	81.13	62.94	69.73	116.68	83.28	89.11	114.18	102.33	124.71	78.62	93.71
HREE	14.32	15.27	10.37	6.24	8.67	5.84	6.25	11.54	23.44	25.96	10.94	9.26	11.76	7.65	9.48
LREE/HREE	7.14	12.85	9.86	10.08	9.36	10.77	11.15	10.11	3.55	3.43	10.43	11.05	10.60	10.27	9.88
LaN/YbN	6.21	12.60	8.95	9.89	9.52	10.47	11.34	10.15	3.50	3.26	10.56	11.78	10.81	10.61	9.58
Eu/Eu*	0.62	0.59	0.63	0.65	0.61	0.65	0.65	0.57	0.68	0.67	0.63	0.61	0.64	0.68	0.64

(3) Redox conditions

In the Huangyingxiang section, the Ni/Co ratios of samples B1–B8 from the Wufeng Formation range from 12.63 to 55.90, indicating anoxic conditions. Upward, the depositional environment evolves into dysoxic conditions, with the Ni/Co value decreasing sharply to 5.56 in the Guanyinqiao Formation. The Ni/Co ratio of the bottom sample B10 from the Longmaxi Formation is 5.84, which also corresponds to a dysoxic water environment. The Ni/Co ratios of samples B11–B15 vary from 20.39 to 48.36, suggestive of anoxic conditions. The V/Cr ratio of the samples in this section is relatively high, ranging from 5.07 to 12.39. According to this index, all the samples from the three formations are in anoxic environments. The average V/Cr for the Wufeng Formation is 6.46, and those of the Guanyinqiao and Longmaxi Formations are 6.05 and 10.21, respectively. Although the redox states of the three formations are anoxic, the Guanyinqiao Formation exhibits weak oxidation water conditions. Although one sample at the bottom of the Longmaxi Formation has a V/V + Ni value of 0.45, indicating an oxic environment, the black shale in the lower Longmaxi Formation is expected to be anoxic. The redox indices (V/V + Ni, V/Cr, and Ni/Co) indicate that the water experienced an anoxic-dysoxic-anoxic process from the Wufeng Formation to the lower part of the Longmaxi Formation in the Huangyingxiang section.

(4) Paleoproductivity

(i) Cooper

In the Huangyingxiang section, the Cu content of sample B9 from the Guanyinqiao Formation is 51 μg/g, significantly higher than that of the Wufeng Formation (4.42–79.3 μg/g; avg. 19.66 μg/g), and the Cu content of the black shale from the Lower Longmaxi Formation is 11.5–65.1 μg/g (avg. 25.83 μg/g), suggesting that the productivity is lower than that of the Guanyinqiao Formation. According to the surface profile, the Cu content in the Huangyingxiang section is significantly lower than that of the Huadi No. 1 well, likely because of weathering. Only one sample of the Guanyinqiao Group is derived from this section. Therefore, the Cu content does not reflect its real productivity, and the Guanyinqiao Formation has the highest productivity.

(ii) Biogenic silica content (Si_{bio})

The Si_{bio} content of the Wufeng Formation in the Huangyingxiang section ranges from 21.21 to 34.27% with an average value of 29.09%. This value suddenly decreases to 5.46% in the Guanyinqiao Formation, indicating that the productivity decreases considerably. The Si_{bio} contents in the lower part of the Longmaxi Formation range from 20.48 to 28.32%, except for sample B10, with a value of 1.98%, and the average value is 24.36%. The variation characteristics of biogenic silica content show that the productivity of the Wufeng Formation in the Huangyingxiang section is the highest. The productivity decreases in the Guanyinqiao Formation, and the productivity of the Longmaxi Formation is high.

(iii) Biogenic barium content (Ba_{bio})

The Ba_{bio} content (826.98–1201.96 ppm; avg. 940.48 ppm) of the Wufeng Formation in this section exhibits a high productivity that decreases to 638.98 ppm in the Guanyinqiao Formation, suggesting a lower productivity than that of the Wufeng Formation. The Ba_{bio} content in the Longmaxi Formation ranges from 655.98 ppm to 1306.96 ppm (avg. 1108.64 ppm), reflecting the highest productivity in the three formations. The Ba_{bio} content can change based on the redox conditions, and thus, the productivity must be comprehensively considered.

The Cu, biogenic silica and biogenic barium contents indicate that the variations in the productivity during the sedimentary period of the Wufeng-Longmaxi Formations in the Huadi No. 1 well are consistent with those of the Huangyingxiang section. The paleoproductivity of the Guanyinqiao Formation is lower than that of the Wufeng and Longmaxi Formations.

(5) Terrigenous input

The variations in the Ti/Al ratios of 15 samples from the Wufeng-Longmaxi Formations in the Huangyingxiang section are similar to those of the Huadi No. 1 well, and it ranges from 0.05 to 0.07. The Zr/Al (avg. 22.53) and Th/Al (avg. 2.20) ratios of the Wufeng Formation are 15.46–29.09 and 1.97–2.78, respectively. The Zr/Al and Th/Al ratios of the Guanyinqiao Formation are 16.34 and 2.07, respectively. The Zr/Al (avg. 14.59) and Th/Al (avg. 2.25) ratios of the lower part of the Longmaxi Formation are 13.68–17.15 and 1.97–2.52, respectively. Although the Zr/Al and Th/Al values of the Wufeng Formation are the highest, the terrigenous inputs of the Guanyinqiao Formation are the lowest. The mean values of Ti/Al, Zr/Al and Th/Al of the Wufeng Formation and Longmaxi Formation and comparable, and it can be argued that the terrigenous inputs do not increase considerably in this section (Tables 4.36, 4.37 and 4.38).

3. The relationships between the organic matter and other factors

Factors controlling the accumulation of marine organic matter include the tectonic movement, sea-level change, redox conditions, terrigenous inputs and paleoproductivity, and different researchers have different opinions on which factor has the most significant influence. Li et al. (2008b) argued that as the ice age reached its crest in the Wufeng period, intense upwelling occurred in the Yangtze area, which promoted the organic production. Black shales of the Longmaxi Formation developed in the early period of transgression. During the initial transgression, coarser-grained siliciclastics were constrained, and the bottom of the sea was an anoxic environment; thus, the organic matter was buried and preserved. Yan et al. (2008, 2009c) pointed out that the organic-rich source beds of the Wufeng-Longmaxi Formations in the Yangtze area were strongly related to the following factors: high organic matter production and burial, rapid rise in the sea level owing to the rapid increase in the temperatures and melting of glaciers between the glacial and postglacial ages, enrichment of phosphorus elements and adsorption of clay minerals during the preservation of organic matter. Chen et al. (2011b) believed that with the accelerated amalgamation between the Yangtze and Cathaysian blocks, a large epicontinental sea with subbasins (or sags) linked to a largely increased landmass developed on the Yangtze Block, forming subbasins within which the water circulation was restricted and oceanic stratification intensified. The enhanced chemical weathering likely increased the nutrient flux into the basins, thereby increasing the primary productivity, which promoted organic accumulation. Zhang et al. (2012) analyzed the sedimentary environment of the Wufeng-Longmaxi formations and concluded that the retained anoxic water environment and low deposition rate were the main factors controlling the development of black shale in the Longmaxi formation. Zhang et al. (2012) noted that the organic-rich source rocks of the Ordovician–Silurian were closely related to the oxygen content in the atmosphere, xerothermic climate and rapid rise in the sea level caused by a rapid increase in the temperatures and melting of glaciers between the glacial and postglacial ages. Zhou et al. (2011), Wang et al. (2014), Jiang (2018), Li (2017), Chen et al. (2018) and He et al. (2018) analyzed the enrichment levels of the major and trace elements and degree of pyritization and concluded that the retained anoxic water environment was the main factor controlling organic accumulation. Qiu et al. (2019) indicated that the black shales of the Wufeng-Longmaxi Formations

Table 4.37 Total organic matter, paleoproductivity, redox conditions and terrigenous input indices of the Huangyingxiang profile, Wulong District, Chongqing City

Samp-le (HYX)	B1	B2	B3	-B4	B5	B6	B7	B8	B9	B10	B11	B12	B13	B14	B15
Formation name				Wufeng formation				Guanyinqiao formation				Longmaxi formation			
TOC	4.85	4.79	4.58	4.33	4.46	3.76	2.96	2.75	2.82	2.98	3.25	3.36	4.23	4.26	4.48
Ba (ppm)	852	1202	951	827	1014	871	927	880	639	656	1168	1103	1307	1108	1310
SiO_2(wt%)	31.29	21.21	26.36	32.84	31.68	34.27	33.32	21.73	5.46	1.98	23.29	24.22	20.48	28.32	25.49
Cu (ppm)	79.3	18.5	10.1	11.5	10.3	4.42	7.22	15.9	51	65.1	11.5	29.4	18.3	15.2	15.5
TiO_2(wt%)	0.271	0.399	0.35	0.21	0.28	0.21	0.22	0.45	0.19	0.17	0.404	0.363	0.46	0.30	0.36
Al_2O_3(wt%)	5.11	8.73	6.92	4.13	4.59	3.93	3.90	7.83	4.07	3.69	7.51	7	8.67	5.64	6.70
Zr (ppm)	76.8	116	94.1	33.8	70.7	45	37	70.4	35.2	33.5	56.2	50.7	64.3	44.1	49
Th (ppm)	7.51	10.5	7.59	4.52	5.2	4.12	4.86	8.16	4.47	4.92	8.48	8.57	10.5	5.87	8.05
V (ppm)	320	435	365	196	186	201	188	288	135	148	823.00	575.00	787.00	442.00	551.00
Ni (ppm)	22.9	47.4	37.8	25.9	21.2	23.4	31.6	59.5	150	180	79.80	82.00	84.00	76.80	91.00
Co (ppm)	0.556	0.848	0.815	2.05	0.597	0.478	1.89	4.19	27	30.8	1.65	2.77	4.12	3.32	3.14
Babio	851.98	1201.96	950.97	826.98	1013.98	870.98	926.98	879.97	638.98	655.98	1167.97	1102.97	1306.96	1107.98	1309.97
Sibio	31.29	21.21	26.36	32.84	31.68	34.27	33.32	21.73	5.46	1.98	23.29	24.22	20.48	28.32	25.49
Ti/Al	0.06	0.05	0.06	0.06	0.07	0.06	0.06	0.07	0.05	0.05	0.06	0.06	0.06	0.06	0.06
Zr/Al	28.39	25.10	25.69	15.46	29.09	21.63	17.92	16.98	16.34	17.15	14.14	13.68	14.01	14.77	13.81
Th/Al	2.78	2.27	2.07	2.07	2.14	1.98	2.35	1.97	2.07	2.52	2.13	2.31	2.29	1.97	2.27
V/V + Ni	0.93	0.90	0.91	0.88	0.90	0.90	0.86	0.83	0.47	0.45	0.91	0.88	0.90	0.85	0.86
Ni/Co	41.19	55.90	46.38	12.63	35.51	48.95	16.72	14.20	5.56	5.84	48.36	29.60	20.39	23.13	28.98
V/Cr	7.98	7.47	7.31	6.53	5.44	6.22	5.68	5.07	6.05	6.46	12.39	10.81	10.98	10.47	10.17

Table 4.38 Parameters for shale gas evaluation and zone selection (Le et al. 2020)

Evaluated items	Referenced standard (%)
Organic carbon content	>2
Thermal maturity	>1.35
Brittle mineral content	>40
Clay mineral content	<40
Porosity	>2
Permeability/nD	>100
Water saturation	<45
Air content/($m^3 \cdot t^{-1}$)	>2
Depth/m	<4500
Thickness of high-quality shale/m	>20
Pressure coefficient	>1.2
Distance from denudation line/km	>7–8
Distance from fault/m	>700
Seismic data	Two-dimensional
Surface conditions	Sufficient area for deployment

were rich in algae, radiolarians, graptolites and other organisms, and the contents of barium, phosphorus, nickel, zinc and other nutrients were high, indicating the high productivity of the ocean surface, which facilitated the formation of abundant organic matter. Notably, research on the relationship between the bentonite deposits in the Wufeng-Longmaxi Formations and enrichment and preservation of organic matter in black shales is limited and no consensus exists. Li et al. (2014a) noted that the number of volcanic ash layers on the centimeter to millimetre scale was positively correlated with the thickness of organic-rich shales deposited in commonly anoxic deep-water environments. In addition, the frequent deposition of volcanic ash on the micrometer scale led to the formation of thick high-quality source rocks. Su et al. (2017) analyzed the relationship between the development characteristics of bentonite and shale reservoir quality of Wufeng-Longmaxi formation marine shales in the Fuling area and concluded that discrete volcanic eruptions likely led to the changes in the redox conditions during deposition. Moreover, the volcanic material emitted by volcanoes provided rich nutrients for seawater, promoted the development of organisms and significantly increased the productivity of the ancient ocean. The moderate volcanic activity helped enrich the organic matter and siliceous minerals.

Wu et al. (2018) analyzed the relationship between paleoproductivity, redox conditions and volcanism within an isochronous stratigraphic framework and concluded that intense and frequent volcanic activity had dual effects on organic-rich shale: Volcanic ash provided a sufficient supply of nutrients, which enhanced the marine productivity. Moreover, the extremely anoxic environment generated owing to the volcanic activity enhanced the burial amount and preservation rate of organic matter. Qiu et al. (2019) considered the black shale of the Wufeng-Longmaxi Formations in South China as an example and performed a comparative analysis of the TOC content, paleoproductivity and redox conditions between the pozzolanic shale and normal sedimentary shale. The authors indicated that the pozzolanic shale did not considerably influence marine areas with high productivity or promote organic matter enrichment. The enrichment of organic matter is influenced by a variety of factors. Considering the Wufeng-Longmaxi black shale, the correlation between the organic carbon content and different factors is examined. The TOC, redox conditions, paleoproductivity and terrigenous input variations from the Wufeng to Longmaxi formations of the Huadi No. 1 well and Huangyingxiang section (Figs. 4.87 and 4.88) indicate that the productivity (Cu, Ba_{bio}, Si_{bio}) and TOC are weakly and positively correlated. The redox indices (V/V + Ni, V/Cr, Ni/Co) are weakly and positively correlated with the TOC, and the correlation coefficient is slightly larger than that for the paleoproductivity. In terms of the terrigenous inputs, the other two indices are not strongly correlated with the TOC, except for the weak positive correlation of Zr/Al. The correlations between the TOC in the sediments and redox indices, productivity and terrigenous inputs, indicate that the influence of the redox indices on the organic matter enrichment is the most notable, followed by those of the productivity and terrigenous inputs (Figs. 4.84, 4.85, 4.86, 4.87, and 4.88).

According to the analysis of the TOC, redox conditions, paleoproductivity and igneous input variations in the Wufeng-Longmaxi Formations in the Huangyingxiang section and Huangdi No. 1 well, the TOC varies considerably from the bottom to the top, influenced primarily by the

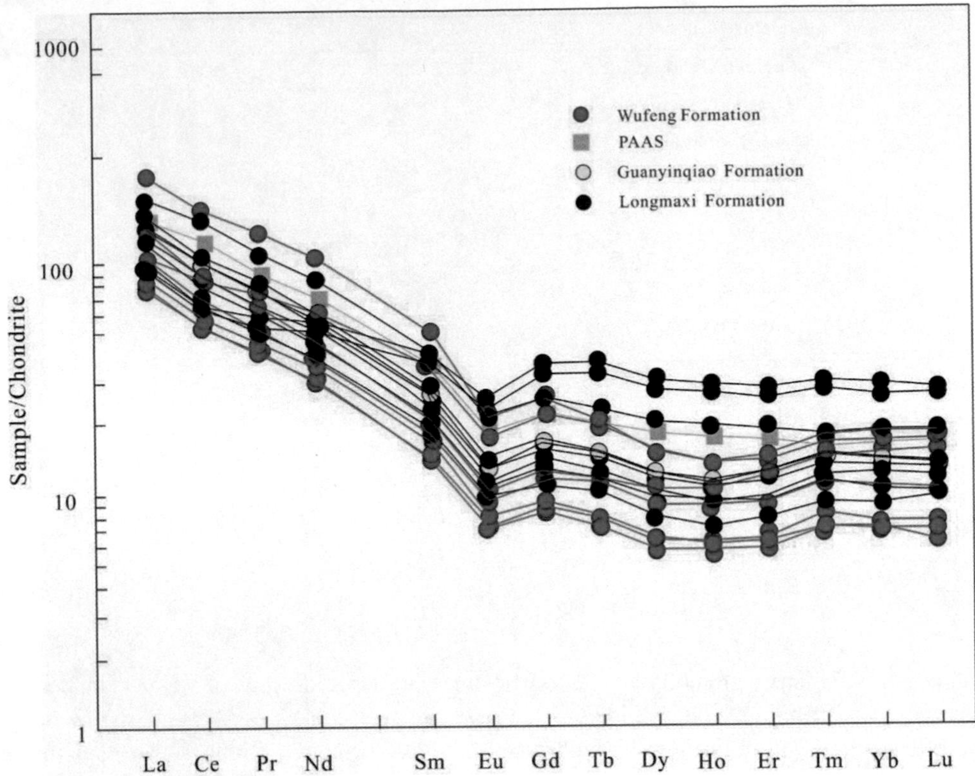

Fig. 4.84 Chondrite-normalized REE distribution patterns of samples from Wufeng-Longmaxi Formations in Huangyingxiang profile (normalization values after Taylor and McLenann 1985)

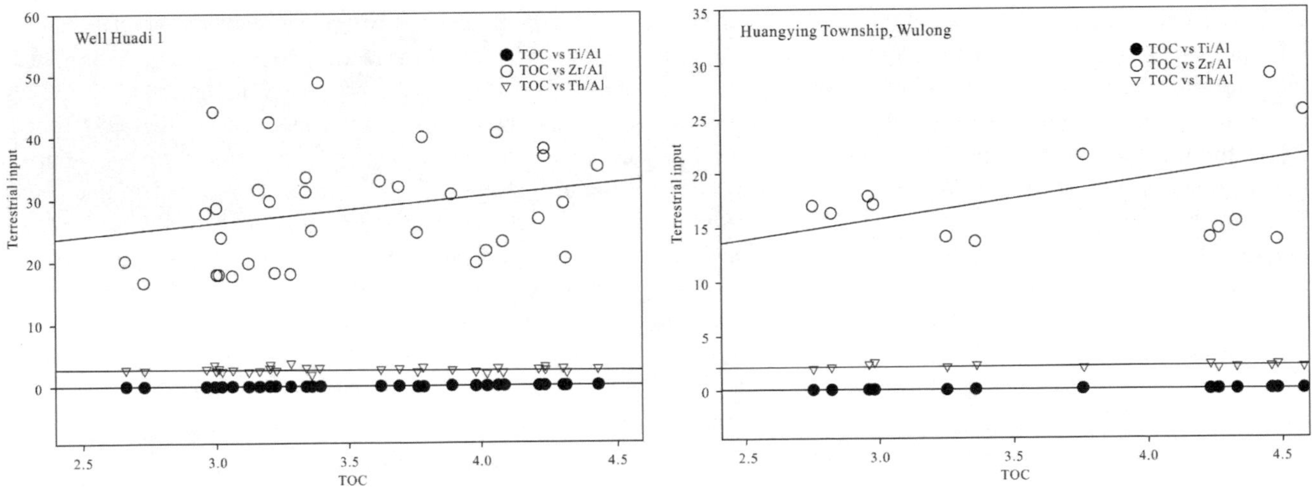

Fig. 4.85 Relationships between the TOC and paleoproductivity of the samples from the Wufeng-Longmaxi Formations in the Well Huadi No. 1 and Huangyingxiang profile, Wulong, Chongqing

sedimentary environment and paleoclimate. The TOC content of the Wufeng Formation increases and later decreases. The Guanyinqiao Formation has a low TOC, but the TOC at the bottom of the Longmaxi Formation is identical to that of the Wufeng Formation: The value increases, remains high for a certain period, and decreases. The redox indices (V/V + Ni, V/Cr, and Ni/Co) indicate that the water experienced an anoxic–oxic-anoxic process. Two sets of black shales were formed in an anoxic environment. The productivity (Cu, Ba_{bio}, Si_{bio}) and terrigenous input indices indicated higher productivity and terrigenous inputs in the black shales of Wufeng and Longmaxi than those for the Guanyinqiao Formation. The correlations between the TOC in the sediments and redox indices, productivity and

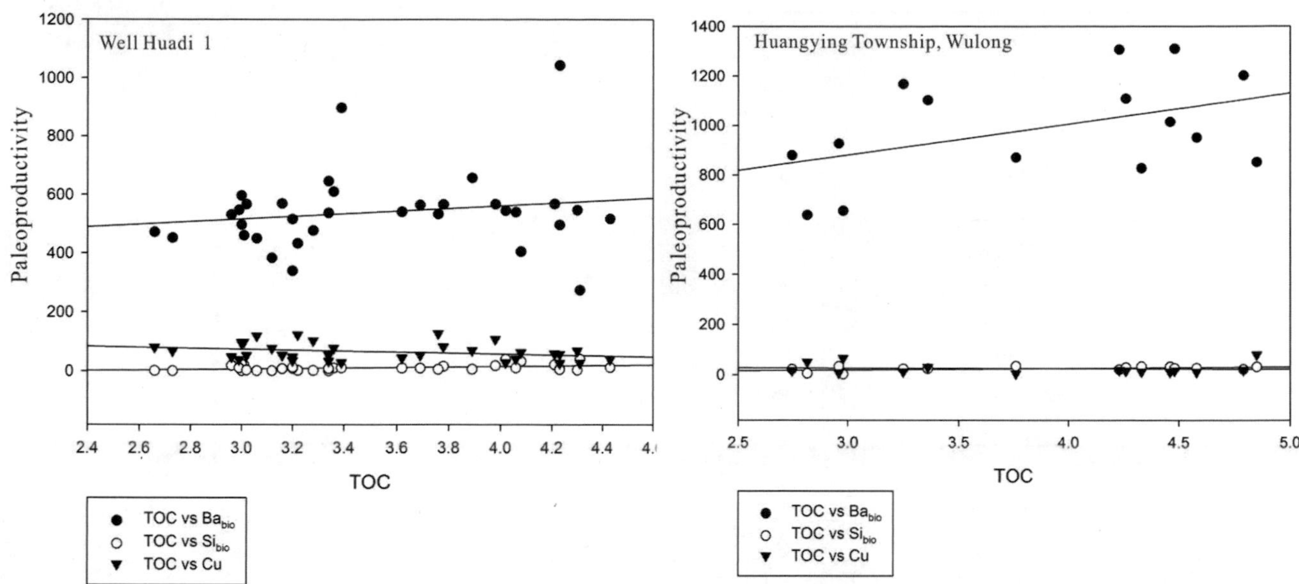

Fig. 4.86 Relationships between the TOC and redox conditions of the samples from the Wufeng-Longmaxi Formations in the Well Huadi No. 1 and Huangyingxiang profile, Wulong, Chongqing

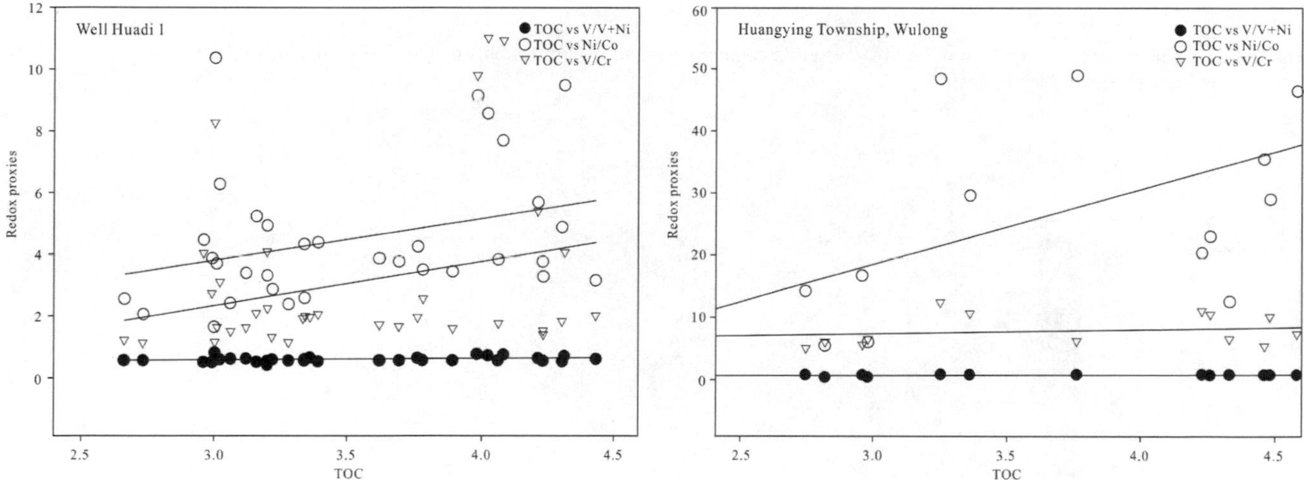

Fig. 4.87 Relationships between the TOC and the terrigenous inputs of the samples from Wufeng-Longmaxi Formations in Well Huadi No. 1 and Huangyingxiang section, Wulong, Chongqing

terrigenous inputs indicate that the influence of the redox indices on the organic matter enrichment is the most notable, followed by those of the productivity and terrigenous inputs.

Overall, the sedimentary environment controls the enrichment of shale gas, and the main factors include an anoxic-reducing environment, rapid transgression, suitable deposition rate, high biological yield and lithofacies development characteristics. The influences can be summarized as follows:

(1) The organic matter is enriched and preserved by the confined anoxic-reducing environment

Owing to the tectonic framework, the study area was in an occluded stagnant environment. According to the geochemical characteristics of black shale in the lower part of the Longmaxi Formation, the sedimentary environment was an anoxic-reducing environment (Figs. 4.89a–c), and the organic matter was well-preserved in the anoxic

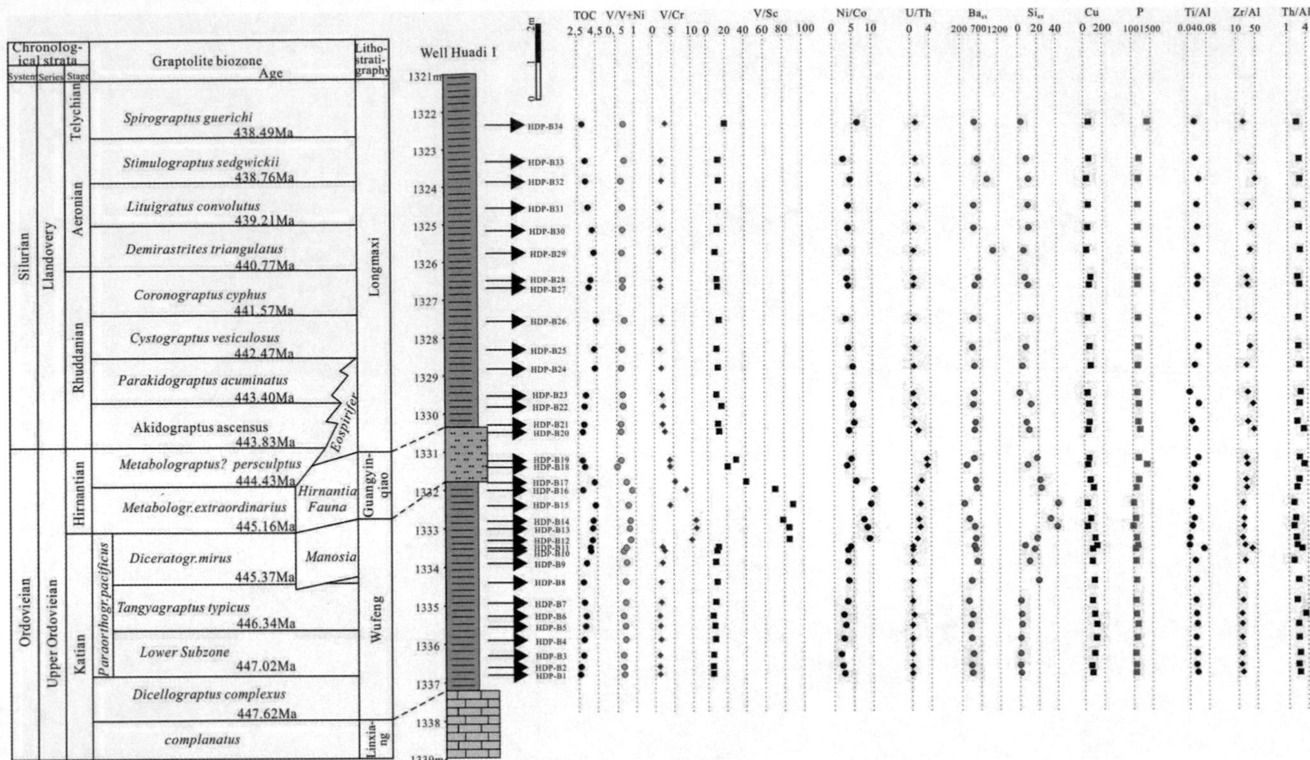

Fig. 4.88 Variations of the TOC, productivity, the redox conditions and terrigenous inputs of the shales and siltstones from Wufeng-Longmaxi Formations in Well Huadi No. 1

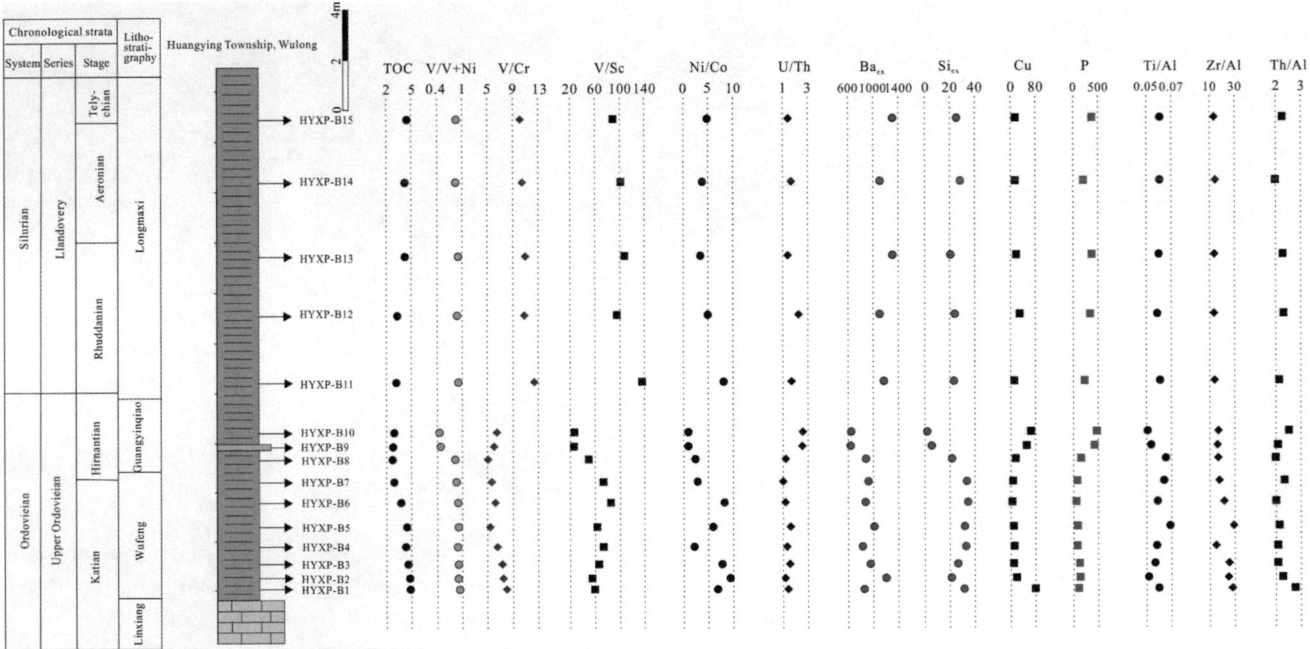

Fig. 4.89 Variations of the TOC, productivity, the redox conditions and terrigenous inputs of the shales and siltstones from Wufeng-Longmaxi Formations in Huangyingxiang profile, Wulong, Chongqing

environment (Zhang et al. 2012a; Li et al. 2008b). Under the marine sedimentary background, undercompensated sedimentary environments, such as shallow-deep-water basins, deep-water shelf basins, deep-water shelf facies and platform depressions, promoted the formation of high-quality marine source rocks (Chen et al. 2006a; Qiaan et al. 2009; Qian et al. 2010b). Furthermore, owing to the small input of terrigenous materials in these sedimentary environments, deep-water environments, biological reproduction on the surface of water, and addition of a large number of marine siliceous sediments, siliceous shale or calcareous shale were easily generated (Fu et al. 2011), which promoted the hydraulic fracturing of shale gas. The Longmaxi Formation of the Silurian period is characterized by a confined and stagnant environment and the presence of siliceous shales, which indicates that the siliceous shales have an organic origin (Wang et al. 2014) rather than clastic origin, as believed by many scholars (Zeng et al. 2011; Liang et al. 2012a; Liu et al. 2011; Zhang et al. 2012a).

Pyrite is abundant in the samples, which may indicate that the bacterial sulfidation of organic matter decreased in deeper sedimentary waters (Beener and Raisewell 1983; Rimmer et al. 2004). The authigenic pyrite is especially developed in this set of strata, mostly in the form of dispersed microcrystalline aggregates, which commonly occur as pellets or roe-shaped, framboidal and biological pseudo-morphs. Huang et al. (2012a) pointed out that this development is suggestive of a restricted reducing environment. In general, the pyrite content is approximately proportional to the TOC, indicating that the sedimentary environment was anoxic and had a low ion concentration (Hackley 2012) (Fig. 4.89d). In addition, Li et al. (2008b) argued that the organic carbon content (TOC) can be used as an indicator of anoxic conditions. The organic carbon content in southern Sichuan and adjacent areas showed an uneven distribution in both the plane and vertical directions, decreased in the vertical direction from bottom to top and gradually decreased in the plane from the basin center to the uplift area. It can be intuitively stated that the anoxic-reducing environment facilitates the enrichment and preservation of organic matter.

Hill et al. (2007) believed that the source beds of the Barnett shale were mainly formed in anoxic seawater with normal salinity and strong upwelling. The development of siliceous rocks can be considered an indicator of upwelling (Lv et al. 2004; Li et al. 2008b). Notably, these rocks are not developed in the Longmaxi Formation. The rocks are distributed only in the western and northern margins of the Yangtze Plate (Liu et al. 2010a, b; Wang et al. 2008) and concentrated at the bottom, indicating that upwelling currents did not develop when the Longmaxi Formation was deposited (Li et al. 2008b; Wang et al. 2008). In addition, in the upwelling active continental margin, the creatures are highly active; therefore, organic-rich source rocks often contain phosphate-rich minerals. However, the analysis showed that the phosphorus element contents of the black shales in the lower part of the Longmaxi Formation and TOC were not positively correlated (Fig. 4.89e), which indicate that the upwelling was not active and did not significantly influence the organic matter production (Wang et al. 2008). A favorable anoxic environment is thus a key factor for the development of high-quality source rocks in the Longmaxi Formation in the Upper Yangtze area (Xiao et al. 2008; Zhang et al. 2012).

(2) The stratified water formed by rapid transgression promotes the enrichment of organic matter

Another important factor influencing the organic matter enrichment of the Longmaxi Formation in southern Sichuan and adjacent areas is the inhibition of terrigenous detritus in the early transgression period (Xiao et al. 2008). When the Longmaxi Formation was deposited, the global climate entered a warm period, and the rapid melting of glaciers led to a rapid increase in the sea level (Li et al. 2008b). Most extensive shale deposits in the Phanerozoic correspond to the global sea-level rise (Arthur and Sageman 1994), and the shales primarily developed at the beginning of transgression rather than at the peak of transgression (Wignall and Maynard 1993). The source rocks in the Middle and Upper Yangtze area mainly developed in transgressive system tracts and were distributed in the middle and lower parts of sedimentary cycles in geological history (Chen et al. 2009; Li et al. 2012b), that is, the early stage of transgression (Li et al. 2008b). In the later stage of transgression, due to the long-term mixing of deep and surface seawater and the injection of terrigenous clastic materials, the anoxic environment at the bottom was destroyed, and the preservation conditions of organic matter deteriorated, as indicated by the decrease in the TOC of the Longmaxi Formation in the Silurian period from the bottom to the top (Li et al. 2008b).

In the Late Ordovician, that is, the Hirnantian period, global glaciation occurred, resulting in the formation of shallow-water grain carbonate or terrigenous clastic sediments of the Guanyinqiao Formation. During the rapid warming of the early Silurian period, the oxygen-rich surface water in the Yangtze region warmed rapidly due to direct solar radiation, while the hypoxic bottom water maintained the paleowater temperature during the glacial period for a considerable period due to the lack of solar radiation (Chen et al. 2006b). Cheng et al. (2013) noted that the sedimentary environment of black shale with benthonic organisms in the lower combination of the Upper Yangtze region was oxic upper water and suboxic or dysoxic bottom water, in terms of the measures of periodic and intermittent hydroenergy. In the oxic upper water, abundant organisms lived and repropagated to provide organic-rich matter for deposits, and in the suboxic or dysoxic bottom water, the decomposition of organic matters was weakened and/or baffled, thereby preserving the organic-rich matter (Chen et al. 2006b; Zheng et al. 2012).

The low temperature of anoxic bottom water rendered it difficult for carbonate minerals to be saturated and participate in chemical precipitation. Therefore, the content of carbonate minerals was lower in the deep-water shelf environment and shallow-water shelf near the center of the basin, which promoted the enrichment of organic matter. In the late stage of the Longmaxi Formation deposition, the bottom water gradually mixed with the upper warm water, which increased the water temperature and carbonate mineral content, leading to the dilution and decreased abundance of organic matter. In addition, the sedimentation of silicon-rich organisms in the upper water promoted the formation of siliceous shale, which facilitated the hydraulic fracturing of shale gas. Therefore, high surface water productivity caused by photosynthesis and anoxic separated basins under a humid climate is a key factor for organic matter enrichment (Wang et al. 2008).

(3) The sedimentation rate controls the dilution of organic matter

Because the stagnant environment in the Yangtze region remained for a considerable period after the Early Silurian period (Zhang et al. 2012a), graptolites developed in the yellow–green shales of the upper part of the Longmaxi formation. In contrast, the black shales of the Silurian developed in an extremely short period of the Early Silurian, indicating that other factors altered or destroyed the early sedimentary balance of oceanic matter.

As a part of the rock composition, the abundance of organic matter is controlled by the content of other components, which is mainly related to the deposition rate (Chen et al. 2006b; Passey et al. 2010). Both geological practice and corresponding laboratory modeling show that low deposition rates limit the preservation of organic matter. However, when the deposition rate is high, the organic matter content per unit volume is significantly diluted and reduced, and thus, an appropriate deposition rate can promote organic matter enrichment (Chen et al. 2006a). Pedersen and Calvert (1990) indicated that the sedimentation rate most conducive to the preservation of organic matter is 20–80 m/Ma, and the paleowater depth is 30–400 m. The sedimentary water must be neutral or slightly alkaline, with pH values ranging between 7.0 and 7.8 (Fu et al. 2008). The early deposition rate of the Longmaxi Formation was low (Zhang et al. 2012a). Feng et al. (1994) believed that the deposition rate of the early Longmaxi Formation was approximately 6–60 m/Ma, which promoted the preservation of organic matter. Organic-rich shales developed in an environment with reduced terrigenous detrital input (Hickey and Henk 2007). In the late Longmaxi Formation, the shallow-water sedimentary environment (Zhang et al. 2012a) witnessed rapid deposition, and the tremendous increase in the detrital sediments resulted in the dilution of organic matter and decrease in the organic matter content. In addition, rocks with low early sedimentation rates and fast late sedimentation rates exhibited the lowest porosity loss (He et al. 2010), which promoted the formation of shale gas reservoirs. The deposition time in the early stage of the Longmaxi Formation was more than half that of the Longmaxi Formation, but the deposition thickness was only 10–30% that of the Longmaxi Formation. The deposition rate of the late Longmaxi Formation was significantly higher than that of the early Longmaxi Formation, and the deposition time was less than half that of the Longmaxi Formation, but the thickness accounted for 70–90% (Zhang et al. 2012a), which promoted the formation of shale gas reservoirs. Therefore, in the shelf environment, the early appropriate deposition rate of the Longmaxi Formation was conducive to the enrichment and preservation of organic matter.

(4) The biological productivity controls the abundance of organic matter

The biological productivity in Paleozoic marine sediments in China is a key factor controlling the abundance of organic matter in sediments (Chen et al. 2006b). The carbon isotopes of black shale kerogen of the Lower Longmaxi Formation in southern Sichuan and adjacent areas range from −31 to −28%, and the organic matter is mainly type I, with a small amount of type II_1, indicating that marine plankton is the main source of organic carbon.

Oceanographic studies show that the Ba accumulation rate is positively correlated with the organic carbon content and biological productivity, and Ba enrichment is suggestive of a high productivity in upper water bodies (Li et al. 2009b). In the Huangchang and Houtan sections in the Middle and Upper Yangtze regions, the Ba content and organic carbon content in the Upper Ordovician and Lower Silurian periods are positively correlated (Li et al. 2009b), and two entities associated with the Silurian Longmaxi Formation organic-rich black shale in southern Sichuan and adjacent areas are also positively correlated (Fig. 4.89f), indicating that the paleoproductivity considerably influences the organic carbon content. Phosphorus is an essential nutrient element for the survival and reproduction of organisms. Phosphorus participates in most metabolic activities of organisms. The distribution of phosphorus in seawater is controlled by biological effects, and the phosphorus contained in the remains of organisms is deposited in the bottom together with the organisms after their death. The sedimentary strata with high phosphorus contents reflect high organic matter contents. Organic-rich source rocks often have phosphate-rich minerals. Therefore, by analyzing the abundance of phosphorus in the sediments, the level of biological productivity in the water can be clarified. The correlation between the phosphorus and organic carbon content in the study area is not obvious (Fig. 4.89e), but when the organic carbon content < 1%, the phosphorus and organic carbon content are correlated, indicating that biological effects influence the enrichment of organic matter. Calvert (1987) noted that the reducing environment does not control the development and distribution of marine organic-rich sediments. High organic carbon abundance in marine sedimentary rocks is the result of high productivity, even in a nonreducing environment. The decomposition of an abundance of organic matter consumes a large amount of oxygen, resulting in anoxic water and the formation of a reducing environment. High biological productivity can also increase the content of brittle minerals (biogenic siliceous and calcareous) in organic-rich shale, thereby increasing the brittleness of shale and enhancing the fracturing effect of shale gas.

(5) The sedimentary environment controls the lithofacies types, mineral compositions and geologic features of shale gas

The mineral composition, organic carbon content and maturity of organic matter are the three most important factors for shale reservoir development (Curtis 2002; Jarvie et al. 2005).

The sedimentary environment affects the organic carbon content and mineral compositions of shale rocks. Zhang et al. (2012a) indicated that the development of organic matter in the Longmaxi Formation in the southern Sichuan Basin was more controlled by the sedimentary environment, and the undercompensatory and hypoxic deep-water environment with a low carbonate mineral content was more conducive to the preservation of organic matter.

According to the petrographic analysis, carbonaceous (calcium-bearing) silty shales, carbonaceous shales and carbonaceous argillaceous limestones are mainly developed in shallow-water shelf environments, and their organic carbon content is relatively high. With the increase in the content of clastic particles and carbonate minerals, the organic carbon content gradually decreases and influences the correlation between the former and organic carbon contents. This observation indicates that the abundance of terrigenous detritus is not conducive to the preservation of organic matter, and the increase in the carbonate mineral content may dilute the abundance of organic matter. The tidal flat environment with shallow water, high oxygen content and rapid warming is not conducive to the preservation of organic matter, and the terrigenous clastic and carbonate mineral content is high. The formation of a large amount of organic carbon is challenging, and thus, the region is not favorable for shale gas. The sedimentary environment affects the types and distribution of lithofacies, and thus, the distribution characteristics of organic matter. Based on the detailed study of the mineral components in southern Sichuan and adjacent areas, the genetic characteristics of clay minerals and siliceous and carbonate minerals reflect the notable influence of sedimentation on the black shale of the Longmaxi Formation in the study area. Therefore, the correlation between the mineral composition and organic matter abundance is attributable to the joint control of the mineral composition and organic matter development characteristics associated with this deposition.

The organic carbon contents of the (calcium-bearing) carbonaceous (siliceous) shale and silt-bearing (calcium-bearing) carbonaceous shale that developed in a deep shelf environment were high, according to the petrographic analysis. The rock type was mainly siliceous shale with low carbonate mineral and high siliceous mineral contents developed in the deep sedimentary water in the early stage of the marine transgression, which promoted the accumulation, exploration and development of shale gas. Carbon-bearing (calcium-bearing) silty shales, carbon-bearing shales and carbon-bearing muddy limestones with higher TOC contents developed in a shallow shelf environment. With the increase in the number of detrital particles and carbonate mineral contents, the organic carbon content gradually decreased and influenced the correlation between the detrital particles and organic carbon content, indicating that the depositional environment rich in oxygen and terrigenous debris was not conducive to the preservation of organic matter, and the increase in carbonate minerals diluted the abundance of organic matter.

The tidal flat environment with shallow water, high oxygen content and rapid warming is not conducive to the preservation of organic matter, and the terrigenous clastic and carbonate mineral content is high. The formation of a large amount of organic carbon is challenging, and thus, the region is not favorable for shale gas. The sedimentary environment affects the types and distribution of lithofacies, and thus, the distribution characteristics of organic matter. Based on the detailed study of the mineral components in southern Sichuan and adjacent areas, the genetic characteristics of clay minerals and siliceous and carbonate minerals reflect the notable influence of sedimentation on the black shale of the Longmaxi Formation in the study area.

Therefore, the correlation between the mineral composition and organic matter abundance is attributable to the joint control of the mineral composition and organic matter development characteristics associated with this deposition.

By examining the organic matter deposition and diagenesis processes, Macquaker et al. (2010) noted that the formation conditions of fine-grained organic-rich shale can be summarized as follows: In a low-energy environment, a large amount of nutrients was supplied by sustained suspension sedimentation, and runoff or upwelling produced overlying water with high organic matter productivity. The primary organic matter was injected into the basin and not diluted by other substances. In bottom water with a long-term anoxic environment, organic matter, which was less affected by microbial degradation, adsorbed on the surface of the clay particles. Sedimentary water with appropriate deposition rates protected organic matter from oxidation or dilution. In general, the confined anoxic-reducing environment, stratified water formed by rapid transgression, appropriate sedimentation rate and high biological yield resulted in the enrichment of organic matter and formation of siliceous shale in the lower part of the Longmaxi Formation, which was conducive to the enrichment and exploitation of shale gas (Fig. 4.90).

4.7.3 Division of Favorable Sedimentary Areas to Shale Gas

The sedimentary facies of the Ordovician Wufeng Formation and Lower Silurian Longmaxi Formation and their relationship with organic matter in Southern Sichuan and its periphery indicate that the sedimentary environment has a fundamental control on the development of shale gas reservoirs. The tidal flat deposits around the uplift are mainly intertidal zone, followed by subtidal zone, showing the characteristics of mixed tidal flat. The sedimentary environment is mainly oxidized, rich in terrigenous clasts and carbonate minerals, with low organic matter content, which is not favorable areas for shale gas. Under the condition of high paleoproductivity, the deep-water shelf sedimentary area in the depositional center is dominated by anoxic reduction environment, which is conducive to the enrichment and preservation of organic matter. Influenced by the rapid warming climate and transgression of the early Silurian, the content of carbonate minerals in the sediments is low, but the biogenic silica content is high, which is conducive to the hydraulic fracturing effect of shale gas. Therefore, the deep-water shelf facies are favorable for shale gas. Carbonaceous siliceous shale has the highest content of brittle minerals, which is the most favorable lithofacies for shale gas. Shallow shelf facies in general can be divided into sandy argillaceous shallow shelf and calcareous muddy shallow shelf. And the sandy argillaceous shallow shelf mainly has lithofacies of calcareous carbonaceous shale, calcareous silty carbonaceous shale and silty shale, carbonaceous shale. These are main sedimentary types of the shallow shelf in the study area, and the development of rock facies types is complicated. Detrital grains, carbonate minerals, siliceous and other biochemical sediments can be deposited, with uneven distribution and different content. And they have high organic content. The microfacies of silty calcareous carbonaceous shale of the deep-water shelf environment deposits are mainly distributed in the south and the west of Sichuan Basin. They are influenced by the limited environment and have relatively high organic matter content, which should be the secondary favorable area for shale gas. Calcareous shale and carbonaceous argillaceous shale are deposited in calcareous muddy shallow shelf environment, showing the characteristics of mixed deposits shelf. The carbonate mineral content is higher, normally between 20 and 30%, and the lithofacies are mainly carbonaceous shale. Affected by high carbonate minerals, organic matter content in the shale is relatively low, usually $< 2\%$, and it is poor for shale gas exploration.

4.8 Shale Gas Area Selected Evaluation

Shale reservoirs are continuous accumulations in which organic-rich shale serves as a source, reservoir and cap rock for continuous gas supply and accumulation (Li et al. 2014). Organic-rich shale generally contains gas but is not homogeneous. Therefore, to find shale gas-rich areas is the key for shale gas exploration, and to obtain industrial gas flow by appropriate engineering measures is the target of shale gas exploitation (Chen et al. 2012). In recent years, the shale gas theory and technology have developed rapidly in China. Additionally, Chinese scholars have made considerable progress through numerous studies. For example, Chen et al. (2011a) and Zeng et al. (2011) compared the geological characteristics of the Lower Paleozoic shale in South China and Barnett shale in North America. An increasing number of scholars believe that marine shale gas in South China is characterized by high thermal evolution, strong structural reworking, complex geostress, deep burial and special surface conditions compared with North America. Therefore, the shale theories and technologies specific to North America cannot be applied in South China (Guo et al. 2012; Wang et al. 2013a, b; Li et al. 2014b). The marine shale gas exploration and evaluation system in South China must be established according to the actual geological conditions referring to evaluation parameters in North America (Zhao et al. 2011; Li et al. 2014b).

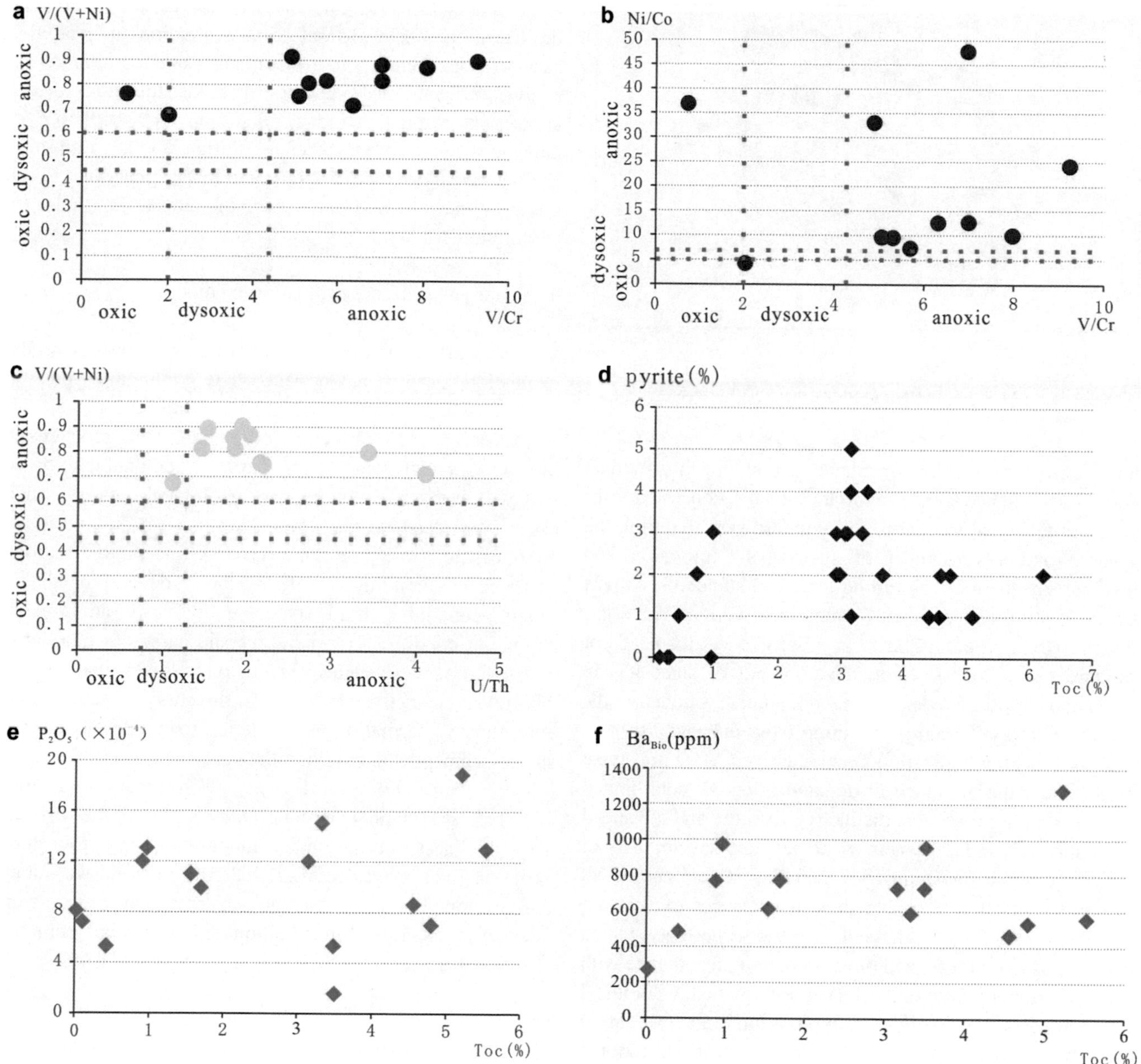

Fig. 4.90 Geochemical characteristics of black shales of Longmaxi Formation

4.8.1 Shale Gas Area Selected Evaluation Parameters

According to "the shale gas resource potential evaluation and favorable area optimization method" made by the ministry of land and resources of oil and gas resources strategic research center and China University of geosciences (Beijing) in August 2011, and "the summary of the shale gas resource potential evaluation and favorable zone selection plan" made by Li et al. (2012c), combined with China's oil and gas exploration situation and characteristics of shale gas resources, the shale gas distribution in China can divide into three parts: prospective area, favorable area and target area (Fig. 4.91). The study in the prospective area focuses on material basis (gas generation capacity) and other parameters; the study in the Favorable areas focuses on shale gas enrichment main factors (gas storage capacity) parameters; the study in the target area and engineering sweet spot area focuses on the parameters including in-situ stress, pressure coefficient, brittleness index and number of natural fractures.

Shale gas exploration in China is still in the stage of ongoing constituency evaluation (Chen et al. 2010; Liu et al. 2009a; Zhang 2010; Li et al. 2010; Liu et al. 2010a; Liu and Wang 2012; Ge and Fan 2013). Based on the successful

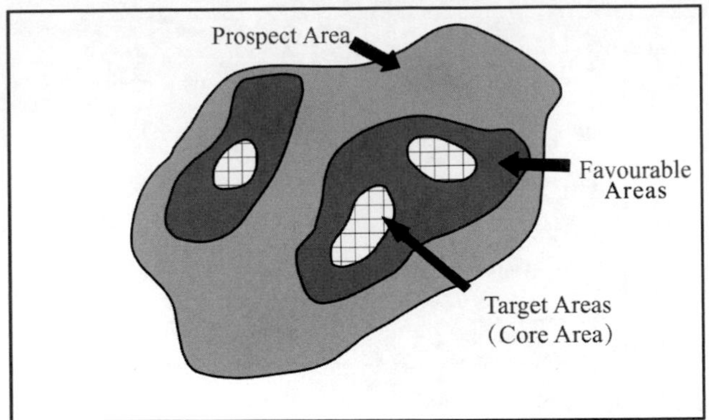

Fig. 4.91 Shale gas distribution area dividing diagram (according to the Ministry of Land and Resources of Oil and Gas Resources Strategic Center, etc. 2011)

development experience of Shale gas in North America, China has attached great importance to the construction of shale gas technical standards and standard system during the research and development of shale gas exploration and development theories, providing technical standards support for the healthy and rapid development of China's shale gas industry (Lehong et al. 2020). In order to meet the needs of standardization for the rapid development of shale gas in China and support the development of shale gas industry, the National Energy Administration approved the establishment of Shale Gas Standardization Technical Committee in August 2013. Under the guidance of demand-oriented, coordinated and supporting principles, distinctive features and advanced application, a standard system of the business development of the whole shale gas industry chain has been established. Combined with the characteristics of shale gas exploration and development in China, the theory and technology system of shale gas exploration and development in accordance with China's national conditions has been established. Geological evaluation standards include 10 exploration target evaluation technology standards, such as geological evaluation, resource evaluation, geological and logging-based data analysis, shale or mudstone microscopic identification, and resource and exploration target evaluation technology. Marine shale gas exploration target optimization method (GB/T 35,110-2017) specifications are used for shale gas evaluation selection and to evaluate the reservoir parameters.

In practical work, this method has certain parameters limitation for shale gas selection evaluation in geological survey stage. Based on the summary and comprehensive analysis of shale gas evaluation methods, this study selects appropriate evaluation indexes and their thresholds to evaluate the shale gas of Ordovician–Silurian Wufeng-Longmaxi Formations in Southern Sichuan and its periphery and divide the prospective areas and favorable areas for shale gas exploration.

1. Shale gas prospect evaluation method

Li et al. (2012c) pointed out that shale gas prospective area evaluation, namely gas-bearing shale evaluation, refers to the area with potential geological conditions for shale gas formation optimized by combining geological, geochemical and geophysical data on the basis of regional geological survey. On the basis of regional geological data, it is necessary to understand the regional structure, sedimentary and stratigraphic development background, find out the regional geological conditions for the development of organic-containing mud shale, preliminarily analyze the formation conditions of shale gas and carry out qualitative and semi-quantitative early evaluation of the evaluation area. The selection method is based on the study of sedimentary environment, formation and structure to divide gas-bearing shale segments, including the organic matter content, organic matter evolution degree, rock mineral composition and shale gas display of shale. After study of the distribution, thickness, burial depth, longitudinal and transverse variation rules of gas-bearing shale, the regions with shale gas development conditions, namely prospective areas were selected by analogy, superposition and synthesis techniques (Table 4.39).

2. Evaluation method for favorable areas of shale gas

Favorable area is selection of the shale gas resource potential assessment in the prospective area (Li et al. 2012c), mainly on the basis of shale distribution, the geochemical indicators, drilling natural gas and hydrocarbon content parameter. After further drilling, shale gas can be acquired in the industrial area.

Table 4.39 Optimization reference index of marine shale gas prospect area in Southern Sichuan Basin and its periphery

Main parameters	Range
TOC (%)	>0.5%, average not less than 1%
Ro (%)	Not less than 1.3%
Depth	100–4500 m
Surface conditions	Plains, hills, mountains, deserts and plateaus
Storage conditions	Regional development and distribution of shale, with general preservation conditions

The selection is based on the spatial distribution of shale, the geological condition investigation, seismic data, drilling and experimental test data, and the characteristics of shale sedimentary facies, tectonic model, shale geochemical index and reservoir characteristics. According to the key parameters of shale, spatial distribution and gas content, the favorable area is further optimized in the prospect area. Selection methods are based on shale distribution, geochemical characteristics, such as the gas-bearing shale gas content, adsorption ability, reservoir space types, reservoir physical property, reservoir fracture and its change rule and to carry out optimization and areas favorable for shale gas resource evaluation by using the superposition of multiple factors, comprehensive geological evaluation and geological analogy.

According to Marine Shale Gas Exploration Target Optimization Method (GB/T 35,110-2017) (Table 4.38), geological characteristics of typical shale gas in Fuling shale gas field and North America (Table 4.40) and geological characteristics of large overpressure shale gas field (Table 4.41), various parameters of favorable shale gas areas can be summarized as follows: The TOC with average value must be > 2%, the thickness of high-quality shale must be > 20 m, the porosity must be > 2%, and the thermal evolution must be at the dry gas stage. According to the determination conditions and criteria for favorable areas or intervals of shale gas in China proposed by analyzing the geological conditions for shale gas formation, accumulation and enrichment in China, progress of exploration and development in China, and comparison with the geological characteristics of shale gas in North America (Dong et al. 2016), and nine parameters: the shale mineral composition, geochemical characteristics, reservoir characteristics, cap rock, rock mechanical properties, resource conditions, gas content, preservation conditions and burial depth (Table 4.42) proposed by the experience of shale gas exploration and development in North America, parameter indices of constituency evaluation, and parameter indices of selected area, and the actual situation of the Ordovician Wufeng Formation-Silurian Longmaxi Formation in the Sichuan Basin (Zhang et al. 2016), favorable shale gas areas, construction production areas and core construction production areas are selected (Table 4.43).

In general, favorable areas for shale gas exploration require thicker gas-bearing shale thickness, higher abundance of organic matter, higher degree of thermal evolution, higher gas content, better preservation conditions and surface conditions suitable for exploration (Table 4.44).

(1) Thick gas-bearing shale: The multistage tectonic modifications of the Wufeng-Longmaxi Formations in southern Sichuan and adjacent areas during geological history render it challenging to preserve the natural gas in organic-rich shale. However, the larger the thickness of organic-rich shale is, the better the self-sealing of the generated natural gas is, and the more favorable the geological conditions for shale gas formation are. At the same time, the thicker the gas-bearing shale is, the more natural gas is generated, and the larger the shale gas is. According to the experience of shale gas exploration and development in the USA, a thickness of 30 m organic-rich shale is selected as the lower limit of evaluation of organic-rich shale thickness in favorable area.
(2) High organic matter abundance: the higher the organic carbon content in shale, the greater the gas produced. Because the adsorbed gas content of Wufeng-Longmaxi Shale in Southern Sichuan and adjacent areas is positively correlated with the organic carbon content, the higher the organic carbon content is always accompanied by the higher the adsorption capacity of shale. At the same time, the content of free gas will increase with the increase of organic matter content. Therefore, the higher the organic carbon content is also accompanied by the higher the gas content of shale. Based on the current data and exploration experience, $\geq$ 2.0% was selected as the favorable area selection criterion in this study.
(3) High thermal maturity: The Barnett shale gas in the USA is formed by the cracking of retained liquid hydrocarbons in the shale in the middle and late stage of thermal evolution, and the enrichment of shale gas is mainly in high maturity (Ro > 1.3%). The thermal maturity of marine organic-rich shale in the Upper Yangtze Block is generally high. However, due to the high maturity, its control effect on shale gas accumulation and enrichment mechanism is not clear. Therefore, Ro < 4.0% is set as the upper limit. The Wufeng-Longmaxi Formations in Southern Sichuan and its periphery are mainly composed of type I and type II 1 kerogen. Therefore, Ro > 1.3% is the lower limit and Ro < 4.0% is the upper limit.
(4) High gas content: The level of gas content is the key to the economic development of shale gas. The higher the gas content is, the better the development prospect and the better the income will be. Considering the special geological characteristics of China, the gas content parameter is temporarily set as $\geq$ 0.5 m^3/t.

Finally, sedimentary structure, surface topography, traffic conditions and engineering conditions are also important parameters for the evaluation of favorable areas. Briefly, the optimization of favorable shale gas exploration areas is a comprehensive analysis based on specific parameters and geological characteristics.

Table 4.40 Comparison of indicators between Longmaxi Formation shale gas in Fuling gas field and typical shale gas in North America (Liu 2015)

Shale name	Marcellus	Haynesville	Barnett	Fayetteville	Eagle ford	Fuling
Shale age	D	J	C	C	K	S
Depth/m	1200–2600	3200–4100	2000–2600	1737	1200–3050	2150–3150
Genetic type	Thermogenic gas	Thermogenic gas	Thermogenic gas	Thermogenic gas	Thermogenic gas	Thermogenic gas
Net thickness/m	15–107(46)	61–97(79)	30–213(91)	6–61(41)	46–91	70–87
TOC	2.0–13.0(4.01)	0.5–4.0(3.01)	3.0–12.0(3.74)	2.0–10.0(3.77)	2.0–8.5(2.76)	1.04–5.89 (3.26)
Ro	0.9–5.0(1.5)	1.2–2.4(1.5)	0.85–2.1(1.6)	2.0–4.5(2.5)	0.8–1.6(1.2)	2.20–3.13 (2.58)
Air content/(m^3/t)	1.70–4.25	2.83–9.34	8.50–9.91	1.70–6.23	5.66–6.23	3.52–8.85 (5.85)
Free gas volume/%	55	75	45	30–50	75	57
Adsorbed gas volume	45	25	55	50–70	25	43
Porosity/%	4.0–12.0(6.2)	4.0–14.0(8.3)	4.0–6.0(5)	2.0–8.0(6)	6.0–14.0	4.0–12.5
Permeability/($\times 10^{-9}\mu m^2$)	0–70(20)	0–5000(350)	0–100(50)	0–100(50)	700–3000 (1000)	< 1000
Siliceous content/%	37	30	45	35	15	31–70.6(44.4)
Clay content/%	35	30	25	38	15	16.6–49.1 (34.6)
% carbonate content/%	25	20	15	12	60	5.4–34.5(10)
Sedimentary environment	Marine	Marine	Marine	Marine	Marine	Marine

4.8.2 Shale Gas Selection Evaluation of Silurian Longmaxi Formation Shale in Southern Sichuan and Its Periphery

1. Prospective area optimization

According to the methods and steps of shale gas investigation in this study, the geological background of the development of black shale of Silurian Longmaxi Formation in southern Sichuan and adjacent areas is mainly the study of sedimentary basin types, as described in Sect. 1. For the study of fine lithofacies palaeogeography, the depositional environment for the favorable hydrocarbon source rocks in Ordovician–Silurian Wufeng-Longmaxi Formations is described in detail in Sect. 4.2. Here, we briefly describe it. During the Ordovician to Silurian, a " shale basin" is developed in the south of Sichuan Basin and adjacent areas. The relatively uplifted highlands around the basin hinder the entry of seawater. Therefore, the hydrogeologic conditions in the basin are restricted and can be only affected by water exchange through the relatively "fixed channels"-'basin-controlled facies". The weak hydrodynamic conditions are favorable for the formation of organic-rich shale - "basic geological conditions of facies controlled oil and gas". The data and maps show that the lithofacies paleogeography matches well with the distribution (thickness contour map) and organic carbon content distribution of organic-rich shale.

After obtaining a certain amount of organic-rich shale ore data and distribution data in the study area, the basic conditions for shale gas prospect optimization can be identified. This study is based on the geological parameters of shale selection evaluation. Therefore, the TOC and Ro are considered the main basis, and the burial depth and preservation conditions are combined to optimize the prospective area in the classification criteria of the prospective area (Table 4.39). The selection method is based on the research of sedimentary facies through the formulation of lithofacies paleogeographic maps and examination of the content of organic carbon (TOC) and degree of thermal evolution of organic matter (Ro) to determine the gas shale section thickness distribution. Moreover, the burial depth and occurrence and fracture development are examined to define the favorable areas of shale gas.

Based on TOC, the TOC of the black shale in the southern Sichuan basin and adjacent areas is almost more than 0.5%. However, in the tidal flat sedimentary area, near the uplift area, the black shale is not developed and the

Table 4.41 Parameters of typical overpressure and high-yield shale gas fields at home and abroad (Honglin Liu et al. 2016)

Parameters	Changning shale gas field	Weiyuan shale gas field	Jiaotianba shale gas field	Barnett	Fayetteville	Marcellus
Depth/m	1500–3000	2500–4500	2000–3500	1980–1600	366–2286	1220–2440
Thickness of high-quality shale/m	46	34	45	31–183	35–61	20–68
Thermal maturity Ro/%	2.8–3.2	2.5–3.0	2.5–3.0	0.8–1.3	1.2–3.0	0.7–2.15
TOC/%	2–4.5	2–6.0	2–5.5	2.5–6.7	2–9.8	2.3–10.5
Siliceous content/%	48	45	50	30–50	30–60	30–60
Total porosity/%	5–6	4–7	5–8	4.5	4–12	3.6–7.0
Permeability ($\times 10^{-9}\mu m^2$)	250–360	300–530	200–552	145–206	200–450	300–900
Pressure coefficient	1.5–2.03	1.3–1.9	1.3–1.7	1.04–1.53	1.1–2.0	1.01–1.64
Gas saturation/%	70	65	80	80	90	75
Test output/($\times 10^4 m^3$/d)	15–30	10–27	20–57	10–35	20–30	10–40
Single-well EUR/($\times 10^8 m^3$)	0.8–1.2	0.8–1.1	0.8–3.0	0.9–2.2	1.0–2.5	0.9–3.1

Table 4.42 Conditions and standards for determining shale gas favorable areas or intervals in China (Dazhong Dong et al. 2016)

Parameters	Standard for area-selection in China	USA		Significance
		standard for area-selection	Lower limit of producing area	
Organic carbon/%	>2.0	>4	>3	The quality and effective range of source rock
Grade of maturity/%	I-II1 > 1.1, II2 > 0.9	>1.4	>1.0	
Brittle minerals such as quartz/%	>40	>20	>40	Reservoir quality
Clay minerals/%	< 30	< 30	< 30	
Porosity/%	>2	>2	1– >2	Potential and prospects
Permeability/($\times 10^{-9}\mu m^2$)	>1	>50	>10	
Air content/(m^3/t)	>2	>2	>1	
Initial daily output of vertical well/($\times 10^4 m^3$/d)	1	4	>0.85	
Water saturation	< 45	< 25	< 35	
Oil saturation	< 5	< 1	低	
Enrichment of resources/($\times 10^8 m^3/km^2$)	>2.0	>2.5	>3	
EUR/($\times 10^8 m^3$/well)	0.3	>0.3	>0.3	
Pressure	Atmospheric pressure-overpressure	Overpressure	Atmospheric pressure - overpressure	Production mode and production capacity
Continuous thickness of effective shale	>30–50	>>30	>30	
Interbed thickness/m	< 1	/	/	
Ratio between sandstone and stratum thickness/%	< 30	/	/	
Lithology and thickness of roof and floor/m	Impermeable rock formation, >10	/	/	
Storage conditions	Stable zone, low degree of transformation	Tectonic stable region	Tectonic stable region	

Table 4.43 Preferred indicators and thresholds for shale gas favorable areas, production areas and core production areas (Zhang et al. 2016)

Parameters	Favorable area	Producing area	Core producing area
Organic carbon/%	>2	>2	>2
Grade of maturity/%	>1.35	>1.35	>1.35
Brittle minerals/%	>40	>40	>40
Clay minerals/%	<30	<30	<30
Porosity/%	>2	>2	>2
Permeability/(mD)	>100	>100	>100
Water saturation/%	<45	<45	<45
Young's modulus/($\times 10^4$ MPa)	2.07	2.07	2.07
Poisson's ratio	0.25	0.25	0.25
Air content/(m^3/t)	>2	>2	>2
Depth/m	<4000	<4000	<4000
Thickness of high-quality shale/m	>30	>30	>30
Pressure coefficient	–	>1.2	>1.2
Tectonic conditions	–	Gently	Gently
Distance from Denudation line/km	–	>7	>7
Distance from Fault/m	–	>1.5	>1.5

organic matter content is very low, which cannot reach the level of organic-rich shale. The highest organic carbon content is > 4% of black shale in the center of the basin enclosed by central Guizhou uplift and central Sichuan uplift, i.e., the vast area from Suijiang and Muchuan to the east to the south of Qijiang, the vast area from Neijiang and Bishan to the south of Xingwen and Xishui, and the narrow area from Luding-Hanyuan to Ebian which is influenced by central Sichuan uplift and Kangdian old land. According to the feedback of exploration and development, the organic carbon content of black shale in Shizhu-Pengshui area in northeast Chongqing is also high, which is consistent with the paleogeography distribution (Fig. 4.92). The thickness of black shale with TOC > 0.5% gradually increases from the uplift area to the center of the basin, and the sedimentary thickness of shuifuan-Jiangjin area is the largest, with an average thickness of > 100 m. In the Luding-Hanyuan area of western Sichuan, the thickness of black shale with TOC > 0.5% is > 60%. Therefore, the organic carbon content, thickness and spatial distribution characteristics of black shale obviously are in conformity with the lithofacies palaeogeographic characteristics. The black shale deposited in the deep-water shelf is characterized by the highest organic carbon content; accordingly, shale thickness is the largest. On the contrary, the shallow shelf shale presents the decreased thickness. The tidal flat black shale is less, and the organic matter content of this rock is the lowest. All the pictures show that depositional environment plays a fundamental role in controlling organic matter enrichment in shale. The study of sedimentary facies and lithofacies paleogeography will provide us with the detailed information of organic matter abundance and the distribution law of organic-rich shale.

The distribution of Longmaxi Formation black shale can be found through outcrop identification and burial depth prediction. Because of the influence of Sichuan and its adjacent area sedimentary tectonic evolution, the Longmaxi Formation black shale accounted for about half of the total distribution area as southwest of zonal distribution in the Southeast of Chongqing, Northern Guizhou and Southeast Sichuan. The Longmaxi Formation is mainly concentrated in the central and Northeastern parts of the region, while the Western part of the Longmaxi Formation is relatively small, forming a long and narrow discontinuous strip. In Southern Sichuan and its periphery, the denudation gradually increases from the Northeast to the southwest, and even the Meso-Neoproterozoic strata are exposed in Ganluo and Emei areas in the west. The burial depth of Longmaxi Formation gradually increases from uplift area to central area. Moreover, because the area mainly presents the characteristics of "synclinal mountain, anticlinal valley", the maximum burial depth of Lonmgaxi Formation is in the core of the anticlinal structure. In the Southern low-slow anticline and the eastern high-steep anticline, near the Rongchang county, the maximum burial depth of Longmaxi Formation in P-1 well is 3856 m, and the maximum burial depth of Longmaxi Formation in YS-2 well and DS-1 well is 3565 m and 3428 m, respectively. The maximum burial depth of Wufeng-Longmaxi Formation in Z-3 well in the south of Lingshui county is 4307 m. The maximum buried depth of Longmaxi

Table 4.44 Shale gas content in black rocks of Longmaxi Formation in Southern Sichuan Basin and its periphery

Scope and Age of study area	Air content (m^3/t)	Average air content (m^3/t)	Remark	Quoted from
Upper Ordovician-Lower Silurian in Sichuan Basin and peripheral area	0.39–2.97	1.19	Gas content less than $0.50m^3/t$ only accounts for 13.60%, gas content of $0.50–1.00m^3/t$ accounts for 39.10%, gas content of $1.00–2.00m^3/t$ accounts for 34.80%, and gas content greater than $2.00m^3/t$ accounts for 13.00%	Nie et al. (2012b)
Longmaxi Formation in Northern Guizhou	0.80–1.40, more than $1.40m^3/t$ at the Southern boundary of Sichuan Basin	–	The conventional oil and gas exploration in Western Sichuan is weak, and the shale gas has not been well-studied	Nie et al. (2012b)
Longmaxi Formation in the Dazhou-Kaixian-Wanzhou area of eastern Chongqing	1.00–1.40	–		
Longmaxi Formation in the Shizhu-Pengshui Area of eastern Chongqing	0.80–1.20	–		
Longmaxi Formation in the Qianjiang-Pengshui-Wulong area of Western Hunan and Hubei	0.80–1.20	–		
Longmaxi Formation, Well Wei 201, 1503.6–1543.3 m	1.70–4.50	–	Accumulative output has reached $29.98 \times 10^4 m^3$	Lin et al. (2012)
Longmaxi Formation, Well Ning 201, 2479–2525 m	1.72–3.50	–		
Longmaxi Formation, Southeast Chongqing	0.42–1.68	0.97		Han et al. (2013)
Longmaxi Formation, Well Ning 201	1.42–2.83	2.08		Wang et al. (2012a)
Longmaxi Formation, Well Yuye 1	1.00–2.10	–		
Longmaxi Formation, Pengshui Block, Southeast Sichuan	–	2.20		
Longmaxi Formation, Weiyuan District	0.30–5.09	1.82		Huang et al. (2012b)
Longmaxi Formation, Changning District	0.90–3.50	1.93		
Wufeng Formation-Longmaxi Formation, Well JY1	0.89–5.19	2.96	On-site core gas content test	Guo et al. (2013)
	1.72–4.66	–	Core isothermal adsorption experiment	
Wufeng Formation-Longmaxi Formation, Changning District	3.50–6.50	–	Core isothermal adsorption experiment	According to internal data of the project, *Evaluation and Optimization of Exploration Blocks for Lower Paleozoic Shale Gas Resources in Southern Sichuan*
	2.40–4.00	–	On-site core gas content test	
Wufeng Formation-Longmaxi Formation, Well Pengye 1	0–4.00	0.50	On-site core gas content test	
	1.79–3.11	–	Core isothermal adsorption experiment	

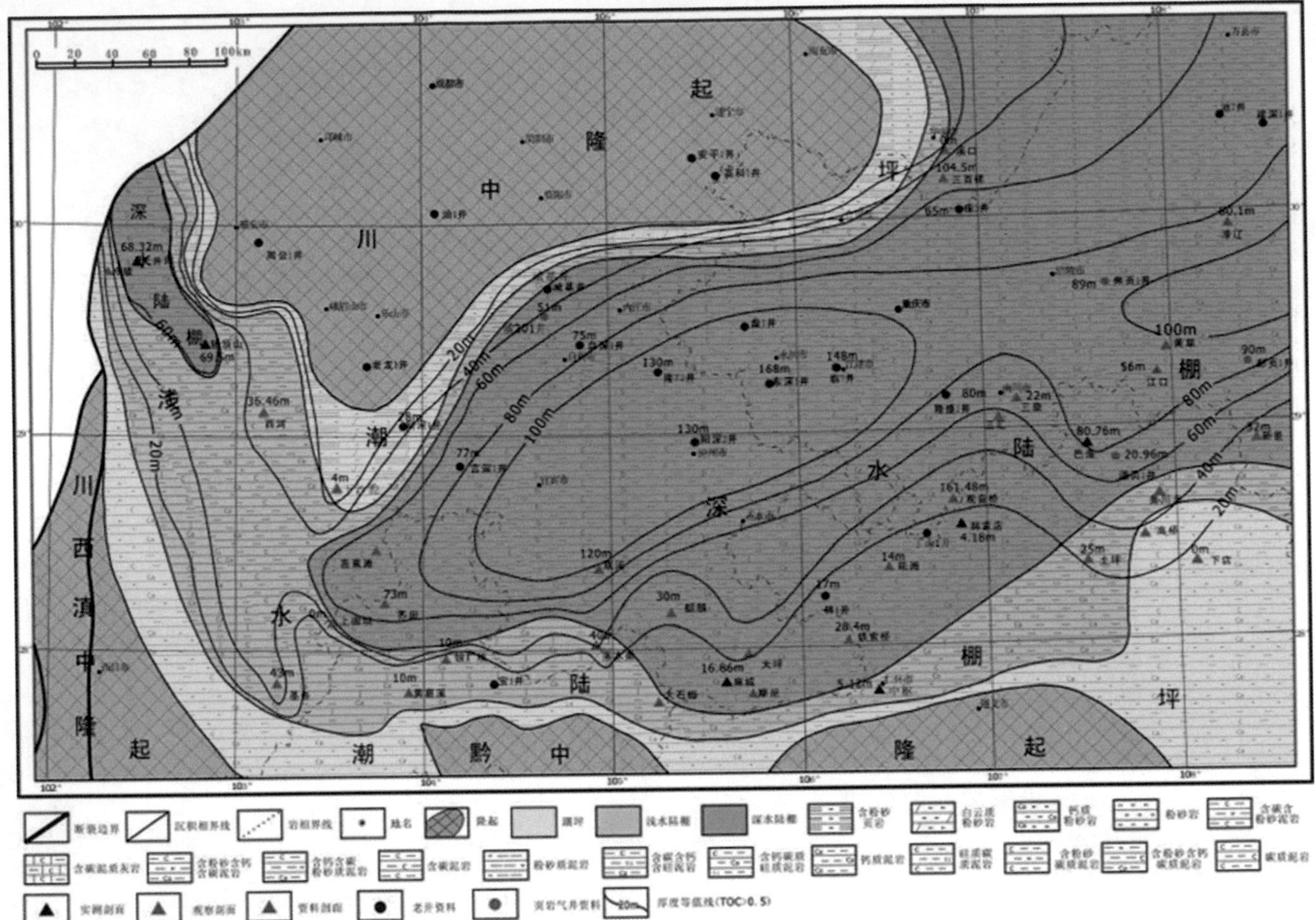

Fig. 4.92 Lithofacies paleogeography and isoline of organic carbon thickness during the deposition of organic-rich shale of Longmaxi Formation in Southern Sichuan Basin and its periphery

Formation is 4900 m in the Qinggang syncline near the Wulong County, 3,400 m at Puzi syncline near Pengshui County, and 3200 m at Tongxi syncline near the Qianjiang District, respectively. The maximum burial depth of Longmaxi Formation is 2761 m and 1518 m for Wusen 1 well and Weiji 1 well located in the middle of low-flat tectonic belt and close to the Central Sichuan Uplift, and 1523 m for Wufeng-Longmaxi Formation in Dingshan 1 well near the Southern denudation area. In general, the Silurian Longmaxi Formation has a large distribution area and burial depth in the south of Sichuan and the central and Northeast of its periphery, and a small distribution area and shallow burial depth in the southwest and Southeast of Sichuan.

In the Southern Sichuan and its periphery, organic matter in the Silurian Longmaxi Formation black shale has a high degree of thermal maturity. The areas with an average organic carbon content greater than 1% and Ro greater than 1.3% are divided into shale gas exploration prospects combined with burial depth characteristics (Fig. 4.93). The distribution of shale gas prospect area in the study area is wide and stable. Compared with the Southern Sichuan-Northern Guizhou and Southeastern Chongqing areas, Western Sichuan is located in high mountainous area with large topographic height difference, many deep faults, strong stratigraphic uplift and denudation, poor surface hydrology and transportation conditions, and low degree of shale gas exploration. Therefore, Western Sichuan is a secondary shale gas exploration prospect area in the study area.

2. Favorable zone optimization

According to the geological and geochemical indicators of shale, on the basis of prospective area evaluation, combined with shale gas content, resource potential and shale storage capacity, and according to the selection evaluation indexes of favorable areas (Table 4.40), the favorable development areas of shale gas of Silurian Longmaxi Formation in Southern Sichuan and adjacent areas are selected. This study mainly applies the geological parameters in shale gas selection evaluation to optimize the favorable area for shale gas, such as thickness, TOC, Ro, mineral composition and total gas content of organic-rich shale, and burial depth,

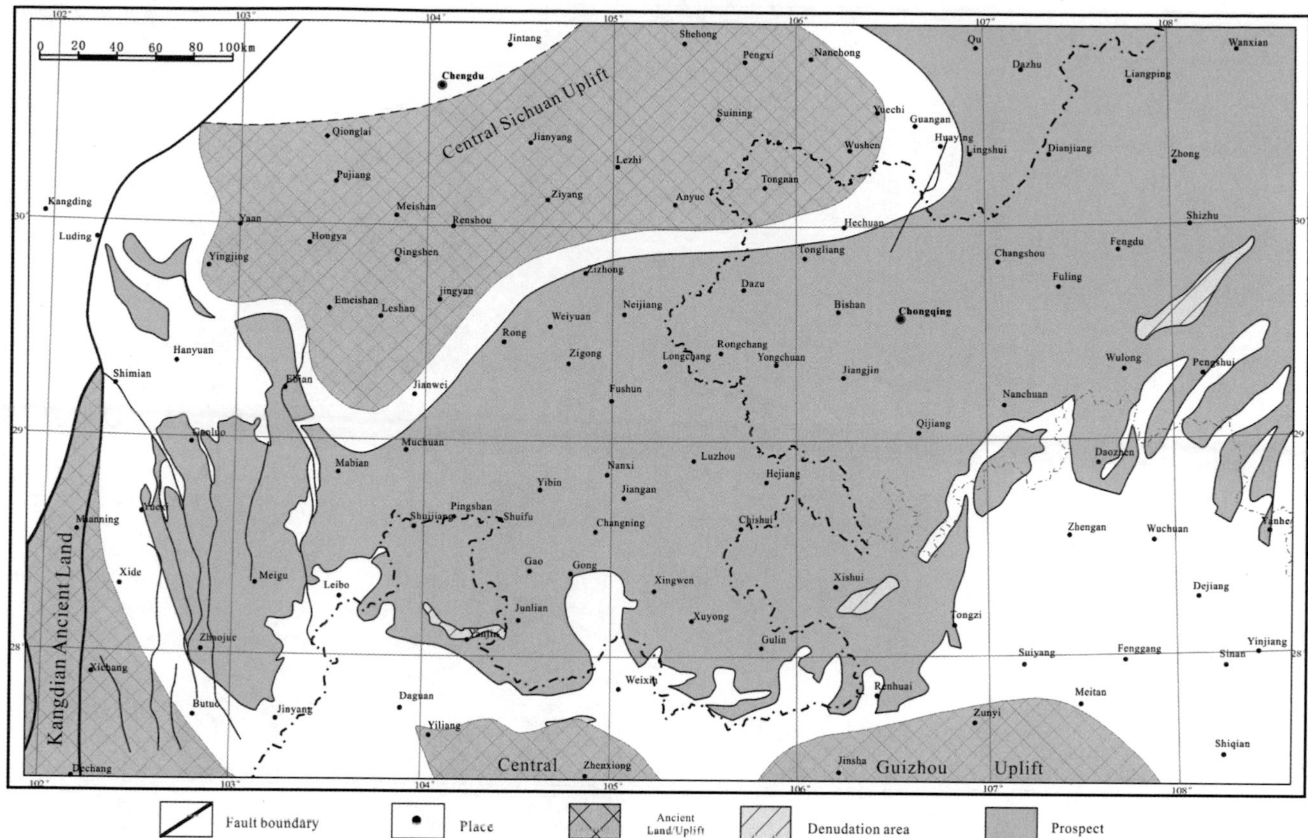

Fig. 4.93 Prospect area distribution for shale gas of black rocks in Longmaxi Formation of Southern Sichuan Basin and its periphery

lower limit of shale area, surface conditions and preservation conditions. Shale gas content research is based on shale gas geological survey wells or exploration wells, through core analysis, isothermal adsorption simulation, as well as logging and logging data analysis, to determine the gas content of gas-bearing shale and its change rule (Li et al. 2011). However, in the stage of shale gas geological survey, the data of shale gas geological survey wells or exploration wells are relatively few, so it is difficult to meet the requirements of shale gas resource potential evaluation. Although gas content is an important indicator for shale gas geological evaluation, resource potential prediction and favorable area selection, it is not the only indicator. Geological characteristics, geochemical characteristics, reservoir physical and chemical characteristics, fluid properties and other factors should be integrated into shale gas geological evaluation (Li et al. 2011). In addition, as discussed above, organic carbon content and mineral composition all affect the gas content of shale. Wang et al. (2013a) pointed out that when TOC is less than 2.0%, the gas content is usually lower than or around 1 m^3/t. When TOC > 2.0% (especially above 2.5%), the gas content of shale increases obviously. Shale gas content is mainly obtained by field analysis of drilling cores and isothermal adsorption test or by formula method. In the process of shale gas regional geological survey, due to the lack of shale gas drilling and data, we mainly use the previous research results for reference. Therefore, on the basis of detailed petrology, organic geochemistry, and a small amount of gas content analysis results combined with previous gas content analysis of Longmaxi Formation in Southern Sichuan and its periphery (Table 4.44) (Nie et al. 2012b; Lin et al. 2012; Wang et al. 2012a; Huang et al. 2012b; Han et al. 2013; Guo and Liu 2013), this study aimed to evaluate the Silurian Longmaxi Formation shale gas favorable area in the study area (Fig. 4.94).

Guided by the evaluation criteria of favorable shale gas areas selection, the favorable shale gas exploration areas of Longmaxi Formation in the study area are preliminarily divided according to the organic carbon content, thermal evolution degree of organic matter and characteristics of stratum burial depth. Controlled by the buried depth, the favorable areas are mainly distributed in the basin and its margin. Based on the geological survey and selection method of shale gas in this paper, the lithofacies palaeogeography map, TOC content and mineral composition superimposed map show the following results: The organic carbon content of carbonaceous and siliceous shale deposited in the deep-water shelf environment is high. When the

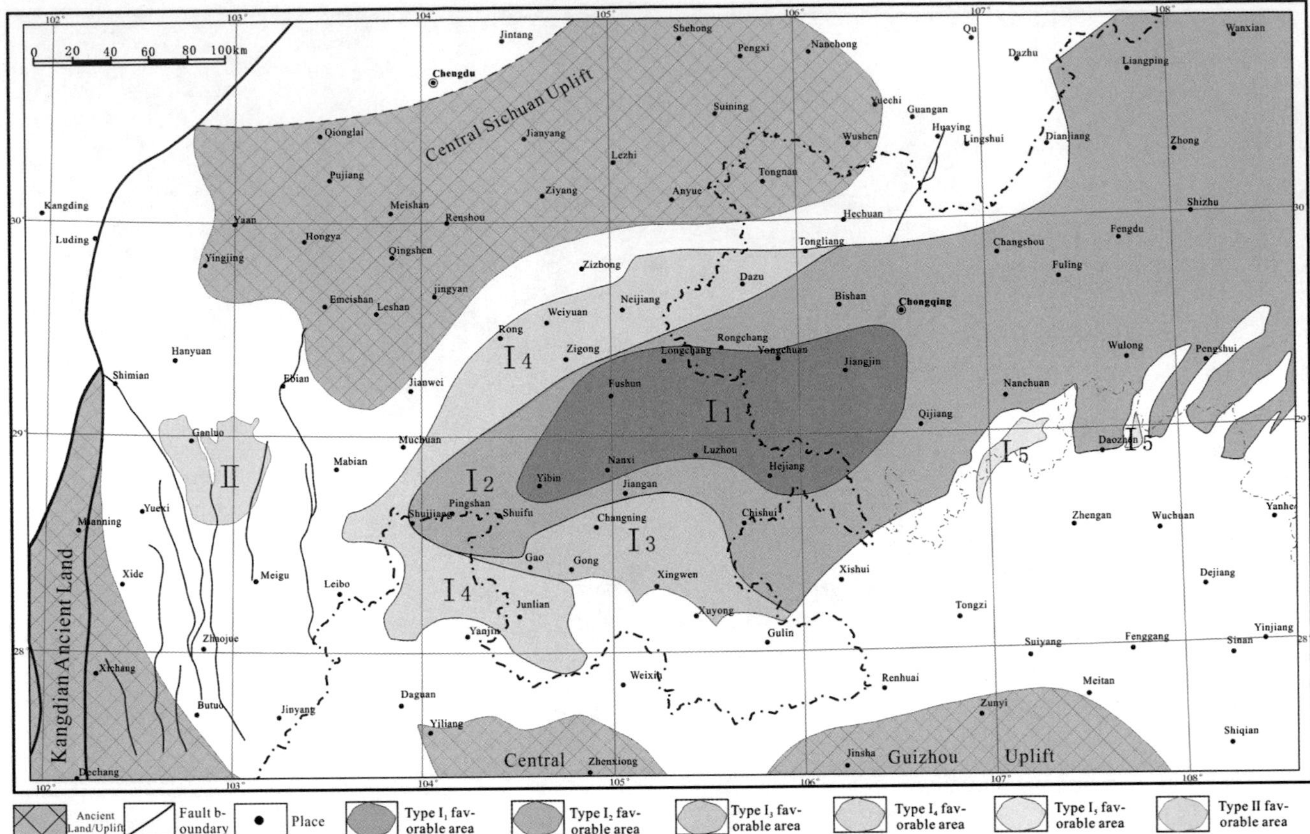

Fig. 4.94 Favorable distribution area for shale gas of black rocks in Longmaxi Formation of Southern Sichuan Basin and its periphery

carbonate mineral content is less than 30%, the higher the quartz content, the lower the clay mineral content is in the shale. Most shale was deposited in the early transgression period and the deep water, so it contains less carbonate mineral but more silica. Carbonaceous silty shale, carbonaceous shale and carbonaceous argillaceous limestone are mainly deposited in shallow-water shelf environment, and their organic carbon content is relatively high. With the increase of the content of detrital inputs and carbonate minerals, the organic carbon content gradually decreases. The positive correlation between the detrital matter and organic carbon content indicates that the abundance of terrigenous detritus is not conducive to the preservation of organic matter, and the increase of carbonate minerals may only dilute the abundance of organic matter. The shallow tidal flat environment with high oxygen content water is not conducive to the enrichment of organic matter. The terrigenous clastic and carbonate mineral content is high, which is not favorable for shale gas (Fig. 4.95).

Considering the influence of mineral components of the Longmaxi Formation on the shale gas exploration and development, the siliceous shale development area is conducive to the enrichment of shale gas. According to the distribution characteristics of prospect area, structure geological backgroud and surface condition, the favorable area can be divided into two types and six subtypes; considering the organic carbon content of the shale, organic-rich shale thickness, and vitrinite reflectance (Ro), the developed deep faults in western Sichuan, and the large topographic height difference, the region can be classified as a class ii favorable area. In this area, the average organic carbon content is > 2%, the thickness of organic-rich shale is > 30 m, the quartz content is 30–50%, the clay mineral content is 40–50% and the carbonate mineral content is 10–30%.

According to different parameters, the south Sichuan and Southeast Chongqing area in the basin and its margin can be divided into five types of favorable areas. Firstly, in the Yibin-Jiangjin area (I_1) at the center of the distribution area, carbonaceous siliceous shale is developed in the deep-water shelf facies, which is most favorable for shale gas enrichment. In this area, the average content of carbonate minerals is < 10%, the average content of brittle minerals such as quartz and feldspar is > 50%, the average content of clay minerals is < 30%, and the organic carbon content is more than 3%, the average content is more than 5%, and the thickness of organic-rich shale (TOC) is more than 1% is more than 80 m. Therefore, it is the most favorable distribution area in the study area. In addition, the terrain of the

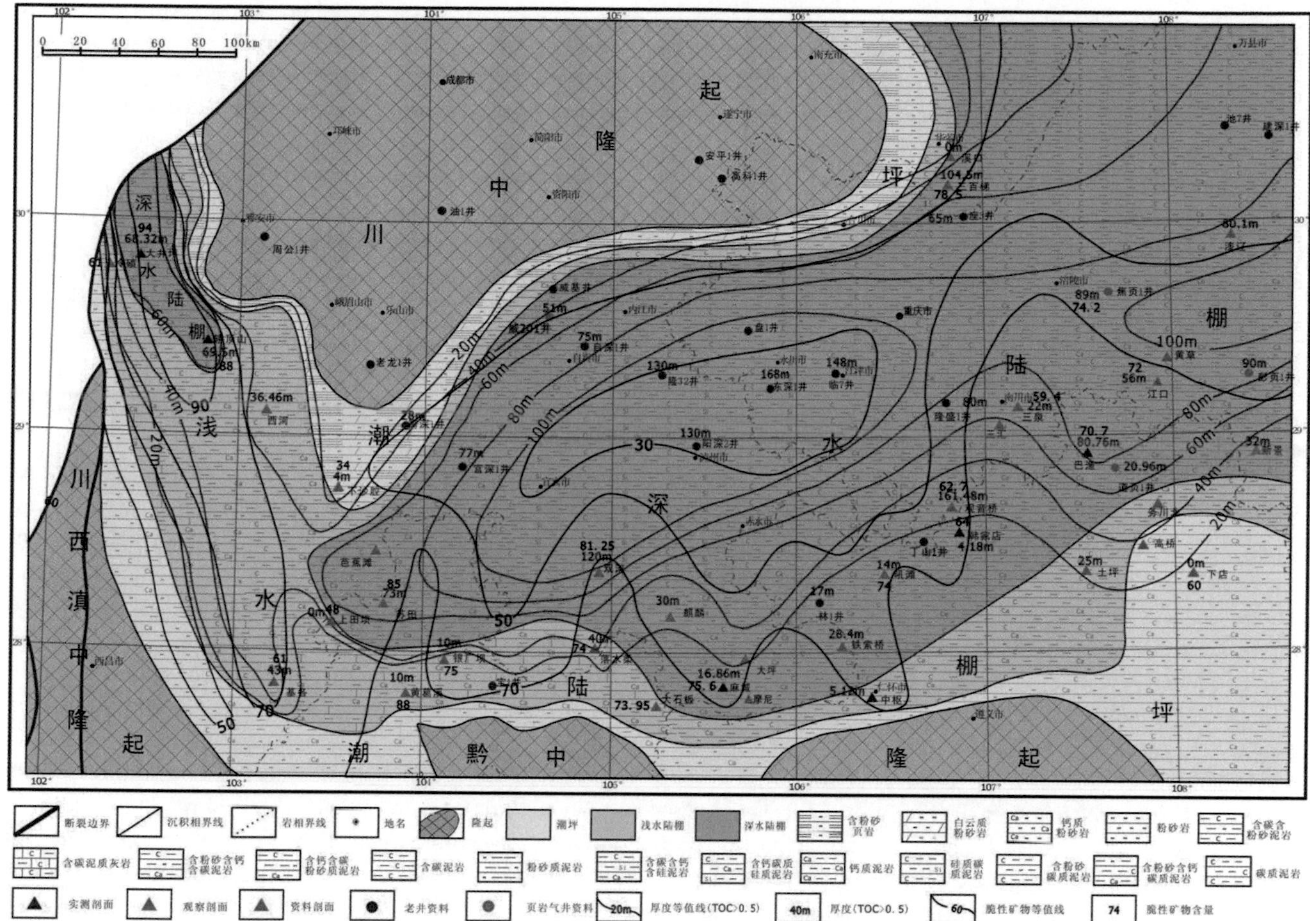

Fig. 4.95 Lithofacies palaeogeography, TOC thickness and brittle mineral composition of Longmaxi Formation organic-rich shale in Southern Sichuan and its periphery

area is slow in the north and steep in the south, and the stratum burial depth is less than 1500 m. The fracture is not developed, and the preservation condition is good. In addition, the terrain in this area is slow in the north and steep in the south, and the stratum buried depth is less than 1500 m. In addition, faults are not developed in this area, which has good preservation conditions.

The vast area of Suijiang-Chongqing-Zhongxian (I_2) at the edge of the I_1 favorable area is the secondary favorable area. From southwest to Northeast, "silty calc-bearing carbonaceous shale + calc-bearing silty carbonaceous shale" and "silty shale + carbon shale" deep-water shelf facies are developed successively. The mineral composition of Longmaxi Formation black shale in this area is similar to that of I_1 favorable area, while the content of organic matter and thickness of organic-rich shale are slightly reduced (TOC > 3% and thickness > 60 m, respectively). (1) The investigation well of shale gas in this area-DY-1 well, the most favorable organic-rich shale interval (TOC > 1.5%) from 570.1 to 597.1 m, thickness 27.0 m; the second favorable organic-rich shale interval (TOC < 1.5%): from 489.0 to 570.1 m, thickness 81.1 m. The average content of brittle minerals such as quartz + feldspar, carbonate minerals and clay minerals in organic-rich shale is 67.56%, 13.58% and 18.86%, respectively. In the 404–423 m gas logging, the maximum total hydrocarbon was 2.95%, and the maximum total hydrocarbon was 1.9%. The total gas content of the four samples is 1.84–2.69m^3/t, indicating a good gas display. It shows that the area has good shale gas exploration potential. (2) The Longmaxi Formation in Fuling area of the high and steep faulted fold belt in eastern Sichuan Basin buried within 4000 m has a stable distribution. The organic carbon of Longmaxi Formation in this area is more than 3.0%; the content of quartz + feldspar and other brittle minerals is more than 50%; the content of carbonate minerals is generally no or less than 10%; nanoscale pores, especially microfractures, are developed. The surface of the area is mainly exposed Jurassic, and the synclinal belt is exposed Triassic. The area is a high distribution area of abnormal pressure, with pressure coefficient of 1.5–2.0 showing good preservation conditions. In the concentrated distribution area of stress field, fracturing has great potential. A breakthrough of shale gas scale has been achieved in Jiaoshiba, Fuling. Jiaoye -1HF has tested 203,000 m^3 and has been drilled with high production and good exploration prospects. (3) In qijiang-Xishui area of Southern Sichuan low-slow anticline structural belt, permian–Triassic system is exposed on the surface with relatively shallow burial depth of 1000–3000 m. The average TOC of Longmaxi Formation in Guanyinqiao area of Qijiang is 3.18%; favorable shale thickness is more than 30 m, kerogen type ii_1 is dominant; Ro is generally greater than 2 and the highest is 2.279, reaching the gas generation stage. The average content of brittle minerals such as quartz + feldspar is more than 40%, carbonate minerals are about 16% and clay minerals account for 40%. Shale in this area is generally dominated by siliceous shale and clay shale; however, carbonate shale is not developed.

Carbonate shales are developed in changning-Xuyong (I_3) area. The average content of quartz + feldspar minerals and clay minerals is > 50% and 30–50%, respectively. The average content of carbonate minerals is 10–30%, and the organic carbon content is mostly more than 2%. This area is located in the low and slow anticlinal structural belt in Southern Sichuan. Most of the area is covered by Triassic-Jurassic strata, followed by Silurian strata, and Ordovician strata are developed in the south. TOC of Longmaxi Formation black shale in Xulong Macheng section is 3.21% on average, Ro is about 2.2%, reaching the gas generation stage. Brittle minerals such as quartz and feldspar are about 35.2% on average, carbonate minerals are 34% on average, and clay minerals are about 30%. The content of carbonate minerals increases, and the content of brittle minerals is relatively high, which is also easy to fracture. The buried depth of Longmaxi Formation in this area is between 2000 and 3000 m, and the fracture is relatively undeveloped, which has good preservation conditions. In Shuanghe section, Changning, the average TOC of Longmaxi Formation black shale is 3.36%, and the favorable thickness is more than 30 m. Quartz + feldspar and other brittle minerals account for 45.2%, about 24.7% of carbonate minerals and 30% of clay minerals. The content of carbonate minerals in Wufeng-Longmaxi black shale increases, and the carbonate shale increases. The overall buried depth of Wufeng-Longmaxi Formation in this area is less than 3500 m, and the faults are relatively undeveloped and indicate the good preservation conditions in this area. The main exploration targets in Changning shale gas exploration area are Lower Cambrian Niutitang Formation and Lower Silurian Longmaxi Formation, and the burial depth of Longmaxi Formation is about 1500–3000 m. A total of 9 wells were drilled in the area, and all of them found shale gas except Ning 206, among which Ning 201–H1 wells produced 50,000 m^3 of gas per day. At present, CNP is drilling a conventional and shale gas exploration well – Yang 1 well in Fulong Village, Zhandong Township, Xuyong County. The well is drilled from the Permian Makou Formation with a drilling depth of 3620 m of which 986 m is drilled into the bottom of Longmaxi Formation. According to the gas measurement and logging curves, 58 m of organic-rich shale was found in the Longmaxi Formation. Through the water invasion experiment of Longmaxi Formation core, the bubbles are obvious and the gas content is high.

The shallow shelf areas between the Western Sichuan basin and the Southern Sichuan basin are characterized by "silty and calc-bearing carbonaceous shale + calc-bearing silty and carbonaceous shale" lithofacies assemblages, with quartz content ranging from 30 to 50%, clay mineral content ranging from 30 to 50% and carbonate mineral content ranging from 10 to 30%, which are also favorable areas. The thickness of organic-rich shale in this region is relatively thin, with organic carbon content > 2% in most cases, and the thickness of organic-rich shale > 30 m, which is mainly distributed in the semi-arc region of Dazhu-Zigong-Junlian (I_4). (1) Weiyuan-Zigong-Luzhou area is located between Weiyuan National Shale Gas Demonstration Zone and Fushun-Yongchuan shale gas cooperation Zone. According to the seismic profile and Zishen 1 well, the organic-rich shale of Longmaxi Formation in this block is well-developed, which is buried deep in the abdomen of the ground within 4500 m, above which is mainly covered by Jurassic-Cretaceous strata. The organic carbon content of black shale is generally more than 3%, the content of brittle minerals such as quartz + feldspar is more than 50%, the content of carbonate minerals is less than 10%, and the content of clay minerals is about 30–50%. (2) In Weiyuan structural belt, by July 2012, CNPC has drilled several shale gas wells, among which W- 201 well, the first shale gas well in China, encountered 43 m organic-rich shale of Longmaxi Formation. After fracturing, Longmaxi Formation has produced 0.3–17,000 m^3/d, among which Longmaxi Formation has produced 495,800 m^3 in total. Secondly, the daily production of Longmaxi shale gas in W-202 and W-201 wells is over 10,000 m^3, which has accumulated 341,400 m^3 and 3.199×10^6 m^3, respectively, showing good exploration prospects. The favorable area between Leibo and Suijiang is distributed in the paleo-uplift structural belt in the southwest of Sichuan Basin. The Longmaxi Formation black shale is relatively stable, with organic carbon content between 1 and 3%, and the distribution area with an average of > 2% is narrow and zoned. The content of brittle minerals is more than 70%, but the content of carbonate minerals is relatively high, between 20 and 40%. The carbonate shale is more developed than the siliceous or clay shale, and the surface is mainly exposed to the Cambrian-Jurassic system. The buried depth is 2000–3000 m, and the fractures are not well-developed, and the preservation conditions are better.

The Southeast corner of Qijiang and Nanchuan as well as the eastern part of Daozhen (I_5) is located in the depression between the mountains. The deep-water shelf facies of "silty carbonaceous shale + a small amount of silty carbonaceous shale" is developed in this area, which is conducive to shale gas enrichment. The organic carbon content of black shale in Longmaxi Formation is higher than 2%, and the thickness of organic-rich shale is higher than 30 m. In addition, the content of carbonate minerals is lower than 10%, the content of brittle minerals such as quartz + feldspar is higher than 50%, and the average content of clay minerals is lower than 50%, which is conducive to the development of shale gas. Considering the influence of strong formation uplift and complex surface conditions, as well as the small distribution area, it is classified as the most unfavorable type I favorable exploration area for shale gas.

In the Yingjing-Hanyuan area between the low uplift tectonic belt in Western Sichuan and the thrust fold belt in Northeast Yunnan, the deep-water shelf facies of "calc-bearing siliceous shale" were developed, which is conducive to the formation of shale gas distribution areas. However, Longmaxi Formation shale in the Jiaodingshan section of the Hanyuan is characterized by high content of organic carbon > 3%, Ro greater than 3%. Quartz and feldspar and other high brittleness mineral content > 50%, carbonate minerals are not evenly distributed, but the content of clay minerals content is so low, usually < 30%, an average of about 15%, low content of clay minerals content. They are disadvantage for shale gas adsorption. In addition, the Western Sichuan region has relatively large terrain elevation difference, relatively developed hidden faults and general preservation conditions. Combined with the gas content of Longmaxi Formation in the study area is low (< 0.5 m^3/t) (Nie et al., 2012b), so it is a non-favorable area for shale gas exploration in the study area.

4.8.3 Shale Gas Area Selection Evaluation of Ordovician Wufeng Formation in Southern Sichuan and Its Periphery

Considering the difference in the material composition between the Wufeng and Longmaxi Formations, favorable exploration areas selection was carried out for both the Wufeng and Longmaxi Formations. The thickness of the Wufeng Formation is low, and the highest thickness in the interior and periphery of the Sichuan Basin is slightly more than 20 m, lower than the required 30 m favorable exploration thickness. Therefore, the selection of favorable exploration areas for the Wufeng Formation is not based on the sedimentary thickness. According to the TOC distribution characteristics, thermal evolution characteristics, burial depth and other factors of organic-rich shale in the Wufeng Formation, the distribution range of the prospective Wufeng Formation is optimized (Fig. 4.96). The areas around the central Sichuan uplift and western Sichuan-Central Dianan-Central Guizhou uplift are outside the prospective areas, and most of the Sichuan Basin and the surrounding area is a potential area for shale gas exploration and research in the Wufeng Formation.

According to the optimal conditions of favorable exploration areas, the favorable exploration areas are selected, and the content of brittle minerals and preservation conditions is superimposed. The favorable exploration areas of the Wufeng Formation are divided into three types: most favorable areas (class i), favorable areas (class ii) and sub-favorable areas (class iii) (Fig. 4.96).

The most favorable area in Wulong-Daozhen-Pengshui (i): This area is mainly distributed in Daozhen syncline and several small synclines around it. The average organic carbon content of Wufeng Formation of Daozheng section and Daoye well 1 is 4.52% and 3.79%, respectively, up to 6.7%, and the hydrocarbon generation potential is good. Kerogen type is mainly type I and type II1. *Ro* is in the range of 2.1–2.6%, which is in the middle and late stage of evolution. The fracture is relatively undeveloped and the preservation condition is good. In terms of rock mineral composition, the clastic minerals of brittle minerals are mainly quartz, accounting for 70.7%, while carbonate minerals are not, and the content of clay minerals is 29.3%. The clay minerals are mainly illite and Aemon mixed layer, and the content of brittle minerals is relatively high and easy to fracture. In addition, the drilling of shale gas in Well Daoye 1, the first shale gas survey well implemented in Daozhen area, shows that the area has good shale gas exploration potential. In summary, Wulong-Daozhen area is the most favorable area for shale gas exploration in Wufeng Formation.

The most favorable area in Weiyuan, Zigong and Luzhou: The area is mainly located between Weiyuan National Shale gas Demonstration Zone and Fushun-Yongchuan shale gas cooperation Zone. According to the seismic profile and Zishen 1 well, the Wufeng-Longmaxi Formation organic-rich shale beds are well-developed in the area, which are buried deep in the ground, and the upper part is mainly covered by Jurassic-Cretaceous strata. The organic carbon content of black shale in Wufeng Formation is more than 3%, the clastic quartz content is more than 50%, and the carbonate minerals are almost absent, and the burial depth is less than 4000 m.

The most favorable Area of Xishui (i): The area is mainly distributed in the low and slow anticline structural belt in Southern Sichuan, with permian–Triassic system exposed on the surface and relatively shallow buried depth of 1000–3000 m. TOC of Wufeng Formation in Qijiang Guanyinqiao section is 6.65% on average and 6.73% on the highest, with good hydrocarbon generation potential. Kerogen type II is dominant, *Ro* 2.235–2.262. In terms of mineral composition, the content of clastic minerals quartz + feldspar is about

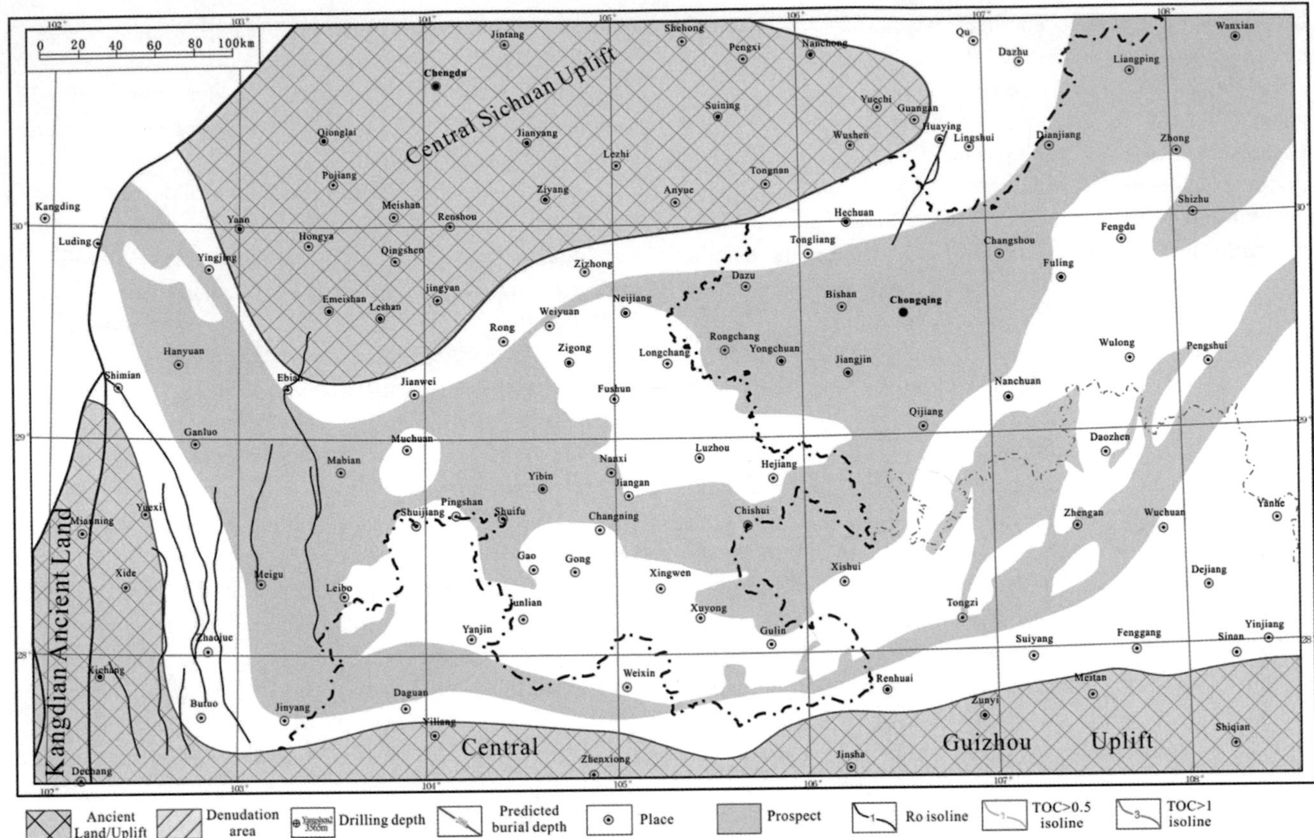

Fig. 4.96 Distribution of prospect area of Ordovician Wufeng Formation in Southern Sichuan and its periphery

54.7, carbonate minerals are less than 10% and clay minerals are about 37%.

The most favorable area of Fuling (i): distributed in the high and steep faulted fold belt of eastern Sichuan Basin, with a buried depth of less than 4000 m and stable distribution; the content of brittle minerals such as quartz and carbonate are more than 50%. Nanoscale pores, especially microcracks, are developed. Jurassic strata are mainly exposed on the surface, and Triassic system is exposed in syncline zone. Here is abnormal high value distribution are with pressure coefficient 1.5–2.0 and good preservation conditions. In the concentrated distribution area of stress field, fracturing has great potential. The exploration degree is relatively low, and a breakthrough of shale gas scale has been achieved in Jiaoshiba of Fuling. Jiaoye 1 HF has been tested with 203,000 m^3, and basically all Wells have been drilled with high production and good exploration prospects.

The Xuyong-Gulin favorable area (ii): This area is located in the low and slow anticlinal belt of Southern Sichuan, and the main structure is Baiyanglin-Dazhai anticline, Da 'anliang syncline and Laduyan anticlinal belt. Most of the area is covered by Triassic-Jurassic strata, followed by Silurian strata, and a small amount of Ordovician strata are developed in the Southern part of the block. The average organic carbon content in Wufeng Formation of Xulong Macheng, Gulin Daping and Xulong Moni sections is generally high, about 4.44% in Xulong Macheng and 5.95% in Gulin Daping area, indicating good hydrocarbon generation potential. Kerogen type II is the main type, Ro is in the range of 1.8–2.3%, which is in the middle and late stage of evolution. In terms of brittle mineral content, debris mineral content accounts for 45.6%. Carbonate minerals account for 30%, and clay minerals account for 24.4%. Compared with the most favorable blocks in Qijiang-Xishui or Wulong-Daozhen, the content of carbonate minerals increases, carbonate minerals begin to develop, the content of brittle minerals is relatively high, and it is easy to fracture, the buried depth is between 2000 and 3000 m, and fractures are relatively undeveloped, and the preservation conditions are satisfactory.

The Changning-Junlian favorable area (II) is located on the basis of a low anticlinal structural belt, with blocks changing. Nine holes have been drilled. The Ning 206 wells exhibit shale gas, including Ning 201-H1 gas wells, with a value of 50,000 m^3/d. To implement the gas recovery project, Gong Xien Luo town and surrounding towns have been selected, which have more than 1300 households running on the provided 450,000 m^3 of high-quality shale gas. The TOC of the Wufeng Formation in the Shuanghe area in Changning is 2.41–4.29%, with an average value of 3.13% and high hydrocarbon generation potential. Kerogen type ii is the

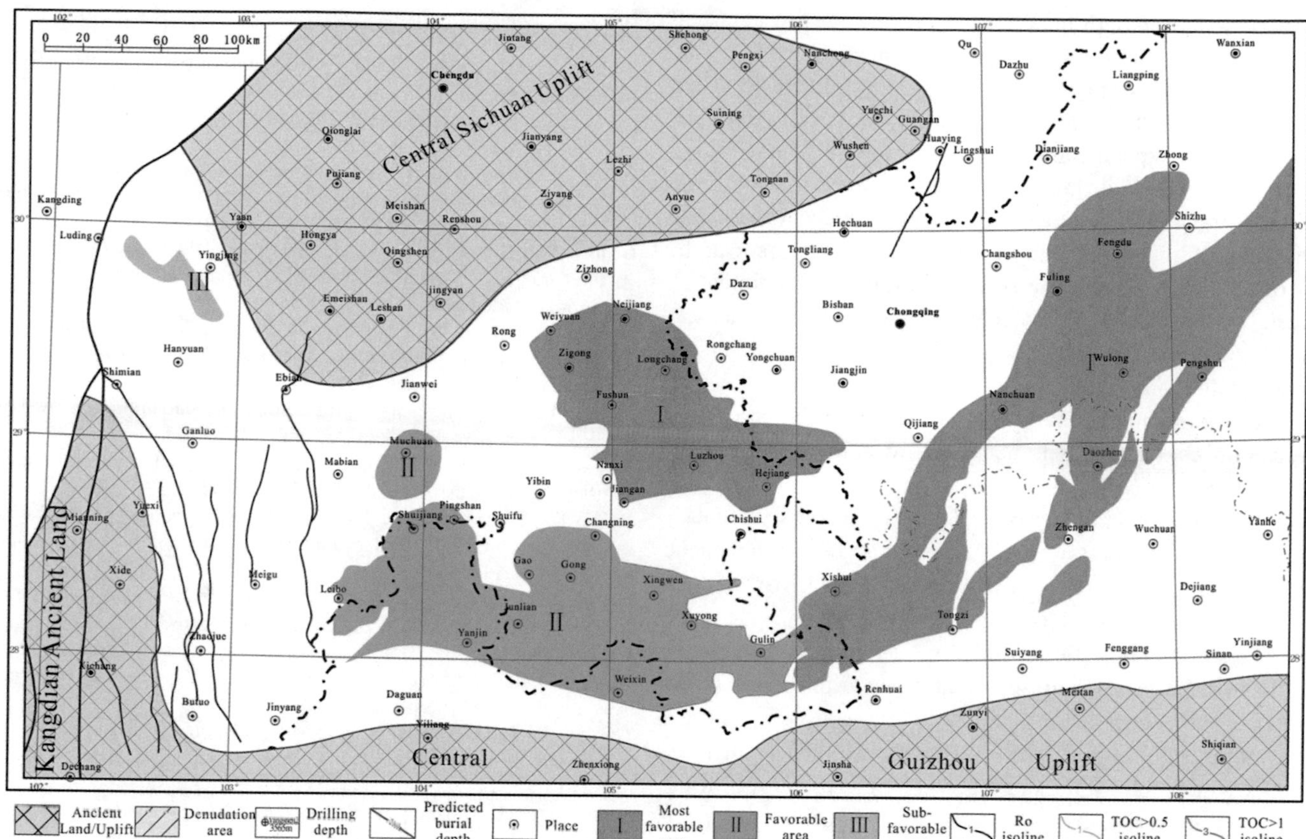

Fig. 4.97 Distribution of favorable area of Ordovician Wufeng Formation in Southern Sichuan and its periphery

main type, and the Ro is 1.730–2.269%, with an average of 1.846, reaching the gas generation stage. In terms of the brittle mineral content, clastic minerals account for 57.5%. Carbonate minerals account for 23.75%, clay minerals account for approximately 18.75%, and carbonate-type minerals have a high content. In addition, the content of brittle minerals is high, the rock is easily fractured and buried within 3500 m, and fractures are relatively undeveloped and have satisfactory preservation conditions.

The Leibo-Suijiang favorable area (ii) is distributed in the paleo-uplift tectonic belt in southwest Sichuan. The Wufeng Formation black shale has a stable distribution, the organic carbon content is 2.36–3.13%, the quartz + carbonate and other brittle mineral content are more than 70%, and the nanoscale pores and microfractures are developed. The surface is mainly exposed in the Cambrian-Jurassic system, with little fracture development and high preservation conditions.

The Yingjing sub-favorable area (iii) is distributed in the western Sichuan lower lung structural belt, western Sichuan and Yunnan between the northeast thrust fold belt, and the Hanyuan car top mountain section. The organic carbon content is 3%, the Ro content is between 3.122 and 3.153, and quartz and feldspar detritus minerals account for approximately 20%. The carbonate mineral content is high (67%). The content in the western Sichuan region is large, the elevation difference indicates fault development, and the preservation condition is ordinary (Fig. 4.97).

The Fuling shale gas field is located in Jiaoshiba town, Fuling District, Chongqing, eastern Sichuan Basin, in south Sichuan and adjacent areas (Fig. 4.95). According to research on the Fuling shale gas field by Guo and Liu (2013), Guo and Zhang (2013), Guo et al. (2014), Zhang et al. (2015), Feng and Mou (2017), He et al. (2016), Fang and Meng (2020) and Wei et al. (2010), the main factor controlling the shale gas enrichment in the Ordovician Wufeng Formation-Silurian Longmaxi Formation in the Jiaoshiba area of Fuling is the high-quality shale in deep-water shelf facies, which facilitates the generation of shale gas and provides a material basis for shale gas enrichment. The abundance of organic matter in shale provides favorable storage space for shale gas accumulation. Satisfactory roof and floor conditions and the formation of independent overpressure storage sites lead to the provision of satisfactory preservation conditions for shale gas enrichment. Guo (2014) and Wang (2015) proposed the theory of "two-element enrichment" and "three-element enrichment" for shale gas in southern marine facies, which indicate that the presence of high-quality shale in the deep-water shelf facies is the basis for shale gas enrichment.

Based on this analysis, we verify that lithofacies paleogeography can be used as a guide for shale gas geological investigation, specifically, "as long as the diagenetic evolution of the Ordovician Wufeng Formation-Silurian Longmaxi Formation in southern Sichuan and adjacent areas is in the middle-diagenetic stage, shale gas reservoirs can be formed". In summary, the search for favorable sedimentary facies (subfacies or microfacies) is the basis of shale gas geological surveys and a key premise for shale gas favorable area evaluation.

For the geological survey of shale gas in the process of selection evaluation, it is necessary to study the structural background, classify the sedimentary facies, identify the enrichment characteristics of shale gas reservoir sedimentary facies and sedimentary microfacies, analyze the diagenetic evolution, select the appropriate parameters such as the line, TOC, Ro and mineral components and evaluate the shale gas. Lithofacies paleogeography, as a comprehensive reflection of facies and sedimentary environments, is the main controlling factor and basic element affecting the development of organic-rich shale and must thus be the basis and key technology for shale gas geological surveys. In the shale gas geology survey phase, the geological conditions of the parameters of the shale gas evaluation must be analyzed. Combined with the actual data, the sedimentary facies (lithofacies paleogeographic) plane exhibition layout must be selected based on the composite organic carbon content contour map, rich organic shale thickness contour map and vitrinite reflectance contour map (main control factors). Next, the mineral components (clay minerals and brittle minerals) must be superimposed with the plane distribution map. Finally, the buried depth map of the study area must be verified and corrected by the gas content data.

References

Algeo TJ, Maynard JB (2004) Trace-element behavior and redox facies in core shales of Upper Pennsylvanian Kansas-type cyclothems. Chem Geol 206(3–4):289–331

Algeo TJ, Lyons TW (2006) Mo–total organic carbon covariation in modern anoxic marine environments: Implicationsfor analysis of paleoredox and paleohydrographic conditions. Paleoceanography 21(21):279–298

Algeo TJ, Tribovillard N (2009) Environmental analysis of paleoceanographic systems based onmolybdenum–uranium covariation. Chem Geol 268(3–4):211–225

Arthur MA, Sageman BB (1994) Marine black shale depositional mechanisms and environments of ancient deposits. Annu Rev Earth Planet Sci 22:499–551

Bai ZR (2012) Sedimentary characteristics of the lower Cambrian Niutitang Fm shale and evaluation parameters of shale gas in Zunyi-Qijiang area. China University of Geosciences, Beijing

Bai BJ, Elgmati M, Zhang H et al (2013) Rock characterization of Fayetteville shale gas plays. Fuel 105:645–652

Barbera G, Ctitelli S, Mazzoleni P (2011) etrology and geochemistry of Cretaceous sedimentary rocks of the Monte Soro Unit(Sicily, Italy): Constraints on weathering. Diagenesis and provenance. J Geol 119:51–68

Beener RA, Raisewell R (1983) Purial of organic carbon and pyrite suffer in sediments over Phanerozoic time: a new theory. Geochim Cosmochim Acta 47:855–862

Bennett RH, Bryant WR, Hulbert MH, et al (1991) Microstructure of fine–grained sediments: drom Mud to Shale. Springer–Verleg, Berlin, Heidelberg

Boles JR, Franks SG (1979) Clay diagenesis in Wilcox sandstones of southwest Texas: Implications of smectite diagenesis on sandstone cementation. J Sediment Petrol 49:55–70

Bowker KA (2003) Recent development of the Barnett Shale play, Fort Worth Basin. W Tex Geol Soc Bull 42(6):4–11

Bowker KA (2007) Barnett Shale gas production Fort Worth Basin: issues and discussion. AAPG Bull 91(4):523–533

Bust VK, Majid AA, Oletu JU et al (2013) The petrophysics of shale gas reservoirs: technical challenges and pragmatic solutions. Pet Geosci 19:91–103

Cai K, Hi LX, Wang CS (2008) The cretaceous clay minerals and paleoclimate in Sichuan Basin. Acta Geol Sin 82(1):115–123

Cai XY, Zhao PR, Gao B et al (2021) Sinopec's shale gas development achievements during the "Thirteenth Five-Year Plan" period and outlook for the future. Oil Gas Geol 42(1):16–27

Calvert SE (1987) Oceanographic controls on the accumulation of organic matter in marine sediments. Geol Soc London Spec Publ 26 (1):137–151

Chalmers GRL, Bustin RM, Bustin AAM (2011) Geological controls on matrix permeability of the Doig-Montney hybrid shale-gas–tight-gas reservoir, Northeastern British Columbia (NTS 093P). Geosci BC Summary Activities 2011:2012–2021

Charvet J, Shu LS, Faure M et al (2010) Structural development of the lower Paleozoic belt of South China: Genesis of an inracontinental orogen. Asian Earth Science 39(4):309–330

Chen B, Guan XQ, Ma J (2011a) The comparison of Paleozoic Shale-gas potential in Upper Yangtze Area with Barnett Shale-gas in North America. J Oil Gas Tech 33(12):23–27

Chen DZ, Wang JG, Yan DT et al (2011b) Environmental dynamics of organic accumulation for the principal Paleozoic source rocks on Yangtze block. Chinese J Geology (scientia Geologica Sinica) 46 (1):5–26

Chen SB, Zhu YM, Wang HY et al (2011c) Characteristics and significance of mineral compositions of Lower Silurian Longmaxi Formation shale gas reservoir in the Southern margin of Sichuan Basin. Acta Petrolei Sinica 32(5):775–782

Chen C (2018) Research on paleoceanography, paleoclimate and formation mechanism of source rock during geologic transition period from late Ordovician to early Silurian in Southern Sichuan Province-Northern Guizhou Province, South China. Geology University of China, Wuhan

Chen X, Chen Q, Zhen Y et al (2018) Circumjacent distribution pattern of the Lungmachian graptolitic black shale (early Silurian) on the Yichang Uplift and its peripheral region. Sci China Earth Sci 48 (9):1198–1206

Chen HD, Huang FX, Xu SL et al (2009) Distribution rule and main controlling factors of the marine facies hydrocarbon substances in the middle and upper parts of Yangtze region, China. J Chengdu Univ Technology (Sci & Tech Edition) 36(6):569–577

Chen SB, Zhu YM, Wang HY et al (2012) Structure characteristics and accumulation significance of nanopores in Longmaxi shale gas reservoir in the Southern Sichuan Basin. J China Coal Soc 37 (3):438–444

Chen JF, Zhang SC, Bao ZD et al (2006a) Main sedimentary environments and influencing factors for development of marine organic-rich source rocks. Marine Origin Petroleum Geol 11(3):49–54

Chen JF, Zhang SC, Sun SL et al (2006b) Main factors influencing marine carbonate source rock formation. Acta Geol Sin 03:467–472

Chen X, Rong J, Fan J et al (2006c) The global boundary stratotype section and point (GSSP) for the base of the Hirnantian Stage (the uppermost of the Ordovician System). Episodes J Inter Geosci 29 (3):183–196

Chen SB, Xia XH, et al (2013a) Classification of pore structures in shale gas reservoir at the Longmaxi Formation in the South of Sichuan Basin. J China Coal Soc 38(5):760–765

Chen WL, Zhou W, Luo P et al (2013b) Analysis of the shale gas reservoir in the Lower Silurian Longmaxi Formatio, Changxin 1 well, Southeast Sichuan Basin, China. Acta Petrologica Sinica 29 (3):1073–1086

Chen SB, Zhu YM, Wang HY (2010) Research status and trends of shale gas in China. Acta Petrolei Sinica 31(4):689–694

Chen XJ, Bao SJ, Hou DJ et al (2021) Methods and key parameters of shale gas resources evaluation. Pet Explor Dev 39(5):566–571

Chen X, Fan JX, Wang WH et al (2017) Stage progressive distribution pattern of the Lungmachi black graptolitic shales from Guizhou to Chongqing, central China. Sci China Earth Sci 47(6):720–732

Chen X, Fan JX, Zhang YD et al (2015) Subdivision and delineation of the Wufeng and Longmaxi black shales in the subsurface areas of the Yangtze Platform. J Stratigr 39(4):351–358

Chen X, Rong JY, Fan J, Zhan, et al (2000) Biostratigraphy of the Hirnantian substage in the Yangtze region. J Stratigraphy 200, 24 (3):169–175

Chen YJ, Deng J, Hu GX (1996) Environmental constraints on the contents and distribution forms of trace elements in sediments. Geology-Geochem 3:97–105

Cheng LX, Wang YJ, Chen HD et al (2013) Sedimentary and burial environment of black shales of Sinian to Early Palaeozoic in Upper Yangtze region. Acta Petrologica Sinica 29(8):2906–2912

Clarkson CR, Solano N, Bustin RM et al (2013) Pore structure characterization of North American shale gas reservoirs using USANS/SANS, gas adsorption, and mercury intrusion. Fuel 103:606–616

Curtis JB (2002) Fractured shale–gas systems. AAPG Bull 86 (11):1921–1938

Daniel JK, Ross DJK, Bustin RM (2009) The importance of shale composition and pore structure upon gas storage potential of gas reservoirs. Mar Pet Geol 26:916–927

Daoudi L, Rocha F, Ouajhain B et al (2008) Palaeoenvironmental significance of clay minerals in Upper Cenomanian-Turonian sediments of the Western High Atlas Basin (Morocco). Clay Miner 43:615–630

Dong DZ, Wang YM, Huang XN et al (2016) Discussion about geological characteristics, resource evaluation methods and its key parameters of shale gas in China. Nat Gas Geosci 27(09):1583–1601

Dow WG (1977) Kerogen studies and geological interpretations. J Geochem Explor 7:79–99

Fan JX, Michael J, Chen X et al (2012) Biostratigraphy and geography of the Ordovician-Silurian Lungmachi black shales in South China. Sci China Earth Sci 42(1):130–139

Fang DL, Meng ZY (2020) Main controlling factors of shale gas enrichment and high yield: a case study of Wufeng-Longmaxi Formations in Fuling area, Sichuan Basin. Pet Geol Exp 42(1):37–41

Fang YT, Bian LZ, Yu JH et al (1993) Sedimentary environment pattern of Yangtze Plate in Wufeng age of Late Ordovician. Acta Sedimentol Sin 11(3):7–12

Faure M, Shu LS, Wang B et al (2009) Intracontinental subduction: a possible mechanism for the Early Palaeozoic Orogen of SE China. Terre Nova 21(5):360–368

Feng JH, Mou ZH (2017) Main factors controlling the enrichment of shale gas in Wufeng Formation-Longmaxi Formation in Jiaoshiba area, Fuling shale gas field. China Petrol Explo 22(3):32–39

Feng ZZ, Wang YH, Liu HJ et al (1994) Chinese sedimentology. Petroleum Industry Press, Beijing

Fu WJ (2000) Influence of clay minerals on sandstone reservoir properties. J Palaeogeography 2(3):59–68

Fu XD, Qin JZ, Teng GE (2008) Evaluation on excellent marine hydrocarbon source layers in Southeast area of The Sichuan Basin-an example from well D-1. Pet Geol Exp 30(6):621–628

Fu XD, Qin JZ, Teng GE (2011) Mineral components of source rocks and their petroleum significance: a case from Paleozoic marine source rocks in the Middle-Upper Yangtze region. Pet Explor Dev 38(6):671–684

Ge XY (2012) Sedimentary facies and lithofacies palogeography of ordovician in Central and Southern Hunan. Shandong University of Science and Technology, TsingTao

Ge ZW, Fan L (2013) Some notable problems about shale gas in the scientific research. Petrol Geol Recov Effi 20(6):19–22

Geology of Sichuan Province (1991) Sichuan Bureau of Geology & Mineral Resources

Guo L (2014) Geological characteristics of the black shale and prospective evaluation of the shale gas in Lower Cambrian of Upper Yangtze Area. Chengdu Univerisity of Technology, Chengdu

Guo QL, Chen XM, Song HQ et al (2013) Evolution and models of shale porosity during Burial process. Nat Gas Geosci 24(3):439–449

Guo TL, Liu RB (2013) Implications from marine shale gas exploration breakthrough in complicated structural area at high thermal stage: taking Longmaxi Formation in Well JY1 as an example. Nat Gas Geosci 24(4):643–651

Guo TL, Zhang HR (2013) Formation and enrichment mode of Jiaoshiba shale gas field, Sichuan Basin. Pet Explor Dev 41(1):28–36

Guo XS, Guo TL, Wei ZH et al (2012) Thoughts on shale gas exploration in Southern China. Eng Sci 14(6):101–105

Guo XS, Hu DF, Wen ZD et al (2014) Major factors controlling the accumulation and high productivity in marine shale gas in the Lower Paleozoic of Sichuan Basin and its periphery: a case study of the Wufeng-Longmaxi Formation of Jiaoshiba area. Geology in China 41(3):893–901

Guo YH, Li ZF, Li DH, et al (2004) Lithofacies palaeogeography of the Early Silurian in Sichuan area. J Palaeogeography (Chinese Edition) 6(1):20–29

Hackley PC (2012) Geological and geochemical characterization of the Lower Cretaceous Pearsall Formation, Maverick Basin, south Texas: a future shale gas resource? AAPG Bull 96(8):1449–1482

Han SB, Zhang JC, Horsfield B et al (2013) Pore types and characteristics of shale gas reservoir: a case study of Lower Paleozoic shale in Southeast Chongqing. Earth Sci Front 20(3):247–253

He L, Wang YP, Chen DF et al (2018) Relationship between sedimentary environment and organic matter accumulation in the black shale of Wufeng-Longmaxi Formations in Nanchuan area, Chongqing. Nat Gas Geosci 30(2):57–72

He XH, Liu Z, Liang QS et al (2010) The influence of burial history on mudstone compaction. Earth Sci Front 17(4):167–173

He ZL, Nie HK, Zhang YY (2016) The main factors of shale gas enrichment of Ordovician Wufeng Formation-Silurian Longmaxi Formation in the Sichuan Basin and its periphery. Earth Sci Front 23(2):008–017

Hickey JJ, Henk B (2007) Lithofacies summary of the Mississippian Barnett Shale, Mitchell 2 T.P. Sims well, Wise County, Texas. AAPG Bull 91(4):437–443

Hill RJ, Jarvie DM, Zumberge J et al (2007) Oil and gas geochemistry and petroleum systems of the Fort Worth Basin. AAPG Bull 91 (4):445–473

Hu CP, Xu DX (2012) Study on shale reservoir evaluation factors. Natural Gas and Oil 30(5):38–42

Huang FX, Chen HD, Hou MC et al (2011) Filling process and evolutionary model of sedimentary sequence of Middle-Upper Yangtze craton in Caledonian (Cambrian-Silurian) [J]. Acta Petrologica Sinica 27(8):2299–2317

Huang JB, Huang HT, Wu GX et al (2012a) Geochemical characteristics and formation mechanism of Eocene lacustrine organic-rich shales in the Beibuwan Basin [J]. Acta Petrolei Sinica 33(1):25–31

Huang JL, Zou CN, Li JZ et al (2012b) Shale gas accumulation conditions and favorable zones of Silurian Longmaxi Formation in south Sichuan Basin, China. J China Coal Soc 37(5):782–787

Huang S, Wang GZ, Zou B et al (2012c) Preferred targets of the shale gas from Silurian Longmaxi Formation in middle-Upper Yangtze of China. J Chengdu Univ Techn (Science & Technology Edition) 39 (2):190–197

Huang SJ, Huang KK, Feng LW et al (2009) Mass exchanges among feldspar, kaolinite and illite and their influences on secondary porosity formation in clastic diagenesis: a case study on the Upper Paleozoic, Ordos Basinand Xujiahe Formation, Western Sichuan depression. Geochimica 38(5):498–506

Huang ZK, Chen JP, Xue HT (2013) Digital core technology–based microstructure characteristics and main controlling factors of continental shale oil: a case study of shale in cretaceous Qingshankou Formation In Songliao Basin. Geol Res 40(1):58–65

IUPAC (International Union of Pure and Applied Chemistry); (1994) Physical chemistry division commission on colloid and surface chemistry, subcommittee on characterization of porous soild (Technical Report). Pure Appl Chem 66(8):1739–1758

Jarvie D M, Hill R J, Pollastro R m. 2005. Assessment of the gas potential and yields from shales: the Barnett shale model [A]. Cardott B J. Unconventional energy resources in the Southern Midcontinent [C]. Norman Oklahoma University Press, 37–50.

Jarvie DM, Hill RJ, Ruble TE et al (2007) Unconventional shale–gas system: The Mississippian Barnett Shale of north–central Texas as one model for thermogenic shale–gas assessment. AAPG Bull 91 (4):475–499

Jarvie DM (1991) Total organic carbon (TOC) analysis, in R. K. Merrill, ed., Source and migration processes and evaluation techniques: AAPG Treatise of Petroleum Geology. In: Handbook of Petroleum Geology, 113–118

Jeong GY, Yoon HI (2001) The origin of clay minerals in soils of King George Island, South Shetland Island, South Shetland Island, west Antarctica, and its implications for the clay–mineral compositions of marine sediments. J Sedimentary 71(5):833–842

Jiang LL, Pan CC, Jinzhong L (2009) Experimental study on effects of minerals on oil cracking. Geochemisca 38(2):165–173

Jiang ZG (2018) The main controlling factors and depotional model of organic matter accumulation in the Wufeng-Longmaxi formations in the Sichuan Basin. Southeast University of Technology, Nan Chang

Jiang ZX, Liang C, Wu J et al (2013) Several issues in sedimentological studies on hydrocarbon–bearing fine–grained sedimentary rocks. Acta Petrolei Sinica 34(6):1031–1039

Jiang ZX (2003) Sedimentology. Petroleum Industry Press, Beijing, 159–164

Jin JL, Zou CN, Li JZ et al (2012) Shale gas accumulation conditions and favorable zones of Silurian Longmaxi Formation in south Sichuan Basin, China. J China Coal Soc 37(5):782–787

Jin N, Li AC, Liu HZ et al (2007) Clay minerals in surface sediment of the Northwest Parece Vela Basin: Distribution and provenance. Oceanologia Et Limnologia Sinica 38(6):504–511

Jin Q (2001) Importance and research about effective hydrocarbon source rocks. Oil & Gas Recov Tech 8(1):1–4

Kamp PC (2008) Smectite–liiete–muscoveite transformations, quarzt dissolution, and silica release in shales. Clays Clay Miner 56(1):66–81

Kang YL, Luo PY, Jiao D et al (1998) Clay minerals and potential formation damage of tight gas–bearing sandstone in Western Sichuan basin. J Southwest Petroleum Institute 60(15):1–5

Khormali F, Abtahi A (2003) Origin and distribution of clay minerals in calcareous arid and semi–arid soils of Fars Province, Southern Iran. Clay Miner 38:511–527

Le H, Chang HG, Fan Y, Chen F, Chen PF (2020) Construction and prospect of China's shale gas technical standard system. Nat Gas Ind 40(4):1–8

Lee YI, Lim DH (2008) Sandstone diagenesis of the Lower Cretaceous sandstone Group, Gyeongsang Basin, Southeastern Korea: Implication for composition and paleoenvironmental control [J]. Island Arc 17:152–171

Lézin C, Andreu B, Pellenard P et al (2013) Geochemical disturbance and paleoenvironmental changes during the Early Toarcian in NW Europe. Chem Geol 341:1–15

Li DH, Li JZ, Huang JL et al (2014a) An important role of volcanic ash in the formation of shale plays and its inspiration. Nat Gas Ind 34 (005):56–65

Li JQ, Gao YQ, Hua CX (2014b) Marine shale gas evaluation system of regional selection in South China: enlightenment from North American exploration experience. Petroleum Geol Recov Eff 21 (4):23–27

Li GP, Zhan RB, Wu RC (2009a) Response of Hirnantia Fauna to the environmental changes before the second phase of late Ordovician mass extinction: example from the Kuanyinchiao Formation at Shuanghe, Southern Sichuan, southwest China. Geol J China Univ 15:304–317

Li SJ, Xiao KH, Wo YJ et al (2009b) Palaeo-environment restoration of Upper Ordovician-Lower Silurian hydrocarbon source rock in Middle-Upper Yangtze area. Acta Petrologica Et Mineralogica 28 (5):450–458

Li J, Yu BS, Guo F (2013a) Depositional setting and tectonic background analysis on Lower Cambrian black shales in the North of Guizhou Province. Acta Sedimentol Sin 31(1):20–31

Li YJ, Liu H, Zhang LH et al (2013b) Lower limits of evaluation parameters for the lower Paleozoic Longmaxi shale gas in Southern Sichuan Province. Scientia Sinica (terrae) 43(7):1088–1095

Li YJ, Liu H, Zhang LH et al (2013c) Lower limits of evaluation parameters for the lower Paleozoic Longmaxi shale gas in Southern Sichuan Province. Science China. Earth Sci 56:710–717

Li J, Yu BS, Liu C et al (2012a) Clay minerals of black shale and their effects on physical properties of shale gas reservoirs in the Southeast of Chongqing: a case study from Lujiao outcrop section in Pengshui, Chongqing. Geoscience 26(4):732–740

Li YF, Fan TL, Gao ZQ et al (2012b) Sequence stratigraphy of Silurian black shale and its distribution in the Southeast area of Chongqing. Nat Gas Geosci 23(2):299–306

Li YX, Zhang JC, Jiang SL et al (2012c) Geologic evaluation and targets optimization of shale gas. Earth Sci Front 19(5):332–338

Li Q, Liu KY, Shao SW et al (2005a) Geological model and genesis mechanism of mudstone fractured oil and gas reservoirs: a case study of mudstone fractured reservoir in the third member of Shahekou area. China University of Geosciences Press, Wuhan, pp 1–29

Li Y, Matsumoto R, Kershaw S (2005b) Sedimentary and biotic–evidence of a warm–water enclave in the cooler oceans of the Latest

Ordovician glacialphase, Yangtze Platform, South China. The Island Arc 14(4):623–635

Li SJ, Xiao KH, Wo YJ et al (2008a) REE geochemical characteristics and their geological signification in Silurian, West of Hunan Province and North of Guizhou Province. Strategic Stud CAE 22 (2):273–282

Li SJ, Xiao KH, Wo YJ et al (2008b) Developmental controlling factors of upper Ordovician-Lower Silurian high quality source rocks in Marine sequenc, South China. Acta Sedimentologica Sinica 26 (5):160–168

Li Y, Wang JP, Zhang YY et al (2008c) The annotation on climate from the Ordovician–Silurian in South China. Prog Nat Sci 18 (11):1264–1270

Li SZ, Qiao DW, Feng ZG et al (2010) The status of worldwide shale gas exploration and its suggestion for China. Geol Bull China 29 (6):918–924

Li YF (2017) Geochemical characteristics and organic matter accumulation of late Ordovician-early Silurian shale in the upper Yangtze platform, and implications for paleoenvironment. Lanzhou University, Lan Zhou

Li YX, Qiao DW, Jiang WL et al (2011) Gas content of gas-bearing shale and its geological evaluation summary. Geol Bull China 30(2–3):308–317

Li ZM, Chen JQ, Shuyun G et al (1997) Sequence stratigraphy and sea level changes of Cambrian in the Southern margin of Ordos Basin. Earth Sci 22(5):479–483

Liang C, Jiang ZX, Yang YT et al (2012a) Characteristics of shale lithofacies and reservoir space of the Wufeng-Longmaxi Formation, Sichuan Basin. Pet Explor Dev 39(6):691–697

Liang T, Yang C et al (2012b) Characteristics of shale lithofacies and reservoir space of the Wufeng-Longmaxi Formation, Sichuan Basin. J Northeast Petrol Univ 39(6):691–697

Liang DG, Guo TL, Bian LZ et al (2009) Some progresses on studies of hydrocarbon generation and accumulation in marine sedimentary regions, Southern China (Part 3): controlling factors on the sedimentary facies and development of palaeozoic marine source rocks. Marine Origin Petroleum Geol 14(2):1–19

Lin LM, Zhang JC, Liu JX et al (2012) Favorable depth zone selection for shale gas prospecting. Earth Sci Front 19(3):259–263

Liu A, Li XB, Wang CS et al (2013) Analysis of geochemical feature and sediment environment for hydrocarbon source rocks of Cambrian in West Hunan-Hubei Area. Acta Sedimentol Sin 31 (6):1122–1132

Liu BJ, Xu XS, Pan XN et al (1993) Crustal evolution and mineralization of the ancient continental deposits in South China. Science Press, Beijing

Liu BJ, Xu XS (1994) Atlas of the Palaeogeography of South China. Science Press, Beijing

Liu BJ (1980) Sedimentology. Geological Publishing House, Beijing

Liu CL, Fan BJ, Yan G (2009a) Unconventional natural gas resources prospect in China. Petroleum Geology and Recovery Efficiency 19 (3):259–263

Liu GS, Xu F, Xu CH, et al (2009b) X-ray diffraction study of clay minerals in Well Ancan 1 of Hefei Basin and analysis of the diagenesis degree. J Hefei Univ of Technology (Natural Science) 32 (12):1911–1915

Liu HL, Wang HY, Liu RH et al (2010a) China shale gas resources and prospect potential. Acta Geol Sin 84(9):1374–1378

Liu W, Xu XS, Feng XT et al (2010b) Radiolarian siliceous rocks and palaeoenvironmental reconstruction for the Upper Ordovician Wufeng Formation in the middle-Upper Yangtze area. Sedimentary Geology Tethyan Geol 30(3):65–70

Liu HL, Wang HY, Sun SS et al (2016) The formation mechanism of over-pressure reservoirand target screening index in south China marine shale. Nat Gas Geosci 27(3):417–422

Liu HL, Wang HY (2012) Adsorptivity and influential factors of marine shales in South China. Nat Gas Ind 32(9):5–9

Liu L, Wang LJ, Yang YZ et al (2012a) Mineral composition and origin of authigenic quartz crystals in black mudstone, in well HX, Qingshankou Formation, Upper Cretaceous, Songliao Basin. J Jilin University (Earth Sci Edition) 42(5):1358–1365

Liu W, Xu XS, Yu Q et al (2012b) Lithofacies palaeogeography of the Late Ordovician Hirnantian in the middle-Upper Yangtze region of China. J Chengdu Univ Tech: Sci Tech Edition 39(1):32–39

Liu W, Yu Q, Yan JF et al (2012c) Characteristics of organic –rich mudstone reservoirs in the Silurian Longmaxi Formation in Upper Yangtze region. Oil Gas Geol 33(3):346–352

Liu RB (2015) Typical features of the first giant shale gas field in China. Nat Gas Geosci 26(08):1488–1498

Liu SG, Ma WX, Luba JS et al (2011) Characteristics of the shale gas reservoir rocks in the Lower Silurian Longmaxi Formation, East Sichuan basin, China. Acta Petrologica Sinica 27(8):2239–2252

Liu Wei XuXS, Feng XT et al (2010) Radiolarian sili- ceous rocks and palaeoenvironmental reconstruction for the Upper Ordovician Wufeng Formation in the middle-Upper Yangtze area. Sedimentary Geology and Tethyan Geology 30(3):65–70

Liu YL, Di SX, Xue XX (1998) Clay minerals from the Lower Cretaccousto Middle Jurassic strata in the Bazhou depression, Eastern Qilian foldedbelt. Journal of Palaeogeography 18(5):10–15

Liu Y (1985) Clay minerals of Late Cretaceous Songliao Basin and their sedimentary environment. Acta Sedimentol Sin 3:131–139

Long H, Wang CH, Liu YP et al (2007) Application of clay minerals in paleoenviroment research. J Salt Lake Res 15(2):456–459

Long PY, Zhang JC, Jiang WL et al (2012) Analysis on pores forming features and its influence factors of reservoir well Yuye–1. J Central South Univ (Sci Technology) 43(10):3954–3963

Loucks GR, Bustin MR, Power IM (2012) Characterization of gas shale pore systems by porosimetry, pycnonmetry, surface area, and field emission scanning electron microscopy/transmission electron microscope image analyses: examples from the Barnett, Woodford, Haynesville, Marcellus and Doig units. AAPG Bull 96(6):1099–1119

Loucks RG, Reed RM, Ruppel SC et al (2009) Morphology, genesis, and distribution of nanometer–scale pores in siliceous mudstones of the Mississippian Barnett shale. J Sediment Res 79:848–861

Loucks RG, Ruppel SC (2007) Mississippian Barnett Shale: Lithofacies and depositional setting of a deep–water shale–gas succession in the Fort Worth Basin, Texas. AAPG Bull 91(4):523–533

Loydell DK (1998) Early Silurian sea-level changes. Geol Mag 135 (4):447–471

Lüning S, Shahin YM, Loydell D et al (2005) Anatomy of a word-class source rock: distribution and depositional model of Silurian organic- rich shales in Jordan and implications for hydrocarbon potential. AAPG Bull 89(10):1397–1427

Luo JL, Morad S, Yan SK et al (2001) Reconstruction of diagenesis of fluvial and lacustrine delta facies sandstone and its influence on reservoir physical evolution: A case study of Jurassic–Upper Triassic sandstone in Yanchang Oilfield. Science in China (Series d) 31(12):1006–1016

Lv BQ, Wang HG, Hu WS et al (2004) Relationship between paleozoic upwelling facies and hydrocarbon in Southeastern marginal Yangtze block. Mar Geol Q Geol 24(4):29–35

Macquaker JHS, Keller MA, Davies SJ (2010) Algal blooms and "marine snow": Mechanisms that enhance preservation of organic carbon in ancient fine-grained sediments. J Sediment Res 80:934–942

Martineau DF (2007) History of the Newark East field and the Barnett Shale as a gas reservoir. AAPG Bull 91(4):399–403

Mastalerz M, Schimmelmann A, Drobniak A et al (2013) Porosity of Devonian and Mississippian New Albany Shale across a maturation gradient: insights from organic petrology, gas adsorption, and mercury intrusion. AAPG Bull 97(10):1621–1643

Merriman RJ (2005) Clay minerals and sedimentary basin history. Eur J Mineral 17:7–20

Milici RC, Ryder RT, Repetski JE (2006) Exploration for hydrocarbons in the Southern Appalachian basin–an overview. Geol Soc America Abst with Prog 38(3):10–11

Mou CL, Ge XY, Xu XS, et al (2014). Lithofacies palaeogeography of the Late Ordovician and its petroleum geological significance in Middle-Upper Yangtze Region. J Palaeogeography (Chinese Edition) 16(4):427–440

Mou CL, Xu XS (2010) Sedimentary evolution and petroleum geology in South China during the Early Palaeozoic. Sedimentary Geol Tethyan Geol 30(3):24–29

Mou CL, Zhou KK, Liang W et al (2011) Early Paleozoic sedimentary environment of hydrocarbon source rocks in the middle-Upper Yangtze region and petroleum and gas exploration. Acta Geol Sin 85(4):526–532

Mu SG, Zhang YM (1994) Reservoirs pores evolution under the control of diagenesis and stage. J Southwest–China Petrol Institute 16 (3):22–27

Nie HK, Bao SJ, Bao SJ, et al (2012a) Shale gas accumulation conditions of the Upper Ordovician–Lower Silurian in Sichuan Basin and its periphery. Oil & Gas Geol 33(3):337–345

Nie HK, Bao SJ, Bo G et al (2012b) A study of shale gas preservation conditions for the Lower Paleozoic in Sichuan Basin and its periphery. Earth Sci Front 19(3):280–294

O'Brien NR, Slatt RM (1990a) Argillaceous rock atlas. Springer–Verlag, New York, p 141

Passey QR, Bohacs KM, Esch WL, et al (2010) From oil-prone source rock to gas-producing shale reservoir-geologic and petrophy characterization of unconventional shale gas reservoir. In: International Oil and Gas Conference

Pedersen TF, Calvert SE (1990) Anoxia vs. productivity: what conrols the formation of organic carbon rich sediments and sedimentary rocks. AAPG Bull 4:454–466

Pollastro RM, Jarvie DM, Hill RJ et al (2007) Adams Geologic frame–work of the Mississippian Barnett Shale, Barnett-Paleozoic total petroleum system, Bend arch–Fort Worth Basin, Texas. AAPG Bull 91(4):405–436

Qian JZ, Fu XD, Shen BJ et al (2010a) Characteristics of ultramicroscopic organic lithology of excellent marine shale in the upper Permian sequence, Sichuan Basin. Pet Geol Exp 32(2):164–171

Qian JZ, Tao GL, Teng GE et al (2010b) Hydrocarbon-forming organisms in excellent marine source rocks in South China. Pet Geol Exp 32(3):262–269

Qian JZ, Teng GE, Fu XD (2009) Study of forming condition on marine excellent source rocks and its evaluation. Petrol Geol Exp 31 (4):366–372, 378

Qin JZ, Tao GL, Tenger, et al (2010) Hydrocarbon–forming organisms in excellent marine source rocks in South China. Petrol Geol Exp 32 (3):262–269

Qiu Z, Zou CN, Wang HY et al (2019) Discussion on characteristics and controlling factors of differential enrichment of Wufeng-Longmaxi Formations shale gas in South China. Nat Gas Geosci 31(2):163–175

Ran B, Liu SG, Sun W et al (2013) Redefinition of pore size characteristics in Wufeng Formation–Longmaxi Formation black shale of Qilongcun section in Southern Sichuan Basin, China. J Chengdu Univ Tech (Science & Tech Edition) 40(5):532–542

Rimmer SM, Thompson JA, Gooodmight SA (2004) Multiple controls on the preservation of organic matter in Devonian-Mississippian marian black shales: Geochemical and petrographic evidence. Palaeogeography, Paleoclimatology, Palaeoecology 215:125–154

Rong JY, Xu, Chen (1987) Faunal differentiation, biofacies and lithofacies pattern of late ordovician (Ashgillian) in South China. Acta Palaeontologica Sinica 26(5):507–526

Rong JY (1984) The ecological formation evidence and glacier activities for the regreession late Ordovician Epoch in Upper Yangtze region. J Stratigr 8(1):19–29

Ross DJK, Bustin MR (2008) Characterizing the shale gas resource potential of Devonian–Mississippian strata in the Western Canada sedimentary basin: application of an integrated formation evaluation. AAPG 92(1):87–125

Ross DJK, Bustin RM (2007) Shale gas potential of the Lower Jurassic Gordondale Member, Northeastern British Columbia, Canada. Bull Can Pet Geol 55(1):51–75

Schoepfer SD, Shen J, Wei H et al (2015) Total organic carbon, organic phosphorus, and biogenic barium fluxes as proxies for paleomarine productivity. Earth Sci Rev 149:23–52

Shen J, Schoepfer SD, Feng Q et al (2015) Marine productivity changes during the end-Permian crisis and Early Triassic recovery. Earth Sci Rev 149:136–162

Sheng XF (1958) Ordovician Trilobites of Southern China. Acta Palaeontologica Sinica 6(2):169–204

Shu LS, Yu JH, Jia D et al (2008) Early paleozoic orogenic belt in the eastern segment of South China. Geol Bull China 27(10):1581–1593

Shu LS (2012) An analysis of principal features of tectonic evolution in South China Block. Geol Bull China 31(7):1035–1053

Sichuan Provincial Bureau of Geology and Mineral Resources (1991) Regional Geology of Sichuan Province[M], Beijing: Geological Publishing House

Slatt RM, Abousleiman Y (2011) Merging quence stratigraphy and geomechanics for unconventional gas shales. Lead Edge 3:274–282

Slatt RM, O'Brien NR (2011) Pore types in the Barnett and Woodford gas shales: Contribution to understanding gas storage and migration pathways in fine–grained rocks. AAPG Bull 95(12):2017–2030

Slatt RM, Rodriguez ND (2012) Comparative sequence stratigraphy and organic geochemistry of North American unconventional gas shales: commonality or coincidence? J Nat Gas Sci Eng 8:68–84

Song ZJ, Zhang ZX, Yu JF, et al (2008) Study on distribution and material sources of clay minerals in surface sediments of the Southern Yellow Sea. J Shandong University Sci Tech (Natural Science) 27(3):1–4

Su WB, Li ZM, Ettensohn FR et al (2007) Distribution of black shale in the Wufeng-Longmaxi Formations (Ordovician–Silurian), South China: Major controlling factors and implications. Earth Sci: J China Univ Geosci 32(6):819–827

Su XH (2017) Sedimentological and geochemical responses of geological events at the Ordovician-silurian boundary-the example from Wufeng-Longmaxi formations in the Qiliao section, Shizhu. Chengdu University of Technology, Chengdu

Su Y, Lu YC, Liu ZH et al (2017) Development characteristics of bentonite in marine shale and its effect on shale reservoir quality: a case study of Wufeng Formation to Member 1of Longmaxi Formation, Fuling area. Acta Petrolei Sinica 38(12):1371–1380

Sui FG, Liu Q, Zhang LY (2007) Diagenetic evolution of source rocks and its significance to hydrocarbon expulsion in Shahejie Formation of Jiyang Depression. Acta Petrolei Sinica 28(6):12–16

Sun YZ (1931) Ordovician Trilobites of Central and Southern China. Acta Palaeontologica Sinica 7(1):1–47

Swanson VE (1961) Geology and geochemidtry of uranium in marine black shales–a review. U. S. Geol Survey Prof Paper 356–C:67–112

Tang Y, Xing Y, Li LZ et al (2012) Influence factors and evaluation meth- ods of the gas shale fracability. Earth Sci Front 19(5):356–363

Taylor SR, McLennan SM (1985) The continental crust: its composition and evolution. Blackwell, Oxford

Thyberg B, Jahren J (2011) Quartz cementation in mudstones: sheet–like quartz cement from clay mineral reactions during burial. Pet Geosci 17:53–63

Tribovillard N, Algeo TJ, Lyons T et al (2006) Trace metals as paleoredox and paleoproductivity proxies: an update. Chem Geol 232(1–2):12–32

Wang HJ, Liu GX (2011) Distribution and evolution of thermal field in middle and Upper Yangtze region. Pet Geol Exp 33(2):160–164

Wang HY, Liu YZ, Dong DZ et al (2013a) Scientific issues on effective development of marine shale gas in Southern China-ScienceDirect. Pet Explor Dev 40(5):574–578

Wang SQ, Wang SY, Man L et al (2013b) Appraisal method and key parameters for screening shale gas play. J Chengdu University of Technology (Science & Technology Edition) 40(6):609–620

Wang XP, Mu CL, Gong YY et al (2013c) Diagenetic evolution and facies of reservoirs in Member 8 of Permian Xiashihezi Formation in the Z30 block of Sulige gasfield. Acta Petrolei Sinica 34(5):883–895

Wang JL, Liu GJ, Wang WZ, et al (2013d) Characteristics of pore–fissure and permeability of shales in the Longmaxi Formation in Southeastern Sichuan Basin. J China Coal Soc 38(5):772–777

Wang QB, Liu RB, Li CY et al (2012a) Geologic condition of the Upper Ordovician-Lower Silurian shale gas in the Sichuan Basin and its periphery. J Chongqing Univ Sci Tech (Natural Sciences Edition) 14(4):17–21

Wang SJ, Yang T, Zhang GS et al (2012b) Shale gas enrichment factors and the selection and evaluation of the core area. Eng Sci 14(6):94–100

Wang YM, Dong DZ, Li JZ et al (2012c) Reservoir characteristics of shale gas in Longmaxi Formation of the Lower Silurian, Southern Sichuan. Acta Petrolei Sinica 33(4):551–561

Wang QC, Yan DT, Li SJ (2008) Tectonic-environmental model of the Lower Silurian high-quality hydrocarbon source rocks from South China. Acta Geol Sin 82(3):289–297

Wang SF, Dong DZ, Wang YM et al (2015a) Geochemical characteristics the sedimentation environment of the gas-enriched Shale in the Silurian Longmaxi Formation in the Sichuan Basin. Bull Mineral, Petrol Geochem 34(6):1203–1212

Wang YC, Liang W, Mou CL et al (2015b) The sedimentary response to Gondwana glaciation in Hirnantian (Ordovician) of the eastern Chongqing and the Northern Guizhou region, South China. Acta Sedimentol Sin 33(2):232–241

Wang YC, Mou CL, Liang W et al (2015c) Sedimentary facies and Palaeogeography of the Northern margin of the Yangzi block during the Hirnantian. Sedimentary Geol Tethyan Geol 35(3):19–26

Wang SJ, Wang LS, Huang JL et al (2009) Accumulation conditions of shale gas reservoirs in Silurian of the Upper Yangtze region. Nat Gas Ind 29(5):45–50

Wang SQ (2013) Shale gas exploration and appraisal in China: problems and discussion. Nat Gas Ind 33(12):13–29

Wang SY, Dai HM, Wang HQ et al (2000) Source rock feature of the South of the Dabashan and Mi–cangshan. Nat Gas Geosci 11(4–5):4–16

Wang YC (2016) Lithofacies paleogeography in Hirnantian stage of late Ordovician in Upper Yangtze region. Chengdu University of Techonology, Chengdu

Wang ZF, Zhang YF, Liang XL et al (2014) Characteristics of shale lithofacies formed under different hydrodynamic conditions in the Wufeng-Longmaxi Formation, Sichuan Basin. Acta Petrolei Sinica 35(4):623–632

Wang ZG (2015) Breakthrough of fuling shale gas exploration and development and its inspiration. Oil Gas Geol 36(1):1–6

Wei XF, Liu ZJ, Wang Q et al (2010) Analysis and thinking of the difference of Wufeng-Longmaxi shale gas enrichment conditions between Dingshan and Jiaoshiba areas in Southeastern Sichuan Basin. Nat Gas Geosci 31(8):1041–1051

Wignall PB, Maynard JR (1993) The sequence sratigraphy of transgressive black shales. Source Rocks Sequencestratigraphy Fram 37:35–47

Wu BH, Yang HN, Li SJ (1993) Mineral composition and sedimentation of sediments in the central Pacific. Geological Publishing House, Beijing

Wu JS, Yu BS, Zhang JC et al (2013) Pore characteristics and controlling factors in the organic–rich shale of the Lower Silurian Longmaxi Formation revealed by samples from a well in Southeastern Chongqing. Earth Sci Front 20(3):260–269

Wu LY, Gu XZ, Sheng ZW et al (1986) Rapid and quantitative evaluation of source rock pyrolysis. Science Press, Beijing

Wu LY, Lu YC, Jiang S et al (2018) Effects of volcanic activities in Ordovician Wufeng-Silurian Longmaxi period on organic-rich shale in the Upper Yangtze area, South China. Pet Explor Dev 45(05):62–72

Wu XH, Liu CW, Ju JC (1997) Study on the differentiation methods of allogenetic, primary and secondary claymineral. J Xianning Normal College 17(3):74–76

Wu YY, Wu SH, Cai ZQ et al (2005) Oil geology. Petroleum Industry Press, Beijing, pp 207–224

Xiao KH, Li SJ, Wang XW et al (2008) Hydrocarbon accumulation features and exploration direction in the Silurian of the Middle-Upper Yangtze Platform. Oil Gas Geol 29(5):589–596

Xie Y, Wang J, Li LX et al (2010) Distribution of thecre–taceous clayminerals in ordos Basin, China and its implication to sedimentary and diagenetic environment. Geol Bull China 29(1):93–104

Xu XS, Liu BJ, Lou XY et al (2004) Marine sedimentary basin analysis and oil and gas resources in central and Western China. Geological publishing house, Beijing

Xu XS, Qiang Xu, Pan GT et al (1996) The evolution of the South China continent and contrast of the global ancient geography. Geological Publishing House, Beijing

Xu XS, Mou CL, Yu Q, et al (2011) Proto-tethys evolution and lower hydrocarbon assemblages in the Middle and Upper Yangtze area. Internal report.

Yan DT, Chen DZ, Wang QC et al (2009a) Geochemical changes across the Ordovician-Silurian transition on the Yangtze Platform, South China. Science China Earth Sci 52:38–54

Yan DT, Chen DZ, Wang QC et al (2009b) Carbon and sulphur isotopic anomalies across the Ordovician-Silurian boundary on the Yangtze Platform, South China. Palaeogeogr Palaeoclimatol Palaeoecol 274:32–39

Yan DT, Wang JG, Wang ZZ (2009c) Biogenetic barium distribution from the Upper Ordovician to Lower Silurian in the Yangtze area and its significance to paleoproductivity. J Xi'an Shiyou Univ (Natural Science Edition) 24(4):16–19

Yan DT, Chen DZ, Wang QC et al (2010a) Large-scale climatic fluctuations in the latest Ordovician on the Yangtze block, South China. Geology 38:599–602

Yan JP, Liu CY, Zhang WG et al (2010b) Diagenetic characteristics of the lower porosity and permeability sandstones of the Upper Paleozoic in the South of Ordos Basin. Acta Geol Sin 84(2):272–279

Yan DT, Wang QC, Chen DZ et al (2008) Sedimentary environment and development controls of the hydrocarbon sources beds: the Upper Ordovician Wufeng Formation and the Lower Silurian Longmaxi Formation in the Yangtze Area. Acta Geol Sin 82 (3):321–327

Young GM, Nesbitt HW (1998) Processes controlling the distribution of Ti and Al in weathering profiles, siliciclastic sediments and sedimentary rocks. J Sediment Res 68(3):448–455

Yu BS (2013) Classification and characterization of gas shale pore system. Earth Sci Front 20(4):211–220

Yu Q, Li YX et al (2011a) Explorative prospect of shale gas of lower Silurian in middle-Upper Yangtze area. Xinjiang Petrol Geol 32 (4):353–355

Yu Q, Mou CL, Zhang HQ et al (2011b) Sedimentary evolution and reservoir distribution of Northern Upper Yangtze plate in Sinian-Early Paleozoic. Acta Petrologica Sinica 27(3):672–680

Yu XH (2008) Sedimentology of hydrocarbon reservoirs in clastic rock series. Petroleum Industry Press, Beijing, pp 87–89

Yuan HR, Nie Z, Liu JY et al (2007) Paleogene sedimentary characteristics and their paleoclimatic implications in the Baise basin, Guangxi, China. Acta Geol Sin 81(12):1692–1697

Yue H, Chang HG, Fan Y et al (2020) Construction and prospect of China's shale gas technical standard system. Nat Gas Ind 40(4):1–8

Zeng XL, Liu SG, Huang WM et al (2011) Comparison of Silurian Longmaxi Formation shale of Sichuan Basin in China and Carboniferous Barnett Formation shale of Fort Worth Basin in United States. Geol Bull China 30(2–3):372–384

Zhan RB, Liu JB, Percival IG, et al (2010a) Biodiversification of Late Ordovician Hirnantia Fauna on the Upper Yangtze Platform, South China. Sci China (Seri D): Earth Sci 40(9):1154–1163

Zhan RB, Liu JB, Percivalig et al (2010b) Biodiversification of late Ordovician Hirnantia fauna on the Upper Yangtze platform, South China. Sci China: Earth Sci 53(12):1800–1810

Zhao Q, Wang HY, Liu RH, et al (2008) Lobal Development and China"s Explo-ration for Shale Gas[J]. Nat Gas Technol Econ, 2 (3):11–14

Zhang CM, Zhang WS, Guo YH (2012) Sedimentary environment and its effect on hydrocarbon source rocks of Longmaxi Formation in Southeast Sichuan and Northern Guizhou. Earth Sci Front 19 (3):136–145

Zhang DW (2010) Strategic concepts of accelerating the survey, exploration and exploitation of shale gas resources in China. Oil Gas Geol 2(31):135–150

Zhang HF, Fang CL, Gao ZX et al (1999) Petroleum geology. Petroleum Industry Press, Beijing, pp 62–70

Zhang H, Gan H (2019) Progress of shale gas constituency evaluation research in China. Complex Hydro Res 12(2):32–35

Zhang JG, Yuan ZW (2002) Formation and potential of fractured mudstone reservoirs. Oil Gas Geol 23(4):336–338

Zhang J, Wang LS, Yang YM et al (2016) The development and application of the evaluation method of marine shale gas in Sichuan Basin. Nat Gas Geosci 27(03):433–441

Zhang RF (1992) Application of scanning electron microscopy to study of mineral change—transformation of feldspar into Cla Y minerals. Sci Geol Sin 1:66–70

Zhang SC, Zhang BM, Bian LZ et al (2005) Development constraints of marine source rocks in China. Earth Sci Front 12(3):39–48

Zhang TS, Kershaw S, Wan Y et al (2000) Geochemical and facies evidence for palaeoenvironmental change during the Late Ordovician Hirnantian glaciation in South Sichuan Province, China. Global Planet Change 20:133–152

Zhang WD, Guo M, Jiang ZX (2011) Parameters and method for shale gas reservoir evaluation. Nat Gas Geosci 22(6):1093–1099

Zhang XL, Li YF, Lv HG et al (2013a) Relationship between organic matter characteristics and depositional environment in the Silurian Longmaxi Formation in Sichuan Basin. J China Coal Soc 38(5):851–856

Zhang ZS, Hu PQ, Shen J et al (2013b) Mineral compositions and organic matter occurrence modes of Lower Silurian Longmaxi Formation of Sichuan Basin. J China Coal Soc 38(5):766–771

Zhang XM, Shi WZ, Xu QH et al (2015) Reservoir characteristics and controlling factors of shale gas in Jiaoshiba area, Sichuan Basin. Acta Petrolei Sinica 36(8):926–939

Zhao CL, Zhu XM (2001) Sedimentary Petrology (3rd edition). Petroleum Industry Press, Beijing

Zhao JZ, Fang CY, Zhang J et al (2011) Evaluation of China shale gas from the exploration and development of North America shale gas. J Xi'an Shiyou Univ (Natural Science Edition) 36(2):1–7

Zhao XY (2009) The impact of clay minerals on oil–gas reservoir. Xinjiang Petroleum Geology 30(4):533–535

Zheng RC, Li GH, Dai CC et al (2012) Basin-mountain coupling system and its sedimentary response in Sichuan Analogous Foreland Basin. Acta Geol Sin 86(1):170–180

Zheng RL (1989) Diagenesis of the quartz sandstones in coal–bearing formations in Shan-Ganning Basin and the evolution of their pore structure. Pet Explor Dev 16(6):31–40

Zhou L, Algeo TJ, Shen J et al (2015) Changes in marine productivity and redox conditions during the late Ordovician Hirnantian glaciation. Palaeogeogr Palaeoclimatol Palaeoecol 420:223–234

Zhou L, Su J, Huang JH et al (2011) A new paleoenvironmental index for anoxic events-Mo isotopes in black shales from Upper Yangtz marine sediments. Sci China Earth Sci 54(7):1024–1033

Zhou MK, Zi WR, Li ZM (1993) Ordovician and Silurian Lithofacies Paleogeography and Mineralization in South China. Geological Publishing House, Beijing

Zou CN, Qiu Z, Poulton SW et al (2018) Ocean euxinia and climate change double whammy drove the late ordovician mass extinction. Geology 46:535–538

Zuo ZH, Yang F, Zhang C et al (2012) Evaluation of advantage area of Longmaxi Formation shale gas of Silurian in Southeast area of Sichuan. Geol Chem Min 34(3):135–142

5 Hirnantian Glaciation

Abstract

The major ice caps in the late Ordovician lay between the base of the *N. extraordinarius* biozone and lower part of the *N. persculptus* biozone, consistent with the low values of CIA, CIW and PIA for the Huadi No. 1 well, reflecting a short-lived cold and dry climate from the top of the Wufeng-Guanyinqiao Formation to the base of the Longmaxi Formation in the Huadi No. 1 well. These observations provide geochemical record evidence for Gondwana glaciation at the end of the Ordovician in South China.

Keywords

Glaciation • Gondwana glaciation • Geochemistry • Hirnantian period • Shale gas

The Late Ordovician-Early Silurian transition was a critical interval in Earth's history, marked by global eustasy, climate cooling (Gondwana glaciation), volcanic eruption, ocean anoxia and biotic mass extinction. These geological events considerably influenced the sedimentation of shales in the Wufeng-Longmaxi Formations, which controlled the formation and distribution of favorable shale gas intervals (Qiu and Zou 2020). In this chapter, we study the Gondwana glaciation.

5.1 The Continental Glacial Activity of Late Ordovician

At the end of the Late Ordovician, large-scale continental glacier activities occurred globally. Geological scholars in France and Algeria first reported the tillites and glacial stria of the Ashgillian stage in the central Sahara. Later, records of continental or marine glacier activities in the late Ordovician, North Africa, South Africa, Southern Europe, Central Europe, West Asia and South America were discovered. The glaciers in this period were part of the Gondwana glacier that surrounded the Antarctic region at that time, centered in North Africa, and extending to South Africa, southern and central Europe, West Asia and South America. Such a large area of glacial activity can reflect the global paleoclimate and paleogeographic characteristics at that time, which has geological significance (Rong and Zhan 1999). In Morocco and Libya, glacier-related strata developed the atypical Hirnantian fauna and were overlain by graptolite strata of the *Parakidograptus acuminatus* zone; thus, these strata were identified in the late Ashgill stage, with a duration of only approximately 50×10^4–100×10^4 a. Although South China was located near the equator at the end of the Late Ordovician, no direct evidence (such as tillites and glacial scratches) has been found for this glaciation in China, and most scholars agree that this glaciation indirectly influenced South China (in the form of sea-level fluctuations, paleoclimate changes, biofacies and paleogeographic characteristics). The formation of ice caps and change in the carbon and oxygen isotope composition in seawater were the characteristic factors for the first pulse of the end Ordovician mass extinction. The average reduction in the global atmospheric and ocean temperatures was 8–10 °C. Most of the sea areas in the late Ashgill turned into cold water areas, which influenced the plankton (graptolites), swimming organisms (trilobites and cephalopods) and benthic organisms (brachiopods and corals) developed in the relatively warm sea areas in the middle of Ashgill. The cold waters with high density in the high latitudes migrated toward the equator, causing cold, deep-water currents with oxygen and nutrients to flow into the waters and ocean currents to overturn. Considerable changes occurred in the ecosystems of the previously formed deeper, oxygen-poor and nutrient-poor waters, resulting in the extinction of the Foliomena fauna, and the subsequent warming of the atmospheric and sea temperatures led to the melting of the ice sheet, which was the characteristic factor for the second pulse of the end Ordovician mass extinction. The rapid rise

C. Mou et al., *Lithofacies Paleogeography and Geological Survey of Shale Gas*, The China Geological Survey Series, https://doi.org/10.1007/978-981-19-8861-5_5

in the global sea level caused seawater to flood the gentle land again, forming a relatively open epigenetic sea and leading to the formation of a global large-scale transgression and anoxic environment. Several toxic water bodies became upwelling currents, resulting in the extinction of Hirnantian fauna (Rong 1999). The Gondwana continental glacier accumulated large amounts of freshwater, leading to global regression, which manifested as different degrees of stratigraphic loss. The important signs of lithofacies and biofacies transformation during this period can be summarized as follows: Black graptolite shale deposition of the Wufeng Formation generally ended in the Yangtze area, and the Guanyinqiao Formation limestone and its shell fauna appeared, which was then superimposed by the Longmaxi Formation shale of the transgression sequence. With the melting of the Gondwana continental ice sheet and beginning of the transgression, the Longmaxi Formation black shales began to be deposited.

Guanyinqiao Formation (late Ashgill) developed Hirnantian Fauna, a cold-water-type fauna mainly inhabiting a part of benthic assemblages 2–3 in the oxygen-rich and nutrient-rich shallow-marine subdomain, with a few types found in the deep-water, oxygen-poor and nutrient-oligotrophic benthic domain (BA4-5). The development of this fauna indirectly confirmed the cold marine environment at the end of the Late Ordovician.

5.2 Geochemical Evidence of Glacial Events in South China: Example of Well Huadi No. 1 in Eastern Sichuan Province

5.2.1 Stratigraphy and Sample Introduction in Well Huadi No. 1

Considering that surface rock samples may be subjected to more weathering, resulting in the migration of some sensitive chemical elements and poor accuracy of the major, trace and rare earth elements data, we selected the shale samples of Wufeng-Longmaxi Formations from the Well Huadi No. 1, in the east of Sichuan basin. Firstly, we finely collected samples from the Wufeng, Guanyinqiao and Longmaxi Formations near the Ordovician–Silurian boundary in well Well Huadi No. 1. The samples were usually 0.5–1 m apart and were compressed to 0.1 m per sample near the boundary. A total of 34 samples were collected in the well.

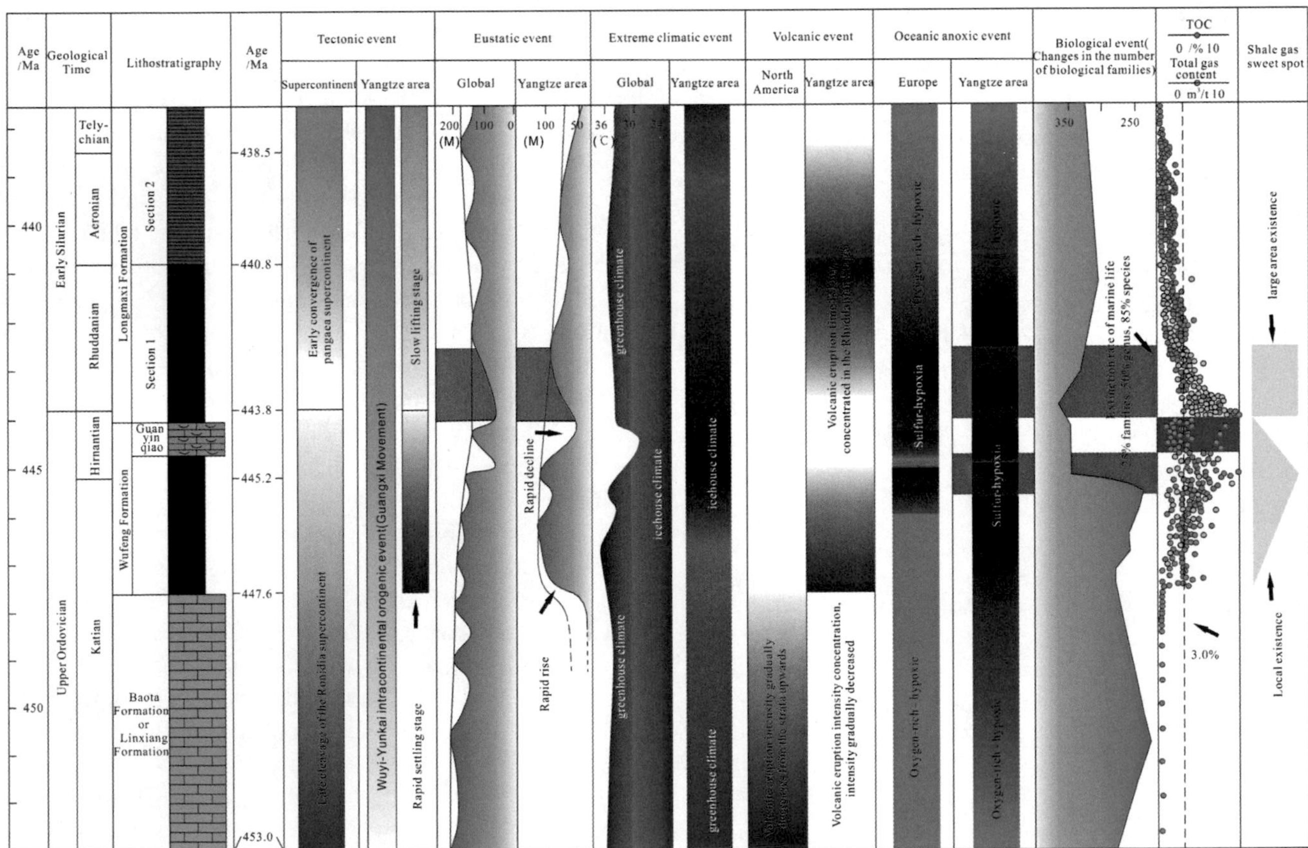

Fig. 5.1 Major geological events during the Ordovician and Silurian transition and characteristics of shale gas sweet-spot intervals of Wufeng-Longmaxi shale in Yangtze area, South China (after Qiu and Zou 2020)

All samples were ground and cleaned with distilled water, then dried and ground to 200 mesh for analytical testing.

The Upper Ordovician-Early Silurian succession in Well Huadi No.1 includes three successive formations: the Wufeng, Guanyinqiao and Longmaxi Formations in ascending order (Fig. 5.1). Wufeng Formation is in conformable contact with the underlying Linxiang Formation. The lithology of Linxiang Formation is mainly gray and dark gray nodular argillaceous limestone with a small amount of brachiopods, trilobites and other organisms. Pyrite aggregates are developed in the limestone, and asphalt can be seen near the bottom. This formation has a depth range of 1337.2–1339.86 and a thickness of about 2.66 m.

The depth of Wufeng Formation ranges from 1331.87 to 1337.2 m, and the lithology is mainly gray-black to black carbonaceous mudstone and carbonaceous siliceous mudstone. Four bentonite beds, about 0.5–1 cm thick each layer, were discovered in the black shales with graptolite and brachiopods fossils. Pyrite veins developed in the bottom of Wufeng Formation and generally parallel to the bedding.

The depth of Guanyinqiao Formation ranges from 1330.37 to 1331.87 m, and the thickness is about 1.5 m. The petrology is dark gray argillaceous siltstone. Silty bands are well developed and distributed in grayish white lamellar on the plane (Fig. 5.2). Lenticular pyrites are commonly developed in the bentonites.

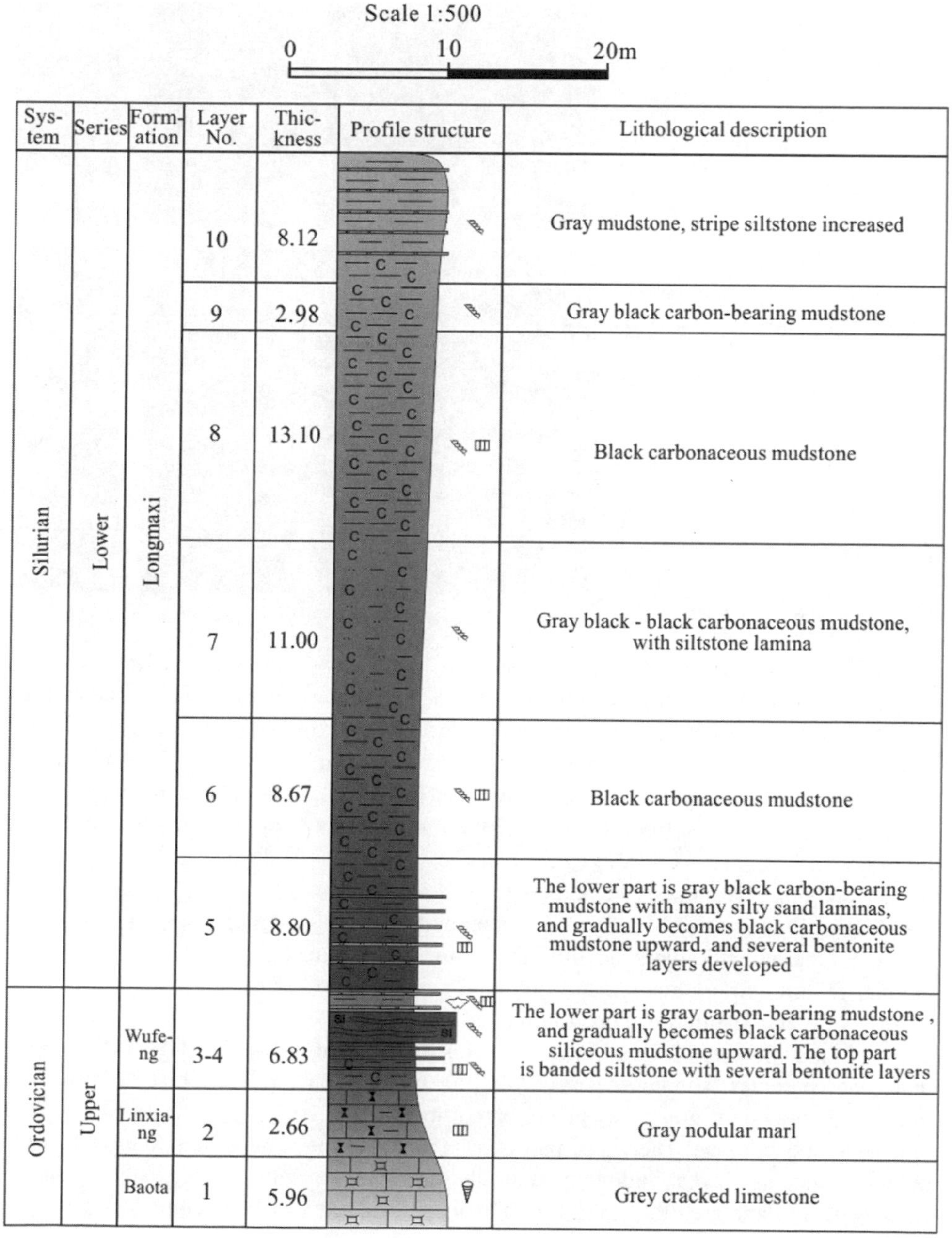

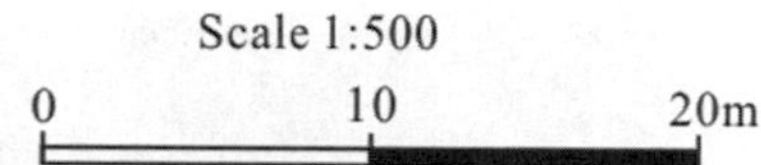

System	Series	Formation	Layer No.	Thickness	Profile structure	Lithological description
Silurian	Lower	Longmaxi	10	8.12		Gray mudstone, stripe siltstone increased
			9	2.98		Gray black carbon-bearing mudstone
			8	13.10		Black carbonaceous mudstone
			7	11.00		Gray black - black carbonaceous mudstone, with siltstone lamina
			6	8.67		Black carbonaceous mudstone
			5	8.80		The lower part is gray black carbon-bearing mudstone with many silty sand laminas, and gradually becomes black carbonaceous mudstone upward, and several bentonite layers developed
Ordovician	Upper	Wufeng	3-4	6.83		The lower part is gray carbon-bearing mudstone, and gradually becomes black carbonaceous siliceous mudstone upward. The top part is banded siltstone with several bentonite layers
		Linxiang	2	2.66		Gray nodular marl
		Baota	1	5.96		Grey cracked limestone

Fig. 5.2 Brief sectional structure of the Wufeng-Longmaxi Formations in the Well Huadi no. 1

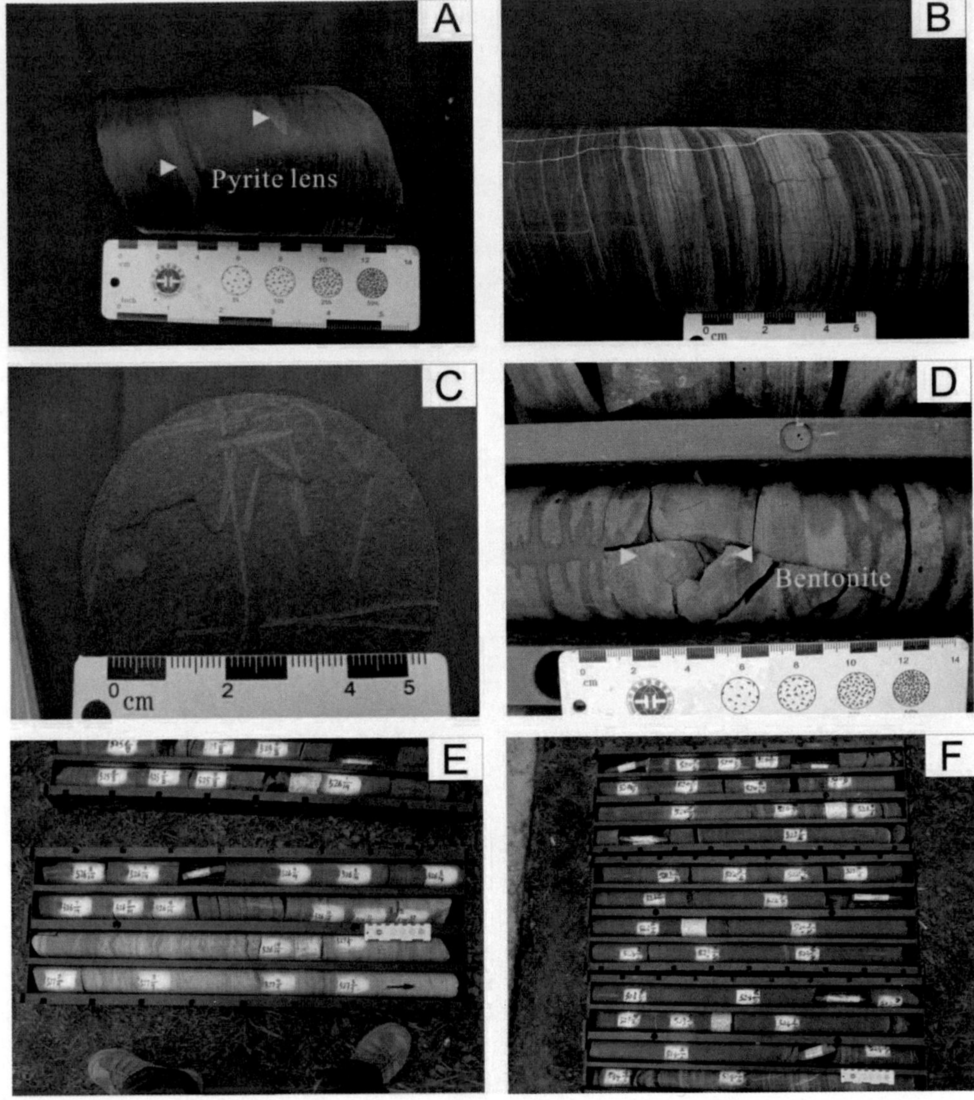

Fig. 5.3 Field photograph characteristics of the Wufeng-Longmaxi Formations in the Well Huadi no. 1. **a** Pyrite occurs as lenses in the black shales of the Wufeng Formation. **b** Laminar structure of siltstones in the Guanyinqiao Formation. **c** Graptolites developed in Longmaxi Formation. **d** Bentonites deposited in Longmaxi Formation. **e** Boundary between the Wufeng to Guanyinqiao Formation. **f** Macrocharacteristics of Guanyinqiao-Longmaxi Formations

The depth range of Longmaxi Formation is 726.56–1330.37 m, and the thickness is about 603.81 m. According to the changes of rock lithology and color, the Longmaxi Formation can be divided into two parts. The lower part of Longmaxi Formation, namely the most favorable black shale section, roughly ranges from 1321 to 1330.37 m, and the lithology of this part is mainly gray black to black carbonaceous shale. Numerous bentonites were discovered in the shales with thickness of 1–2 cm and maximum thickness of 3 cm. Occasionally, there are silty bands with horizontal laminar texture. Graptolites, mainly fine and straight graptolites, were abundantly developed in the shales. The pyrite can be found in Longmaxi Formation as veins and disseminated fine particles. The upper part of Longmaxi Formation mainly contains dark gray, gray-green, carbon-bearing silty mudstone and siltstone. The whole part is manifested by the gradual process of rock lithology and color. The color changes from dark grey to gray-green. The increasing silty contents make the rocks shift from carbon-bearing silty mudstones to mudstones and muddy siltstones rhythm then to siltstones. Near the top numerous limestone, biological limestone interbeds are discovered in the siltstones. The biological particles are mainly crinoids, brachiopods and oolitic particles, and there are wormtrails in siltstone. The changes of lithologic and sedimentary structure characteristics from bottom to top in Longmaxi Formation indicate the process of paleoseawater depth from deep to shallow (Fig. 5.3).

5.2.2 The Lithology of Wufeng-Longmaxi Formations in the Well Huadi No. 1

According to the thin section and X-ray diffraction analyses, the black shales of the Wufeng and Longmaxi Formations have pelitic textures and are composed mainly of clastic

Table 5.1 Mineral compositions of samples from Wufeng-Longmaxi Formations in Well Huadi no. 1 (XRD)

No.	Sample	Clay mineral relative content (%)								Whole rock analysis (%)							
		K	C	I	S	I/S	C/S	Interlaying		Clay miner-als	Quart-z	Orthoclase	Albite	Pyrite	Calcite	Dolomite	Pyroxenes
								I/S	C/S								
1	HDP-B1	2	1	20	/	75	2	8	12	38.7	38.9	0.7	5.6	0.7	6.1	9.3	\
2	HDP-B2	2	1	13	/	82	2	7	12	36.7	35.4	\	5	6.8	1.6	14.5	\
3	HDP-B4	2	1	14	/	82	1	8	14	28.1	41.1	1.3	7	5.5	\	14.2	2.8
4	HDP-B7	2	1	16	/	78	3	7	16	31.5	40.0	\	4	4.1	0.3	20.1	\
5	HDP-B10	1	1	22	/	72	4	6	16	33.6	53.6	\	5	3.5	\	4.3	\
6	HDP-B18	/	/	34	/	66	/	6	/	10.8	85.0	\	1.4	2.8	\	\	\
7	HDP-B19	/	/	25	/	75	/	7	/	71.8	12.5	\	1.5	14.2	\	\	\
8	HDP-B21	2	2	18	/	74	4	10	18	25.8	53.1	1.0	10.3	3.7	2.7	3.4	\
9	HDP-B23	1	2	16	/	78	3	8	13	25.1	47.0	0.8	12.3	6.7	3.6	4.5	\
10	HDP-B25	1	1	15	/	76	7	8	12	28.3	39.5	1.9	11.1	6.4	1.8	11	\
11	HDP-B26	1	3	14	/	76	6	7	15	31	39.7	0.9	10.7	4.1	5	8.6	\
12	HDP-B29	1	2	14	/	80	3	7	13	26.5	36.6	0.7	9.5	2.2	0.7	23.8	\
13	HDP-B30	1	2	13	/	79	5	8	9	32.3	42.8	1.9	10.9	4	2	6.1	\
14	HDP-B31	/	/	11	/	89	/	8	/	46	43.5	\	0.8	9.4	0.3	\	\
15	HDP-B33	1	2	14	/	79	4	9	9	25.3	43.5	1.8	15.2	3.3	1.4	9.5	\
16	HDP-B34	1	2	16	/	77	4	9	10	25.5	33.7	1.3	10.6	3.3	14	11.5	\

K Kaolinite; *C* Chlorite; *I* Illite; *S* Montmorillonite

particles (quartz, feldspar) and clay minerals. The clay minerals with microscale pelitic texture are directionally arranged in the rocks. A part of the clay minerals renders the organic matter as scale aggregates along the bedding, and other minerals fill the small cracks as black bands. Radiolarians can be found in mudstones of the Wufeng Formation, which have an elliptical shape with a diameter of 60–120 μm. The cements are mainly calcite and dolomite, among which dolomites occur as euhedral crystals, while calcite is mostly microcrystalline and distributed in clay minerals. The siltstone of the Guanyinqiao Formation exhibits alternating layers of quartz particles and clay minerals.

The X-ray diffraction (XRD) results of 16 samples from the Wufeng-Longmaxi Formations in the Well Huadi No. 1 (Table 5.1) indicate that quartz is the dominant mineral in the black shales of the Wufeng and Longmaxi Formations. Quartz contents range from 33.6 to 53.6% with an average value of 42.9%, whereas the clay mineral contents have an average value of 31.0% with a range of 25.1–46%. Microcline (0–1.9%), albite (0.8–15.2%), calcite (0–14.1%), dolomite (0–23.8%), and pyrite (0.7–9.4%) contents in the black shales are low. The clay minerals are mainly composed of mixed-layer illite/smectite (72–89%, avg. 78.4%) and illite (11–22%, avg. 15.4%) (Figs. 5.4 and 5.5). The contents of both chlorite and mixed-layer chlorite/smectite are less than 10%, with values of 1–3% and 1–7%, respectively.

The quartz contents in the siltstones of the Guanyinqiao Formation range from 12.5% to 85.0%, and the clay mineral contents range from 10.8% to 71.8%. The clay minerals are dominated by illite (25–34%) and illite/smectite mixed layers (66–75%). The second-most abundant minerals are pyrite (2.8–14.2%) and albite (1.4–1.5%).

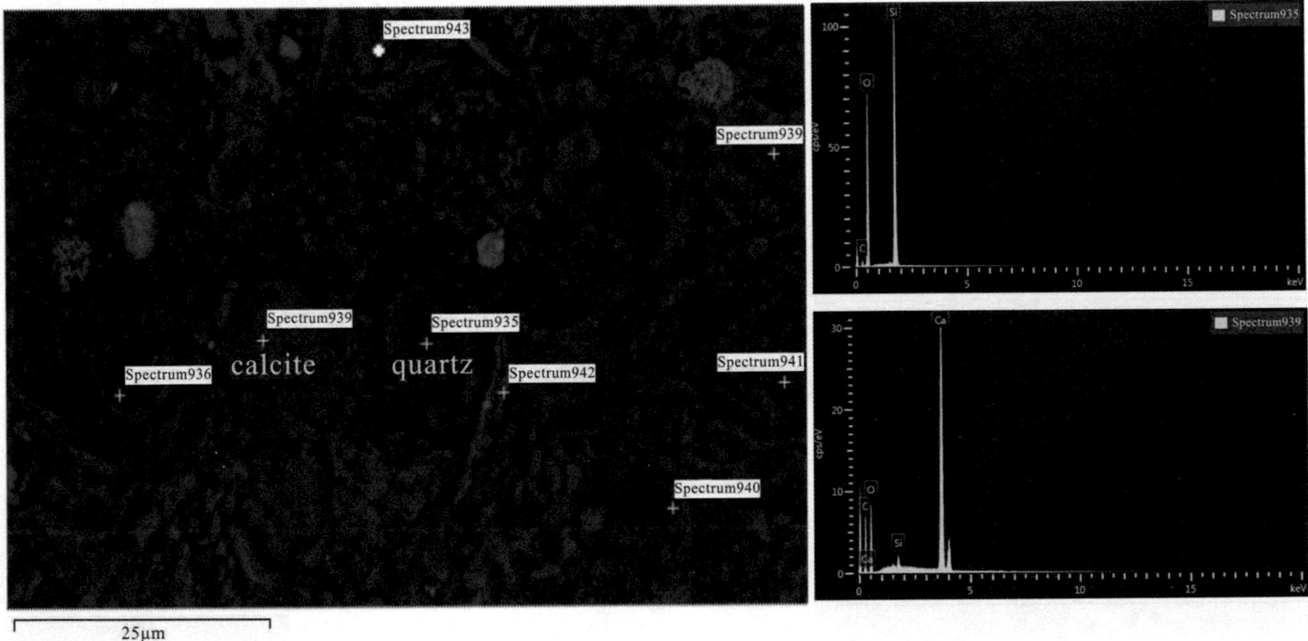

Fig. 5.4 Spectrum diagram of studied sample B36 of Longmaxi Formation

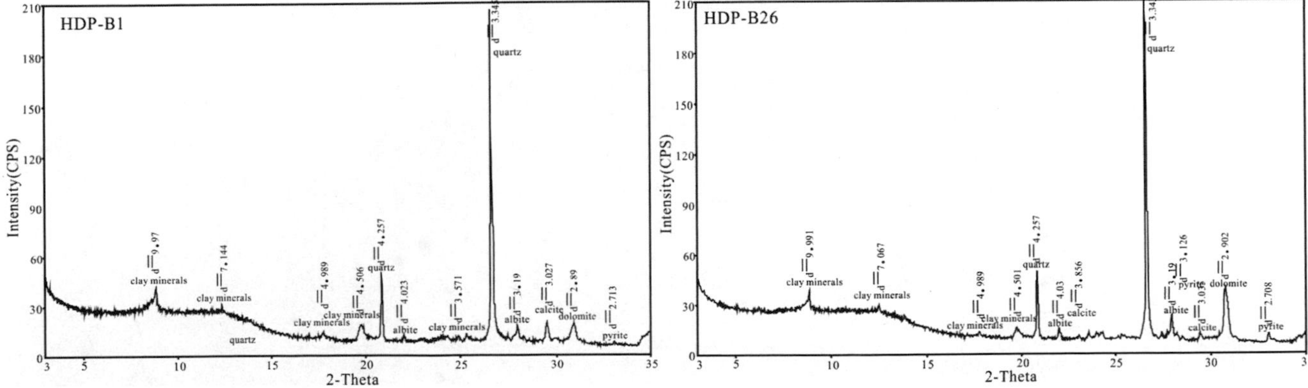

Fig. 5.5 XRD patterns of the black shales of Wufeng and Longmaxi Formations in Well Huadi no. 1

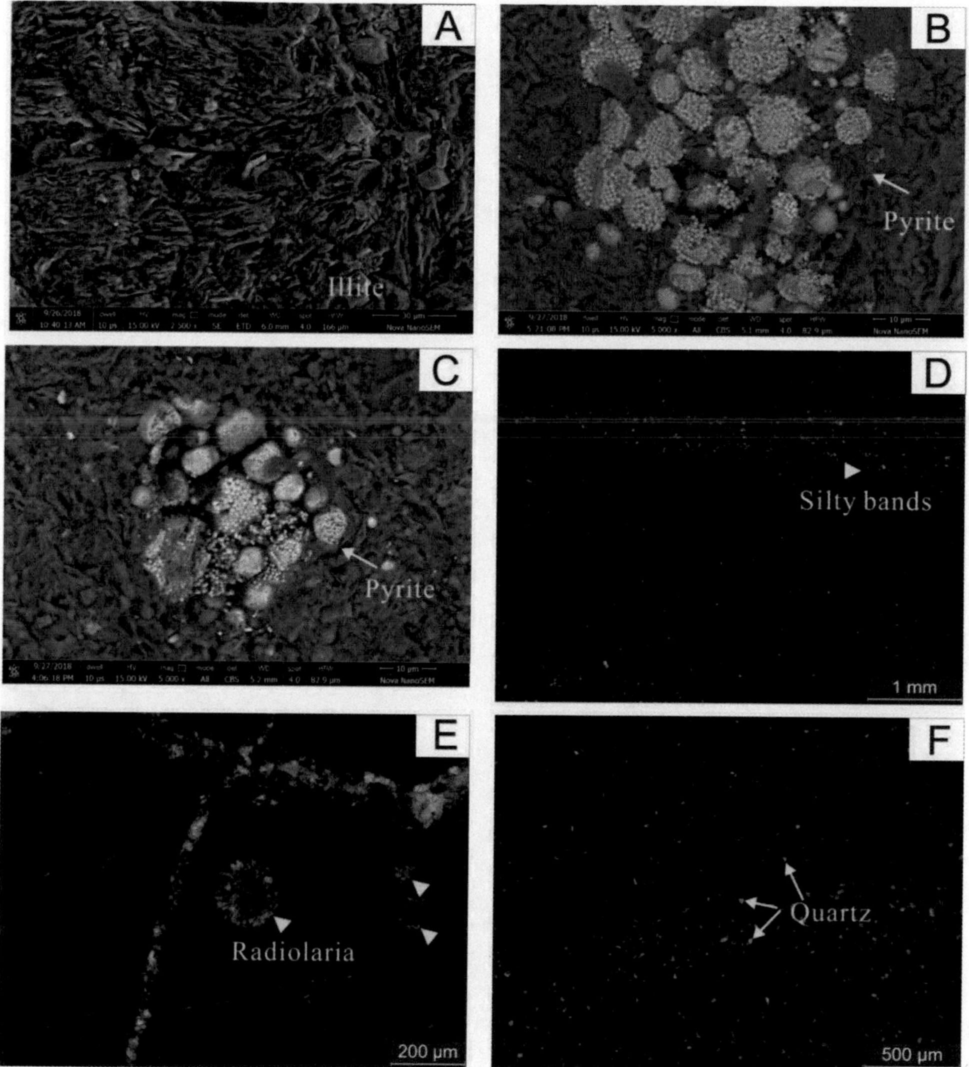

Fig. 5.6 Microcharacteristics of black shales from Wufeng-Longmaxi Formations in Well Huadi No. 1. **a** Scanning electron microscopy image of B17 clay minerals in the black shales of Wufeng Formation; **b** scanning electron microscopy image of the sample B30 in the black shales of Longmaxi Formation: pyrite framboids; **c** scanning electron microscopy image of the sample B15 of Wufeng Formation: pyrite framboids; **d** the horizontal silty bands developed in the black shales of Longmaxi Formation; **e** radiolarian developed in black shales of Wufeng Formation (sample B13); **f** the microscopic characteristics quartz (round-oval shape) in the shales of Longmaxi Formation

The scanning electron microscopy images of the black shales of the Wufeng and Longmaxi Formations show that the pyrites are characterized by euhedral and subhedral fabrics and exist as framboids with a particle diameter of 1–8 μm (Fig. 5.6).

5.2.3 The Geochemical Characteristics of the Wufeng-Longmaxi Formations from the Well Huadi No. 1

The major element concentrations of the 34 samples from Wufeng-Longmaxi Formations in Well Huadi No.1 are listed in Tables 4.47, 5.3. For the black shales and mudstones in the Wufeng Formation, SiO_2 (38.26–91.18%) and Al_2O_3 (1.49–15.48%) are the most abundant oxides. According to the results from the XRD measurements, the mineral compositions of shales from the Wufeng Formation are dominated by quartz and clays, which is in accordance with the chemical compositions. The second most abundant oxides are CaO (0.548%–10.92%), Fe_2O_3 (0.961–7.75%), and K_2O (0.324–4.13%). The concentration of MgO in the shales is 0.554–6.74%. The FeO content of the black mudstones (0.7–2.11%) is slightly lower than the Fe_2O_3 content, while other oxide contents, including those of Na_2O, P_2O_5, TiO_2, and MnO, are lower than 1.0%.

The SiO_2 content of the Guanyinqiao siltstones ranges from 30.09 to 56.61%. The second-most abundant oxide is

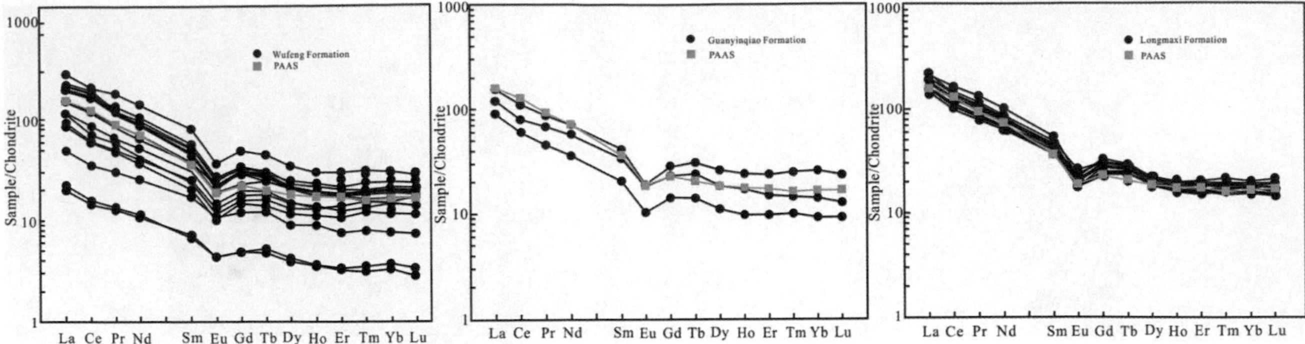

Fig. 5.7 Chondrite-normalized REE distribution patterns of samples from Wufeng-Longmaxi Formations in Huadi no. 1 well (normalization values after Taylor and McLenann 1985)

CaO (8.07–29.81%), the content of which is higher than that of Al_2O_3 (3.22–9.49%). The contents of Fe_2O_3 and FeO are 3.1–4.05 and 0.69–1.23%, respectively. The concentration of MgO is 1.23–3.02%, while the contents of other oxide, such as Na_2O, P_2O_5, TiO_2, and MnO, are equal to or less than 1%.

The shale samples in the Longmaxi Formation are characterized by a dominance of SiO_2 (42.09–65.72%), and Al_2O_3 follows SiO_2 in abundance (8.14–13.64%). The third-most abundant oxide is CaO (2.01–12.22%), with an average value of 5.68%. The contents of Fe_2O_3 (3.40–6.03%, avg. 4.34%), FeO (0.91–2.3%, avg. 1.56%) and K_2O (1.8–3.33%, avg. 2.73%) are close to the average values of the black shales in the Wufeng Formation [Fe_2O_3 (avg. 4.22%), FeO (avg. 2.11%) and K_2O (avg. 2.71%)]. The Na_2O (0.856–1.49%, avg. 1.30%) and MgO (1.57–7.29%, avg. 3.16%) contents are slightly higher than those of the Wufeng Formation shales (avg. 0.52 and 2.00%, respectively). The concentrations of MnO (0.016–0.181%), TiO_2 (0.442–0.621%) and P_2O_5 (0.101–0.281%) are lower than 1%.

Compared with the post-Archean average shale (PAAS) values (Taylor and McLennan 1985), the shales from the Wufeng Formation have slightly higher SiO_2 (avg. 65.92%) and CaO (avg. 2.91%) contents. In addition, the Al_2O_3 (avg. 10.26%) and K_2O (avg. 2.71%) contents are lower than those of the PAAS, which indicates that the content of the clay minerals is low in the samples. Fe_2O_3, MnO, Na_2O, P_2O_5 and TiO_2 are slightly depleted in the shales relative to the PAAS.

The shales from the Longmaxi Formation are enriched in Na_2O and CaO and slightly to strongly depleted in terms of the other major elements (SiO_2, Al_2O_3, Fe_2O_3, K_2O, P_2O_5, TiO_2 and MnO). The SiO_2 (avg. 56.49%) and Al_2O_3 (avg. 11.34%) deficits indicate a decrease in the amount of clay minerals. The enrichment of Na_2O (avg. 1.3%) and CaO (avg. 5.68%) relative to the PAAS can be attributed to the presence of plagioclase and secondary calcite in caliches (Lee 2009), while the depletion of TiO_2 (avg. 0.55%) and K_2O (avg. 2.73%) suggests that phyllosilicate minerals exist in lesser quantities in the shales (Condie et al. 1992; Moosavirad et al. 2011).

The siltstones from the Guanyinqiao Formation are enriched in CaO and slightly depleted in other major elements (SiO_2, Al_2O_3, Fe_2O_3, K_2O, P_2O_5, TiO_2 and MnO). The concentration of P_2O_5 (avg. 0.17%) is similar to that of the PAAS (0.16%).

The total rare earth element contents (ΣREE) from the Wufeng Formation show significant variability (27.17–334.01 $\times 10^{-6}$), with an average of 168.30 $\times 10^{-6}$ (Table 4.48), which is lower than that of the PAAS (184.77 ppm; Taylor and McLennan 1985). The light rare earth element (LREE) contents (21.23 $\times 10^{-6}$–299.83 $\times 10^{-6}$, avg. 149.99 $\times 10^{-6}$) are apparently higher than those of the heavy rare earth elements (HREEs) (3.66 $\times 10^{-6}$–34.18 $\times 10^{-6}$, avg. 18.31 $\times 10^{-6}$), and the mean LREE/HREE ratio is 5.80–11.29. The (La_N/Yb_N) ratio is 6.09–12.07. The Eu/Eu* value is 0.53–0.75. In the chondrite-normalized REE distribution patterns diagram, we can see that the LREE and HREE contents of the samples are slightly lower than those of PAAS. All samples exhibit similar REE distribution curves, showing slight LREE enrichment and relatively flat HREE patterns with weakly negative Eu anomalies. The V-type right-inclining REE distribution model is similar to the granite's, suggesting that the source rocks of Wufeng, Guanyinqiao and Longmaxi Formations are felsic (Fig. 5.7).

5.2.4 The Paleoweathering of the Late Ordovician-Early Silurian Transition in Well Huadi No. 1 and the Implications for the Paleoclimate

During rock chemical weathering processes, chemical weathering increases under humid conditions, causing the leaching of alkali metal and alkaline metal elements, such as Na, K and Ca, and increased concentrations of Al and Si in the residue. The most popular paleoweathering indices used to assess the degree of chemical weathering in the source area are the Chemical Index of Alteration (CIA, Nesbitt and Young 1982), Chemical Index of Weathering (CIW, Harnois 1988) and Plagioclase Index of Alteration (PIA, Fedo et al. 1995).

In the CIA index, new potassium elements are introduced in the potassium metasomatism during diagenesis, resulting in a low CIA value, and thus, the CIA value needs to be corrected. Based on the A–CN–K ternary diagram, Nesbitt and Young (1984, 1989) obtained the weathering profiles and predicted the weathering trendlines of the sediments by analyzing the thermodynamic and kinetic processes of feldspar decomposition and geochemical characteristics of weathering profiles in nature. As shown in Fig. 4-101-(1), solid lines a and b are nearly parallel to the A–CN line and represent the weathering trend without potassic metasomatism. Solid line c represents the weathering trend with potassic metasomatism. In addition, the dotted distance between solid lines 3 and 4 represents the CIA values of the shales with potassic metasomatism. Fedo et al. (1995) pointed that the intersection between the weathering trend without potassic metasomatism and back extension line of the K apex and samples represents the premetasomatized CIA value (the correction value is the solid line distance between lines 1 and 2 in Fig. 4-101-(1); Mou et al. 2019).

The mobility of elements can be evaluated using an Al_2O_3–(CaO^* + Na_2O)–K_2O (A–CN–K) ternary diagram (Nesbitt and Young 1984), where A = Al_2O_3 (mol %), CN = CaO^* + Na_2O(mol%) and K = K_2O (mol%). A–CN–K relationships can be interpreted in terms of the weathering history, paleoclimate and source rock composition (Fedo et al. 1995). The CIA values plotted in an A–CN–K triangular diagram (Fedo et al. 1996) can clarify the differentiation of compositional changes associated with chemical weathering and source rock composition (Nesbitt and Young 1984). As shown in Fig. 5.6, the samples of the three formations plot above the plagioclase-K-feldspar join and show a narrow linear trend that approaches the A apex. The weathering trends lie subparallel to the A–CN joint and do not exhibit any inclination toward the K apex, suggesting that the rocks were not subjected to potash metasomatism during diagenesis. The trendline, when extended backward, intersects the plagioclase-K-feldspar join near the granodiorite and granite fields (potential source rock).

The weathering trends of the samples from the Wufeng-Longmaxi Formations lie subparallel to the A–CN joint and do not exhibit any inclination toward the K apex in Fig. 4-101-(2), suggesting that the rocks in the Huadi No. 1 well were not subjected to potash metasomatism during diagenesis. The CIA_{corr} values calculated from the equation are consistent with the original CIA values (Fig. 5.8, Table 4.2).

The CIA values of the Wufeng-Longmaxi Formations in the Huadi No. 1 well exhibit a certain regularity from the bottom to the top, as shown in Fig. 4-102. The CIA values of black shales from the Wufeng Formation first increase and then decrease from the bottom up, ranging from 66.15 to 72.17 (avg. 68.86). At the top of the formation, the values decrease to a trough, with values between 66.15 and 70.35, and upward in the Guanyinqiao Formation, the CIA values decrease, varying between 59.96 and 62.86 (avg. 61.87). At the top of the Guanyinqiao Formation, the values fall to 59.96. Further upward, at the base of the Longmaxi Formation, the CIA rapidly returns to a high value (62.06). The CIA values of the Longmaxi Formation increase

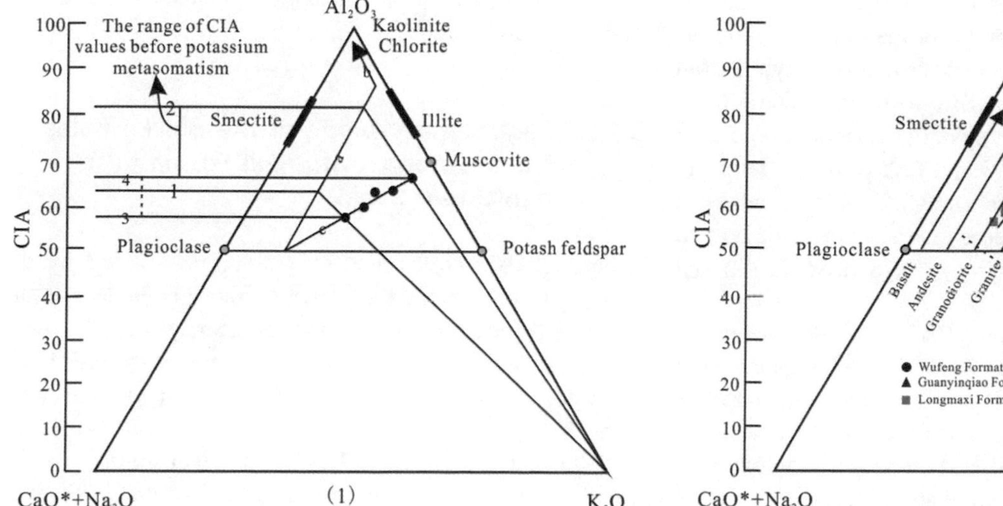
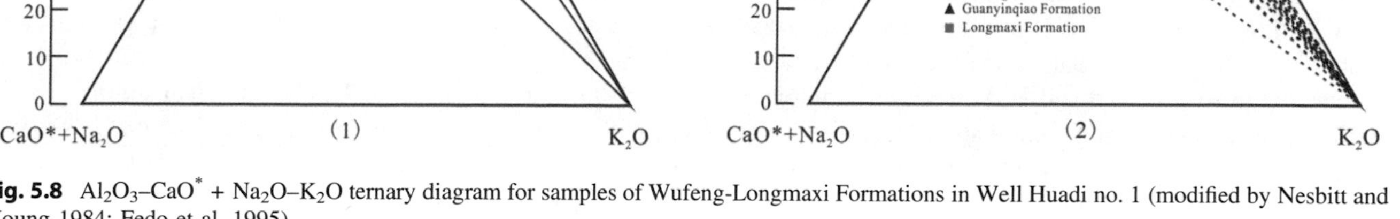

Fig. 5.8 Al_2O_3–CaO^* + Na_2O–K_2O ternary diagram for samples of Wufeng-Longmaxi Formations in Well Huadi no. 1 (modified by Nesbitt and Young 1984; Fedo et al. 1995)

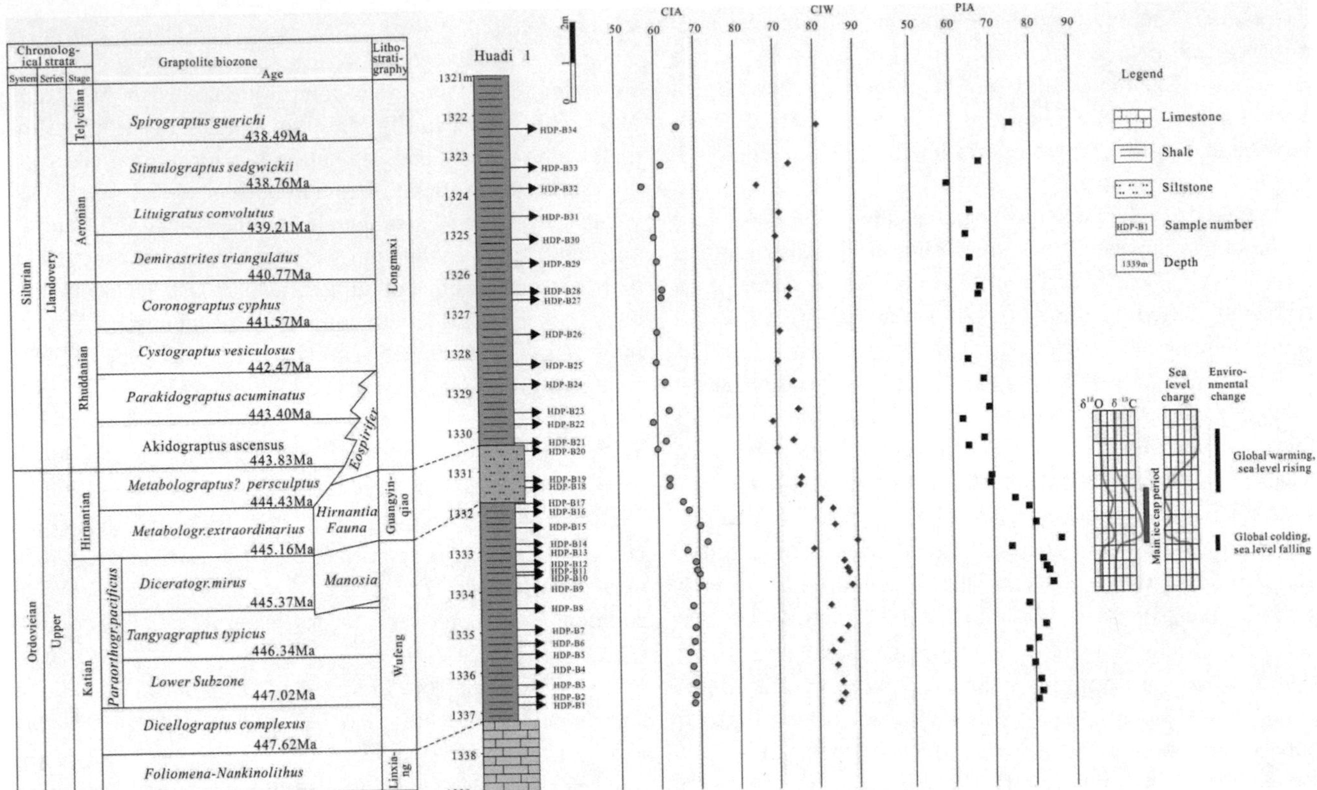

Fig. 5.9 Stratigraphic, lithological correlations and paleoweathering indices of CIA, PIA and CIW of Wufeng and the lower Longmaxi Formations in Well Huadi no. 1 (chronostratigraphic subdivision after Chen et al. 2017; isotope, sea level and environmental change after Brenchley et al. 2006; Harper et al. 2014)

gradually from the bottom up, ranging from 56.64 to 65.59 (avg. 60.87).

The trends of the PIA and CIW in the Wufeng-Longmaxi Formation are the same as those of the CIA. The CIW and PIA values of the lower Wufeng Formation rocks range from 78.98 to 89.96 and 74.46–86.68, respectively. At the top of this formation, the values fall to 80.84–84.35 and 75.38–80.46 and are higher in the strata. In the Guanyinqiao Formation, the CIW and PIA values decrease, varying from 70.10 to 75.98 (avg. 73.91) and from 64.02 to 69.65 (avg. 67.68), respectively. Further upward, at the base of the Longmaxi Formation, the CIW and PIA promptly return to high values (74.19 and 67.92). The CIA and PIA values of the Longmaxi Formation increase gradually from the bottom to the top, ranging from 65.55 to 80.78 (avg. 72.42) and from 59.12 to 74.99 (avg. 66.06), respectively.

Generally, the variations in the CIA value can indicate the paleoclimate (Nesbitt and Young 1984; Young and Nesbitt 1998; Fedo et al. 1995; Yan et al. 2010; Fu et al. 2015; Ge et al. 2019a, 2019b; Ma et al. 2015). Low CIA values (50–60) are suggestive of cold and arid climatic conditions with weak chemical weathering, moderate CIA values (60–80) indicate warm and humid climate conditions with medium chemical weathering, and high CIA values (80–100) indicate hot and humid tropical climatic conditions with intensive chemical weathering. Fluctuations in CIA values indicate that the intensity of chemical weathering changed from moderate to weak and then to moderate, and the climate shifted from warm to cold and then to warm again from the base of the Wufeng Formation to the lower part of the Longmaxi Formation in the Huadi No. 1 well (Fig. 5.9, Tables 5.2 and 5.3).

5.2.5 Sedimentary Mode of the Black Shales from Wufeng-Longmaxi Formations in Sichuan Basin

During the late Ordovician-early Silurian, affected by the Caledonian movement, the Cathaysia block collided with the Yangtze Block and expanded constantly. The Central Guizhou, Central Sichuan and Western Sichuan-Central Yunnan uplifts around the Upper Yangtze area constantly expanded, and the Yangtze area developed into the back-bulge basin confined by the marginal uplifts. As the uplifts expanded, the base of the upper Yangtze basin subsided promptly, and the sea level rose. In addition, frequent volcanic eruptions occurred in the middle of the Katian

Table 5.2 Analytical data of major elements (wt%) of shales and siltstones from Wufeng-Longmaxi Formations in Well Huadi no. 1

Sample	HDP-B1	HDP-B2	HDP-B3	HDP-B4	HDP-B5	HDP-B6	HDP-B7	HDP-B8	HDP-B9	HDP-B10	HDP-B11
Wufeng Formation											
SiO_2	57.02	54.12	56.19	56.27	54.15	53.71	38.26	73.24	69.45	68.59	60.58
Al_2O_3	15.36	15.09	15.48	15.39	14.06	14.23	10.21	8.69	13.35	9.81	13.79
CaO	4.16	4.38	3.1	3.3	4.36	3.68	10.92	1.82	0.647	0.549	2.43
Fe_2O_3	3.3	4.8	5.64	5.3	5.26	6.09	6.92	3.75	3.02	7.75	3.42
FeO	1.19	1.31	0.95	1.05	1.56	1.3	2.11	1.06	1.26	1.23	0.91
Fe_2O_{3t}	4.62	6.26	6.70	6.47	6.99	7.53	9.26	4.93	4.42	9.12	4.43
K_2O	4.09	4.14	4.13	4.03	3.68	3.75	2.87	2.08	3.54	2.6	3.74
MgO	2.59	2.73	2.2	2.48	2.92	3.18	6.74	1.51	1.17	0.919	2.36
MnO	0.057	0.059	0.041	0.044	0.06	0.065	0.177	0.031	0.018	0.018	0.03
Na_2O	0.796	0.721	0.771	0.847	0.841	0.739	0.453	0.534	0.528	0.416	0.608
P_2O_5	0.066	0.069	0.062	0.084	0.099	0.073	0.066	0.08	0.071	0.064	0.1
TiO_2	0.769	0.726	0.761	0.725	0.693	0.684	0.482	0.407	0.619	0.609	0.49
LOI	9.71	9.01	9.62	9.54	10.79	10.79	18.26	7.57	7	8.1	10.35
Total	99.11	97.16	98.95	99.06	98.47	98.29	97.47	100.77	100.67	100.66	98.81
Al_2O_3/TiO_2	19.97	20.79	20.34	21.23	20.29	20.80	21.18	21.35	21.57	16.11	28.14
K_2O/Na_2O	5.14	5.74	5.36	4.76	4.38	5.07	6.34	3.90	6.70	6.25	6.15
CaO*	0.013	0.012	0.012	0.014	0.014	0.012	0.007	0.009	0.009	0.007	0.010
CIA	68.52	68.73	68.81	68.25	67.53	68.64	68.92	68.40	70.53	70.07	69.48
CIAcorr	68.52	68.73	68.81	68.25	67.53	68.64	68.92	68.40	70.53	70.07	69.48
PIA	80.66	81.71	81.26	79.81	78.44	80.69	82.64	78.55	84.55	83.62	82.95
CIW	85.43	86.41	85.92	84.67	83.56	85.41	87.26	83.18	88.49	87.76	87.33
Sample	HDP-B12	HDP-B13	HDP-B14	HDP-B15	HDP-B16	HDP-B17	HDP-B18	HDP-B19	HDP-B20	HDP-B21	
Guanyinqiao Formation											
SiO_2	68.55	91.18	82.35	91.57	76.47	69.01	30.09	56.61	55.44	63.69	
Al_2O_3	8.36	1.57	3.48	1.49	7.69	6.36	3.22	4.82	9.49	13.43	
CaO	0.559	0.943	1.95	0.885	0.548	5.27	29.81	13.71	8.07	2.01	
Fe_2O_3	7.13	1.09	2.06	0.961	2.01	3.28	3.35	3.1	4.05	3.8	
FeO	0.8	0.7	0.98	0.77	0.84	0.88	0.69	1.22	1.23	0.91	
Fe_2O_{3t}	8.02	1.87	3.15	1.82	2.94	4.26	4.12	4.46	5.42	4.81	
K_2O	2.21	0.324	0.879	0.324	2.03	1.61	0.803	1.22	2.11	3.26	
MgO	0.791	0.554	1.13	0.49	0.83	1.38	2.2	1.23	3.02	1.57	
MnO	0.01	0.009	0.02	0.008	0.009	0.052	0.256	0.094	0.143	0.016	
Na_2O	0.394	0.127	0.118	0.084	0.449	0.458	0.315	0.463	1.23	1.42	
P_2O_5	0.079	0.023	0.065	0.032	0.08	0.122	0.256	0.119	0.147	0.117	

(continued)

Table 5.2 (continued)

Sample	HDP-B12	HDP-B13	HDP-B14	HDP-B15	HDP-B16	HDP-B17	HDP-B18	HDP-B19	HDP-B20	HDP-B21
TiO_2	0.299	0.064	0.155	0.058	0.345	0.306	0.142	0.25	0.447	0.548
LOI	10.41	3.54	6.76	3.56	8.96	10.13	25.17	14.26	12.73	8.03
Total	99.59	100.12	99.95	100.23	100.26	98.86	96.30	97.10	98.11	98.80
Al_2O_3/TiO_2	27.96	24.53	22.45	25.69	22.29	20.78	22.68	19.28	21.23	24.51
K_2O/Na_2O	5.61	2.55	7.45	3.86	4.52	3.52	2.55	2.63	1.72	2.30
CaO*	0.006	0.002	0.002	0.001	0.007	0.007	0.005	0.007	0.020	0.023
CIA	69.35	67.11	72.17	70.35	67.63	66.15	62.80	62.86	59.96	62.06
CIAcorr	69.35	67.11	72.17	70.35	67.63	66.15	62.80	62.86	59.96	62.06
PIA	82.14	74.46	86.68	80.46	78.79	75.38	69.38	69.65	64.02	67.92
CIW	86.57	78.98	89.96	84.35	83.88	80.84	75.65	75.98	70.10	73.91

Sample	HDP-B22	HDP-B23	HDP-B24	HDP-B25	HDP-B26	HDP-B27	HDP-B28	HDP-B29	HDP-B30	HDP-B31	HDP-B32	HDP-B33	HDP-B34	PAAS
Longmaxi Formation														
SiO_2	64.54	49.45	53.36	56.3	65.72	59.6	50.91	42.09	62.85	64.25	51.52	58.53	48.02	62.80
Al_2O_3	10.78	13.64	12.93	11.55	11.15	11.12	10.47	8.88	11.32	12.14	8.14	11.43	11.84	18.90
CaO	3.68	6.65	5.8	5.88	2.9	3.91	6.82	12.22	3.45	2.66	9.35	4.63	9.56	1.30
Fe_2O_3	3.4	3.84	5.73	4.58	3.91	4.66	6.03	4.18	3.83	4.23	3.75	4.16	4.59	7.20
FeO	1.32	1.17	1.39	1.94	1.19	1.66	1.93	2.3	1.55	1.68	1.69	1.29	1.82	–
Fe_2O_{3t}	4.87	5.14	7.27	6.74	5.23	6.50	8.17	6.74	5.55	6.10	5.63	5.59	6.61	4.87
K2O	2.48	3.33	3.18	2.7	2.65	2.75	2.58	2.07	2.66	2.85	1.8	2.83	3.13	3.70
MgO	1.72	4.36	2.57	3.03	1.9	2.36	3.98	7.29	2.06	1.87	5.47	2.62	3.42	–
MnO	0.024	0.06	0.038	0.054	0.023	0.033	0.09	0.181	0.024	0.021	0.171	0.044	0.071	2.20
Na_2O	1.47	1.35	1.36	1.47	1.38	1.23	1.14	1.1	1.46	1.49	1.3	1.24	0.856	1.20
P_2O_5	0.125	0.101	0.109	0.12	0.118	0.11	0.109	0.109	0.115	0.118	0.125	0.14	0.281	0.16
TiO_2	0.579	0.495	0.601	0.621	0.581	0.575	0.546	0.448	0.587	0.619	0.442	0.556	0.557	1.00
LOI	9.69	13.58	11.16	10.5	7.58	11.06	12.73	19.89	8.54	7.68	15.33	12.08	15.13	–
Total	99.81	98.03	98.23	98.75	99.10	99.07	97.34	100.76	98.45	99.61	99.09	99.55	99.28	–
Al_2O_3/TiO_2	18.62	27.56	21.51	18.60	19.19	19.34	19.18	19.82	19.28	19.61	18.42	20.56	21.26	18.9
K_2O/Na_2O	1.69	2.47	2.34	1.84	1.92	2.24	2.26	1.88	1.82	1.91	1.38	2.28	3.66	3.08
CaO*	0.024	0.022	0.022	0.024	0.022	0.020	0.018	0.018	0.024	0.024	0.021	0.020	0.014	–
CIA	58.88	62.87	62.00	59.79	60.06	61.26	61.51	60.22	59.55	60.29	56.64	61.51	65.59	–
CIAcorr	58.88	62.87	62.00	59.79	60.06	61.26	61.51	60.22	59.55	60.29	56.64	61.51	65.59	–
PIA	62.58	69.30	67.93	64.06	64.57	66.78	67.16	64.70	63.71	64.86	59.12	67.20	74.99	–
CIW	69.03	75.43	74.29	70.48	71.06	73.32	73.62	71.04	70.21	71.23	65.55	73.69	80.78	–

PAAS Post-Archean Average Shale (Taylor and McLennan 1985)

Table 5.3 Analytical data of rare earth elements (ppm) of shales and siltstones from Wufeng-Longmaxi Formations in Huadi no. 1 well

Sample	HDP-B1	HDP-B2	HDP-B3	HDP-B4	HDP-B5	HDP-B6	HDP-B7	HDP-B8	HDP-B9	HDP-B10	HDP-B11
La	52.8	48.8	53.7	50.20	49.8	52.10	36.20	28.60	38.10	48.80	69.00
Ce	115	108	117	105.0	109.0	117.0	76.60	54.50	72.60	103.0	131.0
Pr	11.80	11.20	11.50	11.10	11.5	12.00	8.06	6.44	8.99	13.30	17.50
Nd	42.60	42.40	42.60	42.10	43.40	46.10	30.60	25.00	33.80	50.50	67.70
Sm	7.26	7.20	6.90	7.08	7.98	8.25	6.04	4.75	5.89	8.78	12.50
Eu	1.18	1.23	1.24	1.30	1.52	1.39	1.29	0.86	1.10	1.53	2.13
Gd	6.26	6.48	6.17	6.37	6.77	7.14	5.89	4.09	4.90	7.00	10.20
Tb	1.09	1.09	1.03	1.08	1.21	1.19	1.14	0.75	0.94	1.12	1.72
Dy	5.37	5.54	5.29	5.59	6.04	5.87	6.41	3.90	5.38	6.12	8.88
Ho	1.09	1.12	1.02	1.12	1.23	1.18	1.33	0.76	1.21	1.29	1.72
Er	2.91	3.03	3.00	3.07	3.40	3.27	3.60	2.01	3.45	4.22	5.00
Tm	0.49	0.51	0.46	0.51	0.51	0.56	0.61	0.33	0.61	0.74	0.79
Yb	3.35	3.38	3.19	3.43	3.67	3.57	4.13	2.15	4.29	5.11	5.14
Lu	0.475	0.478	0.465	0.52	0.531	0.567	0.597	0.314	0.624	0.75	0.733
$\sum$REE	251.67	240.46	253.57	238.47	246.57	260.18	182.49	134.45	181.88	252.26	334.01
LREE	230.64	218.83	232.94	216.78	223.20	236.84	158.79	120.15	160.48	225.91	299.83
HREE	21.03	21.63	20.63	21.69	23.37	23.34	23.70	14.31	21.40	26.35	34.18
LREE/HREE	10.97	10.12	11.29	10.00	9.55	10.15	6.70	8.40	7.50	8.58	8.77
LaN/YbN	11.31	10.36	12.07	10.50	9.73	10.47	6.29	9.54	6.37	6.85	9.63
Eu/Eu*	0.54	0.55	0.58	0.59	0.63	0.55	0.66	0.60	0.63	0.60	0.58
Sample	HDP-B12	HDP-B13	HDP-B14	HDP-B15	HDP-B16	HDP-B17	HDP-B18	HDP-B19	HDP-B20	HDP-B21	
La	21.00	5.43	11.9	4.77	28.3	22.8	28.5	21.4	37.3	50.9	
Ce	37.40	9.86	22.1	8.82	44.8	37.4	49.2	36.8	67.8	101	
Pr	4.88	1.30	2.97	1.22	5.75	4.7	6.52	4.35	8.33	12.6	
Nd	19.40	5.42	12.2	5.06	20.4	17.7	26.8	16.7	33.1	47.2	
Sm	3.89	1.05	2.64	1.11	3.71	3.17	5.18	3.17	6.37	8.22	
Eu	0.75	0.25	0.623	0.253	0.67	0.596	1.05	0.592	1.09	1.28	
Gd	3.68	1.01	2.50	1.01	3.30	3.02	4.78	2.94	5.78	6.74	
Tb	0.68	0.20	0.47	0.18	0.61	0.53	0.893	0.523	1.15	1.09	
Dy	3.82	1.08	2.31	1.01	3.42	2.95	4.7	2.79	6.6	5.55	
Ho	0.80	0.21	0.52	0.20	0.75	0.64	0.974	0.546	1.35	1.11	
Er	2.08	0.56	1.27	0.55	2.32	1.80	2.46	1.6	3.87	3.25	

(continued)

Table 5.3 (continued)

Sample	HDP-B12	HDP-B13	HDP-B14	HDP-B15	HDP-B16	HDP-B17	HDP-B18	HDP-B19	HDP-B20	HDP-B21
Tm	0.35	0.09	0.20	0.08	0.39	0.31	0.369	0.252	0.629	0.489
Yb	2.47	0.631	1.31	0.562	2.58	1.98	2.38	1.55	4.36	3.26
Lu	0.355	0.086	0.186	0.072	0.432	0.29	0.322	0.233	0.598	0.491
∑REE	101.55	27.17	61.19	24.89	117.43	97.88	134.13	93.45	178.33	243.18
LREE	87.32	23.31	52.43	21.23	103.63	86.37	117.25	83.01	153.99	221.20
HREE	14.23	3.86	8.76	3.66	13.80	11.52	16.88	10.43	24.34	21.98
LREE/HREE	6.13	6.04	5.99	5.80	7.51	7.50	6.95	7.96	6.33	10.06
LaN/YbN	6.10	6.17	6.52	6.09	7.87	8.26	8.59	9.90	6.14	11.20
Eu/Eu*	0.60	0.75	0.74	0.73	0.59	0.59	0.65	0.59	0.55	0.53

Sample	HDP-B22	HDP-B23	HDP-B24	HDP-B25	HDP-B26	HDP-B27	HDP-B28	HDP-B29	HDP-B30	HDP-B31	HDP-B32	HDP-B33	HDP-B34	PAAS
La	41.4	47.1	42	44.1	39.2	38.1	37.5	36.9	43.5	41.9	35.1	46.6	52.5	38.2
Ce	76.9	89.8	77.1	78.5	73.7	69	68.3	64.6	77	75.5	60.7	85.5	84.9	79.6
Pr	9.42	11	9.33	9.47	9.01	8.21	8.1	7.8	9.78	9.27	7.45	9.98	10.4	8.83
Nd	36.3	40.6	35.6	36.3	35.4	31.9	31.9	30.5	37.1	35.9	29	39	38.3	33.9
Sm	6.65	7.37	6.43	6.75	6.37	5.71	5.73	5.81	7.02	6.4	5.62	6.96	6.51	5.55
Eu	1.12	1.47	1.21	1.2	1.15	1.03	1.05	1.03	1.16	1.03	1.01	1.25	1.35	1.08
Gd	5.73	6.26	5.56	6.04	5.5	5.02	5.06	4.88	6.2	5.6	4.7	6.32	6.16	4.66
Tb	0.997	1.04	0.968	1.06	0.951	0.885	0.864	0.886	1.02	0.958	0.874	1.09	1.06	0.77
Dy	5.36	5.59	5.13	5.39	4.83	4.65	4.67	4.84	5.55	4.86	4.63	5.83	5.45	4.68
Ho	1.05	1.12	1.01	1.11	0.933	0.927	0.945	0.934	1.12	1	0.89	1.16	1.04	0.99
Er	3.01	3.34	3.03	3	2.67	2.54	2.67	2.56	3.04	2.77	2.47	3.33	3.02	2.85
Tm	0.493	0.542	0.466	0.479	0.379	0.402	0.423	0.413	0.501	0.433	0.418	0.542	0.446	0.41
Yb	3.06	3.38	3.05	3.1	2.61	2.7	2.65	2.72	3.11	2.91	2.62	3.38	2.98	2.82
Lu	0.447	0.534	0.437	0.474	0.375	0.386	0.382	0.382	0.489	0.399	0.369	0.537	0.409	0.43
∑REE	191.94	219.15	191.32	196.97	183.08	171.46	170.24	164.26	196.59	188.93	155.85	211.48	214.53	184.77
LREE	171.79	197.34	171.67	176.32	164.83	153.95	152.58	146.64	175.56	170.00	138.88	189.29	193.96	167.16
HREE	20.15	21.81	19.65	20.65	18.25	17.51	17.66	17.62	21.03	18.93	16.97	22.19	20.57	17.61
LREE/HREE	8.53	9.05	8.74	8.54	9.03	8.79	8.64	8.32	8.35	8.98	8.18	8.53	9.43	9.49
LaN/YbN	9.70	10.00	9.88	10.20	10.77	10.12	10.15	9.73	10.03	10.33	9.61	9.89	12.64	9.72
Eu/Eu*	0.55	0.66	0.62	0.57	0.59	0.59	0.60	0.59	0.54	0.53	0.60	0.58	0.65	0.65

Stage, and volcanic ash fell to the surface of the ocean and provided nutrients for life to flourish. The Yangtze Block ended the carbonate deposition and began to sediment the black shales of the Wufeng Formation by the integrated control of multiple factors. In the early Hirnantian, because of the sustained collision, the basin base continued to decline. However, global glaciation started, ice constantly formed, and the global sea level decreased. Since the falling rate of the basin base was greater than the descent of the sea level caused by the glaciation, and the volcanic eruption events were fewer than before, the increase in nutrients decreased, and the sea level in the Upper Yangtze region constantly increased. Thus, black shales with relatively low TOC values were still deposited in the upper part of the Wufeng Formation in the Yangtze area. In the middle Hirnantian, the compression between the Yangtze and Cathaysia blocks continued, and the basin basement declined. The falling rate of the basin base was less than the descent of the sea level caused by the glaciation, and the relative sea level decreased, depositing the siltstones of the Guanyinqiao Formation. In the late Hirnantian, the sea-level change was the same as that in the early Hirnantian period, and the relative sea level increased. Black shales were deposited in the lower part of the Longmaxi Formation. In the Rhuddanian (early Silurian), the basin basement declined continuously, the glaciation ended, and the global sea level increased. A new round of frequent volcanic eruptions resumed, providing more nutrition for creatures on the surface of the ocean. The combination of the three factors promoted the increase in the sea level, and black shales were deposited in the lower part of the Longmaxi Formation (Fig. 5.10). According to these studies, although both the Wufeng and Longmaxi Formations developed the same black shales, the formation mechanisms were different. In terms of the factors controlling the organic matter enrichment and preservation, apart from the black shale and ancient seawater water environment, the effect of volcanic activity and glaciation cannot be ignored. Tectonic

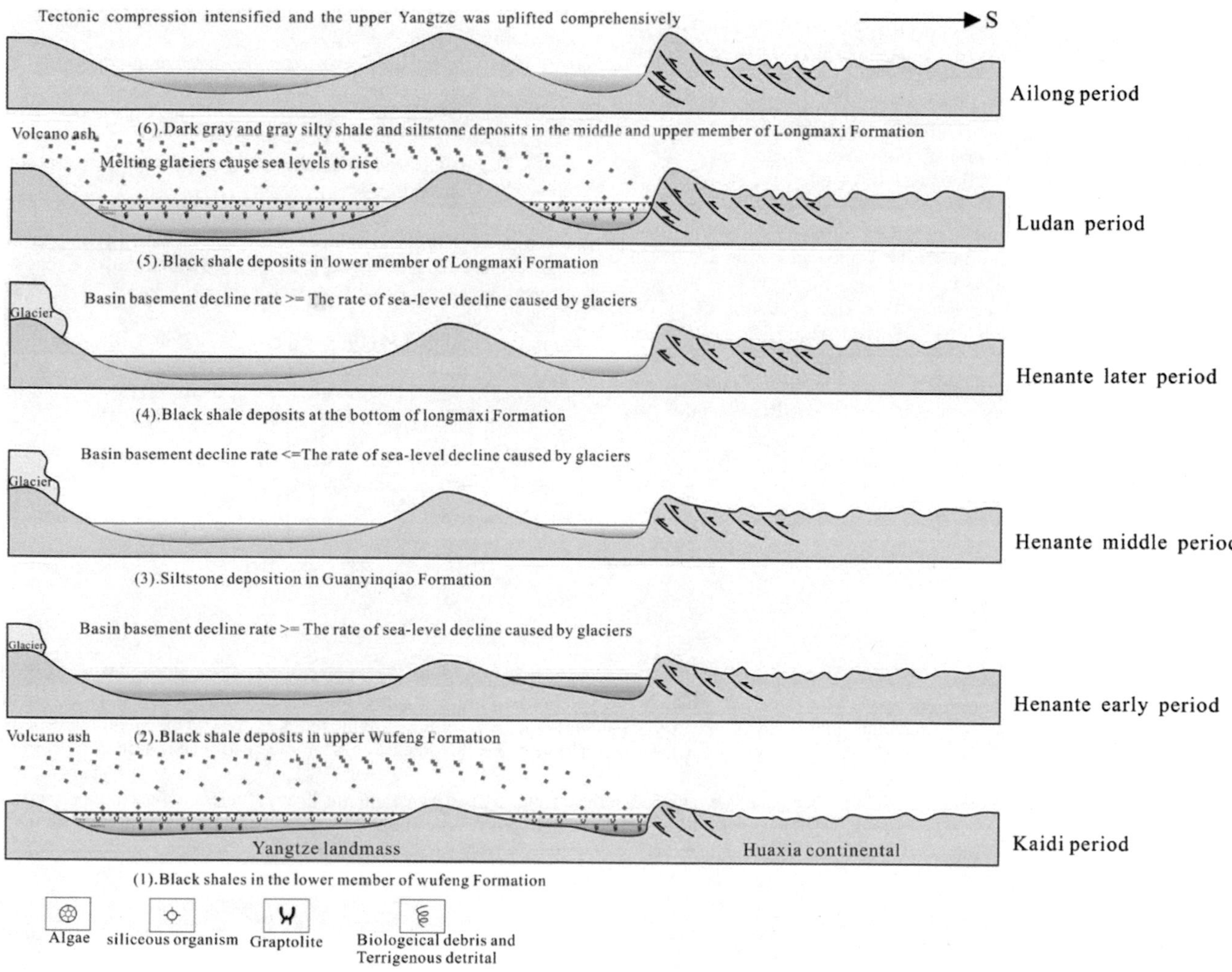

Fig. 5.10 Sedimentary mode of the black shales from the Wufeng and Longmaxi Formations in Sichuan basin

compression resulted in the descending basin basement, and an anoxic water environment was generated. The black shales of the Wufeng Formation were deposited. Notably, the black shales of the Longmaxi Formation were deposited after glacier ablation, and the combination of ablation and tectonic compression led to the increase in sea level and accelerated the formation of an anoxic environment, thereby promoting the burial and preservation of black organic matter.

References

Brenchley PJ, Marshall JD, Harper DAT et al (2006) A late Ordovician (Hirnantian) karstic surface in a submarine channel, recording glacioeustaticsea-level changes: Meifod, central Wales. Geol J 41:1–22

Chen X, Fan JX, Wang WH et al (2017) Stage-progressive distribution pattern of the Lungmachi black graptolitic shales from Guizhou to Chongqing, Central China. Scientia Sinica (Terrae) 47(6):720–732

Condie KC, Boryta MD, Liu JZ et al (1992) The origin of khondalites: geochemical evidence from the Archean to early proterozoic granulitebelt in the North China craton. Precambr Res 59:207–223

Fedo CM, Nesbitt HW, Young GM (1995) Unraveling the effects of Kmetasomatism in sedimentary rocks and paleosols with implicationsfor palaeoweathering conditions and provenance. Geology 23:921–924

Fu XG, Wang J, Chen WB et al (2015) Organic accumulation in lacustrine rift basin: constraints from mineralogicaland multiple geochemical proxies. Int J Earth Sci 104:495–511

Ge XY, Mou CL, Wang CS et al (2019a) Mineralogical and geochemical characteristics of K-bentonites from the Late Ordovician-Early Silurian in South China and their geological significance. Geol J 54:514–528

Ge XY, Mou CL, Yu Q et al (2019b) The geochemistry of the sedimentary rocks from the Well Huadi No. 1 in the Wufeng-Longmaxi Formations (Upper Ordovician-Lower Silurian), South China, with implications for paleoweathering, provenance, tectonic setting and paleoclimate. Mar Pet Geol 103:646–660

Harnois L (1988) The CIW index: a new chemical index of weathering. Sediment Geol 55:319–322

Harper DAT, Hammarlund EU, Rasmussen CMØ (2014) End Ordovician extinctions: a coincidence of causes. Gondwana Res 25:1294–1307

Lee YI (2009) Geochemistry of shales of the upper cretaceous Hayang group, SE Korea: implications for provenance and source weathering at an active continental margin. Sed Geol 215:1–12

Ma PF, Wang LC, Wang CS et al (2015) Organic-matter accumulation of the lacustrine lunpola oil shale, central Tibetan plateau: controlled by the paleoclimate, provenance, and drainage system. Int J Coal Geol 147–148:58–70

Moosavirad SM, Janardhana MR, Sethumadhav MS et al (2011) Geochemistry of lower Jurassic shales of the Shemshak Formation, Kerman Province, Central Iran: provenance, source weathering and tectonic setting. Chemie Dre Erde Geochem 71:279–288

Mou CL, Ge XY, Yu Q et al (2019) Palaeoclimatology and provenance of black shales from Wufeng- Longmaxi Formations in south-Western Sichuan Province: from geochemical records of Well Xindi-2. J Palaeogeogr 21(5):835–854

Nesbitt HW, Young GM (1982) Early proterozoic climates and platemotions inferred from major element chemistry of lutites. Nature 299:715–717

Nesbitt HW, Young GM (1984) Prediction ofsome weathering trends of plutonic and volcanic rocksbased on thermodynamic and kinetic considerations. Geochim Cosmochim Acta 48(7):1523–1534

Nesbitt HW, Young GM (1989) Formation and diagenesis of weathering profiles. J Geol 97:129–147

Qiu Z, Zou CN (2020) Unconventional petroleum sedimentology: connotation and prospect. Acta Sedimentol Sin 38(01):1–29

Rong JY, Zan RB (1999) Ordovician-Silurian Brachiopod Fauna Turnover in South China. Geoscience 13(4):390–394

Rong JY (1999) Ordovician-Silurian Brachiopod Turbover in south china. Geoscience 02:74

Taylor SR, McLennan SM (1985) The continental crust: its composition and evolution. Blackwell, UK, Oxford, Blackwell

Yan DT, Chen DZ, Wang QC et al (2010) Large-scale climatic fluctuations in the latest Ordovician on the Yangtze block, South China. Geology 38:599–602

Young GM, Nesbitt HW (1998) Processes controlling the distribution of Ti and Al in weathering profiles, siliciclastic sediments and sedimentary rocks. J Sediment Res 68(3):448–455

Conclusion

The birth of Fuling shale gas field indicates that the Yangtze plate in Southern China has very good shale gas development conditions. The existing data show that not only Southern China, but also Northwest China, Qinghai Tibet and Central and Eastern China also have great shale gas potential. The ultimate goal of the shale gas geological survey being carried out on a large scale in China is to turn these huge potentials into productivity. In accordance with the general requirements of shale gas geological survey, under the framework of unified guiding ideology and mapping technical rules, this book has comprehensively cleared, collected and consulted more than 10 regional survey data, based on the research results of the two projects of "Research on reservoir forming conditions of combined shale gas in the Southern Sichuan basin and selection evaluation" and "basic geological survey of marine shale gas in the lower Paleozoic of Sichuan Basin". The scientific research report, a large number of scientific research documents and achievement data, and the comprehensive application, mutual verification and constraints of activity theory tectonics, sedimentology, lithofacies paleogeography and related mapping technology, geochemistry, etc. are used to discuss the evaluation methods of shale gas prospect areas; these are favorable areas and target areas. Taking Longmaxi Formation of shale gas development strata in the Southern Sichuan Basin as an example, a series of maps are compiled and their favorable blocks are optimized, basically completing the geological survey of shale gas is the fundamental goal of "looking for prospective areas and favorable areas of shale gas", which provides a scientific basis for further exploration and development of shale gas. Finally, the important understanding and viewpoint of "lithofacies paleogeography research and mapping technology can be the guide (or method) for shale gas geological survey" are determined.

The main achievements are as follows:

(1) The overall system combed the shale gas research status at home and abroad, and the research state of lithofacies paleogeography summarizes the shale gas geological survey work of all kinds of methods and techniques.
(2) Re-understand the definition of shale gas, and put forward that shale gas is the "residual gas" in source rocks that has not been discharged in time, and it exists in the form of adsorbed gas, free gas or dissolved gas, mainly biogas, thermogenic gas or the mixture of both. It is defined that shale, the carrier of shale gas occurrence, must be "source rocks". And it is pointed out that it has typical characteristics such as "the development and distribution of shale gas reservoirs which are not controlled by structure. There are no obvious or fixed boundary trap, and it's only controlled by the gas source area and capping layer of the source rocks itself". And "it has the characteristics of integration of sources and reservoirs, and its hydrocarbon generation residual pores are also the main reservoirs space".
(3) Seven important factors affecting the shale gas enrichment are comprehensively and systematically analyzed. Based on the understanding that "the analysis of factors affecting shale gas enrichment in regional geological survey is actually the analysis of factors affecting the shale gas content", it is pointed out that organic matter content, type and maturity are the basic factors affecting the shale gas enrichment. As a hydrocarbon-generating substance, the former directly controls the existence of the shale gas reservoirs, and its hydrocarbon-generating holes are also the main reservoirs space, which controls the adsorption potential and adsorption gas volume of shale gas. The unique reservoirs space and physical properties of shale itself are mainly influenced by the

C. Mou et al., *Lithofacies Paleogeography and Geological Survey of Shale Gas*, The China Geological Survey Series, https://doi.org/10.1007/978-981-19-8861-5

thermal evolution degree of organic matter. These are the maturity, organic matter content and mineral composition. In-depth researches showed that the content and types of organic matter, rock types and mineral composition characteristics of source rocks were all controlled by sedimentary environment. It is an important understanding that "sedimentary environment is the fundamental factor that determines shale gas enrichment".

(4) Lithofacies paleogeography, as a comprehensive reflection of facies and sedimentary environment, is the main controlled factor and basic factor which is affecting shale gas development. Based on the study of regional sedimentary facies and lithofacies paleogeography, through lithofacies paleogeographic mapping technology, the temporal and spatial distribution and regularity of favorable facies zones of source rocks can be defined. So that the exploration scope of shale gas can be preliminarily delineated. In fact, lithofacies paleogeography research and mapping technology has always been a method for oil and gas exploration. Therefore, this book clearly puts forward for the first time to the important understanding and viewpoint that "lithofacies paleogeography research and mapping are the guide (or key) for shale gas geological survey". This should have important scientific guidance and practical significance for shale gas geological survey in China at present.

(5) Combined with the present situation of shale gas research and exploration in China, three main tasks and objectives of shale gas geological survey in China are clearly put forward, namely ① to find out the characteristics of source rocks, such as lithologic characteristics, sedimentary environment, sedimentary microfacies types and characteristics, organic matter content, mineral composition, etc.; ② make clear the temporal and spatial distribution law of source rocks, including their thickness, buried depth, fine distribution and area, etc.; ③ optimize the prospective areas and favorable areas of shale gas reservoirs. Its fundamental goal is the third point. On this basis of discussing the relationship between lithofacies paleogeography research and shale gas, the important knowledge and viewpoint of "lithofacies paleogeography research and mapping are the guide (or key) of shale gas geological survey" is further determined to guide and play a key technical role in shale gas geological survey. The specific methods and steps of shale gas selection evaluation are as follows: In the process of shale gas geological survey, the selection evaluation should be based on the detailed structural background analysis, and then the analysis of fine sedimentary facies, diagenesis and diagenesis evolution should be carried out to determine the key influencing factors affecting shale gas reservoir enrichment. The appropriate parameter boundaries should be selected. On the basis of fine lithofacies paleogeographic maps, the identified key influencing factors should be superimposed respectively, such as TOC contour map, Ro contour map, mineral component plane distribution map, etc. According to the different evaluation grades of selected areas (i.e., prospective areas, favorable areas and target areas), the range values of key influencing factors can be flexibly selected in real time to delineate the ranges of different selected areas.

(6) Taking Ordovician Wufeng Formation and Silurian Longmaxi Formation shale gas development strata in the South of Sichuan Basin and its periphery as examples, according to the guiding ideology and selection evaluation method put forward above (5) of progress, a series of maps were carefully studied and compiled, and the favorable blocks of Wufeng Formation and Longmaxi Formation shale gas development strata were selected. It is pointed out that in the geological survey of shale gas, no matter how many scientific theories and exploration and development techniques are involved, the understanding of taking sedimentology and paleogeography as the core cannot be deviated. The extensive and comprehensive lithofacies paleogeography can be used as the guide of shale gas geological survey, and the related lithofacies paleogeography mapping technology can be used as the key technical method to realize the task of shale gas geological survey. In this process, lithofacies paleogeography is not only a regional, multi-information and multi-disciplinary comprehensive basic research, but more importantly, it is a method of shale gas geological survey and a method of "finding" shale gas. This understanding of consensus and methodology should be the product of high integration of paleogeography and shale gas exploration and development practice.

At present, the geological survey of shale gas has a low degree of seismic and drilling work in a wide range. So it is impossible to carry out a comprehensive and detailed comparative study, especially the high-precision study of sequence lithofacies paleogeography, which makes the research and understanding of geological characteristics, spatial distribution law and controlled factors of shale gas producing section have certain limitations. Thus, it is affecting the research on the enrichment degree of shale gas reservoirs. The analysis of key geological parameters such as shale gas content and their relationship, as well as their influence on shale gas enrichment, needs a further study. It not only affects the establishment of shale gas evaluation standard system, but also affects the discussion of shale gas selection evaluation methods and technologies. Here, we apologize and please understand that there must be some shortcomings, omissions or deficiencies in the process of comprehensive research and compilation of maps to evaluate and optimize the selected districts. The series of maps compiled in this book and the guiding ideology, methods and techniques may be not fully consider and reflect some important discoveries and new understandings in shale gas geological survey. And the further scientific research and exploration and development. We hope that experts and peers can give valuable opinions and make corrections.

Printed in the United States
by Baker & Taylor Publisher Services